Lecture Notes in Computer Science 12771

More information about this subseries at http://www.springer.com/series/7409

Pei-Luen Patrick Rau (Ed.)

Cross-Cultural Design

Experience and Product Design Across Cultures

13th International Conference, CCD 2021
Held as Part of the 23rd HCI International Conference, HCII 2021
Virtual Event, July 24–29, 2021
Proceedings, Part I

Springer

Editor
Pei-Luen Patrick Rau
Tsinghua University
Beijing, China

ISSN 0302-9743 ISSN 1611-3349 (electronic)
Lecture Notes in Computer Science
ISBN 978-3-030-77073-0 ISBN 978-3-030-77074-7 (eBook)
https://doi.org/10.1007/978-3-030-77074-7

LNCS Sublibrary: SL3 – Information Systems and Applications, incl. Internet/Web, and HCI

This Springer imprint is published by the registered company Springer Nature Switzerland AG
The registered company address is: Gewerbestrasse 11, 6330 Cham, Switzerland

Foreword

Human-Computer Interaction (HCI) is acquiring an ever-increasing scientific and industrial importance, and having more impact on people's everyday life, as an ever-growing number of human activities are progressively moving from the physical to the digital world. This process, which has been ongoing for some time now, has been dramatically accelerated by the COVID-19 pandemic. The HCI International (HCII) conference series, held yearly, aims to respond to the compelling need to advance the exchange of knowledge and research and development efforts on the human aspects of design and use of computing systems.

The 23rd International Conference on Human-Computer Interaction, HCI International 2021 (HCII 2021), was planned to be held at the Washington Hilton Hotel, Washington DC, USA, during July 24–29, 2021. Due to the COVID-19 pandemic and with everyone's health and safety in mind, HCII 2021 was organized and run as a virtual conference. It incorporated the 21 thematic areas and affiliated conferences listed on the following page.

A total of 5222 individuals from academia, research institutes, industry, and governmental agencies from 81 countries submitted contributions, and 1276 papers and 241 posters were included in the proceedings to appear just before the start of the conference. The contributions thoroughly cover the entire field of HCI, addressing major advances in knowledge and effective use of computers in a variety of application areas. These papers provide academics, researchers, engineers, scientists, practitioners, and students with state-of-the-art information on the most recent advances in HCI. The volumes constituting the set of proceedings to appear before the start of the conference are listed in the following pages.

The HCI International (HCII) conference also offers the option of 'Late Breaking Work' which applies both for papers and posters, and the corresponding volume(s) of the proceedings will appear after the conference. Full papers will be included in the 'HCII 2021 - Late Breaking Papers' volumes of the proceedings to be published in the Springer LNCS series, while 'Poster Extended Abstracts' will be included as short research papers in the 'HCII 2021 - Late Breaking Posters' volumes to be published in the Springer CCIS series.

The present volume contains papers submitted and presented in the context of the 13th International Conference on Cross-Cultural Design (CCD 2021) affiliated conference to HCII 2021. I would like to thank the Chair, Pei-Luen Patrick Rau, for his invaluable contribution in its organization and the preparation of the Proceedings, as well as the members of the program board for their contributions and support. This year, the CCD affiliated conference has focused on topics related to cross-cultural experience and product design, cultural differences and cross-cultural communication, as well as design case studies in domains such as learning and creativity, well-being, social change and social development, cultural heritage and tourism, autonomous vehicles, virtual agents, robots and intelligent assistants.

I would also like to thank the Program Board Chairs and the members of the Program Boards of all thematic areas and affiliated conferences for their contribution towards the highest scientific quality and overall success of the HCI International 2021 conference.

This conference would not have been possible without the continuous and unwavering support and advice of Gavriel Salvendy, founder, General Chair Emeritus, and Scientific Advisor. For his outstanding efforts, I would like to express my appreciation to Abbas Moallem, Communications Chair and Editor of HCI International News.

July 2021   Constantine Stephanidis

HCI International 2021 Thematic Areas and Affiliated Conferences

Thematic Areas

- HCI: Human-Computer Interaction
- HIMI: Human Interface and the Management of Information

Affiliated Conferences

- EPCE: 18th International Conference on Engineering Psychology and Cognitive Ergonomics
- UAHCI: 15th International Conference on Universal Access in Human-Computer Interaction
- VAMR: 13th International Conference on Virtual, Augmented and Mixed Reality
- CCD: 13th International Conference on Cross-Cultural Design
- SCSM: 13th International Conference on Social Computing and Social Media
- AC: 15th International Conference on Augmented Cognition
- DHM: 12th International Conference on Digital Human Modeling and Applications in Health, Safety, Ergonomics and Risk Management
- DUXU: 10th International Conference on Design, User Experience, and Usability
- DAPI: 9th International Conference on Distributed, Ambient and Pervasive Interactions
- HCIBGO: 8th International Conference on HCI in Business, Government and Organizations
- LCT: 8th International Conference on Learning and Collaboration Technologies
- ITAP: 7th International Conference on Human Aspects of IT for the Aged Population
- HCI-CPT: 3rd International Conference on HCI for Cybersecurity, Privacy and Trust
- HCI-Games: 3rd International Conference on HCI in Games
- MobiTAS: 3rd International Conference on HCI in Mobility, Transport and Automotive Systems
- AIS: 3rd International Conference on Adaptive Instructional Systems
- C&C: 9th International Conference on Culture and Computing
- MOBILE: 2nd International Conference on Design, Operation and Evaluation of Mobile Communications
- AI-HCI: 2nd International Conference on Artificial Intelligence in HCI

List of Conference Proceedings Volumes Appearing Before the Conference

1. LNCS 12762, Human-Computer Interaction: Theory, Methods and Tools (Part I), edited by Masaaki Kurosu
2. LNCS 12763, Human-Computer Interaction: Interaction Techniques and Novel Applications (Part II), edited by Masaaki Kurosu
3. LNCS 12764, Human-Computer Interaction: Design and User Experience Case Studies (Part III), edited by Masaaki Kurosu
4. LNCS 12765, Human Interface and the Management of Information: Information Presentation and Visualization (Part I), edited by Sakae Yamamoto and Hirohiko Mori
5. LNCS 12766, Human Interface and the Management of Information: Information-rich and Intelligent Environments (Part II), edited by Sakae Yamamoto and Hirohiko Mori
6. LNAI 12767, Engineering Psychology and Cognitive Ergonomics, edited by Don Harris and Wen-Chin Li
7. LNCS 12768, Universal Access in Human-Computer Interaction: Design Methods and User Experience (Part I), edited by Margherita Antona and Constantine Stephanidis
8. LNCS 12769, Universal Access in Human-Computer Interaction: Access to Media, Learning and Assistive Environments (Part II), edited by Margherita Antona and Constantine Stephanidis
9. LNCS 12770, Virtual, Augmented and Mixed Reality, edited by Jessie Y. C. Chen and Gino Fragomeni
10. LNCS 12771, Cross-Cultural Design: Experience and Product Design Across Cultures (Part I), edited by P. L. Patrick Rau
11. LNCS 12772, Cross-Cultural Design: Applications in Arts, Learning, Well-being, and Social Development (Part II), edited by P. L. Patrick Rau
12. LNCS 12773, Cross-Cultural Design: Applications in Cultural Heritage, Tourism, Autonomous Vehicles, and Intelligent Agents (Part III), edited by P. L. Patrick Rau
13. LNCS 12774, Social Computing and Social Media: Experience Design and Social Network Analysis (Part I), edited by Gabriele Meiselwitz
14. LNCS 12775, Social Computing and Social Media: Applications in Marketing, Learning, and Health (Part II), edited by Gabriele Meiselwitz
15. LNAI 12776, Augmented Cognition, edited by Dylan D. Schmorrow and Cali M. Fidopiastis
16. LNCS 12777, Digital Human Modeling and Applications in Health, Safety, Ergonomics and Risk Management: Human Body, Motion and Behavior (Part I), edited by Vincent G. Duffy
17. LNCS 12778, Digital Human Modeling and Applications in Health, Safety, Ergonomics and Risk Management: AI, Product and Service (Part II), edited by Vincent G. Duffy

http://2021.hci.international/proceedings

13th International Conference on Cross-Cultural Design (CCD 2021)

Program Board Chair: **Pei-Luen Patrick Rau,** *Tsinghua University, China*

- Kuohsiang Chen, China
- Na Chen, China
- Wen-Ko Chiou, Taiwan
- Zhiyong Fu, China
- Toshikazu Kato, Japan
- Sheau-Farn Max Liang, Taiwan
- Rungtai Lin, Taiwan
- Wei Lin, Taiwan
- Dyi-Yih Michael Lin, Taiwan
- Robert T. P. Lu, China
- Xingda Qu, China
- Chun-Yi (Danny) Shen, Taiwan
- Hao Tan, China
- Pei-Lee Teh, Malaysia
- Lin Wang, Korea
- Hsiu-Ping Yueh, Taiwan
- Run-Ting Zhong, China

The full list with the Program Board Chairs and the members of the Program Boards of all thematic areas and affiliated conferences is available online at:

http://www.hci.international/board-members-2021.php

HCI International 2022

The 24th International Conference on Human-Computer Interaction, HCI International 2022, will be held jointly with the affiliated conferences at the Gothia Towers Hotel and Swedish Exhibition & Congress Centre, Gothenburg, Sweden, June 26 – July 1, 2022. It will cover a broad spectrum of themes related to Human-Computer Interaction, including theoretical issues, methods, tools, processes, and case studies in HCI design, as well as novel interaction techniques, interfaces, and applications. The proceedings will be published by Springer. More information will be available on the conference website: http://2022.hci.international/:

General Chair
Prof. Constantine Stephanidis
University of Crete and ICS-FORTH
Heraklion, Crete, Greece
Email: general_chair@hcii2022.org

http://2022.hci.international/

Contents – Part I

Cross-Cultural Product Design

Cultural Differences and Cross-Cultural Communication

Contents – Part II

Social Change and Social Development

Contents – Part III

CCD in Autonomous Vehicles and Driving

CCD in Virtual Agents, Robots and Intelligent Assistants

Cross-Cultural Experience Design

A Project-Based Study on User Guidance for Interaction Design

Shuangyuan Cao and Fang Liu[✉]

Xi'an Jiaotong-Liverpool University, Suzhou 215123, China
`fang.liu@xjtlu.edu.cn`

Abstract. This paper studies the usage of user guidance in interaction design based on a digital media art project. A reflection on the project is first conducted to introduce the case. Data on user experience in the project is analyzed to demonstrates the importance of user guidance, and possible optimizations for improving the efficiency of user guidance are discussed. The result suggests that effectively designed user guidance can improve users' experience in an interactive digital artwork, and guidance can be more important for interactions requiring controls unfamiliar with the users. The Fogg Behavior Model is applied to discuss the approaches to optimize the efficiency of user guidance. For motivation, the discussion suggests combining user guidance with narratives, adopting interactive guidance, and rewarding participation in guidance. For ability, designers can simplify the interaction and guidance based on target users and familiarize them with the manipulation.

Keywords: Interaction design · Digital media arts · User guidance · Fogg Behavior Model

1 Introduction

The popularized digital technology has recently engaged numerous artists and designers in creating digital media arts. Digital technology also boosted the demand for interaction design and fusion between digital arts and human computer interaction (HCI) [1, 2]. Because interaction design involves the audience's experience as a part of the artwork, interaction designers should consider efficiently engaging the audience as users for the interaction. User guidance is thus adopted to facilitate the audience's involvement. It guides the audience to participate in the artwork according to the instruction of designers. Interaction designers can also promote the efficiency of interaction by designing the user guidance more efficiently. Well-designed interactive artworks provide immersive experiences to the audience by guiding and facilitating their manipulation [3]. This paper studies user guidance's importance in interaction design and explores possible approaches to improve its efficiency by conducting a case study of a digital media art project named *Mountain Crossing*.

This project is a projection mapping exhibition at Xi'an Jiaotong-Liverpool University, whose target audience is the university's students and faculty members. Projection

© Springer Nature Switzerland AG 2021
P.-L. P. Rau (Ed.): HCII 2021, LNCS 12771, pp. 3–13, 2021.
https://doi.org/10.1007/978-3-030-77074-7_1

mapping is a set of imaging techniques utilized in digital media arts, which project virtual 2D images onto real-world objects [4, 5]. The project *Mountain Crossing* combines projection mapping and interaction design to provide the audience with an immersive experience. It includes a mini-game in the middle of its exhibition to allow the audience to participate in its narrative. However, the user guidance for this project's interaction appears inadequate and reduces the efficiency of the audience's participation. To study the usage of user guidance in interaction design, the paper first introduces and reflects on the project in Sect. 2. Section 3 discusses the importance of user guidance by analyzing the faculty members' experiences, and Sect. 4 explores possible approaches to improve the efficiency of user guidance.

2   Project Reflection

2.1   General Review

Mountain Crossing is an interactive digital media art project based on projection mapping. It advocates environmental protection and tells a story about the conflict between a mountain and humans. The production was completed by a group of six members under the help of tutors. Its exhibition contained two pre-recorded animations and an interactive mini-game. The first animation introduced the conflict between humans and nature, and the second animation showed the ending of the mountain. The mini-game was between the two animations and concentrated on the ending of the humans. Users' performance in the game would determine whether humans can survive the ending.

The project generally shows three features. Firstly, it is based on digital media. The operating system of the project is Microsoft Windows 10. Game engine Unity implemented its interactivity, and projection mapping software MadMapper adjusted video output to fit physical models. Digital media has established its technical foundation. Secondly, the project is interactive. The audience could join in its narrative by using a controller. Their performance would also affect the ending. Thirdly, the project aims at providing an immersive experience, namely immersion. Dawson [6] explained immersion as a state of being engaged in activities both physically and cognitively and excluding unrelated concerns. The exhibition was positioned in an enclosed room to eliminate potential interference. Voice-over and interaction were also considered strategies to improve the immersive experience.

2.2   Problems

A technical problem that occurred in this project is related to software. The projection mapping software occupied the only window focus supported by Windows 10 when it transmitted frames to a projector. The mini-game application cannot detect users' input because input detection required the window focus. To solve this problem, two computers and two projectors were used separately. The second computer specifically ran the interactive application and directly outputted its frames to the second projector.

Another problem was the lack of users in the testing period of interaction design. Edmonds [7] believed that understanding the user is an important concern for designing interactive art. The target audience and potential users of the project were faculty members and students in a university. Nevertheless, only students participated in testing the mini-game. The feedback of the faculty members was not collected timely as references for improving the interaction design. This lack of testing users also led to the inefficiency of user guidance.

2.3 Mini-game

Fig. 1. A typical screenshot from the mini-game's guide video.

The mini-game in the project is a defense game. Rocks are generated from the left and right bounds of the screen. The rocks roll toward a human character and will kill the character if they collide. Users can control an elf to smash these rocks and protect the human. If the human character can reach the left bound of the screen, humans can survive the ending. A Microsoft Xbox One controller is the hardware device that allows users to control the elf. To fully experience the project, two levels of understanding are expected from the audience. The first level is to realize that their performance influences the narrative. The second level is to learn to control the elf properly. These understandings are conveyed to the audience in the user guidance, which is a pre-recorded video displayed before the game starts (see Fig. 1). However, a part of the audience on the exhibition did not reach the second level of understanding.

2.4 Outcomes

The outcomes of producing *Mountain Crossing* can be generalized as five aspects. First, explicit work specification increases the efficiency of cooperation. Inefficiency arose in the early stages owing to unclear work specifications. That situation improved after the division of labor was clarified. Second, communication between group members may be critical in concept development. A divergence occurred in narrative development because group members had different perceptions of environmental protection. Third, interaction design in digital arts is subjective and user-centered. The audience is the users and the subject of interaction with the artist's work. Wright and McCarthy [8] considered user experience an influential concept in designing for HCI and a driver for interaction design. Fourth, the feasibility of combining projection mapping and interaction design was demonstrated. Despite technical issues, the project eventually implemented interactive projection mapping. Existing studies [9, 10] also supported the feasibility of interactive projection mapping projects. Finally, designing user guidance is essential for interaction design. The next section explains this idea methodically.

3   Interaction Design and User Guidance

3.1   Background

User guidance usually instructs users on how to utilize a product effectively. Products that require users' manipulation tend to use user guidance. For example, electrical appliances, software, and instruments may attach an instruction manual. Users' interactivities are also factors in these products' design, and the appropriate behavior for interactivity might be difficult for users to understand, so designers' guidance seems helpful for the product to function effectively. In interaction design, there is a tendency to adopt user guidance as a strategy to guide the audience for a better experience [11–14]. For instance, the research of Rojtberg [14] demonstrated a design of user guidance embodied in the user interface of an interactive augmented reality application. According to Loke and Khut [3], guidance from artists can be important for the audience's experience of an interactive art practice involving bodily participation.

Smith [15] used the term "weakly designed" to describe interactive media that cannot provide different audience members with a similar experience. In the project *Mountain Crossing*, a video shown before the game is the user guidance. Nevertheless, this guidance and the interactivity appeared weakly designed. A part of the audience did not learn the correct controlling when experiencing the mini-game, while others managed to gain the expected two levels of understanding. This section analyzes faculty members' experiences to discuss why the interactivity and its guidance were weakly designed and demonstrate the importance of user guidance for interaction design.

3.2   Method

The mini-game in the testing period is different from the final version for the exhibition, so this section only focuses on data collected on the project's exhibition. Please also note that the word "gaming" especially denotes console gaming in this section because the interactivity is based on the controller of Microsoft Xbox One, a gaming console. Further information about the controller can be found on the support site of Microsoft Xbox [16].

Sample. Seven groups of faculty members attended the exhibition of *Mountain Crossing*, among whom six faculty members participated in the interactive game. For the group not participating in the interactivity, one of the project's members played the mini-game instead of the faculty members. The attending faculty members were mainly staff from Xi'an Jiaotong-Liverpool University, and five groups were mainly formed by members of the School of Film and TV Arts (SoFTA). The six faculty members who have played the game are the sample for this research. Although students also attended the exhibition, their performance was not specifically recorded and is only available as part of a general record of the mini-game.

Materials. The Notepad application of the Huawei smartphone was used to record the performance of participants in the exhibition. Microsoft Word and Excel were used to process the recorded data and produce the table of results.

Procedure. User experience is significant for interaction design [7, 8]. The data were originally recorded as a part of the reflection on producing *Mountain Crossing*, as the behavior varied notably between participants on the exhibition. The data were analyzed to discuss the defect in the interaction design of the project. The topic of the study was afterward decided to be user guidance and interaction design. This analysis also indicated factors leading to the weakly designed interactivity and helped to discuss possible approaches to improve it.

3.3 Result

Table 1. Gaming knowledge and game performance of six faculty members. The numbers of faculty members are arranged chronologically according to their participation in the interactive game. Game performance indicates whether the user understood the proper controlling while experiencing the game.

Number	Gaming knowledge	Game performance
1	High	Yes
2	Medium	No
3	Low	No
4	Low	No
5	Medium	Yes
6	High	Yes

Table 1 shows faculty members' gaming knowledge and game performance. The two faculty members who did not know gaming failed to understand the guidance. The two faculty members knowledgeable about gaming successfully interacted according to the guidance. Among the two faculty members with limited knowledge of gaming, one interacted unsuccessfully, and the other managed to learn the controlling. This result suggests that faculty members with a higher level of gaming knowledge tend to learn the controlling more efficiently. Users' knowledge of console gaming might influence their understanding of the guidance. Furthermore, a part of the target audience cannot participate in the interaction properly due to the weakly designed user guidance.

The performance of students was not specifically recorded. A general record shows that students' game performance had no explicit relation with their gaming knowledge. The attending students tended to enquire about the controls and seek guidance from project members more frequently than faculty members.

3.4 Discussion

Generally, the mini-game has achieved its primary goal of enhancing the immersive experience. It received positive comments from the audience in aspects of graphics, programming, and the sense of entertainment. Although some users did not learn the controls, they still participated in the narrative and were entertained by the effect of interactive projection mapping. There is a trick in the game where a rock suddenly appears at the right border of the screen. This trick also had a dramatic effect on entertainment. Additionally, the design of visuality was successful. Users could identify the state of characters and rocks through their animation, and the controllable character was easy to distinguish.

Nevertheless, the guidance design of this mini-game seems defective. According to Smith [15], weakly designed interactive media provide the audience with more freedom but are incompetent in narratives. For games, narratives help heighten users' sense of involvement, improve their experiences, and motivate them [17]. Users' sense of control over characters considerably affects their engagement in a narrative game [18]. In Mountain Crossing, the narrative is the primary strategy to engage the audience and convey the advocation of environmental protection. Although most users have understood the relation between the game and the narrative, the sense of control could be low for users with limited knowledge about gaming, and their experience would be impaired. The guidance failed to convey the two levels of understanding to all users. The project's design of interaction seemed inadequate in respect to user guidance.

Three factors have led to this inadequacy. First, the feedback from testing users was insufficient. Evaluated user feedback is constructive for interaction design [19]. The faculty members' feedback was not collected before the exhibition. Second, the choice of hardware did not satisfy the target audience. The first faculty member reported that he was more familiar with controllers produced by Nintendo. A student reported that he did not know about the Microsoft Xbox One controller. Third, the game was too difficult. Fogg [20] suggested that designing for behavior change should begin with targeting small behaviors. Simple steps should be designed to develop their proficiency in the basic controls before challenges occur.

This analysis also demonstrates the importance of user guidance for interaction design. Interaction design is highly user-centered. Forbrig [21] believed that maximizing users' experience is the aim of user-centered interaction design. For interactive digital arts, the users are the subject of interaction, but designers produce the interaction. Cognitive differences between users and designers should be considered in user-centered design [22]. User guidance can normalize how the interaction performs. It allows inter-action designers to convey the preferred approach in which the audience can fully experience the interactive artwork. The user guidance in *Mountain Crossing* was weakly designed and unable to provide appropriate guidance for the target audience. Consequently, the user experience was negatively affected. In contrast, effectively designed user guidance can boost user engagement and improve their experience. User guidance is a useful strategy for interaction design.

Sánchez and Espinoza [23] established a Role-Playing Game (RPG) design, whose basic controls "were unanimously accepted" by players without guidance, while multiple-choice selection, a complicated control, needed special guides for most players to perform successfully. A possible explanation for this absence of guidance can be users' application of experience. The basic controlling appears similar to mainstream RPGs, so users may apply their experience learned from other RPGs. Application of experience might also explain gaming knowledge's effect on users' understanding of the guidance in *Mountain Crossing*. Interactions requiring unfamiliar controls of users can attach more importance to user guidance.

3.5 Limitations

Only six samples were recorded on the exhibition of the project. The dearth of samples might influence the accuracy of the result. Additionally, this result is not universal. The general record indicates that the students' performance was different from the faculty members, where the performance seems irrelevant to their gaming knowledge. At the end of the discussion, a hypothesis about users' application of experience is mentioned but still needs evidence. Further studies on examining the relationship between the application of experience and the necessity of guidance can be conducted.

4   Optimizations

The Fogg Behavior Model (FBM) [24] is a design model for persuasive technology. A significant concern of persuasive technology is to change users' behavior through HCI. Existing studies have applied FBM in game design to affect users' behavior outside of the games [25, 26]. Motivation, ability, and triggers are the three principal factors of FBM. In the case of *Mountain Crossing*, the target behavior for users is to accept the guidance and learn the controls. The trigger is the exhibition because the guide video was automatically displayed as part of the exhibition. The motivation is mainly the appeal of interactive projection mapping and the narrative. As for ability, users' gaming knowledge appears to be the ability that impacts their performance most.

WarioWare is a game series known for microgame collections. Nelson and Mateas [27] defined WarioWare-styled games as short games requesting players to complete a simple task in rapid sequence. Gingold [28] suggested that WarioWare effectively conveys each microgame's goal and controls with its fictional representation and assures the diverse microgames to be playable by coherently connecting them by rules. This game series also inspired the design of the mini-game in *Mountain Crossing*. This section explores possible approaches to improve guidance efficiency concentrating on motivation and ability in FBM and WarioWare's guide design.

4.1 Motivation

Motivation is a vital factor in instructional design [29] and game design [30, 31]. WarioWare: Smooth Moves [32] showed two ideas to motivate the users in the guidance. In this game, the guidance occurs as part of each level's background story. The narration will ask the player to perform a specific control, and the central rule of the level's microgames will highlight this control. This design showed a combination of guidance with narratives and involved the players interactively.

Combine Narrative and Guidance. Stories are intrinsic motivators for game players [17]. Interactive digital artworks with narratives may consider embodying guides in their storyline. Guides with a story would appear more explicit and interesting to users. A consistent storyline also allows stronger connections between the guides and other elements.

Interactively Involve Users in Guidance. The guidance can be as interactive as the interaction it instructs. Interactive instructions boost learning better than passive ones [33]. Passive guidance such as video clips or booklets appears unattractive to users and may reduce their motivation to learn the target controls. Interactive guidance also allows users to preview and practice the controls for the interaction to increase their proficiency.

Reward Participation in Guidance. The reward system is a motivational mechanism used in game design [34]. It means to present users with positive feedback to encourage their performance. A reward system can indicate a goal for users and motivate them continuously. Therefore, designers may consider rewarding users for participating in the guidance to promote users' acceptance of it.

4.2 Ability

Fogg [24] suggested that designers should focus on behavior's simplicity to increase users' ability rather than directly improving users' intrinsic ability. The evaluation of users' ability is relative to the interaction's difficulty. To effectively increase users' ability in the interaction, designers can begin with simplifying the interaction.

Simplify the Interaction and Its Guidance. Simplifying the manipulation required to complete the interaction can relatively increase users' ability. It allows the guidance to be more straightforward and understandable because the content to guide becomes easier. The guidance can also be simplified separately. Simplification of guidance can be possibly achieved by filtering unnecessary guides, reducing textual guides' length, or utilizing visual representation.

Consider the Target Users. The simplification should also deliberate the intrinsic ability level of target users to avoid the interaction from being effortless. If users can complete an interaction effortlessly, it might be tedious. Existing studies exemplified difficulty balancing for game design [35, 36]. As users' ability varies, the design of difficulty should consider balancing target users' ability level. Hendrix et al. [36] mentioned that designers could also provide multiple difficulty settings for users to select.

Familiarize Manipulation. Users may apply their experience to guide themselves in an interaction. According to Gingold [28], WarioWare increased its playability by decomposing existing games' activities and mapping them onto players' knowledge. Designers can learn from existing interactive digital artworks to familiarize users with the manipulation of a new interaction. The novelty of the new interaction, however, should not be impaired in the familiarization.

5 Conclusion

This paper examined the usage of user guidance in the context of interaction design by reflecting on a digital art project. The digital art project *Mountain Crossing* implemented interactive projection mapping and was successful concerning user experience and techniques. The mini-game attached to this project was entertaining and received positive feedback from the audience. However, the game's guidance design was defective. The analysis of faculty members' performance indicates that users with limited knowledge about console gaming may have difficulty learning the proper controls. Concentrating on this defect, the importance of user guidance and possible optimizations are discussed.

To conclude, user guidance is useful for interaction design. Efficiently designed user guidance conveys how users can fully experience an interactive digital artwork and improves their experience. Nevertheless, user guidance might be unnecessary when the user is experienced in similar interactions.

There are six suggestions mentioned in the discussion. The discussion on motivation suggests designers to combine the guidance with narratives, interactively involve users in the guidance, and reward users for participating in guidance. Concerning ability, designers can simplify the interaction and guidance deliberating target users' ability level and familiarize them with the manipulation by imitating existing interactive digital art projects.

References

1. Benford, S.: Foreword. In: Candy, L., Ferguson, S. (eds.): Interactive Experience in the Digital Age: Evaluating New Art Practice, pp. v–vi. Springer, Cham (2014). https://doi.org/10.1007/978-3-319-04510-8
2. Candy, L., Ferguson, S.: Interactive experience, art and evaluation. In: Candy, L., Ferguson, S. (eds.) Interactive Experience in the Digital Age: Evaluating New Art Practice, pp. 1–10. Springer, Cham (2014). https://doi.org/10.1007/978-3-319-04510-8_1
3. Loke, L., Khut, G.P.: Intimate aesthetics and facilitated interaction. In: Candy, L., Ferguson, S. (eds.) Interactive Experience in the Digital Age: Evaluating New Art Practice, pp. 91–108. Springer, Cham (2014). https://doi.org/10.1007/978-3-319-04510-8_7
4. Yun, H.R., Kim, D.W., Ishii, T.: A study of digital media art utilizing the contents of the architecture cultural property. Int. J. Asia Digit. Art Des. Assoc. **17**(2), 77–84 (2013). https://doi.org/10.20668/adada.17.2_77
5. Pastor, A.: Augmenting reality: on the shared history of perceptual illusion and video projection mapping (2020). https://arxiv.org/abs/2005.14317

6. Dawson, J.D.: A discussion of immersion in human computer interaction: the immersion model of user experience. Ph.D. thesis, Newcastle University (2016). https://theses.ncl.ac.uk/jspui/handle/10443/3685

7. Edmonds, E.A.: Human computer interaction, art and experience. In: Candy, L., Ferguson, S. (eds.) Interactive Experience in the Digital Age: Evaluating New Art Practice, pp. 11–23. Springer, Cham (2014). https://doi.org/10.1007/978-3-319-04510-8_2

8. Wright, P., McCarthy, J.: The value of the novel in designing for experience. In: Pirhonen, A., Isomäki, H., Roast, C., Saariluoma, P. (eds.) Future Interaction Design, pp. 9–30. Springer, London (2005). https://doi.org/10.1007/1-84628-089-3_2

9. Zhou, Y., Xiao, S., Tang, N., Wei, Z., Chen, X.: Pmomo: projection mapping on movable 3D object. In: Proceedings of the 2016 CHI Conference on Human Factors in Computing Systems, pp. 781–790. Association for Computing Machinery, New York (2016). https://doi.org/10.1145/2858036.2858329

10. Witthayathada, O., Nishio, K.: BUBUU: 3D animation with interactive projection mapping. In: Proceedings of the Annual Conference of JSSD, vol. 66, pp. 442–443 (2019). https://doi.org/10.11247/jssd.66.0_442

11. Markopoulos, P., Shen, X., Wang, Q., Timmermans, A.: Neckio: motivating neck exercises in computer workers. Sensors **20**(17), 4928 (2020). https://doi.org/10.3390/s20174928

12. Sinnig, D., Pitula, K., Becker, R., Radhakrishnan, T., Forbrig, P.: Structured digital storytelling for eliciting software requirements in the ICT4D domain. In: Forbrig, P., Paternó, F., Mark Pejtersen, A. (eds.) HCIS 2010. IAICT, vol. 332, pp. 58–69. Springer, Heidelberg (2010). https://doi.org/10.1007/978-3-642-15231-3_7

13. Wang, Z., et al.: Information-level AR instruction: a novel assembly guidance information representation assisting user cognition. Int. J. Adv. Manuf. Technol. **106**(1–2), 603–626 (2019). https://doi.org/10.1007/s00170-019-04538-9

14. Rojtberg, P.: User guidance for interactive camera calibration (2019). https://arxiv.org/abs/1907.04104

15. Smith, G.M.: What Media Classes Really Want to Discuss. Routledge, Oxon (2011)

16. Get to know your Xbox One Wireless Controller. https://support.xbox.com/en-US/help/hardware-network/controller/xbox-one-wireless-controller. Accessed 25 Jan 2021

17. Huynh, E., Nyhout, A., Ganea, P., Chevalier, F.: Designing narrative-focused role-playing games for visualization literacy in young children (2020). https://arxiv.org/abs/2008.13749

18. El-Nasr, M.S., Milam, D., Maygoli, T.: Experiencing interactive narrative: a qualitative analysis of Façade. Entertainment Comput. **4**(1), 39–52 (2013). https://www.sciencedirect.com/science/article/pii/S187595211200016X

19. Følstad, A., Hornbæk, K., Ulleberg, P.: Social design feedback: evaluations with users in online ad-hoc groups. Hum.-Centric Comput. Inf. Sci. **3**(1), 1–27 (2013). https://doi.org/10.1186/2192-1962-3-18

20. Fogg, B.J.: The new rules of persuasion. RSA J. **155**(5538), 24–29 (2009). https://www.jstor.org/stable/41380568

21. Forbrig, P.: Foreword. In: Kunert, T. (eds.) User-Centered Interaction Design Patterns for Interactive Digital Television Applications, pp. vii–viii. Springer, London (2009).https://doi.org/10.1007/978-1-84882-275-7

22. Mieczakowski, A., Langdon, P., Clarkson, P.J.: Investigating designers' and users' cognitive representations of products to assist inclusive interaction design. Univers. Access Inf. Soc. **12**, 279–296 (2013). https://doi.org/10.1007/s10209-012-0278-8

23. Sánchez, J., Espinoza, M.: Video game design for mobile phones. In: Forbrig, P., Paternó, F., Mark Pejtersen, A. (eds.) HCIS 2010. IAICT, vol. 332, pp. 199–210. Springer, Heidelberg (2010). https://doi.org/10.1007/978-3-642-15231-3_20

24. Fogg, B.J.: A behavior model for persuasive design. In: Proceedings of the 4th International Conference on Persuasive Technology, PERSUASIVE 2009, Article 40. Association for Computing Machinery, New York (2009). https://doi.org/10.1145/1541948.1541999
25. Xi, A.T.Y., Marsh, T.: Identifying triggers within persuasive technology and games for saving and money management. In: Baek, Y., Ko, R., Marsh, T. (eds.) Trends and Applications of Serious Gaming and Social Media. GMSE, pp. 51–70. Springer, Singapore (2014). https://doi.org/10.1007/978-981-4560-26-9_4
26. Lakovic, V.: Crisis management using persuasive technology in a Mobile Game for children. Health Technol. **10**(6), 1579–1590 (2020). https://doi.org/10.1007/s12553-020-00476-9
27. Nelson, M., Mateas, M.: Towards automated game design. In: Basili, R., Pazienza, M.T. (eds.) AI*IA 2007. LNCS (LNAI), vol. 4733, pp. 626–637. Springer, Heidelberg (2007). https://doi.org/10.1007/978-3-540-74782-6_54
28. Gingold, C.: What warioware can teach us about game design. Game Stud. **5**(1) (2005). https://www.gamestudies.org/0501/gingold/
29. Spitzer, D.R.: Motivation: the neglected factor in instructional design. Educ. Technol. **36**(3), 45–49 (1996). https://www.jstor.org/stable/44428339
30. Hironaka, E., Murakami, T.: Game design methodology considering user experience in comprehensive contexts (trial on inducing player to terminate game contentedly by motivation control). In: Ahram, T.Z. (ed.) AHFE 2018. AISC, vol. 795, pp. 390–402. Springer, Cham (2019). https://doi.org/10.1007/978-3-319-94619-1_39
31. Cota, T.T., Ishitani, L., Vieira, N.: Mobile game design for the elderly: a study with focus on the motivation to play. Comput. Hum. Behav. **51**(Part A), 96–105 (2015). https://doi.org/10.1016/j.chb.2015.04.026
32. Nintendo: Intelligent Systems: WarioWare: Smooth Moves. Nintendo, Kyoto (2006)
33. Merrill, M., Li, Z., Jones, M.: Limitations of first generation instructional design. Educ. Technol. **30**(1), 7–11 (1990). https://www.jstor.org/stable/44425441
34. Begy, J., Consalvo, M.: Achievements, motivations and rewards in Faunasphere. Game Stud. **11**(1) (2011). https://www.gamestudies.org/1101/articles/begy_consalvo
35. Hunicke, R.: The case for dynamic difficulty adjustment in games. In: Proceedings of the 2005 ACM SIGCHI International Conference on Advances in Computer Entertainment Technology, pp. 429–433. Association for Computing Machinery, New York (2005). https://doi.org/10.1145/1178477.1178573
36. Hendrix, M., Bellamy-Wood, T., McKay, S., Bloom, V., Dunwell, I.: Implementing adaptive game difficulty balancing in serious games. IEEE Trans. Games **11**(4), 320–327 (2019). https://doi.org/10.1109/TG.2018.2791019

Development of More Concept Words Leads to the Generation of More Idea Sketches

Pei-Jung Cheng[✉]

Department of Advertising, National Chengchi University, 64, Sec. 2, ZhiNan Raod, Wenshan District, Taipei City 11605, Taiwan

Abstract. In this study, we evaluated the performance of designers' idea development using the IDEATOR app as the designers' ideation tool and investigated whether there were differences in behavior linkages and idea sketches among designers from three design fields. The results were as follows: 1. According to our analysis of the video data recorded by designers using IDEATOR, their behaviors included nine behavioral codes separated into three behavior modes (GI, GA, TH) and an error action code; 2. The designers used IDEATOR for ideation and most frequently engaged in GI behavior, especially those in the fields of graphic and product design; 3. The designers who used IDEATOR for ideation had a tendency to frequently develop concept words. In particular, graphic designers were highly dependent on word-type data to perform the ideation process; and 4. The more input concept words on the interface of the IDEATOR mind map employed by the designers, the more idea sketches the designers drew. As they input additional concept words, the designers also produced sketches that fit into multiple lateral thinking categories.

Keywords: Developing HCI expertise and capability worldwide · Design thinking · Ideation

1 Introduction

Many researchers have explored the influences that design support systems have on designers' working process through examination of their design characteristics. For example, Sun, Xiang, Chai, Wang, and Huang [1] proposed the creative segment theory, which postulates that creative segments include inspiration generation, inspiration expression, and visual feedback. They observed that the theory accurately describes the sketching process and developed a sketch-assist system that can structure creative segments. Ozkaya and Akin [2] proposed a requirement design coupling approach that can provide a connection mechanism for requirement-driven design. They modeled a continuous and interactive design process for integrating problem formulation and form exploration to facilitate the architectural design thinking process and enable designers to understand the entire design process from the initial design stage to the stages of construction, maintenance, and completion. Segers, de Vries, and Achten [3] constructed an idea space system to facilitate architectural design thinking. The designers

© Springer Nature Switzerland AG 2021
P.-L. P. Rau (Ed.): HCII 2021, LNCS 12771, pp. 14–26, 2021.
https://doi.org/10.1007/978-3-030-77074-7_2

were inspired by the system's word–image connection to think creatively and consider more perspectives, thus enhancing their work efficiency.

Siangliulue, Chan, Gajos, and Dow [4] conducted an online experiment to explore two mechanisms that offer examples at the appropriate moment. Their results indicated that examples provided on demand can assist designers in developing novel ideas, whereas examples delivered at the wrong time can suppress their idea generation. Golembewski and Selby [5] invented Ideation Decks, a card-based design ideation tool that helps designers effectively examine specific problems by aiding in iterative design explorations. When engaging creatively with each Instance Card, designers critically reflect on the depicted concept in isolation, exploring and gaining insight regarding their internal conceptual models in that instance. Cruz and Gaudron [6] also described an "open-ended objects tool" for use in brainstorming; this tool stimulated participants to reflect on emotions and desires and to establish a participatory atmosphere among themselves.

Ahmed [7] enhanced the reuse of design knowledge by developing index knowledge in the field of engineering design. He observed engineering designers and determined that 24% of them spent most of their time searching for information. Therefore, he asserted that information searches are vital to the design process and accordingly developed a method that enables designers to index design knowledge. Westerman and Kaur [8] examined the retrieval of inspirational images from computer databases, proposing that creative design tasks require the support of information systems for both convergent and divergent processes. Our previous research [9] assumption was that designers are accustomed to looking for resources online by using keywords as a result of Internet technology development. Words have since become recognized as a critical component for researchers in the field of design cognition to understand how designers transform their design concepts. The present study focused on the relevant connection between keyword thinking and design concepts and suggested that the search for resources using keywords should be regarded as a component of "seeing."

As designers decide which keywords to use for their "seeing," they must identify words related to the design task to begin the retrieval process, by which they can find the needed resources and the proper materials, a step known as "thinking first." With the advent and explosion of Internet resources in particular, designers' approaches to finding these resources have dramatically changed. Today, designers can be inspired by the substantial amount of Internet resources available, which is a notable change from designers in the past, who used books as references.

1.1 Creative Idea Generating App IDEATOR

Our preliminary research proposed four modes of association based on designer behavior during ideation [10, 11]. We then developed a creative idea generation tool, IDEATOR [9], to support designers during ideation. The main functions of IDEATOR (Fig. 1) are developed from designers' behavioral modes and resource searching needs during ideation, supporting a designer's formulation of concepts by integrating image searches, and stimulating design actions by displaying all of the images on an image board. Unlike other design thinking support systems, IDEATOR emphasizes recording a designer's word-thinking paths and processes (Function A in Fig. 1). It assists designers with the repeated input, access, and storage of information. Our preliminary research results indicated that IDEATOR enables designers to add their own ideation sketches and brief descriptions while recording each concept, which makes the results of their idea maps similar to a designer's self-reports. During the process, keywords, ideation sketches, and images serve as a stimulus or as the object of "seeing" in the "seeing–moving–seeing" model [12].

A. Idea developing	B. Image searching	C. Image comparing

Fig. 1. Three main functions of IDEATOR (Screenshot sequences from left to right indicate the functions for 'idea developing-mind map', 'image searching', and 'image comparing').

IDEATOR provides designers with copious visual stimulation during idea development (Functions B and C in Fig. 1), which is in accordance with research demonstrating that a designer's mental imagery can be triggered by an abundance of visual stimuli [12–18]. Such mental imagery aids designers in generating new ideas.

However, most research on design support tools has focused on studying designers from a specific design field rather than comparing designer characteristics among various fields. Therefore, the present study investigated and compared the influence of IDEATOR on the idea development of designers in the fields of graphic design, product design, and interior design. Specifically, we examined the differences in behavior linkages and idea sketches among designers from these fields.

2 Research Method

We explored how designers in different fields develop ideas with IDEATOR and the effects that the app has on ideation. We collected data using IDEATOR. Specifically, IDEATOR was adapted to record the designers' idea maps, index reference content, concept words of association, and sketch development. Further details of the research methods are provided in the following sections.

2.1 Design Task, Process and Participants

To ensure consistency, the designers performed a design task assigned by us. Regardless of their field, all of designers had an assigned task for the same café place. Specifically, graphic designers were required to design a logo, product designers were required to design a chair, and interior designers were required to design a bar for a coffee shop named "at Café."

Before the design task was executed, task instructions, pieces of A4 paper for sketching, and an iPad Mini with the IDEATOR app were provided to each designer. Designers were taught how to use IDEATOR and were informed that their behavioral data would be collected and analyzed. The designers were given freedom regarding their working environment and Internet usage.

The designers had 1 week to finish the task and were allowed to finish early. In addition, upon completion, the designers were required to turn off the screen-recording app (Shou.TV mobile game streaming 0.7.13) on their iPad and write down the drawing completion time point for each sketch on the paper (see Fig. 2).

A total of 15 designers (10 male and 5 female) with an average of 3 years of experience were invited to participate. Of them, five were graphic designers, five were product designers, and five were interior designers.

Sketches of participant D4	Sketches of participant G1

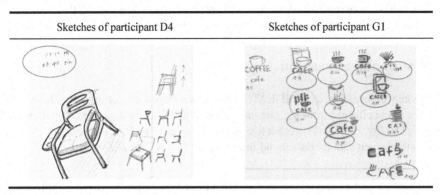

Fig. 2. Idea sketches of designers D4 and G1. Red circles indicate completion time points (Color figure online)

2.2   Data Analysis

For data analysis, each designer's IDEATOR data from the mobile device, screen capture recordings, and developed idea sketches were collected. Video and protocol data analysis was conducted by the researcher and two coders, and the internal consistency of the coding results was tested.

The study used behavior-recording software, The Observer XT was used to collect and analyze video data. First, the researcher and two coders individually marked the change points of the videotaped behavior of the designers according to the behavioral definitions and coding scheme from our previous study [9]. Subsequently, the two coders listed clips that could not be categorized under any behavioral code. Thereafter, the researcher and two coders discussed possible revisions to the behavioral definitions and coding scheme.

The designers' recorded video data using IDEATOR to record their self-ideation process. Their operational behavior did not perfectly match the behavioral codes used in the previous study because of the app's revised interface and functions. For example, for sketching behavior, the behavioral coding in the preliminary study comprised "Creating new sketches" and "Continuing to sketch." Although the research data was provided by the designers, the mobile screen-recording app (Shou.TV mobile game streaming 0.7.13) was unable to record the designers' sketching process, which was done on paper. Therefore, the sketching behavior in this study was classified under the behavioral codes "Sketching on paper (SOP)" and "Drawing on sketch pad (DOSP)."

Our modified behavioral coding scheme is presented in Table 1. There are three behavioral modes. The "Gathering information (GA)" mode includes the three behaviors of "Retrieving information (RI)," "Referring to relevant information (RRI)," and "Referring to the saved data (RSD)." The "Generating ideas (GI)" mode includes the three behaviors of SOP (Sketching on paper), DOSP (Drawing on sketch pad), and "Adding a new branch idea (ANI)." Finally, the "Thinking (TH)" mode includes "Revising a branch idea (RBI)," "Highlighting a branch idea (HBI)," and"Purposeless action (PA)." In addition, because the designers were unfamiliar with IDEATOR, they may have performed an "Error action (EA)," an action that belongs to none of the three modes.

3   Results and Discussion

To explore whether the effect of IDEATOR on the design process differed by a designer's field, the data collected from the graphic, product, and interior designers were denoted G1–G5, D1–D5, and I1–I5, respectively. Each designer had a complete set of data comprising screen capture data, hand-drawn sketches, and IDEATOR data.

3.1   Analysis and Comparison of the IDEATOR Operation Records Segment Encoding in Three Design Domains

According to our analysis using Observer XT, the mean duration of each designer's ideation was 1873 s (or 31 min). Video data on the designers' interface operations were divided into 767 segments according to the behavioral coding scheme (Table 1), and the coding results of all the segments are detailed in Table 2. During ideation, the product designers had the greatest number of IDEATOR segments (M = 75), whereas graphic designers had the fewest IDEATOR segments (M = 33).

ANI (Adding a new branch idea) had the most segments (M = 288, 37.5%) coded under it, followed by PA (Purposeless action, M = 127, 16.6%) and RI (Retrieving information, M = 85, 11.1%). The least prevalent type of behavior was DOSP (M = 12, 1.6%), followed by SOP (Sketching on paper) (M = 17, 2.2%). For the interior designers in particular, the RSD (Referring to the saved data) behavioral segment was more prevalent, even more so than PA (Purposeless action) and RI (Retrieving information); DOSP was also more common than SOP (Sketching on paper). Interestingly, the designers in all three fields seemed not to use IDEATOR's built-in sketch pad to sketch (DOSP), and the graphic designers used this function the least. The interior designers were the group most likely to use a sketch pad to draw sketches; only one interior designer did not use this tool. The remaining four used the sketch pad to sketch (DOSP) with higher frequency and duration than those of designers sketching on paper (SOP).

Table 1. Behavioral coding scheme (revised from the behavior codes of [9])

Behavior mode	Behavior (code)	Definition
Gathering information (GA)	Retrieving information (RI)	Retrieving information on-line for capturing ideas, sketching or drawing; saving the retrieved information in the hard disc to be the reference later.
	Referring to relevant information (RRI)	Referring to the information they have retrieved on-line in advance. Retrieving action is not included in the behavior.
	Referring to the saved data (RSD)	Referring to some saved data that have been retrieved on-line by them in advance.
Generating ideas (GI)	Sketching on paper (SOP)	Creating the new shapes, labels or lines.
	Drawing on sketch pad (DOSP)	Drawing the new shapes, labels or lines on sketch pad.
	Adding a new branch idea (ANI)	Adding an idea in the mind map area of IDEATOR as the new branch to be used or further thinking later.
Thinking (TH)	Revising a branch idea (RBI)	Revising the idea, fixing the words of an idea, adjusting the level of an idea in the mind map area of IDEATOR.
	Highlighting a branch idea (HBI)	Applying different color to a branch idea for highlighting its important or using several colors to those ideas for separating them from each other.
	Purposeless action (PA)	Making a move purposively, such as touching and moving working area back and forth.
None	Error action (EA)	Making a move incorrectly or not accordance with the operational rules of IDEATOR.

When referring to the three modes (GA, GI, and TH), the designers' behavior segments corresponding to the GI mode (Generating ideas, including codes SOP, DOSP, and ANI) occupied 41.3% of the behavioral segments, which was higher than the frequency of behavior segments corresponding to the GA mode (Gathering information, including codes RI, RRI, and RSD) occupied 21.5%. This finding indicates that the frequency of the designers' GI behavior occupied more than two-thirds of all of the behavior segments; that is, the designers most frequently performed GI behaviors during the ideation process.

However, regarding differences among designers of different fields, GI behavior was observed more frequently in the graphic and product designers than were the other two types of behaviors. In particular, the graphic designers' GI behavior accounted for more than half of the total number of behavior occurrences, whereas the product designers' GI behavior accounted for more than 40% of the total number of behavior occurrences. Additionally, among the graphic designers, GA (Gathering information) behavior occurred the least frequently of the three modes, whereas among the interior designers, GA (Gathering information) behavior occurred the most frequently of the three modes.

3.2 Behavioral Relationship Among the Three Design Domains in Ideation

All of the before–after behavior linkages with the designers' sketching behaviors (including SOP and DOSP in IDEATOR) and data referred to before sketching are presented in Table 3. The second column shows the designers' DOSP before–after behavior, with "-" indicating no action after drawing. The third column indicates the type of data that the designers referred to before sketching. For example, designer G1 appears to have engaged in PA (Purposeless action) before drawing the sketch, followed by no other

Table 2. Encoding of designers' behavior segments in three fields

B-mode	GA			GI			TH			None	Total
B-Code	RI	RRI	RSD	SOP	DOSP	ANI	RBI	HBI	PA	EA	Segs.
G1				1		15		3	2		21
G2				1		12		2	3	2	20
G3	1			1		29	11		14	1	57
G4				1		11	1	3	2		18
G5	4		3	1		20	3		10	7	48
Total	5	0	3	5	0	87	15	8	31	10	164
D1	12		4	3		4	2			2	27
D2	14	5	6	1	2	49	7		38	7	129
D3	14	1	10	1		13	4		2	6	51
D4	2	4		1		50	7	11	16	20	111
D5	7	6		1		31	2		5	5	57
Total	49	16	20	7	2	147	22	11	61	40	375
I1	14	2	14	1	2	2	2		14	2	53
I2			10	1		15	10	7	2	1	46
I3	4	2	3	1	1	12	2	5	13	11	54
I4	6		2	1	1	5			3		18
I5	7	1	7	1	6	20		7	3	5	57
Total	31	5	36	5	10	54	14	19	35	19	228
Segs.	85	21	59	17	12	288	51	38	127	71	767
Pct.(%)	11.1	2.7	7.7	2.2	1.6	37.5	6.6	5.0	16.6	9.0	100
Order	3	8	5	9	10	1	6	7	2	4	

behavior after the sketch was completed. G1 reported searching for a word reference before sketching. Among all of the designers, only designer D1 engaged in three periods of sketching behavior when using IDEATOR v.2 to proceed with ideation; the other designers sketched until they were satisfied with the drawing and then exhibited no further behavior. Thus, only the first and second sketches of D1 followed the behaviors of RBI (Revising a branch idea) and RI (Retrieving information).

The most common behavior among the designers before sketching was PA (Purposeless action), followed by RI (Retrieving information); RSD (Referring to the saved data), HBI (Highlighting a branch idea), ANI (Adding a new branch idea), EA (Error action), and RRI (Referring to relevant information) behavior also appeared once. Among these behaviors, PA (Purposeless action) occurred before the graphic designers began sketching, with only designer G2 also engaging in HBI (Highlighting a branch idea) before drawing. The product designers' behavior was different: before beginning to sketch, designers D1 and D5 engaged in RSD (Referring to the saved data), RI (Retrieving information), and RRI (Referring to relevant information); designers D2 and D4 engaged in PA (Purposeless action) and EA (Error action); and designer D3 engaged in ANI (Adding a new branch idea). Among the interior designers, designers I1, I4, and I5 engaged in RI (Retrieving information) and RSD (Referring to the saved data) and designers I2 and I3 engaged in PA (Purposeless action).

Regarding the type of data referenced before beginning their sketches, nine designers (G1–G5, D2, D3, I2, and I3) looked at words and six designers (D1, D4, D5, I1, I4, and I5) looked at pictures. This pattern suggests that product and interior designers tend to refer to words or pictures before sketching, whereas graphic designers tend to refer solely to words before sketching.

By comparing the behavior of the designers with their preferred data reference type, we discovered that the graphic designers usually engaged in PA (Purposeless action) and looked at the words on the mind map before sketching. PA (Purposeless action) was less common among the product designers and interior designers, who usually began their sketches after engaging in RI (Retrieving information), RSD (Referring to the saved data), and RRI (Referring to relevant information) and looking at either the words or the pictures on the mind map. The results of the video data analysis indicated that compared with the traditional behavior adopted by designers of relying on image data for reference before proceeding with the ideation process, the graphic designers in the present study were the most dependent on word references to perform the ideation process.

Table 3. Behavior linkages of before–after drawing sketches and reference data type before drawing

Participant	G1		G2		G3		G4		G5	
	Before	After	Before	After	Before	After	Before	After	Before	After
1st Sketch	PA	--	HBI	--	PA	--	PA	--	PA	--
Reference Type	words		words		words		words		words	
Participant	**D1**		**D2**		**D3**		**D4**		**D5**	
	Before	After	Before	After	Before	After	Before	After	Before	After
1st Sketch	RSD	RBI	PA	--	ANI	--	EA	--	RRI	--
Reference Type	images		words		words		images		images	
2nd Sketch	RI	RI								
Reference Type	images									
3rd Sketch	RI	--								
Reference Type	images									
Participant	**I1**		**I2**		**I3**		**I4**		**I5**	
	Before	After	Before	After	Before	After	Before	After	Before	After
1st Sketch	RI	--	PA	--	PA	--	RI	--	RSD	--
Reference Type	images		words	--	words	--	images		images	

3.3 Designers' Sketches, Referenced Pictures, and Lateral Thinking Categories

The researcher and two coders analyzed the sketches, reference content, and concept words drawn by all the designers during the ideation process. The results revealed that all together, the designers drew a total of 121 idea sketches, saved 100 reference pictures, and input 284 concept words. The categorization of designer G1's sketches and of designer I1's picture references is shown in Fig. 3, and the classification of all of the designers' idea sketches, reference pictures, and concept words is presented in Table 4.

The five graphic, product, and interior designers drew a total of 67, 48, and 6 sketches, respectively. On average, each graphic designer produced 13.4 sketches, each product designer produced 9.6 sketches, and each interior designer produced 1.2 sketches.

Regarding lateral thinking classification, designer G3 demonstrated the highest rate of lateral thinking among the graphic designers, with sketches separated into seven categories. By contrast, designer G1 demonstrated the lowest rate of lateral thinking in the graphic design field, with sketches only separated into three categories. Furthermore, the logo sketches drawn by the graphic designers were mostly variations of the "coffee cup + word" concept, followed by the "type design" concept and the "coffee beans + word" concept. Designer D3 demonstrated the highest rate of lateral thinking among the product designers, with sketches separated into five categories; D2, D4, and D5 demonstrated the lowest rate of lateral thinking among the product designers, with sketches separated into only two categories. In addition, most of the product designers' coffee shop chair sketches were variations of "arc" sketches and "simple line" sketches, followed by the "plant shape" and "coffee bean shape." Among the interior designers, designer I5 had sketches that were separated into two categories of lateral thinking, and all of the other interior designers' sketches were placed in one category of lateral thinking. Most of the coffee shop bar design sketches made by the interior designers were categorized as a "space layout" design.

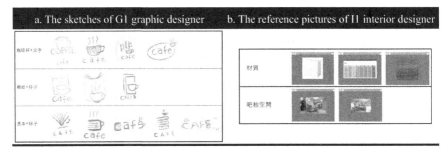

Fig. 3. Sketches of designer G1 and the reference pictures of designer I1

A closer examination of the reference picture analysis results revealed that the five graphic designers had only saved three reference pictures. The average number of pictures saved by each graphic designer was 0.4, which was the smallest average among the three design fields. Only designer G5 saved three pictures that were grouped into two categories. By contrast, the five product designers saved 59 reference pictures in total. The average number of pictures saved by each product designer was 4.2, which was the highest average among the three design fields. Their reference pictures were mostly classified as "arc chair design," followed by "coffee shop space." Finally, the interior designers saved 38 reference pictures altogether. The average number of pictures saved by each interior designer was 3.6, and their reference pictures were mostly classified as "coffee shop space," followed by "material."

In addition, the five graphic designers input a total of 78 concept words for concept word analysis. Among these designers, the average number of input concept words was 17.8 per designer. Most of their concept words were classified as "style," followed by "objects" and then "feeling." The five product designers input a total of 192 concept words. The average number of input concept words was 41.4 per designer, constituting the highest rate of input across the three design fields. Most of their concept words

Table 4. Number of designers' idea sketches, reference pictures, concept words and category in three design fields

Idea Sketches																
Group	Graphic Design					Product Design					Interior Design					
Participant	G1	G2	G3	G4	G5	D1	D2	D3	D4	D5	I1	I2	I3	I4	I5	Ave.
Amount	12	18	16	10	11	8	4	15	11	10	1	1	1	1	2	8.07
Average			13.4					9.6					1.2			
Categories	3	4	7	5	4	3	2	5	2	2	1	2	1	1	1	2.87
Average			4.6					2.8					1.2			
Reference Pictures																
Group	Graphic Design					Product Design					Interior Design					
Participant	G1	G2	G3	G4	G5	D1	D2	D3	D4	D5	I1	I2	I3	I4	I5	Ave.
Amount	0	0	0	0	3	9	22	7	8	13	5	13	4	6	10	6.67
Average			0.6					11.8					7.6			
Categories	0	0	0	0	2	1	7	3	2	8	2	4	3	4	5	2.73
Average			0.4					4.2					3.6			
Mind map Ideas																
Group	Graphic Design					Product Design					Interior Design					
Participant	G1	G2	G3	G4	G5	D1	D2	D3	D4	D5	I1	I2	I3	I4	I5	Ave.
Amount	17	17	26	9	20	5	104	22	48	28	5	13	11	5	25	23.67
Average			17.8					41.4					11.8			
Categories	7	6	5	4	7	3	5	5	7	5	1	4	3	2	5	5.0
Average			5.8					5.0					3.0			

were classified as "requirement and restriction," followed by "style" and then "feeling." Finally, the interior designers input a total of 56 concept words. The average number of input concept words per designer was 11.8, constituting the lowest rate of input across the three design fields. Most of their concept words were classified as "style," followed by "objects" and then "requirement and restriction."

In order to understand whether a correlation existed among the designers' idea sketches, reference pictures, and concept words, all of the designers' idea sketches, sketch categories, reference pictures, reference picture categories, concept words, and concept word categories were subjected to a correlation coefficient test (Table 5). The statistical results revealed that positive correlations between the number of idea sketches and "idea sketch" categories; the number of idea sketches, concept words, and "concept word" categories; the number of "idea sketch" categories and "concept word" categories; the number of reference pictures and "reference picture" categories; and the number of concept words and "concept word" categories for six types of double variables. Negative correlations were also observed among the number of idea sketches, reference pictures, and "reference picture" categories and among the number of "idea sketch" categories, reference pictures, and "reference picture" categories for four types of double variables.

In terms of the results of the correlation coefficient test, we observed a positive correlation between quantity and category among the designers' idea sketches, reference pictures, and concept words. Our most notable finding was that the greater the number of concept words that the designers input on the mental map interface of IDEATOR v.2, the greater the number of idea sketches they drew. With respect to negative correlations, the more reference pictures and categories saved by the designers, the fewer idea sketches they drew. In particular, the more "reference picture" categories they referred to, the fewer sketches they produced, which led to the emergence of fewer "idea sketch" categories. In accordance with this discussion of the relationships between the designers' idea sketches, reference pictures, and concept words, we argue that the designers' concept words may increase idea sketch quantity and expand lateral thinking compared with reference pictures. The results reinforce the importance that word-to-word association has for designers' ideation: when they input more concept keywords, their sketches span across more categories. We will verify these results and increase the number of designers studied in future research.

Table 5. Results of the correlation coefficient test between designers' idea sketches, reference pictures, concept words and category

Item		Sketches	Sketch Category	Picture	Picture Category	Word	Word Category
Sketches	r	1.000					
	p	--					
Sketch Category	r	.826**	1.000				
	p	.000	--				
Pictures	r	-.488*	-.487	1.000			
	p	.034	.033	--			
Picture Category	r	-.570*	-.624**	.850**	1.000		
	p	.013	.006	.000	--		
Word	r	.446*	.187	.319	.309	1.000	
	p	.048	.252	.123	.131	--	
Word Category	r	.746**	.443*	-.172	-.205	.704**	1.000
	p	.001	.049	.270	.232	.002	--

* means P<.05, ** means P<.01

4   Conclusion and Recommendations

In this study, we evaluated the performance of designers' idea development using the IDEATOR app as the designers' ideation tool and investigated whether there were differences in behavior linkages and idea sketches among designers from three design fields. The results were as follows:

1. According to our analysis of the video data recorded by designers using IDEATOR, their interface operation behaviors included nine behavioral codes separated into three behavior modes (GI, GA, TH) and an error action code;

2. The designers used IDEATOR for ideation and most frequently engaged in GI behavior, especially those in the fields of graphic and product design;
3. The designers who used IDEATOR for ideation had a tendency to frequently develop concept words. In particular, graphic designers were highly dependent on word-type data to perform the ideation process; and
4. The more input concept words on the interface of the IDEATOR mind map employed by the designers, the more idea sketches the designers drew. As they input additional concept words, the designers also produced sketches that fit into multiple lateral thinking categories.

This research achieved the expected results in line with the innovative results of the previous preliminary study. In addition, this study provides an effective recording tool for use in future ideation process research. Furthermore, on the basis of the relationships among idea sketches, reference pictures, and concept words, we argue that concept words have a greater impact on idea sketch quantity and lateral thinking than do reference pictures. This assertion will be verified with further research and a larger pool of participants.

Acknowledgement. The author gratefully acknowledges the support provided by the Ministry of Science and Technology under Grant No. MOST 109–2410-H-004–036. Additional gratitude goes to the 15 designers who participated in this study and the two coders, Nian-Chen Cai and Tsai-Ping Chang, who participated in the analysis.

References

1. Sun, L., Xiang, W., Chai, C., Wang, C., Huang, Q.: Creative segment: a descriptive theory applied to computer-aided sketching. Des. Stud. **35**(1), 54–79 (2014)
2. Ozkaya, I., Akin, Ö.: Requirement-driven design: assistance for information traceability in design computing. Des. Stud. **27**(3), 381–398 (2006)
3. Segers, N.M., de Vries, B., Achten, H.H.: Do word graphs stimulate design? Des. Stud. **26**(6), 625–647 (2005)
4. Siangliulue, P., Chan, J., Gajos, K.Z., Dow, S.P.: Providing timely examples improves the quantity and quality of generated ideas. In: Proceedings of the 2015 ACM SIGCHI Conference on Creativity and Cognition, Glasgow, UK (2015)
5. Golembewski, M., Selby, M.: Ideation decks: a card-based design ideation tool. In: Proceedings of the 8th ACM Conference on Designing Interactive Systems, Aarhus, Denmark (2010)
6. Cruz, V., Gaudron, N.: Open-ended objects: a tool for brainstorming. In: Proceedings of the 8th ACM Conference on Designing Interactive Systems, Aarhus, Denmark (2010)
7. Ahmed, S.: Encouraging reuse of design knowledge: a method to index knowledge. Des. Stud. **26**(6), 565–592 (2005)
8. Westerman, S.J., Kaur, S.: Supporting creative product/commercial design with computer-based image retrieval. In: Proceedings of the 14th European Conference on Cognitive Ergonomics: Invent! Explore! London, UK (2007)
9. Cheng, P.-J.: Development of a mobile app for generating creative ideas based on exploring designers' on-line resource searching and retrieval behavior. Des. Stud. **44C**, 74–99 (2016)

10. Cheng, P.-J.: A study on Designers' Searching-retrieving Behaviour in the Ideation Process. (PhD), National Yunlin University of Science &Technology, Unpublished doctoral dissertation (2010)

11. Cheng, P.-J., Yen, J.: Study on searching-retrieving behaviour in designers' ideation process. Bull. Jpn. Soc. Sci. Des. **55**(3), 91–98 (2008)

12. Schön, D.A., Wiggins, G.: Kinds of seeing and their function in designing. Des.Stud. **13**(2), 135–156 (1992)

13. Dorst, K., Cross, N.: Creativity in the design process: co evolution of problem-solution. Des. Stud. **22**(5), 425–437 (2001)

14. Suwa, M., Gero, J., Purcell, T.: Unexpected discoveries and S-invention of design requirements: important vehicles for a design process. Des. Stud. **21**(6), 539–567 (2000)

15. Verstijnem, I., Hennessey, J., Leeuwen, C., Hamel, R., Goldschmidt, G.: Sketching and design creative discovery. Des. Stud. **19**(4), 519–546 (1998)

16. McGown, A., Green, G., Rodgers, P.: Visible ideas: information patterns of conceptual sketch activity. Des. Stud. **19**(4), 431–453 (1998)

17. Goldschmidt, G.: On visual design thinking: the vis kids of architecture. Des. Stud. **15**(2), 159–174 (1994)

18. Herbert, D.: Architectural and Study Drawings. Wiley, New York (1993)

Facial Feature Recognition System Development for Enhancing Customer Experience in Cosmetics

Irene Chiocchia and Pei-Luen Patrick Rau[✉]

Tsinghua University, Beijing, China
{chiocchiai10,rpl}@mail.tsinghua.edu.cn

Abstract. Face recognition has gained increasing attention during recent years thanks to technology development, however, the majority of its application is limited to access and security, while a much wider potential is yet to be exploited. Based on the insight provided by the literature review the present work research objective has been defined as the development of a smartphone application applying neural networks to identify and classify two eye features: color and shape. The main methodology steps include eye shape and color classes definition and selection, dataset collection and preprocessing neural network development, interface flow definition and smartphone application deployment. The result is an integrated and interactive system that is able to make relevant and customized make-up suggestionS to the user, achieving satisfying performances in terms of user-friendliness and accuracy (73.89% and 74.58% for color and shape classification, respectively). The present study proposes three main findings: definition of eye shapes and color classification, development of a neural network system for eye feature classification with high accuracy performances, deployment of a user-friendly smartphone app for personalizing customer experience in cosmetics. Therefore, the main contribution of the present study is expanding potential face recognition applications as well as providing a successful example of customer value creation through creatively applying Face recognition and Neural Networks.

Keywords: Face recognition · Image processing · Smartphone application · Cosmetics · Customer experience · Neural networks

1 Introduction

Face Recognition is a biometric system that exploits the unique facial characteristics of a person's facial traits to provide automatic recognition. Despite the significant technical improvement of the face recognition system in recent years, the performances in terms of precision and effectiveness are still relatively low compared to other biometric techniques. However, facial recognition technology has the potential to capture more personal information about the user (e.g. demographics, ethnicity, attractiveness, state of mind, emotions).

© Springer Nature Switzerland AG 2021
P.-L. P. Rau (Ed.): HCII 2021, LNCS 12771, pp. 27–40, 2021.
https://doi.org/10.1007/978-3-030-77074-7_3

A wide range of face recognition applications has already been developed for access and security control functions. However, these functions exploit limitedly the information that could be collected using face recognition. Conversely, using face recognition and image classification for service customization could further exploit these systems' potential. Service customization through face recognition has been applied in a variety of industry settings. However, some of these applications could be limited due to the low number of customer touchpoints (e.g. hospitality), or the low possibility to customize services by using face-related information (e.g. transportation), or high competitiveness and low space for innovation (e.g. retail, restaurants).

However, in the cosmetic industry, a wider potential could be unlocked.

Indeed, the Cosmetic industry could benefit from facial recognition by gaining information about customers' facial features characteristics to personalize customer experience starting from product screening, selection, and trial, usage, and post-purchase services. Moreover, only a few applications have been developed, mainly focused on the combination of AR and Facial features localization.

In Cosmetics, new value creation opportunities could be created by changing perspective: applying face recognition not to identify people but to learn about their unique features and use this information to provide knowledge to both customers and companies. The level of analysis of human faces changes from high level faces profiling and classification, to low-level features categorization, searching for detailed information about the single features instead of the overall face.

2 Literature Review

Face recognition is a sub-area of pattern recognition research and technology whose objective is to detect and classify human faces. Although the first studies on face recognition systems could be traced back to 1960, performance improvement of such systems has been achieved only during the last decade, thanks to development in pattern recognition, machine learning, and higher computer power efficiency. While the first computer-based systems were based on a set of geometrical features, recently, nowadays a wide set of methodologies are being used, ranging from template matching, or feature extraction by neural and Hopfield-type networks [21].

The face recognition process could be divided into four different phases: acquisition and pre-processing, face detection, face recognition, further processing (expression detection/feature extraction). The latter step refers to the collection of further information and analyzing facial landmarks [15]. For instance, Fuentes-Hurtado [11] has developed a novel methodology to extract and group three facial features (mouths, noses, and eyes) according to their appearance.

Moreover, Different face recognition techniques have different levels of robustness to variations of illumination, rotation, scale, and facial expression which vary in dependence on the attribute to identify [18].

During recent decades, a wide range of technologies has been developed and applied to face recognition. The most important, considering their popularity and performances, are traditional face recognition algorithms (Geometric techniques, PCA.), Neural networks. Moreover, other techniques could be used such as face descriptor-based methods, 3D-based face methods, video-based methods, and Gabor wavelets.

2.1 Traditional Face Recognition Algorithms

The traditional face recognition algorithm, first developed in late 1970, could be further divided into two categories: Local feature approach and Holistic approach.

The Local feature approach, use geometric facial features and geometric distances between them to perform face recognition. Kanade [14] presented an automatic feature extraction method based on ratios of distances with a recognition rate between 45–75%. Cox [7] has introduced a mixture-distance technique that achieves a recognition rate of 95%. One of the main advantages of these techniques is that local appearance features are more stable to local changes such as expression and misalignment compared to holistic techniques. However, they do not have a high degree of accuracy and require considerable computational capacity [22].

The holistic approach refers to the application of holistic texture features to the whole face. The most commonly used algorithms include the following: principal component analysis (PCA), independent component analysis (ICA), linear discriminate analysis (LDA), and linear regression classifier (LRC). Although these methods are relatively easy to apply and low computational cost is demanded, they are highly sensitive to variations in illumination, distortions due to facial expressions, and other disturbances. To deal with such cases, nonlinear extensions have been proposed like kernel PCA (KPCA), kernel LDA (KLDA). Most of these nonlinear methods are based on kernel techniques, which map the input face images into a higher-dimensional space. According to Hassaballah [12], applying kernel-based nonlinear methods do not produce a significant improvement compared to linear methods. Although PCA can outperform many other techniques when the size of the database is small, when the dataset size increases, its performance degrades [4]. Moreover, in most cases, they use unautomated pre-processing methods such as hand-labeling of key facial regions. This reduces the technique efficiency and increases the error rates that could be introduced during the preprocessing phase [23].

2.2 Neural Networks

Artificial neural networks are computer-based systems able to adapt, learn and organize data that has been extensively applied for image classification and face recognition. According to the type of structure, Neural Networks could be classified into different types; one of the most important is Convolutional Neural Networks (CNN).

CNN is defined as a class of neural networks that is specialized in processing grid-like structured data, such as images. What makes this type of network different from the others is the convolutional layer. In a convolutional layer each set of neurons analyses a specific region of the image along 3 dimensions. Neurons belonging to one layer are connected with the subsequent layers of neurons in a feed-forward manner and respond to partly overlapping regions in the visual field.

Laundry [16] conducted a comparative study of CNN, PCA, and KNN, demonstrating the superior performances of CNN over the other techniques: The best results (accuracy of 98.3%) were obtained using the proposed CNN when trained using 320 images, while the other methods' accuracy was between 80% and 70%. Moreover, other successful applications of CNN for face pattern recognition could be found in the literature [1, 2, 13, 17, 20]. CNN models have been also used to identify and locate eyes on grayscale images by searching characterizing features of the eyes and eye sockets [8–10]. Moreover, CNN has been combined with face alignment methods for predicting personal attributes from facial images [3, 18]. The main advantages of CNN are their ability to overcome partial distortion and occlusion, high scalability, high accuracy, and the possibility to use multiple labels for classification. However, the main disadvantage of CNN is the great number of training samples required. If a small dataset is used accuracy performance degrades significantly [19].

3   Methodology

The development of the present face recognition system is constituted by seven main phases: Idea formulation, Eye class selection, Application features and interface flow definition, Image database collection, Neural Network development, App implementation, Testing, debugging and releasing.

3.1   Idea Formulation

Insights from the literature review were used to identify an area of research with the highest unexploited potential. Considering the extremely low number of applications and research articles developed, facial features classification in the cosmetic industry has been selected as the most promising area for further research. Moreover, due to the high efficacy and efficiency, Neural Networks has been selected to perform facial features classification. However, the lack of standardized classifications developed by previous research, make it more reasonable to focus on a single facial feature. Then, based on previous research work and make-up articles, an evaluation of facial features' easiness of classification and importance in cosmetics was performed. The outcome of this analysis suggested that eyes are the most interesting feature for the present research and in particular eyes' color and shape.

Therefore, the research idea has been defined as the development of face recognition and eye classification smartphone applications for the cosmetic industry. The main objective of the application is to provide a useful suggestion about makeup styles and products through eye shapes and color identification and clustering. By using the application, users could take or upload a picture of their eye on which Neural Network model will be run to identify the type of eye shape and color. The outcome of the classification is used to display customized makeup and product recommendations. Additional features of the app will include an online shopping area, personalized profile and history, other suggestions, and tutorials.

When considering the number of eye feature classes, it is important to define a trade-off between accuracy and efficiency. Indeed, a high number of eyes' types of classes may lead to and excessive fragmentation, introducing more error. Conversely, when choosing too few classes, it is not possible to provide a personalized and useful suggestion. By examining this trade-off, three eye color (blue, green, brown) and four eye type (monolid, round, upturned, and downturned) classes were defined.

3.2 Eyes Classes Selection

The majority of the classification methods developed by research is based on geometrical analysis of eyes features on different dimensions: visibility of the upper crease, space between the iris and the lower lid, the orientation of the outward angle of the compared to the centerline, the distance among the inner corner of the eye (Fig. 1).

Considering the degree of visibility of the upper crease four main classes could be defined: monolid (crease is not visible and not exits), hooded (crease is partially visible), visible crease (crease is present and visible, could be slightly covered by the eyelid), protruding eyes (the whole crease is visible and it has a wide dimension) [5]. Crease visibility also impacts how to apply makeup to increase the dimension of the eyes or mitigating excesses. However, determining crease visibility will result in high difficulty when using two-dimension images since the extent of the visibility of the lid could be distorted when using frontal images. Therefore, it appears more reasonable to consider only monolid eyes for present classification, whose features could be discerned by the overall appearance of the eye.

Concerning the outer corner of the eye orientation, three cases could be defined: upturned, central, downturned. Upturned eyes have their outer corner above the centerline defining the horizontal axis of the eyes, while downturned eyes have their outer corner below the centerline. The average case has the outer corners positioned around the horizontal central line. The orientation of the outer corner of the eye is highly relevant in defining how to apply eye makeup products in order to compensate for the preponderance of the upper lid or the lower lid [6]. Moreover, an image processing algorithm could be trained to recognize such eye shape by focusing on the pixels on the outwards area of the eyes.

The space between the iris and the lower lid allows us to define whether a specific set of eyes is round or almond-shaped. Round eyes have a white cornea area visible below the eyelid, while almond eyes have not. A wide range of styles has already been developed for round eyes to emphasize their roundness or mitigate it. Moreover, the possibility to deduce the eye shape by looking at different parts of the eyes, makes round eyes a perfect candidate to be included in the present model.

The distance among inner corners leads to identify mainly two cases: close-set eyes (distance between the inner corners of the eyes is less than an eyeball width apart) and wide-set eyes (the distance is more than an eyeball width). However, to detect this feater a picture including both eyes is required, introducing a higher amount of noise. Moreover, this eye features' on make-up recommendation is limited. For these reasons, these eye types were not included in the classification.

 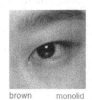 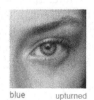

green   downturned   brown   monolid   blue   upturned   green   round

Fig. 1. Eye classes - Examples

3.3   Application Features and Interface Flow Definition

The app design has been developed using Just in Mind prototyping, a prototyping soft-ware enabling the design of wireframes and the possibility to turn them into interactive prototypes. The user experience through the app could be discomposed in different phases, which are not strictly sequential: Login, Eye scan, Eye classification results, Makeup styles, Products, Shopping, Personal profile.

When opening the app, if the user has not logged out, he/she will be redirected to the home interface. Otherwise, a sign-in page will require the user to insert personal information. In the case of a new user, he/she could click on "Sign up" to create a new profile by inserting personal information on a dedicated interface (Fig. 2).

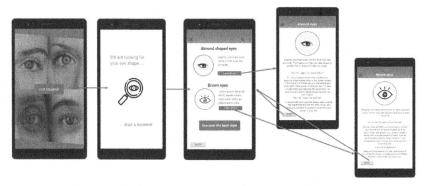

Fig. 2. User Interface Flow – Eye classification (NN function) and Eye classification result

After a new user has been successfully registered, the user will be asked to take or upload a picture of her/his eye on which neural network will be run. Then, using the neural network output, a new user interface will be displayed showing the results about eye shapes, color classes, and a brief description. By clicking on "Learn more", another interface will be displayed showing personalized makeup suggestions.

The user could go back to the previous interface and continue his/her journey by clicking on "Discover makeup styles" (Fig. 3). By clicking on this button, and user interface containing a list of makeup styles will be displayed (between 5 and 15 depending on the eye type). All the makeup styles displayed on the page are considering the specific characteristic of the eye shape and color, to present to the user only the makeup styles that are consistent with his/her profile. Each makeup style will be identified by a title, a brief description, and a picture. Then the user could click on "learn more" to display a detailed description of the makeup, a bigger picture, and the type of eyes for which it is suitable. If the user is interested in learning about how to replicate it, he/she could just click on "show tutorial". Then, another user interface will display the main steps to realize the selected style and the products to replicate the style.

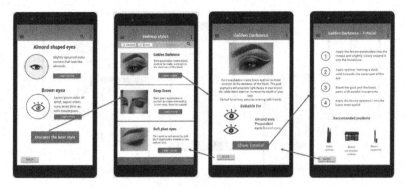

Fig. 3. User Interface Flow – Personalized makeup styles

When clicking on each product a dedicated interface will show detailed information: a big picture supported by a small description and the type of eyes for which it is suitable. The user could choose to purchase the product by clicking on "Shop now" or visualize all makeup styles using that product by clicking on "Show other makeup styles" (Fig. 4).

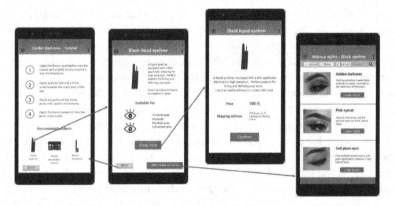

Fig. 4. User Interface Flow – Personalized products styles

A menu button will be present on the upper right corner of every interface page with the following options: personal profile, eye scan, products, makeup styles (Fig. 5).

When clicking on the "products" section on the menu, the user will get access to a list of eye makeup products with brief descriptions. When clicking on a specific product a dedicated page with further details will be displayed, as described before. "makeup styles" leads to a list of makeup styles, which is the same displayed after eye scanning or clicking on "discover new makeup styles" on a product page.

By clicking on "personal profile" it is also possible to visualize the user profile icon and personal information (e.g. name, surname, age, gender, county) and the results from the last eye scanning on the type of eye shape and color. Eye scan allows to get back to the eye scanning function by uploading or taking a picture.

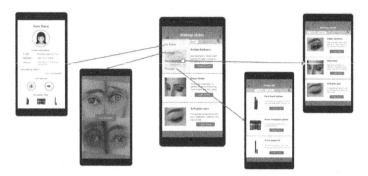

Fig. 5. User Interface – Menu

3.4 Image Database Collection

To enable a precise recognition and classification of facial features, the image database to be used in the study should be composed of colored eye images with good illumination and neutral expression. Being a relatively new research topic, not many databases have been developed for a similar purpose. Therefore, the characteristics of the available datasets (e.g. black and white images, multiple eye images of the same person) are incompatible with the requirement of the present study. Consequently, a new database was assembled by collecting images on a variety of websites (e.g. Pinterest, Google). A total of almost 1631 face images in JPG format were collected, but after sorting and selection, only 1000 eyes images were used for training the model.

After collection, using Anaconda and OpenCV an eye detection function was developed to locate eyes in the picture and crop the image to include only the eye.

After the face detection, and eye detection models have been imported, for each image a loop is run for detecting the position of the face and of the eye in the picture. Then the area around the eye's location is cropped, the picture is aligned considering the orientation of the face, and saved as a jpg file. A sorting phase followed, where all the images with poor illumination, resolution, or difficulty to categorize were discarded. The remaining pictures were categorized and labeled according to eyes color and shape, obtaining a rather uniform distribution among labels.

Figure 6 shows the distribution of the dataset's images across the four shapes and three color classes. The lower number of green and blue eyes for the monoid eye shape is connected to genetic aspects. To compensate for this, a higher percentage of blue and green eyes was retrieved for other eye shapes. As well, downturned and upturned are less common eye shapes than round and monolid, which led to a slightly lower number of samples for the former classes compared to the latter.

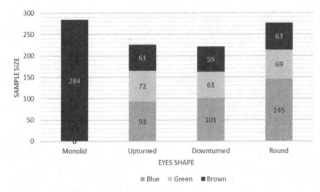

Fig. 6. Dataset distribution

3.5 Neural Network Development

The main tools used to develop the AI function for eye type classification are Tensorflow and AutoML. Tensorflow is an end-to-end open-source library for machine learning. Tensorflow high-level API based on Keras API standard allowed us to define and train neural networks with a variety of layers and characteristics.AutoML provides training services for neural network image labeling models. By providing image labels and the input dataset, through one-hour training a neural network model is trained and can be subsequently implemented on an Android Studio developed app.

The list of eye images represents the input of the convolutional neural network, while the list of colors' and shapes' labels is the desired output of the model.

First, a Neural Network model was developed by using Tensorflow. The model is composed of four convolutional layers (Conv2D) and three fully connected layers (Dense). Between consecutive convolutional layers, a two-dimensional pooling layer is inserted (MaxPooling2D) (Fig. 7).

Convolutional layers are two dimensional with a kernel size of 5×5. Each convolutional layer applies a 5×5 pixel filter to each part of the image returning as output a feature map, which indicates the location of the labeled features. By increasing the number of output filters, feature extraction on a finer scale is performed.

Added after every convolutional layer, pooling layers reduce the dimensionality of the features by converting every 2×2 pixel grid in the image to one pixel. By applying dropout to the output of the pooling layer, a randomly selected fraction of the input is set equal to 0, to prevent overfitting. "Flatten()" is used to connect the convolutional layers to the fully connected layers whose number of neurons decreases progressively.

Then, a Neural Network model is trained using AutoML, showing an accuracy of 72.5% which is 15.5% points higher than the one obtained with Tensorflow (56%); higher performances as probably due to the higher computational capabilities of Google cloud infrastructure, which is exploited for the AutoML training process.

The Neural Network model developed with AutoML is the final model applied to the Smartphone application, both due to higher accuracy, and better integration with the other app functions.

```
#dividing training and testing dataset
X_train, X_test, y_train, y_test = train_test_split(X, y, random_state=42, test_size=0.1)
#creating a model instance with keras
model = Sequential()
#adding a 2d convoltuional layer 5x5 with 16 output filters, adding 2x2 pooling layer and dropout
model.add(Conv2D(filters=16, kernel_size=(5, 5), activation="relu", input_shape=(400,400,3)))
model.add(MaxPooling2D(pool_size=(2, 2)))
model.add(Dropout(0.25))

#adding a 2d convoltuional layer 5x5 with 32 output filters, adding 2x2 pooling layer and dropout
model.add(Conv2D(filters=32, kernel_size=(5, 5), activation='relu'))
model.add(MaxPooling2D(pool_size=(2, 2)))
model.add(Dropout(0.25))

#adding a 2d convoltuional layer 5x5 with 64 output filters, adding 2x2 pooling layer and dropout
model.add(Conv2D(filters=64, kernel_size=(5, 5), activation="relu"))
model.add(MaxPooling2D(pool_size=(2, 2)))
model.add(Dropout(0.25))

#adding a 2d convoltuional layer 5x5 with 64 output filters, adding 2x2 pooling layer and dropout
model.add(Conv2D(filters=64, kernel_size=(5, 5), activation='relu'))
model.add(MaxPooling2D(pool_size=(2, 2)))
model.add(Dropout(0.25))

#add 2 dense connected layers
model.add(Flatten())
model.add(Dense(128, activation='relu'))
model.add(Dropout(0.5))
model.add(Dense(64, activation='relu'))
model.add(Dropout(0.5))
model.add(Dense(2, activation='sigmoid'))


model.compile(optimizer='adam', loss='binary_crossentropy', metrics=['accuracy'])
model.fit(X_train, y_train, epochs=10, validation_data=(X_test, y_test), batch_size=64)
```

Fig. 7. Neural Network code - Tensorflow model

3.6 App Implementation

The app was fully developed using Android Studio, integrating with Firebase, and SQLite database. A total of 15 activities and 5 classes were developed using Kotlin and Java. App development followed a sequential and iterative process. Activities were defined sequentially following the user path and the prototype developed with Just in Mind Prototype. For each function, several debugging phases were carried out to prevent run time errors and tackle exceptions. There are five main app functions: authentication and personal profile, Picture collection and standardization, eye classification, personalized makeup styles, and product navigation.

Authentication and Personal Profile
When opening the application, if the user has not logged out in the previous session, the user id and result of the last eye scanning is retrieved using an instance of SQLite database, and the user is redirected to the personalized makeup style list. When clicking on "Sign In" the information inserted in the text fields are compared with the ones stored on the Firebase database by calling an instance of "Firebase Authentication". In case the authentication is successful the user will be redirected to the personalized makeup list page, if not, an error message will be displayed.

When registering a new user, if all the fields have been correctly filled, the user is inserted in the SQLite database and is redirected to the main AI function.

Besides, when clicking on the menu item "Personal profile", an instance of the SQLite database is created, several queries allow to retrieve the required user information using the current user ID and customizing the information on the interface.

Picture Collection and Standardization

To reduce the noise introduced by variations in illumination, expression, and orientation, an interface before the picture uploading is inserted, asking the user to take off glasses, assume a neutral expression, use a neutral light and include just the eye in the picture. To further reduce noise, regulation of the picture orientation is performed using dedicated functions that deduce the orientation of the picture and correct it.

Eye Classification, Personalized Makeup Styles and Products Navigation

After the image standardization, the two models for eye shape and color classification are applied to a dedicated activity. After the model has been retrieved from the local storage, a new instance is built, and a score threshold value is associated when the model is called on the image. Then a series of if-else clauses allows assigning the right code to the eye type identified, which is memorized in the SQLite database instance of the current user and is passed to the subsequent activity: an overall result interface, whose text and the image will be modified according to the results of the classification, using the numerical code.

Similarly, also the dedicated pages for the description of eye color and shape, as well as the make up styles and product list will be personalized. When an item in the make up/product list is selected the relative position of the item clicked is memorized in a local variable that is passed to the following interface to personalize its content.

3.7 Testing, Debugging and Releasing

After the app has been tested on simulated and real devices and released through a signed release APK (Android Package Kit), which is the file format used by Android operating system for the distribution and installation of mobile applications.

4 Results

The result of the present study is an eye image classification smartphone app, deploying two neural network models. The main criteria used to access the application design and technical performances are accuracy and speed.

The accuracy of the Neural Network model is assessed using several metrics: average accuracy and recall, percentage of false positive, true negative, and accuracy of label prediction. Precision defines the percentage of correct predictions over the overall predictions performed, which is the amount of true positive over the overall assigned labels. Therefore, precision is inversional proportional to the percentage of false-positive, the number of labels predicted that were incorrect (1).

$$\text{Precision} = (\text{True positive})/(\text{True positive} + \text{False positive}) \qquad (1)$$

Recall is calculated as the ratio between the correctly assigned labels and the total number of labels that should have been assigned, which is the amount of true positive over the sum of true positive and false negative (2), thus is inversional proportional to the percentage of false positive. Recall is called sensitivity and could be interpreted as the probability of assigning a relevant label to an object in the dataset.

$$\text{Recall} = (\text{True positive})/(\text{True positive} + \text{False negative}) \tag{2}$$

A trade-off between recall and precision should be defined by selecting a specific scoring threshold, which is directly proportional to precision and inversional proportional to recall. To define the optimal threshold, false positive minimization is set as a priority, while false negative should be set as low as possible to enable the model to assign all relevant labels. For the eye shape model, the value of 0.45 was selected, which leads to a precision of 82.98% and a recall of 41.49%. While for the eye color model, a score threshold of 0.4 leads to obtain 76.47% accuracy, and 52% for recall.

Another important parameter is the average precision, which measures the accuracy of the model across all score thresholds. The value for the eye shape model was 73.55%, while for the eye color model was 74.58%, showing good average results.

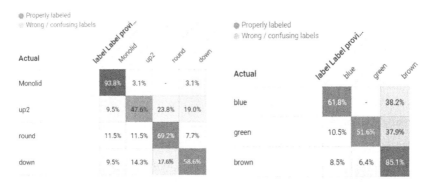

Fig. 8. Confusion matrix for the eyes shape (left) and eye colour (right) model

Confusion matrix rows show predicted results, while columns show the real results (Fig. 8). The percentage on the diagonal indicate the correctly predicted labels, while for a cell in column i and row j, the percentage indicates the incorrectly predicted images with label i, that should have been predicted as j. For the eye shape classification model, excellent results are shown for Monolid eyes (93.8% accuracy) and satisfying results for round eyes (69.2%). While, lower performances are achieved for downturned and upturned eyes (respectively 58.6% and 47.6% accuracy), due to the limited number of images used as a sample.

Considering the eye color classification model, excellent results are shown for brown eyes (85.1%) and satisfying results for blue eyes (61.8%). While, lower performances are achieved for green eyes, with 51.6% accuracy, due to the lower amount of image collected for this class (202) compared to blue 339 and 467 brown eyes. Such a gap is also related to the lower popularity of the green eye as an eye color trait.

The model only takes 1 s to run and 1 more second to display the result on the interface, allowing for a fast and efficient solution.

5 Conclusion and Final Remarks

The present research work is aimed at developing a smartphone app for cosmetics with face features recognition and classification function.

First, an extensive literature study was conducted, exploring in detail the state of the art of both technical methodologies and market applications. Based on the literature review study, a research gap with unique market potential was identified: using face recognition to personalize the customer experience in cosmetics. The originality of the research topic made it necessary to use creatively the available resources and tools, yet a rigorous methodology was used.

Assessing the overall performances of the system, satisfying results have been achieved on user-friendliness, accuracy and speed. Readable, simple, and intuitive interfaces allow the user to easily navigate through the app and access a variety of features. Moreover, high accuracy (73.55% and 74.58% for the color and shape classification models respectively) high speed, low memory, and computational cost contribute to overall excellent technical performances.

Other advantages for the user include the increase in the convenience of getting a professional consultation, the reduction of the commitment to purchase, higher personalization of service both on store and on the app, by collecting information on the user's eye type and behaviour in using the app.

Further research activities on the topic include widening the number and type of classes for eye features, eventually using unlabelled machine learning, increasing the number of samples in the dataset and developing a complete business model.

In conclusion, the originality of the present work lays mainly in the development of an interactive smartphone application that creates value in a specific industry by innovatively using face recognition technology. With the present work, we hope to inspire other inventions in the application of face recognition technology in unexploited areas to create a meaningful and valuable solution with market potential.

Acknowledgement. This work was funded by Tsinghua University Initiative Scientific Research Program 20193080010.

References

1. Alizadeh, S., Fazel, A.: Convolutional neural networks for facial expression recognition (2017)
2. Arya, S., Agrawal, A.: Face recognition with partial face recognition and convolutional neural network. Int. J. Adv. Res. Comput. Eng. Technol. (IJARCET) 7(1) (2018)
3. Balya, D., Roska, R.: Manufactured in the Netherlands. Face and eye detection by CNN algorithms. J. VLSI Signal Process. **23**, 497–511 (1999)
4. Bhele, S.G., Mankar, V.H.: A review paper on face recognition techniques. Int. J. Adv. Res. Comput. Eng. Technol. (IJARCET) **1**(8) (2012)

5. Burton, C.: What's Your Eye Shape? Beautilish (2012)
6. Cardellino, C.: Your Ultimate Guide to Applying Eyeliner on Every Single Eye Shape, Cosmpolitan (2018)
7. Cox, I.J., Ghosn, J., Yianilos, P.N.: Feature-based face recognition using mixture-distance. In: Computer Vision and Pattern Recognition. IEEE Press (1996)
8. Dubey, A.K., Jain, V.: A review of face recognition methods using deep learning network. J. Inf. Optim. Sci. **40**(2), 547–558 (2019). https://doi.org/10.1080/02522667.2019.1582875
9. Eide, A.J., Jahren, C., Jorgensen, S., Lindblad, T., Lindsey, C.S., Osterud, K.: Eye identification for face recognition with neural networks. In: Applications and Science of Artificial Neural Networks II (1996). https://doi.org/10.1117/12.235925
10. El-Sayed, M.A., Khfagy, M.A.: An identification system using eye detection based on wavelets and neural networks. Int. J. Comput. Inf. Technol. **01**(02) (2012)
11. Fuentes-Hurtado, F., Mas, J.A., Naranjo, V., Alcaniz, M.: Automatic classification of human facial features based on their appearance. PLoS ONE (2019)
12. Hassaballah, M., Aly, S.: Face recognition: challenges, achievements and Future directions'. IET Comput. Vis. **9**(4), 614–626 (2015)
13. Kamencay, P., Benco, M., Mizidos, T., Radil, R.: A new method for face recognition using convolutional neural network digital image processing and computer graphics. **15**(4) (2017)
14. Kanade, T.: Picture processing by computer complex and recognition of human faces (1973)
15. Kumar, S., Sigh, S., Kumar, J.: A study on face recognition techniques with age and gender classification (2016)
16. Landry, L., Fute, E., Tonye, E.: CNNSFR: a convolutional neural network system for face detection and recognition. (IJACSA) Int. J. Adv. Comput. Sci. Appl. **9**(12) (2018)
17. Lawrence, S., Giles, C.L., Tsoi, A.C., Back, A.D.: Face recognition: a convolutional, neural-network approach. IEEE Trans. Neural Netw. **8** (1997)
18. Lewenberg, Y., Bachrach, Y., Shankar, S., Criminsi, A.: Predicting personal traits from facial images using convolutional neural networks augmented with facial landmark information (2016)
19. Mohammed, A.A., Sajjanhar, A.: Experimental comparison of approaches for feature extraction of facial attributes. Int. J. Comput. Appl. **38**(4), 187–198 (2016). https://doi.org/10.1080/1206212X.2016.1207427
20. Ranjan, R., Sankaranarayanan, S., Castillo, C.D., Chellappa, R.: An all-in one convolutional neural network for face analysis (2018)
21. Stringa, L.: Eyes detection for face recognition . Appl. Artif. Intell. Int. J. **7**(4), 365–382 (1993). https://doi.org/10.1080/08839519308949995
22. Sutherland, K., Renshaw, D., Denyer, P.B.: Automatic face recognition. In: First International Conference on Intelligent Systems Engineering, Piscataway, NJ, pp. 29–34. IEEE Press (1992)
23. Zahraddeen, S., Mohamad, F.S., Yusuf, A.A., Musa, A.N., Abdulkadir, R.: Feature extraction methods for face recognition. Int. Rev. Appl. Eng. Res. (IRAER) **5**(3) (2016). ISSN 2248-9967

Maintenance Feasibility Analysis Based on a Comprehensive Indicator

Shaowen Ding[1], Teng Zhang[2], Changhua Jiang[1], Bo Wang[1], Yan Zhao[1], Fenggang Xu[1], and Jianwei Niu[2(✉)]

[1] National Key Laboratory of Human Factors Engineering, China Astronaut Research and Training Center, Beijing 100094, China
[2] School of Mechanical Engineering, University of Science and Technology Beijing, Beijing 100083, China
niujw@ustb.edu.cn

Abstract. Maintainability is an important attribute in product design, which has a significant impact on reducing product's whole life cycle costs. Visibility, accessibility, operating space, and comfort are the important factors associated with maintainability in maintenance operations. In this paper, a comprehensive evaluation model is constructed to comprehensively evaluate the main factors that affect maintainability. In order to assess accessibility during operation, we classify the accessibility of maintenance equipment in several different levels based on the human accessibility envelope. By literature research, we divide the human view into several different areas to assess the visibility of the repair operation. We use RULA evaluation method to assess maintenance's comfort in different operating positions and classify them into different levels. For the operational space assessment, we use sweep volumes to evaluate the repair space of the hand during a repair operation. And we also construct an evaluation model base on hierarchical analysis for comprehensive analysis of maintainability. In this paper, we take the carbon ring removal process on an aircraft engine as an example and use our constructed evaluation criteria to assess the serviceability of the repair process.

Keywords: Maintainability · Visibility · Accessibility · Operating space · Hierarchical analysis

1 Introduction

Equipment maintenance is a common problem faced by all machinery and equipment. Maintainability is an important indicator of equipment performance, it reflect the ease of maintenance. Effective maintenance can reduce maintenance costs and improve the maintenance of space utilization. Ergonomics take into account maintenance operations and the maintainability of the equipment should be evaluated. The results of the evaluation provide feedback for the design of the equipment, enabling humans to perform maintenance tasks quickly, comfortably and safely.

© Springer Nature Switzerland AG 2021
P.-L. P. Rau (Ed.): HCII 2021, LNCS 12771, pp. 41–48, 2021.
https://doi.org/10.1007/978-3-030-77074-7_4

For the assessment of maintainability, Regazzoni and Rizzi used digital human technology to simulate maintenance and assembly processes. They evaluated visibility, accessibility and fatigue and then gave an evaluation method based on the simulation system. They apply virtual maintenance simulation to the evaluation of maintenance operation processes [1]. Liang Ma et al. analysis the factors of fatigue during maintenance and classified them into several levels during the maintenance process [2]. Geng et al. evaluate human comfort during maintenance operations. The comfort of the operating position is an important indicator of maintainability [3]. The above scholars focus on unilateral factors that affect maintainability, which lack of comprehensive consideration of indicators. In response to this problem, some scholars have developed assessment guidelines that take multiple factors into account. Slavila et al. evaluated the maintainability during the product design process with the fuzzy theory. They gave qualitative indexes some fuzzy values [4]. Lu and Sun built a virtual product maintainability evaluation model based on fuzzy multiple attributes decision making theory [5]. In order to comprehensively evaluate the maintainability metrics, Xu et al. develop a fuzzy evaluation-based model to guide product design [6].

The research of the above scholars has made great progress in human factors assessment, but there are still some gaps: (I) For operational space assessments, most assessments are based on expert experience or use geometric models, the validity of it is difficult to ensure. In this paper, a spatial scanning method is used to assess the operating space of manual operation, which is more suitable for complex maintenance scenarios. (II) For the assessment of serviceability, most studies focus on single-factor assessments and lacked a holistic approach. In this paper, a hierarchical analysis is used to comprehensively assess maintainability.

2   Method

2.1   Indicators for Evaluating Maintainability

There are many criteria for evaluating the maintainability of equipment, we chose accessibility, visibility, comfort, and operating space as our basic evaluation criteria. Accessibility is used to assess whether the equipment within human reach, visibility is used to assess whether the maintenance equipment is in a blind spot, comfort is used to assess how comfortable people are in their posture when they perform maintenance operations, and operating space is used to assess whether there is any collision between the maintenance personnel and other objects in the environment when they perform maintenance operations.

2.2 Accessibility Assessment

In order to perform an accessibility analysis, it is important to identify the critical areas for accessibility analysis. For the replacement repair, we expect maintenance equipment to be within physical reach. The degree of accessibility can be assess using an envelope ball tool (shown in Fig. 1). The level of human accessibility depends not only on the length of the forearm, but also on the width of the shoulder, the degree of rotation of the shoulder and trunk joints. Accessibility levels can be divided into the following categories. (I) Maintenance worker in a natural position, the equipment is inside the envelope ball, and maintenance personnel can easily reach the equipment. (II) The maintenance equipment is located inside the envelope ball and the maintenance worker needs to change position to reach the object. (III) Maintenance equipment is out of the reach of maintenance workers.

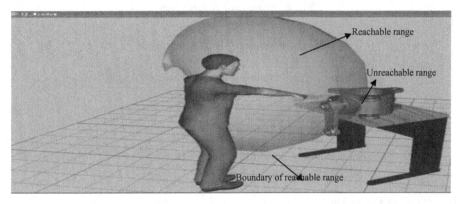

Fig. 1. Reach envelope of a limb

2.3 Visibility Assessment

According to ergonomics theory, a person's field of vision and line of sight can have a great impact on the work, which determine the accuracy and duration of human movements, so it is necessary to quantitatively evaluate the visual factor, which is generally measured in terms of visibility and visualization. When humans are in their natural standing position, Main parameter is the angle of the cone in the field of view of the operating object. $30 < Ang < 45$ is the most comfortable field of view; when $45 < Ang < 90$, the visual perception decreases as the operating object moves away from the center of the field of view [7]. Based on the angle of view of the eye, a view cone can be created in the simulation software (Fig. 2 and Table 1).

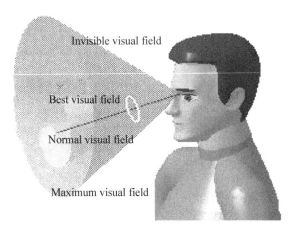

Fig. 2. Vision area of static eyes

Table 1. Maintenance visibility evaluation

Maintenance site location in the cone	Evaluation grade
Best visual field	A
Part in the best visual field	B
Maximum visual field	C
Part or full in invisible field	D

2.4 Comfort Evaluation

Based on the accessibility of maintenance operations, in order to ensure the sustainability and safety of maintenance activities, we evaluate the maintenance worker's posture from an ergonomic perspective, so as to ensure the sustainability and safety of maintenance activities. RULA (Rapid Upper Extremity Assessment) is widely used because the process is simple and feasible without interference. RULA was performed separately for arm, wrist, neck torso and leg analysis [8]. Table 2 shows the range of scores for each segment.

Table 2. Human comfort evaluation

RULA score	Posture evaluation	Evaluation grade
1–2	Don't need to change posture for a long time	A
3–4	Need to change posture after a long time maintenance	B
5–6	Need to change posture from time to time	C
7–8	Need to change posture immediately	D

2.5 Operation Space Evaluation

The purpose of the operating space assessment is to satisfy the maintenance personnel to carry out maintenance task activities, obviously, the larger the maintenance space, the better the maintenance task, however, for the actual maintenance tasks, maintenance space is often very tight and there are more interference problems, through virtual simulation methods we can simulate the operator in different sizes of the maintenance space maintenance operation activities. Maintenance operations are divided into three basic units of maintenance activity: screw, twist, and translate. two types of scan spaces are defined: free SV and constrained SV [9], the former being the maximum range of motion of the repairer's hand, and the latter being the actual sweep space of the repairer's hand. By comparing the surface area and volume of the free and constrained SVs, an indicator of the state of the maintenance space can be calculated.

Here, we define P_s and P_v as the ratio of the surface area of the constrained SV to the surface area of the free SV and the ratio of the volume of the constrained SV to the volume of the free SV respectively:

$$P_s = S_{csv}/S_{fsv} \tag{1}$$

$$P_v = V_{csv}/V_{fsv} \tag{2}$$

Where S_{csv} and S_{fsv} are the surface area of the constraint swept volume and the free swept volume, respectively; V_{csv} and V_{fsv} are the volume of the constraint swept volume and the free swept volume, respectively.

Based on the discussion above, we divide the results of the maintenance space evaluation into four levels. By using the ergonomics requirements of each maintenance operation unit, we can obtain the specific evaluation criteria for the different maintenance operations (Table 3).

Table 3. The evaluation criteria of different maintenance operations

Evaluation level	Ps	Pv
A	>0.85	>0.8
B	0.6–0.85	0.5–0.8
C	0.4–0.6	0.35–0.5
D	<0.4	<0.35

2.6 Ergonomics Comprehensive Evaluation

In order to evaluate maintenance ergonomics, this paper developed a comprehensive solution base on AHP theory [10].

First, we identify the set of evaluation indicators, $U = \{u_1, u_2, \cdots, u_p\}$, u_i represents the i-th evaluation indicator.

Then, we determine the judging level $V = \{v_1, v_2, \cdots, v_p\}$, each level corresponds to a fuzzy subset, and this paper uses four levels for evaluation: A, B, C, D.

After constructing the rank fuzzy subset, it is necessary to quantify each factor of the evaluated matter one by one.

We judge the importance of factors named from R_1 to R_n in the same layer, the judgment matrix R can be constructed.

$$R = \begin{pmatrix} r_{11} & r_{12} & \cdots & r_{1n} \\ r_{21} & r_{22} & \cdots & r_{2n} \\ \vdots & \vdots & \ddots & \vdots \\ r_{n1} & r_{n2} & \cdots & r_{nn} \end{pmatrix}$$

r_{ij} refers to the result of the comparison of r_j and r_j.

The weight of each factor is actually the calculation of the maximum characteristic radix and eigenvector. Finally, we can get the factors weight:

$$W = \{w_1, w_2, \cdots, w_n\}$$

After getting fuzzy matrix and weight of factors, the comprehensive evaluation for single level factors is:

$$B = W * R = (b_1, \cdots, b_n)$$

3 Results

The carbon ring of aircraft engine is an important part of the engine, and the disassembly process requires maintenance workers to perform a series of operations: first, the maintenance worker use the handle to engage the tool, and then, the maintenance personnel hand tighten the bushing until the collet is firmly against the carbon ring stopper. Then, the operator connects the sliding hammer to the handle and removes the assembly from the gear shaft. Finally, the operator unscrewed the sleeve and removed the carbon ring stopper from the tool. Due to lack of operation space, heavy operating tools, it is necessary to analyze ergonomics factors in the assembly process (Fig. 3).

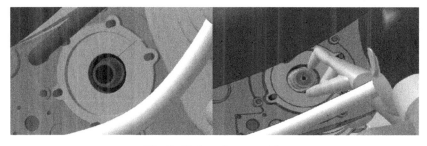

Fig. 3. Carbon ring assembly

Above solution applied to the ergonomic evaluation of assembling carbon ring. First, the evaluation factors set U and alternative set V were created:

$$U = \{\text{Accessibility, Visibility, Comfort, Operation Space}\}$$

$$V = \{A, B, C, D\}$$

The assembly process was simulated in Delmia (https://www.3ds.com/products-ser vices/delmia/). The fuzzy comprehensive evaluation matrix is as follows.

$$R = \begin{bmatrix} 1 & 0.5 & 0 & 0 \\ \frac{2}{3} & 1 & \frac{1}{3} & 0 \\ \frac{2}{3} & 1 & \frac{1}{3} & 0 \\ 0 & \frac{1}{4} & 1 & \frac{3}{7} \end{bmatrix}$$

Then we determine the weights, determine the comparative information of the indicators, and calculate the C, Q, and weights in turn.

$$C = \begin{bmatrix} 0 & 0 & 1 & -1 \\ 0 & 0 & 1 & -1 \\ -1 & -1 & 0 & -1 \\ 1 & 1 & 1 & 0 \end{bmatrix}$$

$$Q = \begin{bmatrix} 1 & 1 & e & e^{-0.4} \\ 0 & 0 & e & e^{-0.4} \\ e^{-1} & e^{-1} & 1 & e^{-1.4} \\ e^{0.4} & e^{0.4} & e^{1.4} & 0 \end{bmatrix}$$

From this we can derive the weights $W = (0.1975, 0.1905, 0.2935, 0.3185)$. Comprehensive evaluation by building a comprehensive evaluation model

$$[0.1975, 0.1905, 0.2935, 0.3185] \begin{bmatrix} 1 & 0.5 & 0 & 0 \\ \frac{2}{3} & 1 & \frac{1}{3} & 0 \\ \frac{2}{3} & 1 & \frac{1}{3} & 0 \\ 0 & \frac{1}{4} & 1 & \frac{3}{7} \end{bmatrix} = (0.2892, 0.3682, 0.2667, 0.0759)$$

4 Discussions

The results of the comprehensive metrics evaluation can help the designer to identify human factors issues when they design the maintenance task. The final score for this maintenance operation was normal, which indicates that the maintenance process could be improved.

By expert evaluation, we obtain an evaluation matrix. We translate the evaluation matrix into weighted values of factors. By analyzing the weights, we found that people have a high demand for operating space and a low demand for operating comfort. Analyzing the reasons, we concluded that it is difficult for maintenance workers to carry out maintenance operations if the operating space is not sufficient. For the comfort indicator, the operator can change the comfort of the operation by adjusting the posture.

There are still 26.7% of evaluation results focusing on "C" grade. The reason for this is the poor visibility of the repair equipment during the disassembly process, which affects the entire repair task.

The purpose of this paper is to explain the methodology for evaluating maintenance operations. The details of the evaluation will not be elaborated on.

5   Conclusion

In this paper, we propose a comprehensive metric evaluation method to assess the maintainability of equipment. We choose accessibility, visibility, operating space, and comfort as the indicators for evaluating maintainability, then, we use a fuzzy evaluation method to comprehensively evaluate the indicators based on the evaluation criteria for each indicator. By analyzing the evaluation results, we can find irrationalities in the maintenance process.

Acknowledgements. This research is supported by Equipment Advanced Research Project (Grant NO. 61400040103), the Open Funding Project of National Key Laboratory of Human Factors Engineering (Grant NO. 6142222190309), the Foundation of National Key Laboratory of Human Factors Engineering (SYFD1700F1801, SYFD180051801), China.

References

1. Regazzoni, D., Rizzi, C.: Digital human models and virtual ergonomics to improve maintainability. Comput.-Aided Des. Appl. **11**(1) (2014)
2. Ma, L., Chablat, D., Bennis, F., et al.: Fatigue evaluation in maintenance and assembly operations by digital human simulation in virtual environment. Virtual Reality (2011)
3. Geng, J., Zhou, D., Lv, C., et al.: A modeling approach for maintenance safety evaluation in a virtual maintenance environment. Comput.-Aided Des. **45**(5), 937–949 (2013)
4. Slavila, C.A., Decreuse, C., Ferney, M.: Fuzzy approach for maintainability evaluation in the design process. Concurr. Eng. Res. Appl. **13**(4), 291–300 (2005)
5. Zhong, L., Sun, Y.: Maintainability evaluation model based on fuzzy multiple attribute decision making theory for virtual products. China Mech. Eng. (2009)
6. Luo, X., Yang, Y., Ge, Z., et al.: Fuzzy grey relational analysis of design factors influencing on maintainability indices. J. Process Mech. Eng. **229**(1), 78–84 (2015)
7. Itti, L., Koch, C.: Computational modelling of visual attention. Nat. Rev. Neuroence **2**(3), 194–203 (2001)
8. Qiu, S., Yang, Y., Fan, X., et al.: Human factors automatic evaluation for entire maintenance processes in virtual environment. Assembly Autom. **34**(4), 357–369 (2014)
9. Zhou, D., Chen, J., Lv, C., et al.: A method for integrating ergonomics analysis into maintainability design in a virtual environment. Int. J. Ind. Ergon. **54**, 154–163 (2016)
10. Saaty, T.L., Kearns, K.P.: The analytic hierarchy process. Anal. Plan. (1985)

Design for the Speculative Future as a Knowledge Source

Fangzhou Dong[1]([⊠]), Sara Sterling[1], Yuzhen Li[2], and Xiaohui Li[2]

[1] Xi'an Jiaotong-Liverpool University, 111 Ren'ai Road,
Suzhou Industrial Park, Suzhou 215123, Jiangsu, China
Fangzhou.Dong@xjtlu.edu.cn
[2] Dalian University of Technology, No. 2 Linggong Road,
Ganjingzi District, Dalian 116024, Liaoning, China

Abstract. Design for the speculative future, as one genre of cultural design, aims at nudging cultural changes through posing questions and arousing discussions. As cultural changes are large-scale and long-term processes, this paper examines the values of design for the speculative future from the designers' perspective in a relatively short term. Design for the speculative future is practiced in two student studios under the topic of Chinese wedding culture, and its values are summarized based on interviews of studio participants and theoretical analysis. It is argued that design for the speculative future acts as a knowledge source for designers to both understand cultural components and adopt a design and critical mindset. The anthropological research of the design topic, as well as the design project communication with broader audiences, offer an opportunity for the designers to understand the cultural components in detail and from multiple perspectives. Tools, approaches, and techniques practiced in design for the speculative future are able to be utilized in other forms of design activities. The critical implications embedded in design for the speculative future encourages designers to critically evaluate their own skill sets and other design projects.

Keywords: Cultural design · Design for future · Knowledge source · Critical · Design tools

1 Research Background

1.1 Culture and Design

Culture and design are both multi-faceted concepts and transformed throughout the years. Evolving from its etymological roots simply meaning "growth" in agriculture and animal husbandry, culture has been recognized as a whole of material and immaterial aspects [2, 32], as a "way of life" [2, 18, 29] transmitting meanings [13] and having an impact on behavior [11, 14, 22, 25]. At the same time, "Design" as a verb and a noun describes an intentional human activity and its outcome of conceiving and implementing for specific purposes [5, 28], in which cultural contexts and cultural meanings are intensively explored [1, 12, 21]. The interpretations of culture and design reveal some

© Springer Nature Switzerland AG 2021
P.-L. P. Rau (Ed.): HCII 2021, LNCS 12771, pp. 49–61, 2021.
https://doi.org/10.1007/978-3-030-77074-7_5

aspects that culture and design are always intertwined. Culture as "a way of life" transfers meanings, which are intensively explored in design activities. This paper researches at the intersection of culture and design and categorizes design in "design for culture" and "cultural design" examining the perspectives that design takes to utilize culture as a resource, as Balsamo [1] states that culture is both a resource and an outcome of design practice. Design for culture describes the practice that aims at solving problems under a certain cultural context and tries to make the designed products/services fit into the cultural environment. It overlaps with the "emerging design" as Manzini [24, p. 52] refers to, as well as design for politics [7] and commercial and responsible design by Tharp and Tharp [30]. On the other hand, cultural design refers to the practice of posing questions, arousing discussions, and thus nudging changes in the cultural context. Instead of trying to make the design fit into the cultural environment, cultural design intends to visualize social-cultural issues under the context, and the designed props might be incompatible with the current cultural environment. It overlaps with a series of design forms including adversarial design [7], critical and speculative design [10], discursive design [31], etc.

1.2 Design for the Speculative Future and Speculative Ethnography

Cultural design focuses on the potential social, political, and cultural issues, creating new cultures and aiming at nudging cultural change via different approaches. For instance, speculative design intends to inspire public discussions on the rational implementation of technologies mainly through provocative images and multimedia scenarios in the exhibition [10]; discursive design is conceptualized to open up conversations, debates, and critical reflections by creating alternative objects in which cultural meanings are disrupted and reorganized [31]. This paper discusses a cultural design approach entitled *design for the speculative future*, further referred to throughout this paper as *DSF* [9], which proposes to emphasize its focus on cultural lifestyles. It is proposed based on the four cones of possible, plausible, probable, and preferable futures illustrated by Dunne and Raby [10, p. 5] and is mapped overlapping with the plausible and preferable cones (see Fig. 1). Situated in the plausible cone indicates that DSF operates within a seeming reasonable at the same time seeming ridiculous area. Meanwhile, the term "speculate" in linguistics has a strong relation with "conjecture" and is often utilized with "critical" implementations in fields such as philosophy [4] and design [10, 23]. Therefore, DSF takes "future" at the center and critically conjectures plausible scenarios and lifestyles. Through communicating those scenarios and props with broader audiences, it is expected to attract public attention, generate discussion, and nudge changes on the selected facet of the cultural ritual.

DSF scientifically reimagines future cultural rituals based on ethnographic explorations of the past and present, with Speculative Ethnography [9] (see Fig. 2) as a tool. Speculative Ethnography is mapped in the "discover" and "define" phase of the British double diamond design process. It starts with "exploring the now" when designers select one or more cultural components and conduct anthropological research to understand the current culture. Then the history and development of the selected cultural components are traced mainly through desk research methods. Trends are able to be generated based on the history and current situation of the selected cultural components, and the future can be speculated following the trend and exaggerating the changing speed. In this way, future scenarios are created based on the developing trend of culture which makes it seem reasonable, on the other hand simplifying the interactions between other factors (e.g. technology and economy) and culture which makes it ridiculous to some extent.

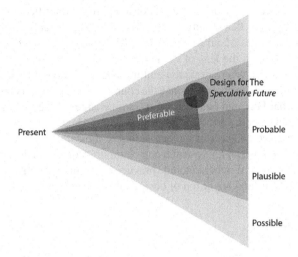

Fig. 1. The future cones illustrated by Dunne and Raby [10, p. 5] and the concept of DSF [9, p. 1884]

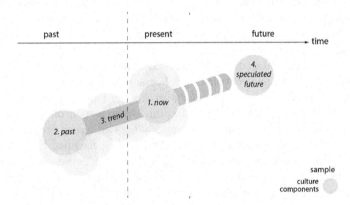

Fig. 2. Speculative Ethnography guideline of "exploring the now", "tracing the past", "generating the trend", and "speculating the future" [9, p. 1888].

1.3 Research Focus

The aim of cultural design and DSF, in theory, is to nudge forward cultural changes by attracting public attention and creating discussions. However, as cultural and behavior changes are large-scale and long-term processes [20, 26], they are hard to notice in relatively short-term research projects. The participation of either online and offline exhibitions [10, 31] or discussion and creation activities [8] can encourage some deeper exploration and culture exchange between participants or viewers, however, it is tricky to measure the impact of such activities in the long run. Nevertheless, cultural design has its values in other aspects, as indicated by Jakobsone [16] with critical design regarded as a source of knowledge that benefits other kinds of design practice. Therefore, this paper focuses on the values of cultural design, DSF in particular, in the relatively short term, examining it through the lens of the knowledge source. Special attention is paid to DSF practice in China, where dramatic cultural changes brought by technology and economic transformation are observed since the country's reform and opening from 1978 [6, 15]. We argue that DSF can be regarded as a knowledge source for Chinese student designers in the field of both cultural and design studies. The benefit for students in the culture field is a detailed and critical understanding of the cultural component resulting from their culture or ethnographic research and communication with the audiences. The valuable practices in the design field include practice of cultural design tools, integration of multiple design tools, and cognition of design and critical mindset.

2   Research Process

This research utilizes a qualitative research approach. Two online DSF studios with the topic of Chinese wedding culture, one of the most significant family events in China, were organized by the authors. There were four student participants (aged 20–22, one male) in the first studio, recruited by One Pear, a design education platform in China, from both industrial design educational background (three students) and non-design backgrounds. The studio lasted for eight weeks for individual DSF projects. The second studio involved thirty-five student participants (aged 19–23, fifteen males) from industrial design (nine students) and environmental art design backgrounds. The studio is held within the design school of a comprehensive university in China. For participants from an industrial design background, it was set as an extracurricular workshop for them to join freely. Students from environmental art design participated in this studio as part of their compulsory course. They were divided into nine groups to conduct DSF group projects over three weeks. In both studios the methodologies of DSF and Speculative Ethnography are introduced to and practiced by the students. In addition, students communicated their design projects with broader audiences and guided group discussions on their projects after the designing process. Participants were interviewed more than two weeks after the studios were over about their experience during the studio and continuous impact on their further understanding and practice of design. Interview data is coded and values of DSF were summarized combining the theoretical analysis. As the studios were held in a Chinese environment and all the coordinators and participants were Chinese native speakers, data was collected, coded, and analyzed in Chinese first, and then translated into English.

3   Example Projects

There were thirteen projects conducted by the participants during the two studios. Three of them are presented in this section as examples to explicate DSF as a knowledge source in both cultural and design aspects. The project descriptions are summarized and translated from the student presentations.

3.1   Personalized Ceremony

This project focuses on Chinese wedding ceremonies. The Chinese wedding ceremony has evolved from a standardized process to personalized one, from family- and rules-oriented ceremony in ancient time, to regularized, revolutionary, and political weddings during the Chinese Cultural Revolution, then to individualization and customization in the contemporary ceremony planning services. Although the current wedding ceremony may contain similar contents, newlyweds do not have to follow the strict procedures as in the past. Ceremonies are planned based on the social and family needs of the couples. It is, therefore, speculated that in the future the wedding ceremony would become so personalized that there are no identical ceremonies among any couples. The following sketches illustrating designs of wedding ceremonies were created by the student designers to visualize the speculated future scenario (see Fig. 3).

Fig. 3. Example wedding ceremony scenes in the speculated future of the personalized ceremony project. Designed and provided by the student designers.

3.2   "Car-ing"

The cultural trend in terms of the whole marriage ritual in the "car-ing" project is interpreted as throughout history wedding customs and steps are combined instead of simply being deleted. Chinese wedding culture originates from "sanshu liuli" (three letters and six etiquettes), some parts of which are inherited and some are combined to fit into the modern marriage context. For instance, the three documents are combined as a marriage certificate, and the pick-up and ceremony phases remain although the specific procedures are combined and simplified. Therefore the designers speculate that in the future all the etiquettes would be combined as one step. The whole marriage ritual would be held in one place in a special wedding vehicle (see Fig. 4), which can provide sufficient space for wedding ceremonies as well as mobility which makes it possible to keep the pick-up procedure.

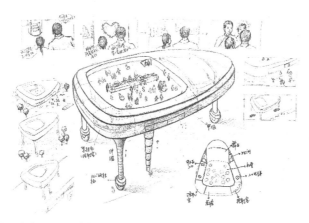

Fig. 4. Speculated future wedding scenarios taking place altogether in the special vehicle. Designed and provided by the student designers.

3.3 AIB Plan

The AIB plan project notices the time period needed from the couple first meeting each other, to the courtship period and then getting married. In the past, couples were matched by the parents' order and on the matchmaker's word, which means they know little about each other when they get married and spend their whole life developing an understanding of each other. Now the couples are able to date and become acquainted with each other for years or months between when they meet and get married. It is interpreted that the time they get to know each other moves ahead, the time spent on it has become shorter, and the extent they know about each other when getting married has become deeper. Following the trends, the future is speculated as they can know each other in three seconds before they meet and get married. Based on the future scenario, they designed the service system to enable the potential couples to know each other in three seconds (see Fig. 5).

Fig. 5. Speculated "user" journey of the future service system design. Designed and provided by the student designers.

4 Findings and Discussion: DSF as a Knowledge Source

4.1 Cultural Knowledge

Although design for culture and cultural design takes culture as a resource based on different ideologies, cultural research is crucial for both forms of design. As a result, through design practices, cultural knowledge is summarized and acquired by student designers. In the case of DSF practice, knowledge is acquired mainly through the anthropological research process in the early stage of the projects and the design communication with broader audiences for cultural exchanges. A comprehensive and critical understanding of Chinese wedding culture, which used to be tiny ideas or scattered gossips of customs, is reported by all the participants after attending the studios. A comprehensive understanding of culture is one of the influences for the personality which impacts behaviors of an individual [14, 17]. As the participants to the studios are aged 19 to 23 who are not married, the DSF practice provides an opportunity for them to be exposed to wedding culture before they are engaged in their own wedding rituals, and to make their wedding decisions on a broad knowledge basis.

Reflexive Ethnographic Research
With the support of Speculative Ethnography as a guideline, the focus of DSF projects is closely connected with the anthropological view of culture as a "way of life" [2, 18, 29]. This indicates that in addition to collecting data about people's behavior, designers also try to collect data about the meanings behind the behavior patterns, as culture is researched in a holistic and contextual sense [13]. Designers research the associated lifestyle related to the selected cultural components adopting anthropological and/or ethnographic methods, and learn about its development in history in order to generate trends and speculate the future.

Studio participants all report their comprehensive and detailed understanding of one or more components of Chinese wedding culture after attending the studio. Before joining the studios, their understanding of wedding culture was mainly based on wedding participation experience, news articles and social media. Born and grown up in China, they regard the practice of wedding rituals as common sense instead of as knowledge. As the designers of the personalized ceremony project describe, they had a negative impression of the wedding since many vulgar customs such as bride price disputes have circulated throughout various media outlets. Other participants also complained about their experiences of attending wedding ceremonies and described wedding ceremonies as 'complicated", "expensive" and "exhausting". During their research sessions in the studio, they try to position themselves neutrally to describe cultural components instead of defining some of the cultural aspects as problems in order to speculate the future. Meanwhile, cultural research reveals meanings of the wedding rituals, which makes participants understand why and how wedding behaviors are formed. The designers of the personalized ceremony project expressed that their understanding of wedding culture changed from "a single ritual and ceremony" to a series of rituals and customs containing meanings that have transformed throughout history impacted by its economic and political environment.

Design Communication

The design communication activities with broader audiences provide a platform for cultural exchange. Culture is interpreted as "the collective programming of the mind that distinguishes the members of one group or category of people from others" [14, p. 6]. Thus culture is often presented on different levels based on the sizes of the identity groups or categories, such as national, regional, organizational, and gender [14, 17]. The cultural research as a DSF process is on a certain group of people due to the time and resource limits of students, and the design communication process involves audiences out of the research group. Therefore, cultural knowledge on different levels is exchanged between the designers and the audiences and the opinions of the audiences add more information to the designers' knowledge range.

When asked in the interviews what the audiences bring to them during the design communication phase, designers all mention "more opinions and possibilities", which might open up conversations of other cultural components. In the discussion of the "caring" project, the description of one audience about his/her expectation of the future wedding ceremony inspired the designers to quickly notice the culture transfer from steampunk, to cyberpunk, then to "artificial intelligence punk" in the future. Designers of the personalized ceremony project expressed that they notice more aspects of Chinese weddings when audiences describe their expectations. Previous to their joining in the studio they identified the wedding as a "problematic" part of culture. The audiences' discussion reminds them that they amplified the negative aspects and ignores many romantic and positive meanings of the wedding. Although after the studios, many participants still hold negative opinions of some wedding customs and they generally do not think wedding rituals can be changed in their generation as those customs are deeply rooted and have historical, economic and political reasons, their positions are chosen based on a broader knowledge.

4.2 Design Knowledge

It is not surprising that the practice of DSF, a design approach alongside the normative design for culture, produces design knowledge for the student designers. Contributions to the students' design learning journeys including practical benefits, which are the practice of cultural design tools and the integration of multiple design tools, and ideological benefits, which provide a new cognition of "design" and a critical mindset.

The Practice of Cultural Design Tools

DSF is supported by Speculative Ethnography, which is surely practiced by the student designers in the studios to imagine plausible future scenarios. The first two steps of Speculative Ethnography, exploring the now and tracing the past, provide a logical guideline for cultural design research, while the last two steps, generating the trend and speculating the future involve cultural interpretations from designers' perspective and require creativity. The practice of Speculative Ethnography in the DSF project thus is a practice of balancing scientific research and creativity in design. Speculative Ethnography is not only practiced as a tool for designers to speculate the future but also a communication strategy to convey the speculated future to broader audiences. Describing both what and why the future scenarios would be like following the "past, now, trend and future" order to the audiences is able to draw more attention to the cultural components the project tries to address instead of the rationality and possibility of the speculated future. On the other hand, although Speculative Ethnography is proposed for DSF, it can be utilized in broader cases. The concept of research following chronological thinking can be utilized in any form of design, as design is a future-oriented activity that explores future ways of living [27] regardless of the distance of its vision.

In the after-studio interviews, participants all mentioned that Speculative Ethnography provides a framework for their wild imagination. Some designers of the AIB plan and "car-ing" project had conducted speculative design projects before joining the studio. They describe their former speculative design process as "tianma xingkong" which means they can freely imagine the future without any constraints. They are often inspired by fictional books, movies and artworks. Speculative Ethnography limits their imagination to some extent, as the speculated future has to follow the generated trend grounded in culture, however, makes them feel their design more reasonable, and offers inspiration when they are out of creative ideas. As one participant states, "suddenly there are many restrictions for my imagination, it is hard to come up with design ideas, this is a small difficulty. But later I can continue to produce ideas even with more restrictions". Some participants report that they utilize Speculative Ethnography in other design projects and creative activities. One designer of the "car-ing" project uses the term "addicted" to describe his/her practice of Speculative Ethnography after the studio. As he describes:

> "I am addicted to this thinking mode when I finally understand it. Its result can be very different and interesting. Walking on the side of the road I see something at any time, and I just spend a few seconds thinking about it following the logic of Speculative Ethnography, it's a lot of fun."

A participant in the AIB plan project shares his/her final year project in the interview and states that he/she delivered the project presentation following the chronological order suggested by Speculative Ethnography. The project designs a private capsule bathroom in public areas as a standardized facility for smart cities. Instead of displaying problems and solutions, the participant first describes the future culture where this design is positioned. It is described that based on cultural research on the past and present, three trends are generated, and the capsule bathroom is designed to fit into the future with these trends. "I choose to use the concept of Speculative Ethnography when preparing my presentation because I want to be more logical", said the participants, "I plan to use it often in my future presentations. It makes me feel supported by many theories".

Integration of Multiple Design Tools

DSF and cultural design only define their aims in a broad sense and are open to all types of outcomes. The outcomes of DSF can either be a speculated service system or transportation or posters, etc. Therefore approaches, tools and techniques in all senses of design might be adopted in the DSF projects depending on specific cases. Speculative Ethnography is mapped in the discover and define phase in the British double diamond design approach to help designers build the speculation. As DSF is created for public communication, designers need to adopt appropriate tools to visualize and materialize the speculation for effective communication. For instance, the process of cultural exploration from the past to the present and speculation of future is suggested to be visualized for a clear presentation; the speculated future can be depicted by either design fiction [3], or scenario and storyboard, or both. Props or diegetic prototypes [19] may need to be designed to support the speculated scenarios, thus tools and skills in the field of product design, transportation design, service design might be utilized. This requires participants to have a good command of multiple design tools in order to flexibly choose and integrate appropriate tools in the DSF project.

All the projects conducted during both design studios in this research utilizes data visualization to explain to the audiences how the future is speculated, as stated in the former sections, to draw more attention to the cultural aspects instead of the rationality of the speculated future. The speculated future scenarios are displayed via scenarios and storyboards by all the projects, as scenarios and storyboards through sketch or PhotoShop are the most accessible methods for participants from a design background. Some projects produce videos to demonstrate the procedures for the wedding ceremony. A large range of approaches is adopted to demonstrate their prop design, as their props are in varieties of forms including bank service, clothes changing installation, transportation, wedding scenes, etc. For instance, the AIB plan project displays a "customer" journey map to communicate how future couples are able to know everything about each other in only three seconds (see Fig. 5); the "car-ing" project build a computational 3D model of their vehicle design to show its semantic meanings (see Fig. 4).

Cognition of Design and Critical Mindset

DSF and cultural design are regarded as new ideologies of design for the student designers especially in China. The practice of DSF broadens their cognition of "design" from something which acts solely as a problem-solving activity to one that supports creative explorations in juxtaposition with logic-centric praxes. Critical thinking, in a broader sense, is embedded in DSF and cultural design. Critiques implied in DSF and cultural design projects are not targeted only to the cultural components that the project addresses, but also to the cognition of design [16]. The design outcomes of DSF and cultural design projects produce alternative possibilities to the cultural components [10]. Similarly if we see "design" as a cultural component, the practice of DSF produces alternative possibilities to the culture of design itself. Only when student designers are exposed to a broad range of possibilities can they make their decisions for their future. DSF practice enables them to realize the social responsibility of designers. One participant states that "commercial design is of course important (I have to make a living), but I must also consider doing something interesting and valuable for the society, so that it is not in vain that I insist on choosing design as my career".

The DSF studio experience guides the participants to the venue of being critical to everything, including the wedding culture and the design activity. With a reference to the Cultural knowledge section, participants' perceptions of Chinese wedding culture have expanded from limited aspects of physical objects and ritual processes to multiple and critical integrations. In the field of design, the integration of multiple design tools in DSF projects encourages participants to review the tools and skills they master, notice any weak parts and practice them in the future. Participants generally are not satisfied with the final visual effect of their projects and intended to improve their sketching and visual communication design skills. The designers of the personalized ceremony project realized that the visual effect of their speculated scenarios, the wedding ceremony sketches (see Fig. 3), does not convey their speculation to their satisfaction. They speculate that there would be no identical ceremonies of any couples in the future, however, in their sketches the ceremonies look similar to most wedding ceremonies in the current situation - the T-shape stage, traditional Chinese furniture, romantic pink color, etc. "We tried to sketch each ceremony differently," said the designers, "but we cannot imagine a real unique ceremony. Also we are afraid that if we sketch the ceremony so uniquely it will not look like a wedding. We need to keep learning to improve our sketch skills, and also improve our creativity." In other design projects after the studios, participants start to challenge the "user needs" listed by themselves or their teammates in order to avoid design with bias. As one participant describes, "I start to critically think about the so-called user needs before I create something for them, to evaluate whether the needs are valuable to meet".

5 Conclusion

This paper examines DSF, design for the speculative future, through the lens of regarding it as a knowledge source. By collecting and analyzing qualitative data of two online DSF studios, findings are displayed that knowledge in the realms of both culture and design is acquired by student designers in engaging with this methodological tool. Through historical and participant ethnographic research of some cultural components combined with the design communication process with broader audiences, a comprehensive and critical cognition of culture is formed by the participants. As a cultural design tool, Speculative Ethnography is practiced in the DSF studios as an approach to speculating the future and communicating design concepts with broader audiences. This tool was also utilized in other forms of design by the participants after attending the studio, revealing the success of DSF in sparking student designers' creative thinking processes. The DSF project also involves practice of approaches and tools in other areas of design such as product design and service design. The practice of these tools trained the student participants' ability to balance scientific research and creativity in design, as well as to integrate multiple design tools and approaches depending on specific cases. Critical and reflective thinking is conveyed to the student participants through DSF practice, in a way that enabled them to think reflectively about their own place in the design process, and understand that their own positions may impact the nature of how they engage with realizing the meaning of user needs as well as other formative factors. As students start to reflect their own design skill sets and other design projects after the studios, DSF is a tool that may act as a bridge between the areas of present- and future-making, as well as a catalyst for enabling creative thinking.

References

1. Balsamo, A.: Keyword: DESIGN. Int. J. E-Learn. Educ. Technol. Digit. Media **1**(4), 1–10 (2010)
2. Bennett, T.: Culture: A Reformer's Science. Sage Publications, London (1998)
3. Bleecker, J.: Design Fiction: A Short Essay on Design, Science, Fact and Fiction - Near Future Laboratory. Near Future Laboratory (2009)
4. Broad, C.D.: Critical and speculative philosophy. Contemp. Br. Philos. Pers. Statements 77–100 (1924)
5. Buchanan, R.: Design and the new rhetoric: productive arts in the philosophy of culture. Philos. Rhetoric **34**(3), 183–206 (2001)
6. Chen, S.: Economic reform and social change in china: past, present, and future of the economic state. Int. J. Polit. Cult. Soc. **15**(4), 569–589 (2002)
7. DiSalvo, C.: Adversarial Design: Design Thinking Design Theory. MIT Press, Cambridge (2015)
8. Dong, F., Sterling, S.: Audience matters: participatory exploration of speculative design and Chinese wedding culture interaction. In: Proceedings of IASDR Conference 2019, Manchester (2019)
9. Dong, F., Sterling, S., Schaefer, D., Forbes, H.: Building the history of the future: a tool for culture-centred design for the speculative future. In: Proceedings of the Design Society: DESIGN Conference, pp. 1883–1890 (2020)
10. Dunne, A., Raby, F.: Speculative Everything: Design, Fiction, and Social Dreaming. MIT (2013)
11. Eckhardt, G.: Culture's consequences: comparing values, behaviors, institutions and organisations across nations. Aust. J. Manag. **27**(1), 89–94 (2002)
12. Flusser, V.: The Shape of Things: A Philosophy of Design. Reaktion, London (1999)
13. Geertz, C.: The Interpretation of Cultures: Selected Essays. Basic Books (1973)
14. Hofstede, G., Hofstede, G., Minkov, M.: Cultures and Organizations: Software of the Mind. McGraw-Hill (2010)
15. Hu, X., Chen, S., Zhang, L., Yu, F., Peng, K., Liu, L.: Do Chinese traditional and modern cultures affect young adults' moral priorities? Front. Psychol. **9** (2018)
16. Jakobsone, L.: Critical design as a resource adopting critical mind-set. Des. J. **22**(5), 561–580 (2019)
17. Karahanna, E., Evaristo, J., Srite, M.: Levels of culture and individual behavior: an integrative perspective. J. Glob. Inf. Manag. **13**(2), 1–20 (2005)
18. Katan, D.: Translation as intercultural communication. In: Munday, J. (ed.) The Routledge Companion to Translation Studies, pp. 74–92. Routledge, London and New York (2009)
19. Kirby, D.: The future is now: diegetic prototypes and the role of popular films in generating real-world technological development. Soc. Stud. Sci. **40**(1), 41–70 (2009)
20. Klasnja, P., Consolvo, S., Pratt, W.: How to evaluate technologies for health behavior change in HCI research. In: Proceedings of the CHI 2011 Conference on Human Factors in Computer Systems. ACM, New York (2011)

21. Latour, B.: A cautious Prometheus? A few steps toward a philosophy of design (With Special Attention to Peter Sloterdijk). In: Hackney, F., Glynne, J., Minton, V. (eds.) Networks of Design: Proceedings of the 2008 Annual International Conference of the Design History Society (UK), pp. 2–10. Universal-Publishers, Boca Raton (2009)
22. Linton, R.: The Cultural Background of Personality. Appleton-Century (1945)
23. Malpass, M.: Between wit and reason: defining associative, speculative, and critical design in practice. Des. Cult. **5**(3), 333–356 (2013)
24. Manzini, E.: Design culture and dialogic design. Des. Issues **32**(1), 52–59 (2016)
25. O'Driscoll, J.: Face, communication and social interaction. In: Bargiela-Chiappini, F., Haugh, M. (eds.) Journal of Politeness Research. Language, Behaviour, Culture **7**(1) (2011)
26. Rapp, A.: Design fictions for behaviour change: exploring the long-term impacts of technology through the creation of fictional future prototypes. Behav. Inf. Technol. **38**(3), 244–272 (2018)
27. Sanders, E., Stappers, P.: Probes, toolkits and prototypes: three approaches to making in codesigning. CoDesign **10**(1), 5–14 (2014)
28. Simon, H., Laird, J.: The Sciences of the Artificial. MIT Press (1996)
29. Sparke, P.: An Introduction to Design and Culture: 1900 to the Present, 3rd edn. Routledge (2013)
30. Tharp, B., Tharp, S.: Discursive Design: Critical, Speculative and Alternative Things. The MIT Press, Cambridge (2018)
31. Tharp, B., Tharp, S.: Discursive design basics: mode and audience. In: Proceedings of Nordic Design Research Conference (2013)
32. Williams, R.: Keywords: A Vocabulary of Culture and Society, New Oxford University Press, New York (2015)

Effects of Players' Social Competence on Social Behaviors and Role Choice in Team-Based Multiplayers Online Games

Ka-Hin Lai, Bingcheng Wang, and Pei-Luen Patrick Rau[✉]

Department of Industrial Engineering, Tsinghua University, Beijing, China
rpl@tsinghua.edu.cn

Abstract. Recently, social interaction between gamers has caught the attention in sociology and psychology. The purposes of this study are: (1) to explore the relationship between gamer background information and social competence with player social behavior and role choice in team-based multiplayer online games. (2) To verify the relationship which found in the exploration. Therefore, an online survey (N = 177) and an experiment (N = 37) were conducted to find out how the background and social competence of player affects the gamer social behavior and role choice. The results showed that: social competence and player background have an effect on social behavior and role-choice in game. Specially, conflict management has a significant positive effect on negative communication frequency and game ranking has a significant positive effect on leadership. Additionally, KDA (Kill/Death/Assists) and cooperation behavior have a significant positive on winning. Finally, author gave some suggestion to future research and gamer based on the conclusion.

Keywords: Online game · Social competence · Social behaviors · Role choice

1 Introduction

Nowadays, advancement in mobile phones have enable people to access online games more easily. People are spending more time and consuming more money in online games. A latest report [1] pointed out that the average revenue per user currently amounts to 19.54 US dollars. The report also reveals that revenue in the online games segment reaches 17,141 million US dollars in 2020. User penetration is 11.8% by 2020 and is expected to hit 13.0% by 2024. In global comparison, China, as an emerging economy, earned the largest share in online game with 4,249 million dollars in 2020. Currently, many gamers prefer to team-based multiplayers online game (TMOG) than massively multiplayer online role-playing games (MMORPG) due to the TMOG playing time is shorter than MMORPG. Released in December 2015 by Raptr most people play PC client in the game, the number of League of Legend (LOL) gamers accounted for 22.92%, the number of Counter-Strike: Global Offensive gamers accounted for 6.88% in second place and the number of DOTA 2 gamers accounted for 5.09% in the fourth. TMOG has become one of the most popular game types now.

© Springer Nature Switzerland AG 2021
P.-L. P. Rau (Ed.): HCII 2021, LNCS 12771, pp. 62–79, 2021.
https://doi.org/10.1007/978-3-030-77074-7_6

The prosperity of the TMOG has also witness the growing willingness of players' social behaviors in games. Since TMOG is a team-oriented games, players need to cooperate and communicate with each other to win the games, which involves various social behaviors. In addition, online games provide players with convenient and diverse virtual environment for social interaction. In such environment with higher participation and interaction, people can communicate with others, transforming the real social relationship into a virtual social relationship, thereby providing a new channel to meet human interpersonal communication needs [2]. A report on Chinese players' behavior [3] indicates that about half players (51.2%) use voice to chat with other players and most of them are female players. Male players, on the other hand, are willing to communicate with other players for better cooperation and victory in games. Social behaviors have become an integral part of online games. However, in-game social behaviors are not always positive as destructive social behaviors and cyberbullying amongst players, such as verbal aggression and sexual harassment, destroyed their gaming experience [4, 5]. Especially for adolescents, such destructive social behaviors exposure will adversely affect their mental development [6, 7]. To counter this phenomenon, many TMOGs have provided the function to detect and report teammates' destructive behaviors. As such, research on the in-game social behavior should be one of the valuable research topics in the field of psychology and games.

A considerable amount of literature has shown that one's social behaviors are closely associated with his or her social competence [8–11]. Some researchers have also found close relationship between social competence with in-game behaviors [12–14], but many specific behaviors in TMOG have not yet been covered. Hughes (2017) have concluded a focus group to find out various social behavior in TMOG: trolling, role choice, leadership, autonomy/cooperation, teaching, raging, building positive rapport and enemy support. But how these behaviors are influenced by players' social competence has been a largely under explored domain. To fill this gap, we conducted a two-phase research to find out that the relationship between player's social competence and their behavior in TMOG. In phase I, we conduct of survey to explore the relationship between competence, gamer backgrounds and social behavior in TOMGs. In phase II, we carried an empirical research to further investigate the impact of particular social competence, i.e., conflict management, on in-game social behaviors.

This research will aid game developers to understand the user's in-game behaviors and to build more appropriate user models, so as to design a more reasonable and effective mechanism to reduce the occurrence of destructive social behaviors in games, thereby improving the players' gaming experience. For players, the research can help them consciously improve their social skills, thereby changing their in-game social behavior and earning them the respect and understanding from other players.

2 Related Works

2.1 Team-Based Multiplayers Online Games

Online games are popular entertainment for people due to the interactivity and anonymity of the Internet. Online virtual environment has become a place for people to get rid of real stress and to reduce anxiety [15]. Players achieve the goal in a virtual game environment to get involvement, positive mood, pleasure, and joy [16]. Multiuser dungeons (MUDs) are one early type of online multiplayer computer game that combined role-playing and chat rooms. Since then, many derivatives continue to attract numerous players (Hsu and Lu, 2004). Massive multiplayer online role-playing games (MMORPGs) is one of the kinds and it allow players to communicate with each other in the more realistic network game environment, and players develop social relationships with their virtual identity in the game world.

Unlike MMORPGs, players in team-based multiplayer online games (TMOGs) do not need to spend much time to earn experience points and use those points to reach character "levels". The gamers play in a fair environment. Players now be able to access online games at the same time and enjoy simultaneous interaction with other people through the network. The timeliness allows players to interact socially with others [17]. Interpersonal interaction is one of the most crucial characteristics of optimal computer game experiences [18]. Social interaction in the online games is important for winning, because players compete in groups against other groups and information sharing within groups is critical. There are 4 types of interpersonal interaction in online games: cooperation, transaction, competition and communication. Cooperation means players work in teams to achieve the game goals. Transaction means players can exchange the virtual items with others. Competition means there will be competitors in online games. Communication means players can use words, emoticons or voice to communicate with other players so that increasing interactivity of the game.

Some research showed that social interaction of online game has a negative effect on players. Online games will increase the sense of insolation in younger players [19], or weaken gamer's offline relationship [20, 21], or make players become gaming addicts [22].

2.2 Social Behavior and Role-Choice in Online Games

Traditionally, researchers have summarized that the social behavior can be categorized into anti-social behavior and pro-social behavior [23]. Kiesler [24] suggested that social behavior can be affected by the two personality dimensions: dominance and affiliation. dominance has a positive effect on prosocial behavior [25] and has a positive effect on antisocial behavior, such as ridicule, verbal attacks and pranks [26, 27]. Affiliation has a positive effect on pro-social behavior such as encourage, anti-bullying [28, 29] while has a negative effect on anti-social behavior such as bullying [29].

In game research, scholars also examined similar conclusion that affiliation personality of gamers has negative effect on anti-social behavior, and has a positive effect on prosocial behavior; while dominance of players has both positive effect on prosocial behavior and anti-social behavior [12].

Additionally, researchers have used automated behavioral reports from video games to determine associations between personality traits and behaviors, the result showed that player character traits can be measured by five dimensions: extraversion, agreeableness, conscientiousness, emotional stability, and openness to experience [14]. And players with extraversion prefer to team-based activities. Other researchers found that male players sometimes would think that online games are not suitable for female players and they would receive some sexual harassment or verbal aggression from man. And that will affect the female players experience and pleasure during gaming [4].

In MMORPG, according to the player's consciousness, imagination and expectation of characters, players can act different roles in the game. The motivation of role choice can be considered in three dimensions: achievements (including skills, competition, etc.), sociality (including social, relationship and collaboration, etc.) and obsessions (including exploration, role playing, appearance, etc.) [30]. Scholars pointed out that player's performance such as sociality, exploration, killing, achievements and level upgrade can affect player's role choice. Specially, sociality has a positive effect on social behaviors.

In TMOG, however, the characters are different from MMORPG's. It is unnecessary for players to choose the same role in every round but based on specific requirements (own expectation or teammates demand) to choose role. The relationship between players' character is more like a team rather than the traditional relationship in MMOPRG such as marriage. Therefore, in TMOG, the character role is not permanent. There are two type roles in general: carry role is an active, dominant, high damage output role; support role is a passive, utility-oriented role (support positions) [12]. Scholars have found out several social behaviors: trolling, role choice, leadership, autonomy/cooperation, teaching, raging, building positive rapport and enemy support. These social behaviors can be considered in three dimensions: neutral social behaviors, positive social and negative social behaviors. In the virtual environment, people don't need to worry about their behavior will affect real life and the role preference in the network is unconstrained by reality. Thus, character performance and role-choice of the individual in network are different from those of the real world [12].

2.3 Social Competence and Interpersonal Competence

Social Competence (SC) is the ability that help individual to promote interpersonal relationships and can be divided by five dimensions: cooperation, assertion, responsibility, empath and self-control [31]. Researchers defined social competence from other perspectives and classified SC into two dimensions: emotional and social. Each dimension included expressivity, sensitivity, and control. emotional and social represented non-verbal and verbal competence (Riggio, 1989). People with low social competence often have poor perceive ability in social interactions. They misunderstand other people's words, emotion, and that leading depression. Additionally, the researchers pointed out that children with social anxiety only get anxiety in some social situation rather than their low social skills. Therefore, instead of simply training children's social skills, children should reduce their social anxiety and change their wrong social cognition.

There are some empirical studies between social competence and online game. Researchers explored the relationship between SC and player's participation. They founded that player's participation has negative effects on emotional expressivity, social expressivity, social sensitivity, and social control; players with higher participation though that they lack expression ability [32–34]. However, some scholars pointed out that player engagement has a positive effect on emotional expressivity, emotional control and social expressivity. It does not agree with the above research results. The authors suggested that the sample was different from the previous study, and the sample was limited to white participants. Finally, the form of self-evaluation may lead to different correlations [35].

3 Phase I: Exploring the Influence of Players' Social Skills on Their Social Behavior and Role Preference

3.1 Questionnaire Design and Administration

In phase I, we conduct of survey to investigate the relationship between players' social competence and their social behavior in TOMGs.

The questionnaire used in the survey consisted of four parts: players' demographics, their social competence, and in-game social behaviors. Demographics includes players' gender, age, educational background, number of TMOG played, favorite TMOG, first time TMOG played, game ranking, cost, frequency per week, and gaming locations.

To assess social behavior, five items were conducted including mood (Scales on a five-point scale ranging from 1 "positive" to 5 "negative"), leadership (Scales on a five-point scale ranging from 1 "does not apply at all" to 5 "does fully apply"), autonomy/cooperation (Scales on a five-point scale ranging from 1 "autonomy" to 5 "cooperation"), carry role preference and support role preference (Scales on a five-point scale ranging from 1 "does not apply at all" to 5 "does fully apply").

Social competence is measured by interpersonal Competence Questionnaire (ICQ) [36]. In our survey, we used the Chinese version ICQ [37] which consisted of 30 items were conducted on a 5-point scale from 1 "total disagreement" to 5 "total agreement" including initiation, negative assertion, disclosure, emotional support and conflict management (e.g. "When angry with a companion, being able to accept that s/he has a valid point of view even if you don't agree with that view" in conflict management dimension). We also added an additional item under each dimension asking player whether their social competence changed after playing the games (e.g. "Do you think you get more initiative when interaction with others after playing the games?").

We also included a reverse question to assure respondents were paying attention to the survey in general. If a respondent has a high degree of positive correlation between the reverse questions and its corresponding question, it is judged that the participant has not completed it carefully. This sample is invalid and will be eliminated in the following analysis.

Participants in phase I were recruited through a social network platform (WeChat). To be considered for this study, participants must have TMOG experience (such as LOL, DOTA2, Storm Hero and Overwatch). After finishing the questionnaire participants received 10–15 RMB as a reward. Of the 227 consented participants, 50 participant records were eliminated due to non-TMOG gamers and failure to accurately respond to the validation questions. At last total 177 questionnaires could be used for data analysis.

3.2 Result and Analysis

Reliability of the Questionnaire

Reliability was measured through Cronbach's alpha, and a value greater than 0.7 was considered acceptable [38]. Some researchers also point out that alpha above 0.6 was also acceptable for exploratory research [39]. The Cronbach's alpha of the ICQ in our research was 0.815, which is considered good under the 0.7 criteria. For individual construct, the Cronbach's alpha of initiation, negative assertion, disclosure, emotional support and conflict management were 0.782, 0.757, 0.697, 0.808, 0.661. The disclosure and conflict management are a little bit below 0.7 but still above 0.6.

Descriptive Statistics

Regarding the demographics of the sample participants in Table 1, 82.5% of the respondents were male, and 17.5% were female. Most of the respondents (69.5%) were 21–25 years old. Young people were the largest group of online-game players, and 54.2% of the respondents played online games at dormitories. Most of the respondents (76.8%) only played two type TMOGs, and 67.8% of respondents played TMOGs under 10 h per week.62.7% of respondents thought their game ranking above the average.

Table 1. Demographics of respondents

Demographics	Category	Frequency	Percentage
Gender	Male	146	82.5%
	Female	31	17.5%
Age	<20	28	15.8%
	21–25	123	69.5%
	>25	26	14.7%
Education	Under high school	4	2.3%
	Colloge	6	3.4%
	Bachelor	85	48.0%
	Above Master	82	46.3%
The number of TMOG	≤2	136	76.8%
	≥3	41	23.2%
Game ranking	Below the average	66	37.3%
	Above the average	111	62.7%
Cost	≤100	59	33.3%
	100–500	64	36.2%
	>500	54	30.5%
Frequency per week	≥10	120	67.8%
	10–20	46	26.0%
	>20	11	6.2%
Location	Net bar	33	18.6%
	Home	42	23.7%
	Dormitory	96	54.2%
	Working place	6	3.4%

Influence of Demographics on In-Game Social Behaviors
Given that the collected data on social behaviors did not follow normal distribution, a nonparametric test analysis of variance was conducted on emotion, leadership and autonomy/cooperation, carry role performance and support role performance regarding gamer demographics. For two independent samples (gender, game ranking and the amount of TMOGs), Mann-Whitney U tests were used. Results revealed that there was a significant effect of game ranking on leadership ($Z = 4.212$, $p < 0.05$) and autonomy/cooperation ($Z = 2.266$, $p < 0.05$). There was no significant difference in social behavior and role choice on gender and amount of TMOGs played. For three or more independent sample (age, education, cost, frequency, location), Kruskal-Wallis tests were used. Results revealed that there was a significant effect of leadership on age (Chi-square $= 8.84$, $df = 2$, $p < 0.05$), a significant effect of emotion on frequency (Chi-square $= 13.52$, $df = 2$, $p < 0.05$). It demonstrated that older group (>20 yrs.) produced significantly higher leadership than younger group did and higher frequency group (>20h per week) produced significantly negative emotion than other groups did. There was no significant difference in social behavior and role choice on gender and amount of TMOGs. There was no significant difference in social behavior and role choice on other groups.

Influence of Social Competence on In-Game Social Behaviors
Next, we examined the effects of social competence on in-game social behaviors. Table 2 shows the correlations between social competence and social behavior, role preference for the current study. Some notable interrelations include: Initiation and Negative Assertion both were positive significantly related to Autonomy/cooperation, leadership and carry role preference; Disclosure was negative significantly related to Emotion; Conflict Management was positive significantly related to support role preference and was negative significantly related to emotion and leadership; Emotional Support was negative significantly related to emotion and was positive significantly related to Autonomy/cooperation. It demonstrated that higher Disclosure, Conflict Management and Emotional Support produced more positive emotion in TMOGs; players with higher Initiation, Negative Assertion and Emotional Support will prefer cooperation in games, also they prefer carry role; with higher Initiation and Negative Assertion and lower Conflict Management will prefer leading their teams in game.

Table 2. Correlations between interpersonal competence and social behavior, role preference

	Initiation	Negative assertion	Disclosure	Conflict management	Emotional Support
Emotion	0.114	0.075	**−0.160***	**−0.199***	**−0.162***
Autonomy/cooperation	**0.194****	**0.184***	0.093	0.100	**0.183***
Leadership	**0.166***	**0.244***	0.136	**−0.221***	−0.103
Carry role	**0.206****	**0.272****	−0.019	0.059	0.032
Support role	0.106	−0.061	0.004	**0.170***	0.002

*$p < 0.05$. **$p < 0.01$

Figure 1 shows the relationship between Conflict Management, gamer's backgrounds and social behaviors. As we can see, Conflict Management was positively related to support role, was negatively related to emotion and leadership. In addition, age, ranking and frequency.

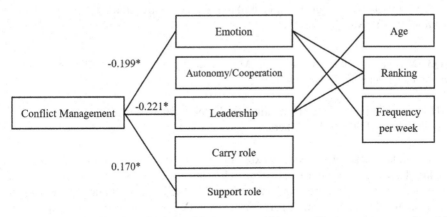

Fig. 1. Phase II: Verification of relationship between ranking, conflict management and social behaviors, role-choice.

3.3   Research Model and Questions

In phase I, we explored the relationship between player backgrounds, social competence and social behavior, role choice preferences. The result showed that conflict management was significantly negative related to leadership and was positively related to support role preference. In addition, game ranking, age and the frequency of gamers have different effect on the social behavior. But there is a limitation in phase I research. The questionnaire survey in phase I is memory-based and self-reported, which increased the probability of biased responding. For this purpose, a behavior observation experiment based on actual social gaming behavior to establish strong evidence for criterion-related validity was conducted. In phase II study, we carried an experiment investigate the relationship between ranking, conflict management and social behaviors, role-choice. It was hypothesized that gamer's ranking and player's conflict management in interpersonal competence affect online player social behavior (neutral, positive or negative social behaviors) and role-choice (carry and support role).

RQ1. Does Conflict Management have different effect on gamers' social behavior or role-choice?

- H1.1. Players with lower Conflict Management prefer leading their team in TMOGs.
- H1.2. Players with higher Conflict Management have more positive emotion and less negative emotion in TMOGs.
- H1.3. Players with higher Conflict Management prefer support role.

RQ2. Does game ranking of players have different effect on gamer social behavior or role-choice?

– H2.1. Players with higher game ranking prefer leading their team in TMOGs.
– H2.2. Players with higher game ranking prefer carry role.
– H2.3. players with higher game ranking prefer cooperation in their team than autonomy in TMOGs.

RQ3. Does game ranking and Conflict Management of players have different effect on gamer social behavior or role-choice?

– H3. players with higher game ranking and lower Conflict Management prefer leading their team in TMOGs than others.

To validate these hypotheses, an experiment was conducted to observe player actual social behavior and role choice in League of Legends. The reasons of selecting this game were mentioned in literature review and currently LOL is the most popular game in the world so that the recruitment of participants would be more convenient. In experiment progress, communication behaviors of participants were recorded and then participants were interviewed and filled out with a questionnaire about the social behaviors of the game.

3.4   Methodology

Participants
To be considered for this study, participants must have been 20 years or older and frequency less than 20 h a week at the time of survey completion to reduce effect of participants' complex background on social behavior. Additionally, participants were limited to gamers who had played the online game League of Legends. Finally, 37 participants were selected to participate in this behavior experiment. Table 3 shows descriptive statistics of the participants.

Task
The participants were asked to play the classic mode in Summoner's Rift (a map) for ranking match. Because of ranking match related to player rank, they would pay more attention to winning so that can promote the communication between players. In order to not affect overall communication, players needed to play by themselves and communicate to strangers rather than play with their friends together.

Independent Variables
There are two independent variables in this experiment, Conflict Management and game ranking.
   Conflict Management: the level of Conflict Management is based on the interpersonal competence questionnaire, and 18 participants have higher scores on Conflict Management while 19 participants have lower scores.

Table 3. Descriptive statistics of the participants.

Demographics	Category	Frequency	Percentage
Gender	Male	33	89.2%
	Female	4	10.8%
Age	20–22	22	59.5%
	23–24	10	26.0%
	>25	5	13.5%
Game years	<1	4	10.8%
	1–3	14	37.8%
	>3	19	51.4%
Frequency per week	≤10	17	45.9%
	10–20	15	40.6%
	>20	5	13.5%

Game ranking: There are 7 tiers (Bronze, Silver, Gold, Platinum, Diamond, Master and Challenger) and each tier from Bronze to Diamond is divided into five divisions. The names of each division is a roman numeral between V (5) and I (1). Higher numbers correspond to lower skill level). From the current game ranking distribution in China, about 30% of players are Bronze tier, about 40% gamers are Silver tier and the 30% remaining are above Gold tiers. Therefore, this experiment chose participants above Gold III as high-ranking players while below Silver II as low-ranking players.

Dependent Variables

In the experiment, the dependent variables were collected mainly from two aspects: Firstly, actual social behavior in game including positive social behavior (the frequency of positive communication), negative social behavior (the frequency of negative communication), neutral social behavior (tagging info to teammate), role-choice and game result.

- Positive communication frequency: players send positive messages in chat box (such as encourage, teaching, building positive rapport) and the positive communication frequency is the number of positive messages divided by the game time.
- Negative communication frequency: players send negative messages in chat box (such as anger or raging, blaming teammate) and the negative communication frequency is the number of negative messages divided by the game time.
- Neutral communication frequency: players tag a signal to teammate including "danger", "on my way", "enemy missing" or "assist me" to their teammates and the neutral communication frequency is the number of tagging divided by the game time.
- Role Choice: the role (carry or support) in the game, 1means carry role and 2 means support role.
- Game result: 1 means defeat and 2 means victory.

- Secondly, after the experiment, a self-assessment 5-point scale questionnaire was conducted to investigate participants' feeling about their game performance, including Leadership, Autonomy/cooperation, Encourage, Teaching, Give up on themselves and Raging.
- Leadership: degree to which a player choice to "call the shots" or to follow the commands of others.
- Autonomy/cooperation: degree to which a player's level of autonomy in game, as determined by their tendency to follow the team's strategy or to proceed with their own agenda.
- Encourage: the degree to which a player's tendency to encourage other players.
- Teaching: the degree to which a player's tendency to share knowledge with other players in order to help improve the others' gameplay.
- Give up on themselves: the degree to which the tendency for a player to give up the match prematurely before defeated.
- Raging: the degree to which the tendency for a player to express excessive verbal aggression towards other players.

Procedure

The experiment was carried out in a human-computer interaction laboratory. Participants played LOL in lab and were recorded. Figure 2 showed that the scenario when participants were playing online game. Author used Open Broadcaster Software (OBS) to record player's facial express and gaming screen for counting the communication frequency. Before the experiment, participants were told that the experiment purpose, tasks and some matters needing attention and then log their own account to play LOL. During the experiment, participants played one ranking match and were encouraged to communication to their teammates and they could interact with teammates at chat box; after

Fig. 2. Experiment scenario of phase II study.

the experiment, author recorded the role-choice, game time, result and then interviewed participants for a few questions about the social behaviors and sent a questionnaire to them.

3.5 Result and Analysis

First, we checked the normality of the dependent variables, which turns out that all the dependent variables are non-normally distributed. Thus, a nonparametric test analysis of variance was conducted on these dependent variables regarding Conflict Management and Ranking. For this two-independent sample, Mann-Whitney U tests were used. The results are shown on Table 4 and Table 5.

On Conflict Management, results revealed that there was a significant effect of Negative frequency ($p < 0.05$) and Raging ($p < 0.05$) and There was no significant difference in other dependent variables ($p > 0.05$). On Ranking, results showed that there was a significant effect of Neutral frequency ($p < 0.05$) and Leadership ($p < 0.05$) and there was no significant difference in other dependent variables ($p > 0.05$). It demonstrated that high ranking group (above Gold III) produced significantly higher leadership and neural frequency than low ranking group did, and higher conflict management group produced significantly lower negative frequency and raging than low conflict management group did.

Table 4. Result of conflict management of the dependent variables.

Variables	Conflict management		Mann-Whitney U	Z	p
	High	Low			
Positive frequency	0.16	0.15	155.5	−0.47	0.64
Negative frequency	0.04	0.12	99.0	−2.23	0.03*
Neutral frequency	0.21	0.2	169.0	−0.61	0.96
Role Choice	1.33	1.53	138.0	−1.17	0.33
Game Result	1.39	1.47	156.5	−0.51	0.66
Leadership	3.33	3.16	156.0	−0.47	0.66
Autonomy/Cooperation	3.94	3.79	149.0	−0.70	0.52
Encourage	3.67	3.32	149.0	−0.69	0.52
Teaching	3.17	3.21	166.0	−0.16	0.89
Give up on themselves	1.56	2.00	153.0	−0.62	0.60
Raging	1.61	2.53	101.0	−2.24	0.03*

In the observation, this paper found that not only player social behaviors can be affected by the conflict management and ranking, but also can be influenced by gamers' performance. Therefore, this paper did a correlation analysis for this additional factor: one is the rate of participation, which is the ratio of player killing number to their team

Table 5. Result of game ranking on dependent variables

Variables	Conflict management		Mann-Whitney U	Z	p
	High	Low			
Positive frequency	0.18	0.14	127.5	−1.24	0.28
Negative frequency	0.06	0.09	141.0	−0.84	0.42
Neutral frequency	0.27	0.16	74.0	−2.89	0.00**
Role Choice	1.25	1.57	114.0	−1.93	0.10
Game Result	1.50	1.38	148.0	−0.71	0.55
Leadership	3.69	2.90	101.50	−2.12	0.04*
Autonomy/Cooperation	3.88	3.86	167.00	−0.03	0.99
Encourage	3.63	3.38	153.50	−0.46	0.66
Teaching	3.56	2.90	112.50	−1.76	0.09
Give up on themselves	1.69	1.86	168.00	0.00	1.00
Raging	1.88	2.24	138.00	−0.97	0.37

killing number. Another is the KDA (Kills/Deaths/Assists), which is the ratio of player kill number to player death number. The result of Pearson correlation between these two factors and social behaviors was showed in Table 6.

Table 6. Correlations between rate of participation, KDA and social behaviors

	Rate of participation	KDA	Role choice	Game result
Positive frequency	0.22	−0.20	−0.04	−0.17
Negative frequency	−0.10	−0.35**	0.23	−0.29*
Neutral frequency	0.02	0.29	−0.03	0.29
Role Choice	−0.72	−0.09	1	0.09
Game Result	0.00	0.66**	0.09	1
Leadership	0.23	0.25	−0.04	0.15
Autonomy/Cooperation	−0.02	0.21	0.17	0.34**
Encourage	0.37**	0.19	−0.03	0.17
Teaching	0.17	0.29	−0.14	0.23
Give up on themselves	−0.28*	−0.34**	−0.17	−0.21
Raging	−0.13	−0.43**	0.03	−0.35**

*p < 0.05. **p < 0.01

It was revealed that the correlations between rate of participation, KDA and social behaviors for the current study. Some notable interrelations include: rate of participation was positive significantly related to encourage ($p < 0.01$) and negative significantly relate to give up on themselves ($p < 0.05$), KDA was negative significantly related to negative frequency ($p < 0.01$), give up on themselves ($p < 0.01$) and raging ($p < 0.01$) and positively related to game result ($p < 0.05$). Game result was negative related to negative frequency ($p < 0.05$) and raging ($p < 0.01$), positive related to autonomy/cooperation ($p < 0.01$). It demonstrated that higher autonomy/cooperation and KDA, and lower negative communication, raging produced more victory in TMOGs.

From the analysis above, we can find that emotion has significant different in conflict management, namely low conflict management group has more negative communication and raging in game. Leadership has a significant difference in ranking group, which means that high ranking group has more leadership in TMOGs.

4 Discussion

In phase I, some game background will affect the player's social behavior, such as in different game ranking, higher ranking players prefer to lead and cooperating with their teammate in game. The reason is that when a player reaches a certain high ranking, they understand how to win, so they will be more proactive in cooperating and leading their teams for victory; in different age, players older than 20 will be more active in leading their team. Players under the age of 20 are more reserved and will not be active in leadership and communication. Gamers who play game more than 20 h a week have more negative emotion and behaviors because in TMOGs, players need to consume an amount of energy and then their emotion control will be weaken. Additionally, social behaviors on gender have no significant difference ($p > 0.05$), but the p value was close to significant standard. The possible reason is the number of female samples is much smaller than male's and therefore, in the future research can continue to explore the influence of gender on social behaviors.

Players with higher Initiation or Negative Assertion prefer to command their team in game. The reason is leadership requires the abilities that players to take the initiative to communicate with their teammate, and to reject some wrong suggestions. Player with lower Conflict management also prefer to command their team in game. It is hard to control their emotions when they encounter a difficult game situation so that players with higher Conflict management can control their mind in different situations so that they will not take the initiative to communicate with teammate. Players with higher Disclosure, Emotional Support and Conflict Management have more positive emotion and behaviors in game. These players can understand other feelings and at the appropriate time to release their emotions and they are positive in reality, therefore can also be reflected in the game. This is similar to previous researchers' conclusions: people with affiliation personality also exhibit more pro-social behaviors in the game (Chelsea M. Hughes, 2017).

In phase II, there was no significant difference in Leadership among players with different Conflict Management groups. Possible reason is that players would not display usual behaviors in the unfamiliar experimental environment and leadership would be affected by other factors such as the game's situation, player's performance etc., However, players with low Conflict management significantly have more raging and negative communication than those with high Conflict management. Interestingly, there were no significant difference on the positive communication frequency and possible reason is that players were observed in laboratory so that they would friendly interact with other.

Players with higher ranking prefer to command in game than lower ranking groups thus the hypothesis 2.1 was confirmed. Frequency of tagging a signal was also significantly different in ranking groups. High ranking players are more clearly that how to help the team for victory, so they in the game no matter if the situation will take the initiative to command. On the contrary, low ranking gamers need to speed an amount of energy in controlling their character and they have no extra time and energy to communicate with others. There was no significant difference in the autonomy/cooperation thus hypothesis 2.3 was not confirmed. The possible reason is they played TMOG with strangers so that participants would prefer to follow their own agenda.

There were no opposite results between phase I and phase II. In phase II, lower Conflict Management group have higher negative communication frequency and raging and higher ranking group have higher leadership, which was consistent with the results of the phase I. Players with higher Conflict Management prefer to support role and follow direction in phase I but there were no significant different in phase II. Some results have not been verified due to, on the on hand, the social behaviors of players can be affected by other factors (such as KDA, rate of participation); on the other hand, player behaviors in laboratory may be different from the behaviors in player familiar environment. Moreover, based on the experiment analysis and interviews after experimenting this paper summarized 3 factors that influence the player's communication intention in game: 1) the overall communication environment. Subjects would decrease their communication intention when nobody reply to them, even though they tried to take the initiative to talk to their teammate and finally players stopped the communication. 2) Player performance. Subjects argued that with high KDA or rate of participation they would communicate to teammates more effectively. If players with poor KDA or rate of participation, their teammates would ignore subjects' communication and even have negative effects (such as raging or verbal aggression). In other words, players with better performance are more persuasive in the game. 3) Gaming situation. Participants believe that when in an inferior gaming situation, players would increase negative communication with each other, such as questioning his teammates, verbal aggression and raging. However, when players in a superior situation, they would decrease their communication (both negative and positive). In addition, several subjects believed that primary and middle school students have more negative emotions in game. Because teenagers' minds are not yet mature so that negative factors such as violence in games can affect them more easily, making them behaved anti-socially.

5   Conclusion

This study is an elementary exploration of gamer interpersonal competence and social behaviors of TMOGs. Through the questionnaire and observational experiment, this paper summarized the conclusions, limitations and directions of future research.

In Phase I, we found that interpersonal competence in some dimensions is significantly related to social behaviors in game, such as Initiation is positively related to leadership and Conflict Management is positively related to support role preference.

In Phase II, we provided an observational experiment and found that leadership has a significant difference in ranking group and negative communication frequency has significant difference in Conflict Management group. Additionally, we found that game result is positive significantly related to autonomy/cooperation, KDA, and was negative significantly related to negative communication, raging.

There are several limitations to the study. Firstly, the sample size of questionnaire is insufficient, and most subjects are university students, so conclusion unavoidably has limitations and future research could contain different samples, such as different age groups or cross-cultural groups. Secondly, the experiment was conducted in the laboratory environment. This environment was unfamiliar to participants and they need to be recorded in laboratory. So they would have some unusual social behaviors in game or some usual social behaviors of gamers did not happen. Participants could be observed in their familiar environment in future research. Finally, participants needed to play with strangers in phase II so that they would decrease their communication and future study could design the experiment that gamers play with acquaintances to observe their social behaviors.

The results of the present research open the doors for practical application of interpersonal competence's effect on in-game social behavior. With such an influence of interpersonal competence on in-game behavior, it is possible that real-world interpersonal competence interventions may also alter social behavior in online gaming. Conversely, perhaps interventions implemented through games might result in changes in real-world behaviors. As the accessibility of online game increasing, more and more teenagers (especially primary and middle school students) were exposed to the virtual environment and they could be affected by others easily. Thus, gaming companies might also use this scale to help identify players who might pose a threat to the social health of the gaming community and to protect young teenagers from negative social behaviors. Additionally, the behaviors examined in this study have notable impacts on a player's (and team's) success in the game (e.g., positive cooperation resulting in increased performance, and negative communication or raging resulting in decreased performance). This result may find use in the field of e-sports psychology and performance enhancement.

References

1. Statista: Digital Media Report 2019 - Video Games (2019)
2. Chen, V.H.-H., Duh, H.B.-L., Phuah, P.S.K., Lam, D.Z.Y.: Enjoyment or engagement? role of social interaction in playing massively mulitplayer online role-playing games (MMORPGS). In: Harper, R., Rauterberg, M., Combetto, M. (eds.) ICEC 2006. LNCS, vol. 4161, pp. 262–267. Springer, Heidelberg (2006). https://doi.org/10.1007/11872320_31
3. iResearch: 2015 Chinese game user behavior research report (2015)
4. Fox, J., Tang, W.Y.: Sexism in online video games: The role of conformity to masculine norms and social dominance orientation. Comput. Human Behav. 33, 314–320 (2014). https://doi.org/10.1016/j.chb.2013.07.014
5. Tang, W.Y., Fox, J.: Men's harassment behavior in online video games: personality traits and game factors. Aggress. Behav. 42, 513–521 (2016)
6. Calvete, E., Orue, I.: The impact of violence exposure on aggressive behavior through social information processing in adolescents. Am. J. Orthopsychiatr. 81, 38 (2011). https://doi.org/10.1111/j.1939-0025.2010.01070.x

7. Gentile, D.A., Lynch, P.J., Linder, J.R., Walsh, D.A.: The effects of violent video game habits on adolescent hostility, aggressive behaviors, and school performance. J. Adolesc. **27**, 5–22 (2004). https://doi.org/10.1016/j.adolescence.2003.10.002

8. Fraser, M.W., et al.: Social information-processing skills training to promote social competence and prevent aggressive behavior in the third grades. J. Consult. Clin. Psychol. **73**, 1045 (2005). https://doi.org/10.1037/0022-006X.73.6.1045

9. Rose-Krasnor, L.: The nature of social competence: a theoretical review. Soc. Dev. **6**, 111–135 (1997)

10. Seevers, R.L., Jones-Blank, M.: Exploring the effects of social skills training on social skill development on student behavior. Online Submission **19** (2008)

11. Van den Eijnden, R.J., Meerkerk, G.-J., Vermulst, A.A., Spijkerman, R., Engels, R.C.: Online communication, compulsive internet use, and psychosocial well-being among adolescents: a longitudinal study. Dev. Psychol. **44**, 655 (2008)

12. Hughes, C.M.: A measure of social behavior in team-based, multiplayer online games: the Sociality in Multiplayer Online Games (SMOG) scale. Comput. Human Behav. **69**, 386–395 (2017)

13. Kowert, R., Domahidi, E., Quandt, T.: The relationship between online video game involvement and gaming-related friendships among emotionally sensitive individuals. Cyberpsychol. Behav. Soc. Netw. **17**, 447–453 (2014)

14. Yee, N., Ducheneaut, N., Nelson, L., Likarish, P.: Introverted elves & conscientious gnomes: the expression of personality in world of warcraft. In: Proceedings of the SIGCHI Conference on Human Factors in Computing Systems, pp. 753–762. ACM (2011)

15. Liu, D., Li, X., Santhanam, R.: Digital games and beyond: what happens when players compete? Mis Q. **37**, 111–124 (2013)

16. Csikszentmihalyi, M.: Beyond Boredom and Anxiety. Jossey-Bass, San Francisco (1975)

17. Laurel, B.: Computers as Theatre. Addison-Wesley, Boston (2013)

18. Lewinski, J.S.: Developer's Guide to Computer Game Design. Wordware Publishing Inc., Plano (1999)

19. Orleans, M., Laney, M.C.: Children's computer use in the home: isolation or sociation? Soc. Sci. Comput. Rev. **18**, 56–72 (2000)

20. Kraut, R.E., Rice, R.E., Cool, C., Fish, R.S.: Varieties of social influence: the role of utility and norms in the success of a new communication medium. Organ. Sci. **9**, 437–453 (1998). https://doi.org/10.1287/orsc.9.4.437

21. Kraut, R., Patterson, M., Lundmark, V., Kiesler, S., Mukophadhyay, T., Scherlis, W.: Internet paradox: a social technology that reduces social involvement and psychological well-being? Am. Psychol. **53**, 1017 (1998). https://doi.org/10.1037/0003-066X.53.9.1017

22. Grüsser, S.M., Thalemann, R., Griffiths, M.D.: Excessive computer game playing: evidence for addiction and aggression? CyberPsychol. Behav. **10**, 290–292 (2007). https://doi.org/10.1089/cpb.2006.9956

23. Clarke, D.: Pro-Social and Anti-Social Behaviour. Routledge, London (2003)

24. Kiesler, D.J.: The 1982 interpersonal circle: a taxonomy for complementarity in human transactions. Psychol. Rev. **90**, 185 (1983). https://doi.org/10.1037/0033-295X.90.3.185

25. Carlo, G., Okun, M.A., Knight, G.P., de Guzman, M.R.T.: The interplay of traits and motives on volunteering: agreeableness, extraversion and prosocial value motivation. Pers. Individ. Differ. **38**, 1293–1305 (2005). https://doi.org/10.1016/j.paid.2004.08.012

26. Lee, K., Ashton, M.C., Shin, K.-H.: Personality correlates of workplace anti-social behavior. Appl. Psychol. **54**, 81–98 (2005). https://doi.org/10.1111/j.1464-0597.2005.00197.x

27. Parkins, I.S., Fishbein, H.D., Ritchey, P.N.: The influence of personality on workplace bullying and discrimination. J. Appl. Soc. Psychol. **36**, 2554–2577 (2006)

28. Graziano, W.G., Habashi, M.M., Sheese, B.E., Tobin, R.M.: Agreeableness, empathy, and helping: a person × situation perspective. J. Pers. Soc. Psychol. **93**, 583–599 (2007). https://doi.org/10.1037/0022-3514.93.4.583
29. Tani, F., Greenman, P.S., Schneider, B.H., Fregoso, M.: Bullying and the big five: a study of childhood personality and participant roles in bullying incidents. School Psychol. Int. **24**, 131–146 (2003)
30. Yee, N.: Motivations for play in online games. CyberPsychol. Behav. **9**, 772–775 (2006). https://doi.org/10.1089/cpb.2006.9.772
31. Elliott, S.N., Gresham, F.M.: Social skills interventions for children. Behav. Modif. **17**, 287–313 (1993)
32. Griffiths, M.D.: Computer game playing and social skills: a pilot study. Aloma: Revista de Psicologia, Ciències de l'Educació i de l'Esport **27**, 301–310 (2010)
33. Liu, M., Peng, W.: Cognitive and psychological predictors of the negative outcomes associated with playing MMOGs (massively multiplayer online games). Comput. Human Behav. **25**, 1306–1311 (2009). https://doi.org/10.1016/j.chb.2009.06.002
34. Loton, D.J.: Problem video game playing, self esteem and social skills: an online study (2007)
35. Kowert, R., Oldmeadow, J.A.: (A) Social reputation: Exploring the relationship between online video game involvement and social competence. Comput. Human Behav. **29**, 1872–1878 (2013). https://doi.org/10.1016/j.chb.2013.03.003
36. Buhrmester, D., Furman, W., Wittenberg, M.T., Reis, H.T.: Five Domains of Interpersonal Competence in Peer Relationships. J. Pers. Soc. Psychol. **55**, 991 (1988)
37. Wang, Y.-C., Zou, H., Qu, Z.-Y.: The revision of interpersonal competence questionnaire in middle school students (2006)
38. DeVellis, R.F.: Scale Development: Theory and Applications. Sage publications, Thousand Oaks (2016)
39. Nunally, J.C., Bernstein, I. (eds.): Psychometric Theory. McGraw-Hill, New York (1978)

A Study on the Influence of Intercultural Curation on the Brand Loyalty of Cultural Creative Park Based on the Experiential Marketing Theory

Yun-Chi Lee[1]([⊠]), Tien-Li Chen[2]([⊠]), Chi-Sen Hung[3]([⊠]), and Shih-Kuang Wu[4]([⊠])

[1] Doctoral Program in Design, College of Design, National Taipei University of Technology, Taipei, Taiwan
[2] Department of Industrial Design, College of Design, National Taipei University of Technology, Taipei, Taiwan
chentl@mail.ntut.edu.tw
[3] Department of Communication Design, National Taichung University of Science and Technology, Taichung, Taiwan
[4] Department of Distribution, National Chin-Yi University of Technology, Taichung, Taiwan

Abstract. The aim of this study focuses on the effective indicators of improving customer's brand loyalty towards cultural and creative industries park through curating events and the curation strategies of the park. This study is based on the well-known experiential marketing theory from professor Schmitt (1999) focusing on the discussion of how cultural creative parks create visitors' experience and build up brand loyalty through transnational authorization of international exhibitions.

The methodology of this study includes: 1. In-depth interviews with the 24 Taiwanese co-exhibit illustrators of their intercultural illustration exhibition participation experience and impression to further evaluate the results of the exhibition; 2. Questionnaire surveying which was based on the population variable statistics, experiential marketing experiences, brand loyalty, and questions related to the aforementioned facets.

The result indicates that a fruitful exhibition experience can significantly enhance visitors' brand loyalty towards the park. For Taiwanese co-exhibit illustrators, the experience of co-exhibiting with famous international illustrators can deepen the understanding of different cultures contexts of the exhibited works. This exhibition also demonstrated a mode of experiencing learning and exchanging. In addition, the framework of this study has been verified through the Pearson Correlation Coefficient analysis and the outcomes have shown positive correlations between Strategic Experiential Modules (SEMs) versus the experiential value, the experiential value versus brand loyalty, and experiential marketing versus brand loyalty. The outcome of regression analysis indicates that different experiential marketing strategies and experiential value have a noticeable impact on the brand loyalty towards the park.

Keywords: Cultural creative industries parks · Experiential marketing · Brand loyalty · Exhibition curation (curation)

© Springer Nature Switzerland AG 2021
P.-L. P. Rau (Ed.): HCII 2021, LNCS 12771, pp. 80–99, 2021.
https://doi.org/10.1007/978-3-030-77074-7_7

1 Preamble

1.1 Backgrounds and Motives

In Taiwan, Cultural and Creative Industries is defined as "industries that originate from creativity or accumulation of culture which through the formation and application of intellectual properties, possess potential capacities to create wealth and job opportunities, enhance the citizens' capacity for arts, and elevate the citizens' living environment" (Development of the Cultural and Creative Industries Act 2010). This act confirms the development of cultural and creative industries and values the tangible and intangible asset by preservation, research, innovation, and designs to carry out the values of Taiwan traditional culture. Domestically speaking, cultural creative industries are expected to improve citizens' cultural qualities; while internationally-wise, the pursuit is the unique cultural value and the communication with the world as well as developing the economy of culture.

The Cultural and Creative Industries Park (CCIP) serves as a critical strategy for the development of the cultural creative industry. The establishment of cultural creative parks in each region assimilates to sowing the seeds of cultural creative economy development; and the clustering effect of regional cultural industries will form into "blocks" of industry chains that boost local economy growth. The establishment of cultural creative parks has accumulated the energy of this industry and the clustering effect carries the crucial purpose of cultural creative industry economic values. (Ministry of Culture 2020).

The responsibility of CCIP, instead of serving as an average exhibition and event space, has the mission of promoting regional cultural and creative industries; thus, the operating strategy should be considered even more seriously. In a practical perspective, the consideration of attracting businesses, artists, designers, cultural industry workers, and the general public to gather and exchange has become ever so critical. This research will discuss the art and culture exhibitions of CCIP, focusing on how successful curation and execution communicates the exhibition theme and content to the customers; and further boost the brand loyalty CCIP.

1.2 Research Purpose

To understand how does CCIP attract the public and the professionals to the park and improve the brand loyalty by curating art and culture exhibition, the public held and operated "Cultural Heritage Park, Ministry of Culture" was chosen for case study, focusing on the exhibition- "the Joint Exhibition of Bologna Italy Illustrators"—held during August 2020. This study was based on the theory of experiential marketing and experiential value, targeting participants and the joint illustrators to see whether there were positive effects on prak's brand loyalty after participating the exhibition. The result of this study will provide cultural creative parks with the effectiveness of curating future exhibitions and with the increase of brand loyalty, while revitalizing CCIP and its development in industries' economy.

The purpose of this study listed as follows:

(1) In-depth interview methods were adopted to understand the benefit on Taiwanese illustrators after partaking in the joint international cultural exhibition.

(2) Based on the exhibition's "experiential marketing" and experiential value," the discussion was then focused on the improvement of brand loyalty of public visitors and related effects.

2 Literature Review

2.1 Development of Culture and Creative Industries

During the mid and late twentieth century, globalization has caused a great impact on the economy worldwide. Many countries have turned to innovative products and services that emphasize local culture; and this has been an economic strategy that demonstrates a nation's cultures and tourism features. The Executive Yuan of Taiwan proposed the "Challenge 2008 National Development Plan" in response to the worldwide trend; and the sub-plan "Cultural and Creative Industry Development Plan" has defined "Cultural and Creative Industries" as "accumulation of culture which through the formation and application of intellectual properties, possess potential capacities to create wealth and job opportunities, enhance the citizens' capacity for arts, and elevate the citizens' living environment" (Development of the Cultural and Creative Industries Act 2010) which confirms the development cultural and creative industries. In 2009, the "Development of the Cultural and Creative Industries Act" (DCCIA) was promulgated to navigate the industry development and to emphasize the creation of art, cultural preservation, combination of culture and technology, and improving citizen's cultural literacy and spreading culture and art to the public. Taiwan's ambition of developing cultural and creative industries does not limit to the industry fields which are defined by DCCIA; the expectations are the integration of cultural strategy to each industry field and become a key to industry transformations and adding value to the industry.

2.2 Development and Operation of Cultural Heritage Park of Ministry of Culture

CCIP is a critical strategy for the development of cultural creative industries; and by the establishment of CCIP in each place, the clustering effect of the cultural industry will be seen. After the year 2000, Taiwanese society gradually realize the concept of cultural creative industries; since then, the discourses and actions of cultural heritage preservation become more mature and began to thrive as well as the preservation, maintenance, revitalization, archiving, and passing on of the historic sites, historical buildings, cultural sites, customs, techniques, and tangible and intangible cultural heritages.

During 2003 to 2007, Council for Cultural Affairs, former Ministry of Culture, launched the "Coordination of the Cultural and Creative Industry Development Plan;" and the establishment measures of "Creativity and Culture Special Section" has chosen historic buildings such as, Taipei old distillery, Taichung old distillery, Chiayi old distillery, Hualian old distillery, and Tainan North Gate Warehouses for the demonstration of Taiwan's development of cultural and creative industries and for information exchanges. The five mentioned parks are bases for promoting cultural creative industries that bring commercial and culture together and construct cultural industries that have Taiwan features. These historical buildings are the cultural and creative industry fields for revitalization and reutilization; also, these are the manifestations of how Taiwan focuses on the cultural heritages.

The five cultural creative parks of MOC are operated in different structures that include public and private cooperation, the Operation-Transfer, the Build-Operation-Transfer, the Reconstruction-Operation-Transfer, and parks operated solely by the public section. Taichung Cultural Heritage Park (CHP,) is the only public held and operated park and the demonstration park of the MOC cultural heritage preservation policy. CHP spans over an area of 5.6 hectares, which is the largest park among the five parks, and has the most well preserved historical buildings and old distillery equipment. There are 16 buildings that have been registered as historical buildings, and 12 average buildings. The space of this park is arranged into four major sections, which are administration and incubation section, cultural creative development section, performing sections, and business section (Cultural Heritage Park 2020).

2.3 Experiential Marketing

In Taiwan, a number of cultural creative companies have chosen customer experience, education, and interaction as the focus of operation. It is expected that through the experience process, customers will gain knowledge, techniques, and recreation; moreover, it is also expected to improve the brand identification and loyalty. In todays' commercial activity, internet media, information transmission, and technology development, have created even more options for marketing planning; and operations such as, exhibition events, online or offline community events, self-operated media, VR, AR, Apps…etc. can be the media that creates different experiences.

Schmitt (1999) have provided the marketing personnel with clearer thinking aspects based on the individual psychology and behavior of customers. He suggested that five facets that cause the difference in experiential experiences, are sense, feel, think, act, and relate. The five facets are the Strategic Experiential Modules (SEMs) presented in Table 1. It is considered that if the modules could be properly put to use, it could strategically stimulate customer's senses, reach customers' minds, and influence their behavior.

2.4 Experiential Value

The customer behavior research of Taylor et al. (2018) has discussed the influence of experiential value on customers' behavior. They discovered that a sense of value towards a brand is based on the experience of consuming process. The experiential value has an impact on customer's satisfaction, brand image, and eventually brand loyalty. Mano and Oliver (1993) have categorized the experiential value into intrinsic value and extrinsic value. The intrinsic value consists of the pleasure and playfulness which are generated through the process of experiencing tasks; while extrinsic value is a sense of value that is generated after the completion of an experience task. Babin and Darden (1995) also considered that the intrinsic value is formed out of the pleasure and fun during the experience process instead of the completion of a task; while extrinsic value is obtained often through the completion of task. Besides, Holbrook (1994) has taken the "Activity" as a viewpoint, providing "Reactive value" and "Active Value". The former represent customer's evaluation, attitudes on the experienced object and the understanding of the

Table 1. The Study Variable 1: strategic experiential modules

Variables	Facets	Purposes	Approaches
Study Variable 1: Experiential marketing	Sense	Create cognitive feeling experiences to add value to the event	To understand how to stimulate senses through sight, sound, touch, taste, and smell, generating behavior patterns and creating cognitive experience
	Feel	To Trigger inner emotions, feelings, and thoughts	To understand what are the ways to stimulate customer's inner emotions and boost their active participation and consumption
	Think	By triggering sustaining thinking, customers are constantly participating till the cognitive transformation occurs	The appeal to intellect is to trigger customers' interest through new events, stimulating concentrated thinking or dispersing thinking so that during the question solving process, the experience is then created
	Act	Strategically invite the customers for conduct physical actions	By creating a customer's physical experience, the purpose is to reexamine daily life habits through the experience in order to create changes in lifestyles
	Relate	Connect an individual customer's ideal ego with the others and relate with society culture	The relate marketing facet include sense, feel, think, and act that focus on the personal psychological link of a customer; and draw customers closer to the brand experience and to relate with societal context

Table 2. The Study Variable 2: experiential value evaluation facets

The Study Variable 2: experiential value		
	Active value	Reactive value
Intrinsic value	**Facet 1: Playfulness** Customers may shortly escape the reality feelings through playfulness by the instant pleasure generated through the experience process	**Facet 2: Aesthetics** The Aesthetics value focuses on the attractiveness from visual stimulation. During the experiencing process of consuming, the aesthetic generates from the environment or services and activities that are impressive
Extrinsic Value	**Facet 3: Customer Return On Investment** By the expenditure of money and time on experiencing activities, this facet refers to the mentality of the relatively obtained benefit. which are the rewards resulted from the exchange	**Facet 4: Service Excellence** Service Excellence comes from all that brand's promises; from the customer's perspective, it is the subjective evaluation on the provided services and quality

brand; while "Active Value" is the reaction and evaluation of customers after they participate in the experience and the active value increases with the customers' satisfaction of the participated experience.

In an online shopping environment study, Mathwick et al. (2001) integrated the aforementioned theories of "Extrinsic versus Intrinsic Value" and "Active versus Reactive Value". In the study, the authors have developed an equation, the Experiential Value Scale (EVS), which is used for evaluating experiential values. The four facet in the EVA are: Customer Return On Investment (CROI,) Aesthetics, Playfulness, and Service excellence; it is presented in the following Table 2. The method of this study was designed based on these evaluation facets.

2.5 Brand Loyalty

Jacoby and Chestnut (1978) believes that brand loyalty include complicated customer beliefs, affect, and behavior intentions. There is no doubt that brand loyalty must originate from previous experience of a brand and the customer would then have preference for the brand and finally brand loyalty is then formed. During the experience process, the more positive emotions, the more increase in brand loyalty to the customer. Likewise, CCIP's services and events should effectively satisfy the customer's demands on the participation in art and culture events; specifically, the satisfaction both intellectually and emotionally will increase brand loyalty towards CCIP and could achieve the mission of vitalizing the economy of art and culture industries.

Though the brand loyalty of the customers generate the commitment and persistence, the apparent behaviors might not necessarily be the demonstration of the degree of loyalty. It might be driven by the convenience of product and services, pricing, peer pressures, and other factors. Hallowell (1996) and Oliver (1999) collected various characteristics of customer's brand loyalty and defined it with two clear indicators which are "Conative Loyalty" and "Action Loyalty".

Conative Loyalty: Oliver (1999) defined this as the positive and preferred attitude towards a certain brand that has become a long term commitment that the customers are willing to spread positive word of mouth. Brand loyalty is not a random behavior, instead, it is constructed by experience, comparison, and identification of a customer and eventually the commitments to the brand are generated. Therefore, conative loyalty tends to be more psychological-wise, as a brand should have a certain degree of value to the customer.

Action Loyalty: This refers to the customer's behavior of willing to repeatedly pay for a certain product or service. This action originates from the satisfied experience of obtaining products or services of a brand, even if paying higher cost than other brands, the customer remains willing to purchase and to recommend to others (Oliver 1999).

With all the mentioned discourses on brand loyalty, this study adapted Oliver's (1999) theory as the basis for evaluating customers' brand loyalty. Two facets, the "Conative versus Action Loyalty," are selected for further research. The definitions of the two facets are presented in Table 3.

Table 3. The Study Variable 3: brand loyalty facets

The Study Variable 3: brand loyalty facets	
Facets	Definition
Conative loyalty	Psychological phase of a customer and the commitment towards a brand, originated from the customer's sense of value
Action loyalty	Actions taken by customers, especially the willingness of paying a certain cost for the brand's service or product

3 Methodology

3.1 The Study Framework and Hypotheses

This study explores the experiential marketing strategy of CHP for arts and cultural exhibition curation, and examines the benefits of enhancing customers' brand loyalty in CHP. In response to the purpose of the research, two phases were designed in this study, including the first phase: adopting in-depth interviews to understand the experience evaluation and participation benefits of the Taiwanese co-exhibit illustrators; the second phase: surveying by questionnaires to Understand visitors' experiential marketing and feelings of experiential value, and analyze the relationship between the various facets of brand loyalty.

1. Framework

Focusing on the second part of the study, the literature review covers the experiential marketing, experiential value, and brand loyalty and the framework were presented in Fig. 1. The discussion emphasizes the interrelation of demographic variables, experiential market, experiential value, and brand loyalty. Suggestions on CHP's curation of art and culture events were presented in the last part.

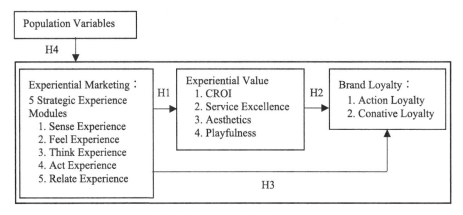

Fig. 1. Framework of this study

2. Hypotheses

Based on the purpose and the framework, the hypotheses of this study are as follows:

H1: Experiential Marketing has a positive influence on Experiential Value.

H2: Experiential Value has a positive influence on Brand Loyalty.

H3. Experiential Marketing has a positive influence on Brand Loyalty.

H4. Demographic Statistic Variables have discernible influence on Experiential Marketing, Experiential Value, and Brand Loyalty.

3.2   Questionnaire Design and Surveying

(1) Questionnaire Design

In order to make the design of the questionnaire conform to the content validity, the questionnaire topics used in this research are developed by theories and aspects of experience marketing, experience value, brand loyalty, etc. After the questionnaire questions are developed, three university teachers who teach in the Applied Chinese Department, Business Design Department, and Marketing Department will edit and confirm the questions to improve the expert validity of the overall questionnaire (Tables 4, 5 and 6).

Table 4. The question design of experience marketing measurement

Variables	Evaluation facets	Topic design
Experiential marketing	Sense experience	I think the content of this exhibition is abundant I think this exhibition attract my various senses I think the design of this exhibition enables me to feel the theme of the exhibition
	Feel experience	I think this exhibition makes me pleasant I think the layout of this exhibition makes it easy to visit I think the content of this exhibition (such as works, exhibition planning) is professional
	Think experience	After visiting the exhibition, it allows me to have more imagination about the theme of this exhibition Through this exhibition, I have new ideas for art activities Through this exhibition, I can learn new knowledge
	Act experience	I am satisfied with the interactivity in the exhibition I think this exhibition makes me want to take photos in the exhibition space I am willing to share the experience and exhibition information of this exhibition with my friends and family
	Relate experience	The theme of this exhibition is related to issues that I care about in life The knowledge learned in this exhibition can be applied to my life I think this exhibition aroused my understanding of society and culture

Table 5. The question design of experience value measurement

Variables	Evaluation facets	Topic design
Experiential value	CROI	I think the content of this exhibition is practical to me
		Due to various costs, this exhibition made me feel worthwhile
		I think the content of this exhibition meets my expectations
	Service excellence	I think the planning of this exhibition is perfect
		I think the content quality of this exhibition is worth recommending
		I am satisfied with the service quality of this exhibition
	Aesthetics	I think this exhibition has a sense of design
		I think such an exhibition can enhance my artistic atmosphere
		I think the atmosphere of this exhibition is very attractive to me
	Playfulness	I think this exhibition makes me feel a temporary escape from daily life
		This exhibition is interesting to me
		This exhibition is innovative to me

Table 6. The question design of brand loyalty measurement

Variables	Evaluation facets	Topic design
Brand loyalty	Action loyalty	After visiting this exhibition, I will revisit the Cultural Park
		After visiting this exhibition, I will pay attention to the future exhibition information
		After visiting this exhibition, I won't share information about the cultural park with friends and relatives
		I would like to learn more about the works of Cultural Heritage Park
	Conative loyalty	There are some inconveniences in the exhibition, which greatly affects my evaluation of the park
		The activities and exhibitions in the park are quite distinctive
		The environmental facilities, styles and styles in the park make me feel very distinctive
		Overall, I am very satisfied with the Cultural Heritage Park

(2) Implementation and Analysis of Questionnaire Survey

This study conducted a paper questionnaire survey in the exhibition area during the exhibition period (July 21 to August 23, 2020). People who have completed the visit were invited to complete the survey. A total of 285 questionnaires were collected, and invalid questionnaires that were omitted and deleted were excluded. There are 274 valid questionnaires in total, and the recovery rate is 96.1%. In order to test whether the content of the questionnaire is consistent, this study first uses Cronbach's Alpha reliability analysis to examine the internal consistency of the questionnaire. Reliability analysis results show that the overall questionnaire reliability is 0.967, and the reliability values of each measurement dimension are all greater than 0.7 (Hee 2014), indicating that the analysis results of this questionnaire have good reliability.

(3) Participatory Observation and Case Study Introduction: Bolognal Illustration Exhibition the "Illustration For Children. Italian Excellence"

This international art and culture exhibition was organized by the Cultural Assets Bureau of the Ministry of Culture and was authorized by the International Children's Book Fair in Bologna, Italy, to exhibit "ILLUSTRATION FOR CHILDREN. ITALIAN EXCEL-LENCE". Bologna Children's Book Fair is a world-renowned professional book fair, with more than 1,400 exhibitors every year and more than 300,000 visitors from 80 countries around the world, bringing together publishers, illustrators, literary brokers, writers, and packagers, Printers, audiobook people, booksellers, librarians, teachers and students... and other people from different fields to participate (Bologna Children's Book Fair 2020). This exhibition was officially authorized by BolognaFiere S.p.A. It aggregated the works of 18 Italian illustrators and displayed a total of 72 masterpieces. In addition to the works of Italian illustrators, in order to promote the exchange of Taiwan's illustration industry, this exhibition also invited 55 illustration works of 32 Taiwanese illustrators to be exhibited together.

The location for this transaction was held in the 1916 Cultural and Creative Workshop in the Cultural Heritage Park. There were static exhibition areas (Italy exhibition area, Taiwan exhibition area), dynamic exhibition area, projection interaction area, experience interaction area, picture book reading area, and book shopping area. There were several sections and surrounding activities (library carts, markets), etc., multiple exhibitions and interactive area content, providing people of different ethnic groups to visit, relax, and experience. The exhibition period dated from July 21 to August 23, 2020. It was a free public welfare exhibition without ticket sales. According to official statistics, a total of 15,000 people visited during the exhibition.

(4) In-depth Interview: Co-Exhibited Taiwanese Emerging Young Illustrators
In order to strengthen the qualitative discourse of this study, 24 Taiwanese illustrators who participated in this exhibition were invited to conduct in-depth interviews during the study. The interview period was September to October 2020. In addition, interviews were conducted in terms of revealing self-branding, enhancing professional knowledge, promoting cultural exchange vision, increasing contacts in related industries, experience value of participating in exhibitions, and overall benefit evaluation. The interview process was recorded by audio recording, and verbatim manuscripts were used to analyze the content of the interview afterwards, so as to provide relevant suggestions for future international art exhibitions.

4 Data Analysis

4.1 Hypotheses Correctness Analysis

After conducting the questionnaire survey in this study, the Cronbach's Alpha reliability analysis was conducted to confirm the internal consistency of the questionnaire design in order to obtain good reliability analysis results. Pearson correlation coefficients were also adopted as a means of analysis of the correlation between experience marketing (strategic experience module), experience value, and brand loyalty. The analysis results are shown in Table 7 Table 8 and Table 9. The analysis results show that "experience marketing (strategic experience module)" has a significant positive correlation with "experience value"; "experience value" and "brand loyalty" have a significant positive correlation; and "experience marketing (strategy) Experience module)" and "brand loyalty" also have a significant positive correlation. The analysis results show that the relationships established by the various hypotheses in this study are positively interrelated and are consistent with the previous discussed theoretical basis, and all hypotheses are correct.

Table 7. Correlation analysis of experiential marketing (strategic experience modules) and experiential (H1)

	CROI	Service excellence	Aesthetics	Playfulness
Sense experience	0.758**	0.872**	0.833**	0.776**
Feel experience	0.838**	0.867**	0.831**	0.771**
Think experience	0.720**	0.771**	0.704**	0.846**
Act experience	0.747**	0.845**	0.848**	0.943**
Relate experience	0.666**	0.603**	0.515**	0.694**

Notes: "*" means the p value < 0.05, "**"means p value < 0.01

4.2 Verification of Hypotheses

(1) Experience Marketing (Experience Marketing Module) Regression Analysis of Experience Value Prediction (H1)

This paragraph analyzes the experience value versus experience marketing, using the five facets of experience marketing to perform regression analysis on the four dimensions of experience value. The analysis results are shown in Table 10, Table 11, Table 12, and Table 13. First observe the standardized R-squared value of each regression analysis which could examine the explanatory strength of the entire regression model. It can be seen that the standardized R-squared value of each table can reach more than 0.8, which shows that each experience marketing dimension can reach more than 80% of the explanatory strength on experience value.

Further observations of the t test value of each table, and observation of significance. It can be seen from Table 10 that "sense experience", "feel experience", and "relate experience" all have a significant impact on "consumer return on investment". In terms of standardized Beta, emotional experience has the highest predictive strength on customers investment returns. It can be inferred that as visitors arouse more emotions from the

Table 8. Correlation analysis of experiential value and brand loyalty (H2)

	Action loyalty	Conative loyalty
CROI	0.711**	0.729**
Service excellence	0.659**	0.724**
Aesthetics	0.628**	0.716**

Notes: "*" means the p value < 0.05, "**"means p value < 0.01

Table 9. Correlation analysis of experiential marketing and brand loyalty (H3)

	Action loyalty	Conative loyalty
Sense	0.602**	0.675**
Feel	0.613**	0.729**
Think	0.610**	0.642**
Act	0.701**	0.714**
Relate	0.659**	0.532**

exhibition experience, the more they will be able to enhance a sense of value of customer investment returns.

It is seen in Table 11 that "sense experience", "feel experience", and "act experience" have a significant impact on "service excellence". From the standpoint of standardized Beta, the "sense experience" is the highest, which shows that if the sense experience of actual contact can be improved in the exhibition activities, it will be able to increase the value and satisfaction of visitors with the superior service.

It can be seen from Table 12 that the four dimensions of "sense experience", "feel experience", "act experience", and "relate experience" all have a significant impact on "aesthetics". From the perspective of standardized Beta, among them, action experience is the most predictive for the improvement of aesthetic value. In other words, if the planning of the exhibition can provide more interactive and participatory content for visitors, it will improve visitors with a higher sense of aesthetic value.

It can be seen from Table 13 that "sense experience", "think experience", and "act experience" can have a significant impact on "playfulness". From the standardized Beta point of view, the "action experience" is the highest, which can explain: the exhibition provides various content that allows visitors to actually participate, interact, and experience. In addition to the above-mentioned value that can strengthen the "aesthetics", it can also enhance the value of "playfulness".

It can be inferred from the results of various analyses that the hypothesis proposed by this study: "Experience marketing has a significant positive impact on experience value" is affirmative.

Table 10. "Experiential marketing (strategic experience modules) regression analysis of "CROI"

Dependent variables	Independent variables	Beta	Unstandardized coefficients		t	Sig.
			B	Std. Error		
CROI	Sense experience	0.120	0.128	0.058	2.194	0.029*
	Feel experience	0.540	0.541	0.055	9.842	0.000**
	Think experience	−0.014	−0.014	0.053	−0.263	0.793
	Act experience	0.094	0.088	0.049	1.807	0.072
	Relate experience	0.267	0.258	0.042	0.000	0.000**

Adjusted R Square: 0.802/Sig. = 0.000

Notes: "*" means the p value < 0.05, "**"means p value < 0.01

Table 11. The "Experiential Marketing (Strategic Experience Modules) Regression Analysis of "Service Excellence"

Dependent variables	Independent variables	Beta	Unstandardized coefficients		t	Sig.
			B	Std. error		
Service excellence	Sense experience	0.340	0.363	0.042	8.553	0.000**
	Feel experience	0.296	0.298	0.040	7.449	0.000**
	Think experience	0.064	0.064	0.039	1.649	0.100
	Act experience	0.331	0.311	0.035	8.773	0.000**
	Relate experience	0.004	0.004	0.031	0.122	0.903

Adjusted R Square: 0.885/Sig. = 0.000

Notes: "*" means the p value < 0.05, "**"means p value < 0.01

Table 12. The "experiential marketing (strategic experience modules) regression analysis of "Aesthetics"

Dependent variables	Independent variables	Beta	Unstandardized coefficients		t	Sig.
			B	Std. error		
Aesthetics	Sense experience	0.294	0.277	0.043	6.464	0.000**
	Feel experience	0.276	0.244	0.040	6.052	0.000**
	Think experience	−0.004	−0.003	0.039	−0.083	0.934
	Act experience	0.529	0.438	0.036	12.236	0.000**
	Relate experience	0.136	0.116	0.031	3.736	0.000**

Adjusted R Square: 0.850/Sig. = 0.000

Notes: "*" means the p value < 0.05, "**"means p value < 0.01

(2) Regression analysis of experiential value on predictive power of brand loyalty (H2)

In this section, experiential value is analyzed for brand loyalty. Four facets of experiential value are taken as independent-variants, and action loyalty and cognitive loyalty are taken as variants for regression analysis one by one. The analysis results are presented in Table

Table 13. The "experiential marketing (strategic experience modules) regression analysis of "playfulness"

Dependent variables	Independent variables	Beta	Unstandardized coefficients		t	Sig.
			B	Std. error		
Playfulness	Sense experience	0.084	0.096	0.034	2.820	0.005**
	Feel experience	0.027	0.029	0.032	0.917	0.360
	Think experience	0.275	0.293	0.031	9.428	0.000**
	Act experience	0.663	0.666	0.028	23.366	0.000**
	Relate experience	−0.012	−0.012	0.025	−0.491	0.624
Adjusted R Square: 0.935/Sig. = 0.000						

Notes: "*" means the p value < 0.05, "**"means p value < 0.01

14 and Table 15. The analysis results showed that the standardized R-squared values of each regression analysis could reach 0.652 (Table 14) and 0.714 (Table 15), respectively, and all reached significant levels. It can be seen that the two facets of experiential value can reach 65.2% and 71.4% explanatory strength on brand loyalty.

Then the t-test values were observed from the two tables, and the significance was observed. It is seen in Table 14 that "CROI" and "playfulness" have a significant impact on "action loyalty". From the perspective of standardized Beta, the "CROI" is higher than the "Playfulness", which indicates that with the increase of the value of the return on investment of the exhibition, visitors will have a higher result of action loyalty.

It can be seen from Table 15 that the three dimensions of "CROI", "aesthetics" and "playfulness" have a significant impact on "conative loyalty". Also from the perspective of standardized Beta, "CROI" is also the most explanatory factor among the three dimensions, which also indicates that if visitors can get a higher perception of the value of return on investment, it will also promote them to produce a higher result of attitude loyalty.

Based on the analysis results, it can be inferred that hypothesis 2, "experiential value has a significant positive effect on brand loyalty", is valid.

(3) Experience Marketing (Experience Marketing Module) Regression Analysis of Brand Loyalty Prediction (H3)

This paragraph analyzes the brand loyalty of experience marketing, taking the five facets of experience marketing as independent variables, taking behavioral loyalty and attitude loyalty as dependent variables and conducting the regression analysis one by one. The regression analysis results show that the standardized R-squared values can reach 0.677 (Table 16) and 0.708 (Table 17), and both reach significant levels. It indicates that the

Table 14. Multiple regression analysis of "experience value" to "action loyalty"

Dependent variables	Independent variables	Beta	Unstandardized coefficients		t	Sig.
			B	Std. error		
Action loyalty	CROI	0.455	0.419	0.075	5.623	0.000**
	Service excellence	0.023	0.021	0.085	0.245	0.807
	Aesthetics	0.067	0.069	0.091	0.763	0.446
	Playfulness	0.240	0.206	0.080	2.572	0.011*
Adjusted R Square: 0.652/Sig. = 0.000						

Notes: "*" means the p value < 0.05, "**"means p value < 0.01

Table 15. Multiple regression analysis of "experience value" to "conation loyalty"

Dependent variables	Independent variables	Beta	Unstandardized coefficients		t	Sig.
			B	Std. error		
Conative Loyalty	CROI	0.312	0.290	0.069	4.194	0.000**
	Service Excellence	0.131	0.121	0.079	1.539	0.125
	Aesthetics	0.226	0.238	0.084	2.821	0.005**
	Playfulness	0.176	0.155	0.074	2.087	0.038*
Adjusted R Square: 0.714/Sig. = 0.000						

Notes: "*" means the p value < 0.05, "**"means p value < 0.01

experience value marketing can reach 67.7% and 70.8% of the explanatory strength of brand loyalty.

Further observation on the t test value and its significance. Table 16 shows that the facet of "marketing experience" and "relate experience" can have a significant impact on "action loyalty". The standardized Beta values of the two are similar, and the "relate experience" is slightly higher, which indicates that the exhibition planning by providing visitors with more diverse behavioral experience content and providing more experiences that have higher relevance content to the visitors themselves will produce higher behavioral loyalty results.

Table 17 shows that the dimensions of "feel experience" and "act experience" have a significant impact on "conative loyalty". Among them, the standardized Beta of "feel experience" is higher than that of "action experience", which shows that if the emotional appeal in exhibition planning can arouse the psychological emotions of visitors, it will result in higher attitude and loyalty.

A comprehensive observation of the analysis results shows that the third hypothesis proposed in this study: "Experiential marketing has a significant positive impact on brand loyalty" is validated.

Table 16. Multiple regression analysis of "experience marketing" to "action loyalty"

Dependent variables	Independent variables	Beta	Unstandardized coefficients		t	Sig.
			B	Std. error		
Action loyalty	Sense experience	0.138	0.135	0.075	1.809	0.072
	Feel experience	0.136	0.126	0.070	1.782	0.076
	Think experience	−0.090	−0.082	0.068	−1.207	0.229
	Act experience	0.354	0.299	0.062	4.461	0.000**
	Relate experience	0.360	0.321	0.054	5.913	0.000**

Adjusted R Square: 0.677/Sig. = 0.000

Notes: "*" means the p value < 0.05, "**"means p value < 0.01

Table 17. Multiple regression analysis of "experience marketing" to "conative loyalty"

Dependent variables	Independent variables	Beta	Unstandardized coefficients		t	Sig.
			B	Std. error		
Conative loyalty	Sense experience	0.084	0.083	0.073	1.147	0.252
	Feel experience	0.373	0.348	0.069	5.068	0.000**
	Think experience	0.046	0.042	0.066	0.637	0.524
	Act experience	0.315	0.274	0.061	4.506	0.000**
	Relate experience	0.046	0.041	0.053	0.784	0.434

Adjusted R Square: 0.708/Sig. = 0.000

Notes: "*" means the p value < 0.05, "**"means p value < 0.01

(4) Variance Analysis on the Influence of Demographic Variables on Experiential Value, Brand Loyalty. (H4)

To verify the research hypothesis 4: "Demographic Statistic Variables have discernible influence on Experiential Marketing, Experiential Value, and Brand Loyalty" This paragraph chooses T-test or single factor variance analysis, targeting the seven demographic variables including gender, residence, and occupation, age, marital status, and monthly income, and conducted the average verification analysis one by one on the five facets of experience marketing, the four facets of experience value, and the two facets of brand loyalty. If the statistics reach a discernible level, it will be analyzed by the Duncan posterior analysis. Limited to the pages, this study presents only the items with noticeable differences in Table 18.

The statistical results show that among the disguised demographics, gender, residence, marital status, and monthly income have no significant impact on experience marketing, experience value, and brand loyalty.

Table 18 shows that the "occupation category" of the visitors has an significant impact on facets such as "sense experience", "feel experience", "act experience", "CROI", "service excellence", "beauty", and "playfulness;" while the occupation from the result shows that the visitors in field of "Design/Cultural Creative/Art" have higher evaluation average compare to other occupations. This shows that the cultural exhibition events were especially preferred by participants of art and cultural fields and that boosts these groups of people with higher conative loyalty.

The "age" of the visitor has a significant impact on the six facets including the "sense experience", "feel experience", "act experience", "service excellence", "aesthetics", and "playfulness". The results of the posterior verification shows that the average number of exhibition evaluations given by people aged "21 to 40" and "41 to 60" is significantly higher than that of "below 20" and "61 years of age". This result shows that people from mature young to middle-aged have a higher evaluation of this art exhibition and are more dependent on it.

The "educational level" of visitors has a significant impact on the four facets of "think experience", "act experience", "service excellence" and "playfulness". Among them, when the students of "Elementary/Junior High School" and "High School/High Vocational College" are visitors, they will give significantly higher evaluation scores to the content of the exhibition activities, especially for the thinking, action experience, and exhibitions in the experience activities. The services of the exhibition and the playfulness in it can particularly evoke these two groups and a higher evaluation was given.

The above-mentioned three demographic variables of occupation, age, education level, etc., can have a significant impact on experiential marketing, experiential value, and brand loyalty, and such research results may be helpful to future exhibition planning. It is specifically an important reference for integrating marketing and event appeals, and it is expected that higher exhibition benefits and evaluation will be obtained. As for the research hypothesis four, "the demographic changes are significantly different from experience marketing, experience value, and brand loyalty." Only some of them have significant impact on variables so the research hypothesis four is only partially verified.

4.3   Discussion on International Cultural Curation Qualitative Benefits

This exhibition was hosted by the Cultural Assets Bureau of the Ministry of Culture and invited professional teams from local universities for operation. The 1916 Cultural and Creative Workshop in the Cultural Assets Park was chosen as the venue. The Cultural Heritage Park is a century-old historical building. In order to be able to revitalize historical buildings and demonstrate the respect for culture and creativity industries, the 1916 Cultural and Creative Workshop was established in 2016 to invite people of all fields of

Table 18. Result of single-way variance analysis of demographic variables for each facet (Listed only those with significant levels)

Demographic (independent variables)	Facets (dependent variables)	F	Significance
Occupation	Sense Experience	3.199	0.005**
Occupation	Feel experience	4.090	0.001**
Occupation	Act experience	4.368	0.000**
Occupation	CROI	3.221	0.005**
Occupation	Service excellence	3.091	0.006**
Occupation	Aesthetics	3.808	0.001**
Occupation	Playfulness	3.532	0.002**
Occupation	Conative loyalty	2.261	0.038*
Age	Sense experience	2.749	0.043*
Age	Feel experience	2.827	0.039*
Age	Act experience	3.583	0.014*
Age	Service excellence	3.003	0.031*
Age	Aesthetics	4.201	0.006**
Age	Playfulness	2.907	0.035*
Education	Think experience	2.383	0.039*
Education	Act experience	2.621	0.025*
Education	Service excellence	2.396	0.038*
Education	Playfulness	2.813	0.017*

Notes: "*" means the p value < 0.05, "**" means p value < 0.01

Taiwan's cultural and creative industries such as, cultural heritage preservation, education, design, etc. Artists, designers, youth creative teams, etc. are invited to station in, and a total of 32 businesses are stationed. This multinational art and cultural exhibition chooses to exhibit in CHP that shows the emphasis on the exchange and activation of the local art and cultural industries. Therefore, it not only exhibits works from Italian illustrators, but also invites 32 Taiwanese illustrators for joint exhibitions. The 32 businesses in the 1916 workshop also organized related activities and marketing campaigns in response to the theme of the Bologna Illustration Exhibition to expand the benefits of the exhibition. After the exhibition was completed, this research began to interview 24 illustrators who participated in the exhibition. The interview period was from September to October 2020. The viewpoints of these illustrators participating in the exhibition were compiled into this research proposal for future exhibition references.

First of all, for the exhibition planning, due to the nature of public interest, it has attracted numerous public and media attention. During the exhibition, there were up to 15,000 visitors, and the number of media integrated marketing and exposure reached 37. In addition, before, during, and after the exhibition, the curatorial unit systematically planned to cooperate with the illustrator, and carried out activities on the social media that was operated by the illustrator. It was an important marketing strategy to effectively gather the attention of fans of each illustrator. To the emerging new illustrators, it was also an important opportunity to enhance self-brand, so this exhibition activity in the park was highly praised. In addition, the series of activities handled by the exhibition are quite diverse, especially planning opening press conferences, exchange meetings, illustrator sharing lectures, etc., so that illustrators can communicate with each other. Not only could the illustrators exchange opinions on techniques, but also discussions related to cultural literacy, creative ideas, brand management, etc. and these have promoted diversified cultural exchanges and increased industry connections. From the discussion of the process of joint exhibition and exchange of transnational works, it is evidential that the introduction of internationally renowned art exhibitions can effectively enhance the visibility and positioning of the park. From the participating illustrators and the public, what they appreciate is not only the story of the painting, but from the connotation of illustration stories to see the background, educational environment, cultural connotation and social atmosphere of illustrators from all over the world. Through the management of international art exhibitions, this type of cultural exchange benefits can be created, and the visitors will be enhanced as it is also an important source of nutrients for cultural literacy.

5   Conclusion and Suggestion

The mission and value bestowed on CCIP(or CHP) should not only be used as a general exhibition venue, but also should be considered with a broader picture of the operational positioning, operating policy, and operating strategy, as well as the disruptive benefits of the local cultural and creative industries. In Taiwan, among the five cultural and creative industry parks under the jurisdiction of the Ministry of Culture, only the CHP located in the Taichung area is operated by the public with the Cultural Asset Bureau stationed in for management and the overall operation of the park; CHP has an important role demonstrating cultural policies which is made by MOC. This study chooses the transnational authorized exhibition, the "Illustrations for Children. Italian Excellence" as case study. Based on theories of experiential marketing, experiential value, brand loyalty, the discussion focuses on the promotion of customer's brand loyalty towards CHP by the curation of art and cultural exhibitions and creating visitors experiences.

This research chooses questionnaire survey method and in-depth interview method to verify the four research hypotheses and achieve the purpose of this study. The results show that all the hypotheses are verified with affirmative results, which will be helpful for the future planning of art exhibitions and integrated marketing in the cultural creative parks.

References

Development of the cultural and creative industries act. Ministry of Culture, Taiwan: https://law. moj.gov.tw/. Accessed 30 Oct 2020

Ministry of Culture, Taiwan. https://www.moc.gov.tw/. Accessed 23 Oct 2020

Cultural Heritage Park: Ministry of Culture: https://tccip.boch.gov.tw/. Accessed 23 Oct 2020

Bureau of Cultural Heritage: Ministry of Culture. https://www.boch.gov.tw/. Accessed 25 Oct 2020

Schmitt, B.H.: Experiential marketing. J. Mark. **15**(1), 53–67 (1999)

Taylor, S., DiPietro, R.B., So, K.K.F.: Increasing experiential value and relationship quality: an investigation of pop-up dining experiences. Int. J. Hosp. Manag. **74**, 45–56 (2018)

Holbrook, M.B.: The Nature of Customer Value: An Axiology of Service in the Consumption Experience. In: Rust, R., Oliver, R.L. (eds.) Service Quality: New Direction in Theory and Practice, pp. 21–71. Sage Publications, Thousand Oaks (1994)

Mano, H., Oliver, R.L.: Assessing the dimensionality and structure of the consumption experience: evaluation, feeling, and satisfaction. J. Consum. Res. **20**, 451–465 (1993)

Babin, B.J., Darden, W.R.: Consumer self-regulation in a retail environment. J. Retail. **71**(1), 47–70 (1995)

Mathwick, C., Malhotra, N., Rigdon, E.: Experiential value: conceptualization, measurement and application in the catalog and internet shopping environment. J. Retail. **77**(1), 39–56 (2001)

Jacoby, J., Chestnut, R.W.: Brand Loyalty: Measurement and Management. New York Press, New York (1978)

Hallowell, R.: The relationships of customer satisfaction, customer loyalty and profitability: an empirical study. Int. J. Serv. Ind. Manag. **7**, 27–42 (1996)

Oliver, R.L.: When consumer loyalty. J. Mark. **63**, 33–44 (1999)

Hee, O.C.: Validity and reliability of the customer-oriented behaviour scale in the health tourism hospitals in Malaysia. Int. J. Caring Sci. **7**(3), 771–775 (2014)

Bologna Children's Book Fair (2020). https://www.bookfair.bolognafiere.it/en/focus-on/illustrators/how-to-take-part/1041.html. Accessed 12 Oct 2020

The Core Values and Methodology
of Cross-Cultural I-Sustainability Design
Thinking

Yu-Chao Liang[1]([✉]), Chao Liu[2], Hao Chen[2], Ding-Hau Huang[3], and Wen-Ko Chiou[1]

[1] Department of Industrial Design, Chang Gung University, Taoyuan City, Taiwan
`wkchiu@mail.cgu.edu.tw`
[2] Graduate Institute of Business and Management, Chang Gung University,
Taoyuan City, Taiwan
[3] Institute of Creative Design and Management, National Taipei University of Business,
Taoyuan City, Taiwan

Abstract. In order to create sustainable products, this study presents the concept of I-sustainability design and its eight core values (flow, mindfulness, sustainability, aesthetics, ergonomics, health, quality, technology). At the beginning of the development of new products, linking product shape, product technical function and sustainable development, analyzing how to clear the product value of I-sustainability, how to excavate potential I-sustainability value of the product, and finally analyzing and evaluating the I-sustainability of the product value, so as to satisfy the customer demand and social sustainable development of the creative products. In order to maximize the I-sustainability value of products, designers can analyze the attributes of eight value opportunities that represent the I-Sustainability Design Thinking (ISDT) core values. By playing the role of the above attributes in product design, product design can achieve breakthrough progress, thus creating sustainable products. Combining the methods of science and design can help to develop more innovative, better integrated and truly transformational initiatives for sustainability. We suggest DT as a suitable approach to ensure that sustainability aspects are already taken into account in the initial phase of organizational innovation processes, i.e. when generating and evaluating SDGs ideas. Furthermore, ISDT provides a differentiated basis underpinning the assumptions of several researchers that a DT-based approach is an essential, yet often overlooked asset when addressing sustainability challenges.

Keywords: SETBS (social-economy-technology-commercial resources-sustainability) factor · I-sustainability design · Product value opportunity gap · I-sustainability value analysis · I-sustainability core values

1 Introduction

With the development of society and the progress of science and technology, people's requirements for product design have developed to a new stage, which is embodied in

© Springer Nature Switzerland AG 2021
P.-L. P. Rau (Ed.): HCII 2021, LNCS 12771, pp. 100–114, 2021.
https://doi.org/10.1007/978-3-030-77074-7_8

the following aspects: the design object moves from single machine to system; The design requirement is to move from single target to multi-target; The field of design moves from a single field to a multi-field. Product update speed is accelerated; Products from free development to planned development; The development of computer has put forward new design requirements (Davis and Tomoda 2018). In addition, the continuous emergence of new materials and new processes makes the replacement cycle of products shorter day by day. Therefore, new features have appeared in modern product design, and stricter requirements have been put forward for the function, reliability and benefit of products. Among these features, 70% depend on the design time, that is, the conceptual design time of products (Cooper 2013).

The excessive industrialization of the West, and the economic development for the purpose of profit, will gradually cost the earth's resources, and lead to the disaster of mankind (Perra et al. 2017). The Book of Changes (I Ching) is an ancient human wisdom in the East. By observing the movement of celestial bodies, the changes of creation and growth of all things on earth, we can understand the fundamental principle behind it, and the creation and way of dealing with the world in which man coexists with heaven and earth, and heaven, earth and man are integrated. It is also expressed in simple illustrations and words and passed on to later generations (TenHouten and Wang 2001). In the 21st century, the international community has woken up to the need for coexistence between human beings and the natural environment, and for balanced and sustainable development. In 2015, the United Nations proposed a plan for sustainable development of the eye Sustainable Development Goals (SDGs). However, how to make the whole mankind aware and consistent remains to be promoted (Naidoo and Fisher 2020).

In the past, most design studies considered to meet the material needs of human beings. By starting from the various material needs of human beings, innovative solutions were sought for various issues and more possibilities were created (Brown 2008). Little research has been done on the design of human spiritual needs. Through the use of the product, we can obtain spiritual pleasure and good emotional experience, or in the use of the product, we can make the human become more and more like the original human, realize the full potential of human. Not only do people want to use a product to complete or perfect a job, but they also want a product to bring about a spiritual experience and connect that experience to some kind of personal dream (Clausen and Pohjola 2013). Therefore, it is necessary to connect the product design with the value concerned by the target users, so as to abandon the past design idea of "form is subordinate to function", but to adopt the design idea of both form and function, and design products with I-sustainability (Verganti 2011). That how to do this, this study focus on how to clear the product value, qualitative analysis the influencing factors of product value opportunity, finally carries on the analysis, evaluation, and combining with reasonable product modeling and functional characteristics, formed to meet customer needs and expectations of breakthrough products, establish its competitiveness in the market.

2   The ISDT

2.1   I Ching

The Qian and Kun diagrams of the Book of Changes can best represent the core, value and method of I-sustainability design thinking. Qian diagrams means heaven revolves and the gentleman to unremitting self-improvement. It can not only describe the way of heaven, the gentleman or the principles of human growth and career, or even can be applied to any living things. We can apply I-sustainability design thinking methods and principles to help an enterprise grow and succeed. In particular, the process of designing an innovative product or service is positive, persistent and consistent (Yuduo et al. 2011). The Kun diagrams is a metaphor for the earth's compassion. To be successful, it is necessary to be as tenacious and practical as the Kun diagrams, but also as kind and motherly as the earth breeds things! For example, from the point of view of I-sustainability design, the innovation of products and services should be based on the spirit of serving the society and designing for human welfare (TenHouten and Wang 2001).

2.2   Sustainable Development

Sustainable development is development that meets the needs of current generations without jeopardizes the ability of future generations to meet their own needs. They are an inseparable system that not only aims to achieve economic development, but also protects the natural resources and environment such as the atmosphere, fresh water, oceans, land and forests on which human beings depend for their survival, so that future generations can enjoy sustainable development and a happy life (Elliott 2012). Sustainable development is not the same as environmental protection. Environmental protection is an important aspect of sustainable development. The core of sustainable development is development, but it requires economic and social development under the premise of strict control of population, improvement of population quality, protection of environment and sustainable use of resources. Development is the premise of sustainable development; Man is the centrosome of sustainable development. Sustainable development is the real development. So that future generations can enjoy sustainable development and live in peace and contentment (Silvestre and Țîrcă 2019). It includes two important concepts: the concept of needs, especially the basic needs of the peoples of the world, which should be considered with special priority; The concept of limitations, technological conditions and social organizations impose limitations on the ability of the environment to meet immediate and future needs (Sachs et al. 2019).

2.3   Design Thinking

Design thinking is a problem solving methodology based on people's needs. It seeks innovative solutions to various issues and creates more possibilities based on people's needs. Tim Brown, president of IDEO Design Company, believes that design thinking is people-oriented design spirit and method, considering people's needs and behaviors, as well as the feasibility of technology or business (Brown 2008). There was a design way of thinking and communication, which was different from the scientific and academic way of thinking and communication, and different from the scientific and academic way of application to solve various problems (Archer 1979). Dorst and Cross proposed that during the design process, new solution ideas can lead to a deeper or alternative understanding of the background of the problem, thereby triggering more solution ideas (Dorst and Cross 2001). Meinel et al. described the five phases of the design innovation process: defining the problem, finding and benchmarking, conceiving, building and testing (Lindberg et al. 2011). Ralph proposed that design thinking has always been at the core of user-centered design and the main method of designing human-computer interface, and is also at the core of recent software development concepts (Ralph 2015).

However, western design thinking mostly considers the material needs of human beings, and there are few studies to discuss the design thinking of the spiritual needs of human beings. Through the use of the product, we can obtain spiritual pleasure and beautiful emotional experience, or during the use of the product, we can make the human become more and more like the original human, realize the full potential of human.

2.4 Situational Story Design Method

The design method of the so-called situation story is the obvious study of the current product design. The situational story method is a design method to describe how technology can help users in the future life through visualization (Nardi 1992). Traditional systematic design methods, mostly from the perspective of designers, carry out functional design by exploring the relationship between objects and objects, while ignoring the fact that designers and users have different understandings of products (Norman 1989). The law of situation story is to simulate the use situation of future products through an imaginary story in the process of product design and development.

In the simulation process, the relationship between the user's characteristics, events, products and environment variables is considered. And constantly in the form of visual and practical experience, guide the staff involved in the design and development, from the perspective of users, to explore the idea of the products, through the use of situational simulation, to explore, analyze the interaction between people and products, and to judge whether comply with the design theme, while testing product ideas, meets the users' potential demand (Kishita et al. 2020). Therefore, situational storytelling is basically a user-centered design. The content in the situational story method must contain visual elements, because they are the main tool to express the design idea, and also the important element to express the interactive relationship between people and products. Situational story is a tool to support the designer to implement the design process, and it can be involved in the whole process, whether before the design begins, after the concept development, or in the formative development and evaluation stage at the later stage of the design (Park et al. 2020). The greatest contribution of situational stories is a predictive description of possible future events (Uwasu et al. 2020).

2.5 I-Sustainability Design Thinking

A changing world: The book of I Ching records the earth's rotation and revolution around the sun in the universe, as well as the eternal rule changes in the four seasons of the year, spring, summer, autumn and winter, which constitute the birth and growth of all things on the earth. As the wisest of all creatures, human beings should take the mindfulness and rationality of the unity of "heaven, earth and man" in product design, coexist with nature in time, and take sustainable development as the goal. This is the best way for human beings to coexist with each other on earth (Wilhelm et al. 1967)!

In the first two hexagrams in the book of I Ching, the "Qian Diagram" represents the rotation and revolution of the earth around the great sun, which represents the positivity of the heavenly movement and the phenomenon of endless circulation. It metaphors all living things in the world, from birth, growth, prosperity, decline, and subsequent regeneration, endless circulation and sustainable development. In the process of "I-sustainability design", it represents the innovative design and development of products and services. From creativity to commercial feasibility, it can be transformed into "dragons" through the process of rigorous divergence, convergence of qualitative research, qualitative research and experimental verification. We should be alert to our "highlight dragon's shame" and plan for the second generation of products as soon as possible, continuing the ancestral line, inheriting and developing forever. Finally, "no leader in a host of dragons" appears to caution us in the "I-sustainability design" group when it comes to the importance of group collaboration without a leader (Lu 2008)!

"Kun Diagram", terrain Kun, virtue carrying things, represents the earth like a mother, with the maternal nature of "kindness" and "Virtue", nurturing the growth of all things on the earth. In the universe, the coincidence between the sun and the earth, the distance between the sun and the earth, the inclination of the earth's axis and the common rotation axis, making it 365 days a year, just forming the earth of spring, summer, autumn and winter, which is suitable for nourishing all things (Cheng 1988). The combination of yin and Yang of heaven and earth can be introduced into the I-sustainability design. The "Qian Diagram" is a metaphor for the firmness and perseverance of the heaven, the practice and positivity of the innovative design. While the "Kun Diagram" is also a metaphor for the mindfulness that designers must have to serve human beings and design to improve human life and living environment. In addition, "Qian Diagram" and "Kun Diagram" supplement each other, and the earth can nourish the growth of all things by absorbing the energy of the sun (Ma 2005)!

In the two Diagram of Qian and Kun, the dominant and subordinate changes of the Yang and Yin also imply the "orientation" and "qualitative" of the main changes in the first part of the "double diamond" method of "I-sustainability design", and the "position" and "qualitative" of the latter part (Clune and Lockrey 2014).

In the center of the eight Diagram, the Taiji image of balance of Yin and Yang is a symbol of the eternal movement and sustainable balance of the heavenly movement. It is also a metaphor for the "I-sustainability design". With human wisdom, it is necessary to consider the sustainable development of the earth and the human beings on the earth in terms of innovation and design (Lui 2005).

In the eight trigrams of Taiji in the book of I Ching, the three trigrams represent the changes and rules of the unity of heaven, earth and man. The eight trigrams represent the heaven, the earth, the thunder, the electricity, the mountains, the rivers, the water and the fire that human beings face. In the natural environment, there are eight main phenomena of natural and physical science, which Human beings live in. How to use human wisdom and I-sustainability design thinking to achieve sustainable development with the earth is the purpose and goal of "I-sustainability design" (TenHouten and Wang 2001)!

3   ISDT Core Values

A product is valuable if it is useful, easy to use, and attractive. Value can be broken down into specific product attributes that support product availability, ease of use, and desirability, and it is these attributes that link the functional characteristics of a product to value. Because a product creates an experience for the user, the better the experience, the more valuable the product is to the user (Vechakul et al. 2015). Ideally, a product can realize a dream by telling the user to solve a problem or complete a task in a more enjoyable way, thus identifying a set of eight opportunities to enhance the value of the product, known as the ISDT core values. The eight value categories are: flow, mindfulness, sustainable development, aesthetics, ergonomics, health, quality, and core technology.

3.1   Flow Experience

Flow is a kind of feeling in which one's mind is completely invested in some activity. When flow is generated, there will be a high sense of excitement and fulfillment. The conceptualization of flow originates from the field of peak experience. In 1968, Maslow began to study peak experiences. The peak of all experiences is self-actualization. The conclusion is that flow experiences are available to all (Fritz and Avsec 2007), although little is known about these experiences. For example, you don't know what it takes to have a peak experience. Nevertheless, Maslow's research confirms that these experiences do exist, may be available to all, and benefit human development.

The concept of flow comes from the study of optimal human experience (Seligman 2002). Flow is the state of mind in which one is completely immersed in the current activity, such as reading or running (Csikszentmihalyi and Nakamura 2010). They described the conditions required to achieve flow: activities with clear goals, a balance between challenges and skills, and clear and real-time feedback on effectiveness and progress. Flow is described as optimal experience in positive psychology, where the feeling of enjoyment and the experience of flow are considered to be an optimal state of well-being and have been shown to increase human happiness (Fritz and Avsec 2007). Csikszentmihalyi and Nakmura (2010) describe it as a state of high attention to the task at hand, with a self-dynamic state. They describe the subjective elements of experience as including enjoyment, reduced self-awareness and altered perception of time. In his phenomenological analysis of flow, Elkington defined it as the dynamic flow nature of experience following the optimal function of consciousness. Elkington described the main components needed to experience flow: in the face of a task that can be completed, focus, clear goals and real-time feedback, which can be controlled, because of concentration, forgetting worries and achieving a state of selflessness (Elkington 2010).

3.2 Mindfulness

Mindfulness therapy was founded by Kabat-Zinn, emeritus professor at the University of Massachusetts. Mindfulness is a kind of self-awareness that focuses on the present and is completely open. It does not require a self-critical attitude, but instead embraces curiosity and acceptance to receive every thought in the heart and mind, namely, the emphasis on facing the present and being aware (Kabat-Zinn 1994). Mindful practitioners live very clearly in the present moment but are not involved in it (Bishop et al. 2004). Mindfulness techniques have been effective in treating a range of mental health difficulties (Baer 2003) and promoting well-being (Brown and Ryan 2003). As a result, mindfulness is increasingly being used as a form of psychological intervention derived from eastern Buddhist meditation practices that draw on concepts related to all things and are thought to be related to spiritual commerce (Baer 2003). Carmody and Bae (2007) found that regular meditation cultivated mindfulness skills in daily life, thereby improving psychological function and increasing happiness. Similarly, Brown and Ryan reported that mindfulness is negatively correlated with negative emotions such as anxiety, depression, and anger, and positively correlated with happiness (Brown and Ryan 2003).

3.3 Sustainable Development Goals

The SDGs or Global Goals are a collection of 17 interlinked goals designed to be a "blueprint to achieve a better and more sustainable future for all". The SDGs were set in 2015 by the United Nations General Assembly and are intended to be achieved by the year 2030. They are included in a UN Resolution called the 2030 Agenda or what is colloquially known as Agenda 2030. The 17 SDGs are: (1) No Poverty, (2) Zero Hunger, (3) Good Health and Well-being, (4) Quality Education, (5) Gender Equality, (6) Clean Water and Sanitation, (7) Affordable and Clean Energy, (8) Decent Work and Economic Growth, (9) Industry, Innovation and Infrastructure, (10) Reducing Inequality, (11) Sustainable Cities and Communities, (12) Responsible Consumption and Production, (13) Climate Action, (14) Life Below Water, (15) Life On Land, (16) Peace, Justice, and Strong Institutions, (17) Partnerships for the Goals. Though the goals are broad and interdependent, two years later the SDGs were made more "actionable" by a UN Resolution adopted by the General Assembly. The resolution identifies specific targets for each goal, along with indicators that are being used to measure progress toward each target. The year by which the target is meant to be achieved is usually between 2020 and 2030. For some of the targets, no end date is given (DU 2016).

3.4 Aesthetics

Product aesthetics, not only refers to the pure form, but rather refers to a number of factors constitute the product modeling (function, material, technology, etc.) of the integrated embodiment, is the organic combination of science and art, it mainly includes: function, comfort, specifications, structure, material, process, shape beauty, color beauty and harmonious beauty alone (Forsey 2016).

3.5 Ergonomics

Ergonomics is a subject that studies various factors such as anatomy, physiology and psychology in a certain working environment, studies the interaction between human, machine and environment, and studies how to consider people's health, safety, comfort and work efficiency in work, family life and leisure time. Human factors engineering provides theoretical knowledge and design basis for product design, which enables designers to follow rules in operation during specific design, reduce time consumption, and reduce labor intensity, and put more energy into solving human-machine problems. The three attributes of value opportunity in human factor engineering are convenience, safety and comfort (Bridger 2008).

(1) Convenience: A product must be easy to use in both physical and cognitive aspects. A product should be able to operate within the natural motion of the human body. For a product that has direct contact and interaction with the user, the size and shape of its components should be logically organized for identification, contact, grasp and operation (Bhise 2011).
(2) Safety: The use of a product should not only be convenient for users, but also have safety, that is, the operator or user, within the scope of consistent use, the user should be protected from harm (Sagot, Gouin, & Gomes 2003).
(3) Comfort: In addition to convenience and safety, a product should be comfortable to use and should not cause unnatural physical and psychological fatigue (Helander 2003).

3.6 Health

Health means that a person is in good physical, mental and social condition. The meaning of modern health is not only a traditional refers to the body without the disease, according to "the world health organization" explanation: health not only refers to a person not have a disease or weak, but refers to a person's physical, psychological and social health state, this is a relatively complete scientific concept of modern about health. The meaning of modern health is diversified and extensive, including physiology, psychology and social adaptability, among which social adaptability ultimately depends on the condition of physical and psychological quality. Mental health is the spiritual pillar of physical health, and physical health is the material basis of mental health (Liu et al. 2020b). Good emotional state can make physiological function in the best state, otherwise it will reduce or destroy some function and cause disease. Changes in physical conditions may lead to corresponding psychological problems. Physical defects and diseases, especially chronic diseases, tend to produce upset, anxiety, depression and other negative emotions, leading to various abnormal psychological states. As a person of body and mind, body and mind are two closely interdependent aspects (Organization 2016).

3.7 Quality

The final value opportunity is quality, which refers to the accuracy of manufacturing, the combination of materials, the bonding process, etc. Although it is related to technology, the emphasis here is on the product processing itself, the expected result of the processing. Quality value opportunities include two attributes: manufacturing process and durability.

(1) Manufacturing process: With the surface process, the product should meet the appropriate tolerance requirements to ensure its performance (Haines-Gadd et al. 2018).
(2) Durability: The performance changes with time, and the product appearance must be kept constant within the expected product life (Haines-Gadd et al. 2018).

3.8 Core Technologies

Aesthetics and personality aim at modeling design, while core technology and quality value opportunity aim at technical factors. Technology is an indispensable part of product design. The most basic is that the product should be able to operate normally, realize the wishes of users, and bring them pleasant feelings and experiences. The core technology of the product mainly includes the reliability of the product and the usability of the product (Baxter 1995).

(1) Usability: The most basic feature of a product is the basic function it contains, that is, the feature of the product to meet the most basic use requirements of users. However, whether it is the core technology or the traditional technology, when a designer designs a product, he must first ensure that the product can achieve the most basic functions (Palmer 2002).
(2) Reliability: During the use of the product, due to the passage of time, the product will inevitably fail, but the product involved must be able to ensure the normal work of the product within the specified life span and normal use environment, so as to realize its value (Kuo et al. 2001).

4 ISDT Methodology

4.1 Mandala Drawing Orientation

Design is a creative activity and a philosophy of life that explores communication from individuals to groups. Mandala warm-up activity aims to combine the flow theory of Western psychology and Eastern meditation activities, combined with Mihaly Csikszentmihalyi's flow and Jung's psychoanalysis and the concept of Mandala, through the study of the collective unconscious of specific groups, specific age groups, to educate designers (Liu et al. 2020a).

Through the activities of cooperative painted mandala, designers can not only satisfy their needs of expression, but also position themselves in a safe space of spirit, while developing empathy and revealing the experience of self-communication with others. The meditative effects of mandala are equivalent to mindfulness meditation practices used in brain neuroscience research. In Jung's Psychoanalysis, he also made important comments on the inner spiritual perfection of the mandala, which various studies have shown mandala drawing have a significant impact on creativity and the promotion of empathy (Liu et al. 2019).

4.2 Macro Forces

The value proposition of innovative products and services is generated by taking the background factor of Social economy (SE), the Core Technology (T) as the dominant factor, and the favorable Business resources of enterprise operation in line with the goal of sustainable development (SDGS) (Cagan et al. 2002).

Social Economic and Technology Forces Divergent. Using yellow 3M sticky notes, cross-field team members brainstorm to create together. According to the above mentioned SET, more than 50 influencing factors are divergent respectively, and then they are ranked in order of group and importance.

Social Economic Convergent TOP10. Select more than 10 Top10 Forces, and list the project names of the important influencing factors of each group with pink sticky notes.

Technology Convergent. Select more than 10 Top10 Forces, and list the project names of the important influencing factors of each group with pink sticky notes.

Business Resources and Sustainable Development Goals. *Business Resources Divergent and Convergent*
Using yellow 3M sticky notes, cross-field team members brainstorm to create together. According to the above mentioned Business Resources, influencing factors are divergent respectively, and then they are ranked in order of group and importance. Select the important influencing factors of Business Resources, and list the project names of the important influencing factors of each group with pink sticky notes.

Sustainable Development Goal
Three major projects and their sub-items are selected from the 17 SDGS goals of the 17 UN development goals for sustainability.

SE.T.B.S Potential Opportunity Gap (POG) Synthesis and Product Ideas Generation. From the above ranking of the 10 important factors of the potential opportunity gap in SE.T.B.S, the first important factor and the correlation between the two opportunity gaps were integrated into the background of 3–4 items for each supporting innovative proposal concept project.

Product Idea and Potential Opportunity Gap Statement. The above integrated SE.T.B.S potential opportunity gap (POG) is arranged and plotted as below.

4.3 Micro Forces

It is micro-user-oriented and highlights the key needs and creative ideas of products in the application context according to the user's product application context of people, things, times and places (Campbell 1992).

Scenario Story Approach. According to the product opportunity gap and proposal conception generated by the above-mentioned macro factors, the typical user representative and the use situation story of the representative are set to describe the demand conditions and the factors to be satisfied under the use situation of the product.

Character Mapping. According to the product conception direction and background generated by the above macro, the attributes of the target group of the product and the Range of each attribute are set. The representative of attribute difference is selected within the scope of this definition. In terms of general experience, select 6–10 event story situations that represent differences and situations.

Scenario Sketch Brainstorming Development. Situational story sketch conception development: the team members select different user representative roles respectively, and think and draw the situation and situation of the product that the role needs or applies. The Critical issues and Key ideas in the scene were introduced to the group members. The group members participated with empathy, and brainstormed to spread the scene pictures according to the situation, so as to spread more demands and creative content, and pasted the scene pictures with yellow sticky notes.

Top10 Themes. The contents of Critical issues and Key Ideas obtained by spreading the yellow sticky notes were concluded into the innovation proposal and Top10 Themes by Grouping and Ranking.

Value Proposition. Based on the above Top10 Themes, solutions were worked out and the value proposition of this study was put forward.

Prototype Development. According to the user's application situation of the product, the model is used to simulate the use process and way to verify the applicability of the product in the use situation.

Through the value propositions, two proposals were selected for model making and simulation verification, and the product proposal with better simulation value was selected.

Prototype Simulation and Evaluation. The system selected above is manufactured into a product entity for verification.

Value Opportunity Analysis. Value opportunity analysis can be applied to tangible products, services, and product-service systems. All of the value opportunity categories can be applied to services, and some of these keywords can become more directly related to product and service design terms.

Business Model Generation. Business model planning: according to the nine important factors of commodity planning strategy, enterprise product development professional team, brainstorming divergence and convergence with 3M yellow sticky notes, to produce business model and main content in line with the conditions of the enterprise.

5   Discussion

This study through the analysis of the present situation of the early stage of the product design research, summed up the I-sustainability design pattern is a new and system of the early stage of the product design analysis theory. There are defects and inadequacies in the traditional method and theory of preliminary analysis of product design. Product designers should carefully analyze the social, economic, technological and sustainable development factors in a given period, abandon the previous single pre-design analysis theory, and propose new design principles and evaluation criteria to determine the most appropriate product opportunity gap according to the analysis results. The value of I-sustainability design thinking should be considered deeply by the design industry (Shapira et al. 2017).

DT's approach to problem framing appears helpful to set the appropriate innovation scope for SDGs. Its strong focus on users and stakeholders fosters the development of SDGs that meet actual user needs. With its focus on iterative experimentation, DT makes it possible to assure positive sustainability effects while reducing the risk of innovation failure. Furthermore, ISDT provides a differentiated basis underpinning the assumptions of several researchers (Buhl et al. 2019; Fischer 2015; Shapira et al. 2017) that a DT-based approach is an essential, yet often overlooked asset when addressing sustainability challenges.

Integrating design and sustainability science hold much value for transforming our social–ecological system to achieve SDGs. However, there are some significant challenges in doing so. Many aspects of a research through design project cannot be pre-determined (Moloney 2015) which provides challenges for traditional research grants. Design approaches to achieving SDGs can be advanced by expanding opportunities for publishing creative explorations and visioning (Wiek and Iwaniec 2014).

6 Conclusion

In the process of product design, designers can through the analysis of the factors of SETBS, grasp the value of a product opportunity gap, on this basis, fully considering the I-sustainability value of products, at the same time, combined the technology of product modeling and the role in the modern products, make maximize embody the I-sustainability value of the product design. In order to maximize the I-sustainability value of products, designers can analyze the attributes of eight value opportunities that represent the ISDT core values. By playing the role of the above attributes in product design, product design can achieve breakthrough progress, thus creating sustainable products.

Combining the methods of science and design can help to develop more innovative, better integrated and truly transformational initiatives for sustainability. We suggest DT as a suitable approach to ensure that sustainability aspects are already taken into account in the initial phase of organizational innovation processes, i.e. when generating and evaluating SDGs ideas.

References

Archer, B.: Design as a discipline. Des. Stud. 1(1), 17–20 (1979)
Baer, R.A.: Mindfulness training as a clinical intervention: a conceptual and empirical review. Clin. Psychol. Sci. Pract. 10(2), 125–143 (2003)
Baxter, M.: Product Design. CRC Press, Boco Raton (1995)

Bhise, V.D.: Ergonomics in the Automotive Design Process. CRC Press, Boco Raton (2011)

Bishop, S.R., et al.: Mindfulness: a proposed operational definition. Clin. Psychol. Sci. Pract. **11**(3), 230–241 (2004)

Bridger, R.: Introduction to Ergonomics. CRC Press, Boco Raton (2008)

Brown, K.W., Ryan, R.M.: The benefits of being present: mindfulness and its role in psychological well-being. J. Pers. Soc. Psychol. **84**(4), 822 (2003)

Brown, T.: Design thinking. Harv. Bus. Rev. **86**(6), 84 (2008)

Buhl, A., et al.: Design thinking for sustainability: why and how design thinking can foster sustainability-oriented innovation development. J. Clean. Prod. **231**, 1248–1257 (2019)

Cagan, J., Cagan, J.M., Vogel, C.M.: Creating Breakthrough Products: Innovation from Product Planning to Program Approval. Ft Press, Upper Saddle River (2002)

Campbell, R.L.: Will the real scenario please stand up? ACM SIGCHI Bull. **24**(2), 6–8 (1992)

Cheng, C.-Y.: On harmony as transformation: Paradigms from the I Ching. In: Harmony and Strife: Contemporary Perspectives, East & West??, pp. 225–248 (1998)

Clausen, T.H., Pohjola, M.: Persistence of product innovation: comparing breakthrough and incremental product innovation. Technol. Anal. Strateg. Manag. **25**(4), 369–385 (2013). https://doi.org/10.1080/09537325.2013.774344

Clune, S.J., Lockrey, S.: Developing environmental sustainability strategies, the Double Diamond method of LCA and design thinking: A case study from aged care. J. Clean. Prod. **85**, 67–82 (2014)

Cooper, R.G.: Where are all the breakthrough new products? Using portfolio management to boost innovation. Res.-Technol. Manag. **56**(5), 25–33 (2013)

Csikszentmihalyi, M., Nakamura, J.: Effortless attention in everyday life: a systematic phenomenology. In: Bruya, B. (ed.) Effortless Attention: A New Perspective in the Cognitive Science of Attention and Action, pp. 179–189. MIT Press (2010)

Davis, C., Tomoda, Y.: Competing incremental and breakthrough innovation in a model of product evolution. J. Econ. **123**(3), 225–247 (2018). https://doi.org/10.1007/s00712-017-0568-y

Dorst, K., Cross, N.: Creativity in the design process: co-evolution of problem–solution. Des. Stud. **22**(5), 425–437 (2001)

Du, C.R.: Sustainable development goals (2016)

Elkington, S.: Articulating a systematic phenomenology of flow: an experience-process perspective. Leisure/Loisir **34**(3), 327–360 (2010)

Elliott, J. (2012). *An introduction to sustainable development*: Routledge.

Fischer, M.: Design it! solving sustainability problems by applying design thinking. GAIA Ecol. Perspect. Sci. Soc. **24**(3), 174–178 (2015)

Forsey, J.: The Aesthetics of Design. Oxford University Press, Oxford (2016)

Fritz, B.S., Avsec, A.: The experience of flow and subjective well-being of music students. Horiz. Psychol. **16**(2), 5–17 (2007)

Haines-Gadd, M., Chapman, J., Lloyd, P., Mason, J., Aliakseyeu, D.: Emotional durability design nine—a tool for product longevity. Sustainability **10**(6), 1948 (2018)

Helander, M.G.: Forget about ergonomics in chair design? Focus on aesthetics and comfort! Ergonomics **46**(13–14), 1306–1319 (2003)

Kabat-Zinn, J.: Catalyzing movement towards a more contemplative/sacred-appreciating/non-dualistic society. Paper presented at the Meeting of the Working Group (1994)

Kishita, Y., Mizuno, Y., Fukushige, S., Umeda, Y.: Scenario structuring methodology for computer-aided scenario design: an application to envisioning sustainable futures. Technol. Forecast. Soc. Chang. **160**, 120207 (2020)

Kuo, W., Prasad, V.R., Tillman, F.A., Hwang, C.-L.: Optimal Reliability Design: Fundamentals and Applications. Cambridge University Press, Cambridge (2001)

Lindberg, T., Meinel, C., Wagner, R.: Design thinking: a fruitful concept for it development? In: Meinel, C., Leifer, L., Plattner, H. (eds.) Design Thinking, pp. 3–18. Springer, Heidelberg (2011). https://doi.org/10.1007/978-3-642-13757-0_1

Liu, C., Chen, H., Liu, C.-Y., Lin, R.-T., Chiou, W.-K.: Cooperative and individual mandala drawing have different effects on mindfulness, spirituality, and subjective well-being. Front. Psycho. 11, 2629 (2020)

Liu, C., Chen, H., Liu, C.-Y., Lin, R.-T., Chiou, W.-K.: The effect of loving-kindness meditation on flight attendants' spirituality, mindfulness and subjective well-being. Healthcare 8, 174 (2020)

Lu, S.: I Ching and the origin of the Chinese semiotic tradition. Semiotica 2008(170), 169–185 (2008)

Lui, I.: The Taoist I Ching. Shambhala Publications, Berkeley (2005)

Ma, S.S.: The I Ching and the psyche-body connection. J. Anal. Psychol. 50(2), 237–250 (2005)

Moloney, J.: Planning a thesis: ways, means and tactics for research through design. In: Christopher, M., Jan, S., Simon, T. (eds.) Perspectives on Architectural Design Research: What Matters, Who Cares, How, pp. 135–137. Spurbuchverlag, Baunach (2015)

Naidoo, R., Fisher, B.: Reset Sustainable Development Goals for a Pandemic World. Nature Publishing Group, Berlin (2020)

Nardi, B.A.: The use of scenarios in design. ACM SIGCHI Bull. 24(4), 13–14 (1992)

Norman, D.A.: The Design of Everyday Things. Doubleday-Currency, New York (1989)

World Health Organization: World health statistics 2016: monitoring health for the SDGs sustainable development goals. World Health Organization (2016)

Palmer, J.W.: Web site usability, design, and performance metrics. Inf. Syst. Res. 13(2), 151–167 (2002)

Park, K.-Y., Park, H.-K., Hwang, H.-S., Yoo, S.-H., Ryu, J.-S., Kim, J.-H.: Improved detection of patient centeredness in objective structured clinical examinations through authentic scenario design. Patient Education and Counseling (2020)

Perra, D.B., Sidhu, J.S., Volberda, H.W.: How do established firms produce breakthrough innovations? Managerial identity-dissemination discourse and the creation of novel product-market solutions. J. Prod. Innov. Manag. 34(4), 509–525 (2017). https://doi.org/10.1111/jpim.12390

Ralph, P.: The Sensemaking-coevolution-implementation theory of software design. Sci. Comput. Program. 101, 21–41 (2015)

Sachs, J.D., Schmidt-Traub, G., Mazzucato, M., Messner, D., Nakicenovic, N., Rockström, J.: Six transformations to achieve the sustainable development goals. Nat. Sustain. 2(9), 805–814 (2019)

Sagot, J.-C., Gouin, V., Gomes, S.: Ergonomics in product design: safety factor. Saf. Sci. 41(2–3), 137–154 (2003)

Seligman, M.E.: Positive psychology, positive prevention, and positive therapy. In: Handbook of Positive Psychology, vol. 2, no. 2002, pp. 3–12 (2002)

Shapira, H., Ketchie, A., Nehe, M.: The integration of design thinking and strategic sustainable development. J. Clean. Prod. 140, 277–287 (2017)

Silvestre, B.S., Țîrcă, D.M.: Innovations for sustainable development: moving toward a sustainable future. J. Clean. Prod. 208, 325–332 (2019)

TenHouten, W.D., Wang, W.: The eight trigrams of the Chinese I Ching and the eight primary emotions. Asian J. Soc. Psychol. 4(3), 185–199 (2001)

Uwasu, M., Kishita, Y., Hara, K., Nomaguchi, Y.: Citizen-participatory scenario design methodology with future design approach: a case study of visioning of a low-carbon society in Suita city, Japan. Sustainability 12(11), 4746 (2020)

Vechakul, J., Shrimali, B.P., Sandhu, J.S.: Human-centered design as an approach for place-based innovation in public health: A case study from Oakland, California. Matern. Child Health J. 19(12), 2552–2559 (2015)

Verganti, R.: Designing breakthrough products. Harv. Bus. Rev. **89**(10), 114–116 (2011)

Wiek, A., Iwaniec, D.: Quality criteria for visions and visioning in sustainability science. Sustain. Sci. **9**(4), 497–512 (2014). https://doi.org/10.1007/s11625-013-0208-6

Wilhelm, R., Baynes, C.F., Jung, C.G.: The I Ching: Or, Book of Changes. Princeton University Press, Princeton (1967)

Yuduo, L., Yi, Q., Donghua, W., Yao, L.: Implications of I Ching on innovation management. Chin. Manag. Stud. **5**, 394–402 (2011)

Sound Signal Sensitivity of Subjective Auditory Features

Jin Liang[1], Xin Wang[1(✉)], Tuoyang Zhou[1], Zhen Liao[1,2], Lei Liu[3], Yang Yu[1],
Liang Zhang[1], Chi Zhang[1], Zhanshuo Zhang[1], and Xiaoyi Li[1]

[1] China Institute of Marine Technology and Economy, Beijing 100081, China
[2] Tsinghua University, Beijing 100084, 266400, China
[3] Beijing Institute of New Technology Applications, Beijing 100094, China

Abstract. In order to evaluate the sound signal sensitivity of subjective auditory features, understand the process of human sound signal recognition, guide machine recognition of sound signal and noise reduction, this study constructed eight sound signals by changing waveform composition, frequency and amplitude of the sound signals. Twelve subjects were asked to evaluate the subjective auditory features of eight sound signals, such as loudness, sharpness, associativity, sweetness and pleasure, as well as the differences of each sound signal with others. It showed that the sound signals were different in sharpness, associativity, sweetness and pleasure, but only associativity could predict the difference of the sound signals, other subjective auditory features could not effectively distinguish of eight sound signals. However, a single subjective characteristic of the sound signals might not sufficiently discriminate them, two or more subjective perceptual features of sound signals might effectively classify them, with subjective sharpness and associativity being the most sensitive features to sound signals discrimination.

Keywords: Sound signal · Sensitivity · Subjective auditory features

1 Introduction

The sound signal is a kind of mechanical wave, which produces a vibration from objects or medium in a certain area. It could be measured by basic acoustic parameters such as sound pressure, intensity, frequency, period, wavelength, speed, particle velocity. However, people often evaluate sound signals through subjective perception features [1], not above acoustic parameters, when they identify sound signals subjectively, such as loudness, sharpness [2]. Different subjective auditory features have different sensitivity to sound signals [3] and have different effects on sound signal recognition [4], which will lead to different accuracy of sound signal recognition [5]. It could be used to guide the optimization of sound signal recognition algorithms and models and improve the accuracy of sound signal analysis. At the same time, noise reduction based on human subjective perception sensitivity of sound signal could also get a better construction effect [6].

© Springer Nature Switzerland AG 2021
P.-L. P. Rau (Ed.): HCII 2021, LNCS 12771, pp. 115–126, 2021.
https://doi.org/10.1007/978-3-030-77074-7_9

2   Method

2.1   Sound Signals

In order to evaluate the sensitivity of different subjective auditory features to sound signals, understand the process of human sound signal recognition, guide sound signal machine recognition and noise reduction, this study constructed eight sound signals by changing waveform composition, frequency and amplitude[7] of the sound signals as shown in Table 1, asked subjects to evaluate the subjective auditory features of the sound signals. There are two groups (3, 7; 5, 6, 8) of sound signals with high similarity, that is, the sound signals of each group are very similar, but there is a big difference between the two groups of sound signals, and other sound signals are also different from these two groups of sound signals. The constructing sound signals with different differences could be used for the subsequent analysis of sound signal sensitivity of subjective auditory features.

Table 1. Sound signal

Signal	Composition	Dominant frequency	Amplitude
1	White Noise	–	45
2	Signal+Disturbance	200	80
3	Signal+White Noise	900	70
4	White Noise	–	50
5	Signal+Disturbance+White Noise	200	80
6	Signal+Disturbance+White Noise	200	81
7	Signal+White Noise	900	65
8	Signal+Disturbance+White Noise	200	79

2.2   Evaluating Method

This study interviewed a number of experts with strong sound signal recognition ability. Based on the interview results, the process of sound signal recognition was analyzed and extracted the subjective auditory features of sound signals related to sound signal recognition: loudness [8], sharpness [8], associativity, sweetness [9] and pleasure [9], as well as the differences of each sound signal with others [10]. Then there were twelve subjects were asked to evaluate the subjective auditory features of eight sound signals by these subjective auditory features of the sound signals. After the subjective auditory features of eight sound signals were obtained, the following analysis were conducted:

(1) The frequency spectrum of eight sound signals was used for the analysis of the composition, frequency and amplitude features of the sound signals, so as to verify the similarity and difference of the physical features of the eight sound signals [11].

(2) The difference analysis of the subjective auditory features (loudness, sharpness, associativity, sweetness and pleasure) of eight sound signals was carried out to evaluate the degree of the eight sound signals' difference in the subjective features.

(3) The correlation analysis between every subjective auditory feature and the differences of sound signals was analyzed respectively to evaluate the correlation between the subjective auditory features of sound signals and sound differences.

(4) With loudness, sharpness, associativity, sweetness and pleasure as independent variables, and differences of sound signals as dependent variables, regression analysis was carried out to analyze the prediction effect of subjective auditory features of sound signals on sound difference.

(5) Cluster analysis on sounds' difference was carried out with loudness, sharpness, associativity, sweetness and pleasure as clustering features respectively as well as two or more subjective auditory features of sound signals were used as clustering features, which is used to evaluate the discriminative power of the subjective auditory features of sound signals on the differences of sound signals.

3 Result

3.1 Differences of the Eight Sound Signals in the Spectrum

The spectrum of eight sound signals was analyzed from the perspective of composition, frequency and amplitude features, and verifies the similarity and difference of the physical features of the eight sound signals (the spectrum of eight sound signals is shown in Fig. 1).

3.2 Differences of the Eight Sound Signals in the Subjective Auditory Features

There was significant difference on the sharpness ($F(7, 3) = 31.01, p < 0.01, \eta^2 = 0.99$), associativity ($F(7, 63) = 3.20, p < 0.05, \eta^2 = 0.26$), sweetness ($F(7, 63) = 3.04, p < 0.01, \eta^2 = 0.25$) and pleasure ($F(7, 63) = 4.36, p < 0.001, \eta^2 = 0.33$) of eight sound signals. The further analysis has shown that the subjective auditory features of sound signals with small physical differences were not significantly different, while the subjective auditory features of sound signals with large physical differences are significantly different, but the subjective auditory features of sound signals with large physical features were not significantly different in all subjective features. These results indicated the sound signals had shown different features in sharpness, associativity, sweetness and pleasure, but a single subjective characteristic of sound signals might not sufficiently discriminate them.

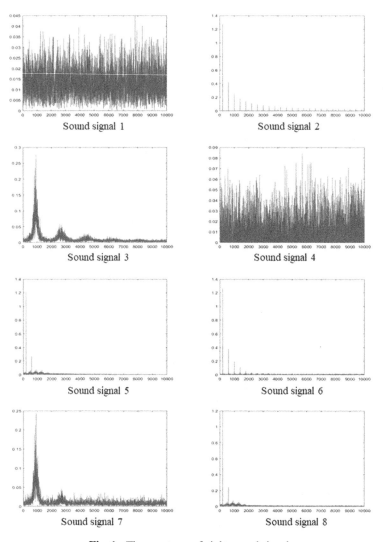

Fig. 1. The spectrum of eight sound signals

3.3 Correlation Between the Differences of the Eight Sound Signals and Their Subjective Auditory Features

The relation between the differences of the eight sound signals and the sound subjective auditory features should be revealed through correlation analysis. Therefore the correlation analysis was conducted on between the differences and the subjective auditory features of eight sound signals, such as loudness, sharpness, associativity, sweetness, and pleasure to guide the selection of subjective auditory features which were sensitive to the difference of the sound signals. It has shown that only associativity was highly correlated with the difference of the sound signals ($r = 0.748$, $p < 0.05$), the loudness, sharpness, sweetness and pleasure of sound signals were not significantly correlated with the difference. However, loudness was highly correlated with sharpness ($r = 0.870, p < 0.01$), sweetness was significantly correlated with pleasure ($r = 0.934, p < 0.01$). This high correlation between subjective auditory features of sound signals implied any one of the two highly correlated indicators could be selected as the sensitive feature in sound signal recognition.

3.4 The Prediction Effect of Subjective Auditory Features of the Sound Signals on Their Differences

The Prediction Effect on the Difference of the Sound Signals Based on a Single Subjective Auditory Feature

Linear regression analysis was carried out and a linear regression model was constructed by taking the loudness, sharpness, associativity, sweetness and pleasure as independent variables respectively, and the difference of the sound signals as the dependent variable.

The prediction model of the loudness for sound signal difference was $y = 0.426x + 2.902$, $R^2 = 0.23$, $P = 0.23$, which prediction effect is poor.

The prediction model of the sharpness for sound signal difference was $y = 0.227x + 4.353$, $R^2 = 0.233$, $P = 0.226$, which indicated that the prediction of sharpness feature to signal difference is poor.

The predictive model of the associativity for sound signal difference was $y = 0.496x + 3.47$, $R^2 = 0.486$, $P < 0.05$, which might predict the signal difference.

The prediction model of the sweetness for sound signal difference was $y = 0.041x + 5.309$, $R^2 = 0.23$, $P = 0.892$, and the prediction of the model is poor.

The prediction model of the pleasure for sound signal difference was $y = -.008x + 5.43$, $R^2 = -.0.167$, $P = 0.979$, and the prediction of the model was poor.

The Prediction Effect on the Difference of the Sound Signals Based on Multiple Subjective Auditory Features

The stepwise regression was carried out by taking loudness, sharpness, associativity, sweetness and pleasure of eight sound signals as the independent variables, and the difference of the sound signals as the dependent variable. The results have shown that only the associativity entered into the prediction model, which was $y = 0.496x + 3.47$ ($R^2 = 0.486$, P < 0.05), while the prediction of loudness, sharpness, sweetness and pleasure was poor. These results implied only associativity could predict sound signals, while loudness, sharpness, sweetness and pleasure were poor predictors for the differences of the sound signals. But the associativity also could not well predict the difference of the sound signals well which had a low R^2. These results indicated that the five perceptual features of loudness, sharpness, associativity, sweetness and pleasure could not effectively distinguish of eight sound signals.

3.5 Discrimination of the Subjective Auditory Features on the Difference of the Sound Signals

Sound Signal Discrimination Based on a Single Subjective Auditory Feature

The subjective auditory features of the eight sound signals, such as loudness, sharpness, associativity, sweetness and pleasure, were used as input variables for cluster analysis, result has shown that:

Although loudness could cluster the similar sound signals 3,7 or 5,6,8 into one category, they could also cluster the different sound signals 2,4 and sound signals 3,7 into one category, and sound signal 1 and sound signals 5,6,8 into one category, which meant that the loudness could reflect the similarity and difference of the sound signals to a certain extent. However, it was not accurate enough, which would lead the loudness could not accurately distinguish all eight sound signals. Therefore, loudness was not the sensitivity feature of sound signals.

The sharpness could cluster sound signal 3 and sound signal 7 into one category, sound signal 5, 6 and 8 into one category, which were consistent with the similarity of the sound signals. However, the sharpness feature could also cluster sound signal 4 and sound signal 3 and 7 into one category, and sound signal 1 and 2 and sound signal 5, 6 and 8 into one category, which meant that although sharpness could reflect the similarity and difference of the sound signals to a certain extent, they were not accurate enough, which might lead subjects could not accurately identify all eight sound signals based on the sharpness only. Therefore, sharpness was not the sensitivity feature of sound signals.

The results of cluster analysis based on associativity were not consistent with the similarity of sound signals, which meant that the associativity could not sensitively reflect the difference of the sound signals, which might lead the subjects could not effectively distinguish all eight sound signals based on the associativity only. So it's not a sensitive characteristic of the sound signals.

The cluster analysis based on sweetness has shown sound signal 3 and sound signal 7 could be clustered into one category, sound signal 5, 6 and 8 could be clustered into one category, which were consistent with the similarity of the sound signals. However, the sweetness could also cluster sound signal 4 and sound signal 3 and 7 into one category, and sound signal 1 and sound signal 5, 6 and 8 into one category, which meant that although the sweetness could reflect the similarity and difference of the sound signals to a certain extent, it was not accurate enough, which might lead the subjects could not accurately distinguish all eight sound signals based on the sweet only. Therefore, the sweetness was not the sensitivity feature of sound signals.

The results of cluster analysis based on pleasure degree were inconsistent with the differences of sound signals, which meant that the pleasure could not sensitively reflect the differences of the sound signals, which might lead the subjects could not effectively distinguish all eight sound signals based on the pleasure only. Therefore, the pleasure was not the sensitivity feature of the sound signals.

Sound Signal Discrimination Based on Two Subjective Auditory Features

The results of cluster analysis based on loudness and sharpness of the sound signals have shown that sound signal 3 and sound signal 7 could be clustered into one category, and sound signal 5, 6 and 8 could be clustered into one category, which were consistent with the similarity of sound signals. However, loudness and sharpness could also cluster sound signal 1 and sound signal 5, 6 and 8 into one category, which meant that although the loudness and sharpness could reflect the similarity and difference of the sound signals to a certain extent, it was not accurate enough, which might lead the subjects could not accurately distinguish all eight sound signals based on the loudness and sharpness.

The results of cluster analysis based on loudness and associativity have shown that sound signal 3 and sound signal 7 could be clustered into one category, and sound signal 5, 6 and 8 could be clustered into one category, which were consistent with the similarity of sound signals. However, loudness and associativity could also cluster sound signal 1 and sound signal 5, 6 and 8 into one category, which meant although loudness and associativity could reflect the the similarity and difference of the sound signals to a certain extent, they were not accurate enough, which might lead the subjects could not accurately distinguish all eight sound signals based on loudness and associativity.

The results of cluster analysis based on loudness and sweetness have shown sound signal 3 and sound signal 7 could be clustered together, sound signal 5, 6 and 8 could be clustered together, which were consistent with the similarity of the sound signals. However, loudness and sweetness could also classify sound signal 4 and sound signal 3 and 7 into one category, and sound signal 1 and sound signal 5, 6 and 8 into one category, which meant that although the loudness and sweetness could reflect the similarity and difference of the sound signals to a certain extent, they were not accurate enough, which might lead the subjects could not accurately distinguish all eight sound signals based on the loudness and sweetness.

The results of cluster analysis based on loudness and pleasure were inconsistent with the similarity of sound signals, which meant that the loudness and pleasure of sound signals could not reflect the difference and similarity of sound signals, which might lead the subjects could not accurately identify all eight sound signals based on loudness and pleasure.

The results of cluster analysis based on the sharpness and associativity showed that sound signal 3 and sound signal 7 could be clustered together, and sound signal 5, 6 and 8 could be clustered together, which were consistent with the similarity of sound signals. These results meant that the sharpness and associativity of sound signals could reflect the similarity and difference of the sound signals. So we could construct a joint index based on the sharpness and associativity for sound signal recognition, and take this index as the sensitivity features of the sound signals.

The results of cluster analysis based on the sharpness and sweetness showed that the sound signal 3 and sound signal 7 could be clustered together, and the sound signal 5, 6 and 8 could be clustered together, which were consistent with the similarity of the sound signals. However, the sharpness and sweetness could also cluster the sound signal 1 and sound signal 5, 6 and 8, which meant that although the sharpness and sweetness could reflect the similarity and difference of the sound signals to a certain extent, they were not accurate enough, which might lead the subjects could not accurately distinguish all eight sound signals based on the sharpness and sweetness.

The results of cluster analysis based on the sharpness and pleasure showed that sound signal 3 and sound signal 7 could be clustered into one category, and sound signal 5, 6 and 8 could be clustered into one category, which were consistent with the similarity of the sound signals. However, the features of sharpness and pleasure could also cluster sound signal 1 and sound signal 8 into one category, and signal 2 and signal 5 and 6 into one category, which meant that although the sharpness and pleasure could reflect the similarity and difference of the sound signals to a certain extent, they were not accurate enough, which might lead the subjects could not accurately distinguish all eight sound signals based on sharpness and pleasure.

The results of cluster analysis based on the associativity and sweetness showed that they could cluster sound signal 3 and sound signal 7 into one category, and sound signal 5, 6 and 8 into one category, which were consistent with the similarity of the sound signals. However, sound signal 1 and sound signal 3 and 7 could also be clustered into one category based on the associativity and sweetness, which meant although associativity and palatability could reflect the similarity and difference of the sound signals to a certain extent, they were not accurate enough, which might lead subjects could not accurately identify all eight sound signals based on the associativity and sweetness.

The results of cluster analysis based on the associativity and pleasure degree were inconsistent with the similarity of sound signals, which meant that the perceptual features of associativity and pleasure of sound signals could not sensitively reflect the difference and similarity of the sound signals, which might lead the subjects might not accurately identify all eight sounds based on associativity and pleasure.

The clustering analysis results based on the sweetness and pleasure showed that sound signal 3 and sound signal 7 could be clustered together, and sound signal 5, 6 and 8 could be clustered together, which were consistent with the similarity of the sound signals. However, the sweetness and pleasure could also cluster sound signal 1, 4 and sound signal 3 and 7, which meant that although the of sweetness and pleasure could reflect the similarity and difference of the sound signals to a certain extent, they were not accurate enough, which might lead the subjects could not accurately distinguish all eight sound signals based on the sweetness and pleasure.

Sound Signal Discrimination Based on Three Subjective Auditory Features

The results of cluster analysis based on loudness, sharpness and associativity showed sound signal 3 and sound signal 7 could be clustered together, and sound signal 5, 6 and 8 could be clustered together, which were consistent with the similarity of sound signals. However, loudness, sharpness and associativity also could cluster sound signal 1 and sound signal 5, 6 and 8, which meant that although the loudness, sharpness and associativity of the signal could reflect the similarity and difference of the sound signals to a certain extent, they were not accurate enough, which might lead the subjects could not accurately distinguish all eight sound signals based on the loudness, sharpness and associativity.

The results of cluster analysis based on loudness, sharpness and sweetness showed sound signal 3 and sound signal 7 could be clustered together, and sound signal 5, 6 and 8 could be clustered together, which were consistent with the similarity of sound signals. However, loudness, sharpness and sweetness could also cluster sound signal 1 and sound signal 5, 6 and 8, which meant that although loudness, sharpness and sweetness could reflect the similarity and difference of the sound signals to a certain extent, they were not accurate enough, which might lead the subjects could not accurately distinguish all eight sound signals based on loudness, sharpness and sweetness.

The results of cluster analysis based on loudness, sharpness and pleasure were inconsistent with the similarity of sound signals, which meant that the loudness, sharpness and pleasure of sound signals could not reflect the similarity and difference of the sound signals, so the subjects could not distinguish sound signals accurately based on loudness, sharpness and pleasure.

The results of cluster analysis based on loudness, associativity and sweetness showed these features could cluster sound signal 3 and sound signal 7, sound signal 5, 6 and 8, and also could distinguish sound signal 1, 2 and 4. The similarity of classification results were consistent with the similarity of sound signals, which meant that the subjects could better perceive the loudness, associativity and sweetness of sound signals. Therefore, we could build a joint index based on the loudness, associativity and sweetness for the subjects to identify the sound signals, and took the index as the sensitivity features of the sound signals.

The results of cluster analysis based on loudness, associativity and pleasure showed these features could cluster sound signal 3 and sound signal 7, sound signal 5, 6 and 8, and distinguish sound signal 1, 2 and 4. The classification results were consistent with the similarity of sound signals, which meant that the subjects could better perceive the loudness, associativity and pleasure of sound signals. Therefore, we could build a joint index based on the subjective loudness, associativity and pleasure for the subjects to identify the sound signals, and took the index as the sensitivity features of the sound signals.

The results of cluster analysis based on loudness and sharpness has shown sound signal 3 and sound signal 7 could be clustered together, and sound signal 5, 6 and 8 could be clustered together, which were consistent with the similarity of sound signals. However, loudness, sweetness and pleasure could also be clustered together with sound signal 4 and sound signal 3 and 7, and 1 and sound signal 5, 6 and 8, which meant that although loudness, sweetness and pleasure of sound signals could reflect the similarity and difference of the sound signals to a certain extent, they were not accurate enough, which might cause the subjects could not accurately identify all eight sound signals based on loudness, sweetness and pleasure.

The results of cluster analysis has shown the sharpness, associativity and sweetness could cluster the sound signal 3 and sound signal 7 into one group, and sound signal 5, 6 and 8 into one group. The classification results could distinguish sound signal 1, 2 and 4. The classification results were consistent with the similarity of sound signals, which meant that the sharpness, associativity and sweetness of sound signals could classify the sound signals. Therefore, a joint index could be constructed based on sharpness, associativity and sweetness for sound signal recognition, and the index could be used as the sensitivity features of sound signals.

The results of cluster analysis based on the sharpness, associativity and pleasure showed these features could cluster sound signal 3 and sound signal 7, sound signal 5, 6 and 8, and could distinguish sound signal 1, 2 and 4. The classification results were consistent with the similarity of sound signals, which meant that the sharpness, associativity and pleasure could classify the sound signals. Therefore, a joint index could be constructed based on sharpness, associativity and pleasure for sound signal recognition, and the index could be used as the sensitivity features of sound signals.

The results of cluster analysis based on the sharpness, sweetness and pleasure showed these features could cluster sound signal 3 and sound signal 7 into one category, and sound signal 5, 6 and 8 into one category, which were consistent with the similarity of sound signals. However, the sharpness, sweetness and pleasure could also cluster sound signal 1 and sound signal 5, 6 and 8 into one category, which meant although sharpness, sweetness and pleasure could reflect the similarity and difference of the sound signals to a certain extent, they were not accurate enough, which might lead the subjects could not accurately distinguish all eight sound signals based on sharpness, sweetness and pleasure.

The cluster analysis results based on the associativity, sweetness and pleasure showed that sound signal 3 and sound signal 7 could be clustered together, and sound signal 5, 6 and 8 could be clustered together, which were consistent with the similarity of sound signals. However, the features of associativity, pleasant degree and pleasant degree could also cluster sound signal 1 and sound signal 3 and 7, which meant although associativity, sweetness and pleasure of sound signals could reflect the similarity and difference of the sound signals to a certain extent, they were not accurate enough, which might lead the subjects could not accurately distinguish all eight sound signals only based on associativity, sweetness and pleasure.

Sound Signal Discrimination Based on Four Subjective Auditory Features

The results of cluster analysis of loudness, sharpness, associativity and sweetness has shown sound signal 3 and sound signal 7 could be clustered together, and sound signal 5, 6 and 8 could be clustered together, which were consistent with the similarity of sound signals. However, loudness, sharpness, associativity and sweetness could also cluster sound signal 1 and sound signal 5, 6 and 8, which means that although the loudness, sharpness, associativity and sweetness of sound signals could reflect the similarity and difference of the sound signals to a certain extent, they were not accurate enough, which might lead the subjects were unable to accurately identify all eight sound signals based on the of loudness, sharpness, associativity and sweetness.

The results of cluster analysis based on loudness, sharpness, associativity and pleasure has shown they could cluster sound signal 3 and sound signal 7, sound signal 5, 6 and 8, and could distinguish sound signal 1, 2 and 4. The classification results were consistent with the similarity of the sound signals, which meant that the loudness, sharpness, associativity and pleasure could better reflect the similarity and difference of the sound signals. Therefore, a joint index could be constructed based on the of loudness, sharpness, associativity and pleasure of sound signals, and the index could be used as the sound signal recognition sensitivity features.

The results of cluster analysis based on loudness, sharpness, sweetness and pleasure has shown sound signal 3 and sound signal 7 could be clustered together, and sound signal 5, 6 and 8 could be clustered together, which were consistent with the similarity of sound signals. However, loudness, sharpness, sweetness and pleasure would also cluster sound signal 1 and sound signal 5, 6 and 8, which meant that although the loudness, sharpness, sweetness and pleasure of sound signals could reflect the similarity and difference of the sound signals to a certain extent, they were not accurate enough, which might lead the subjects could not accurately distinguish all eight sound signals based on the loudness, sharpness, sweetness and pleasure.

The results of cluster analysis based on loudness, associativity, sweetness and pleasure has shown they could cluster sound signal 3 and sound signal 7 into one category, sound signal 5, 6 and 8 into one category, and could distinguish sound signal 1, 2 and

4. The classification results were consistent with the similarity of sound signals, which meant that the subjects had good understanding of loudness, associativity, sweetness and pleasure of sound signals. Loudness, associativity, sweetness and pleasure could better reflect the similarity and difference of the sound signals. Therefore, a joint index could be constructed based on the loudness, associativity, sweetness and pleasure of sound signals, and the index could be used as the sound signal recognition sensitivity features.

The results of cluster analysis based on the sharpness, associativity, sweetness and pleasure has shown these features could cluster sound signal 3 and sound signal 7, sound signal 5, 6 and 8, and could distinguish sound signal 1, 2 and 4. The classification results were consistent with the similarity of the similarity of sound signals, which meant that a joint index could be constructed based on the sharpness, associativity, sweetness and pleasure of sound signals, and the index could be used as the sensitivity features of sound signals.

Sound Signal Discrimination Based on Five Subjective Auditory Features

The results of cluster analysis based on loudness, sharpness, easiness of associativity, sweetness and pleasantness has shown these features could cluster sound signal 3 and sound signal 7 into one category, sound signal 5, 6 and 8 into one category, and could distinguish sound signal 1, 2 and 4. The classification results were consistent with the similarity of sound signals, which meant a joint index used to identify sound signals could be constructed based on the loudness, sharpness, associativity, sweetness and pleasure of sound signals. The index could be used as the sensitivity features of sound signals.

4   Conclusion

(1) There was a significant difference on the sharpness, associativity, sweetness and pleasure of eight sound signals, but only associativity features could predict sound signals, while loudness, sharpness, sweetness and pleasure were poor predictors of differences of sound signals. The comprehensive analysis results have shown that the five perceptual features of loudness, sharpness, associativity, sweetness and pleasure could not effectively predict the difference of eight sound signals.

(2) If two or more subjective perceptual features of sound signals were used, these sound signals might be effectively classified. The subjects' subjective sharpness and easy associativity were the most sensitive features to sound signals.

References

1. Burns, T., Rajan, R.: A mathematical approach to correlating objective spectro-temporal features of non-linguistic sounds with their subjective perceptions in humans. Front. Neurosci. **13**, 794 (2019)
2. Winkler, I., Tervaniemi, M., Huotilainen, M., et al.: From objective to subjective: pitch representation in the human auditory cortex. Neuroreport **6**(17), 2317 (1995)
3. Perrott, D.: Judgments of sound volume: Effects of signal duration, level, and interaural characteristics on the perceived extensity of broadband noise. J. Acoust. Soc. Am. **72**(5), 1413–7 (1982)

4. Ando Y.: Subjective preference of the sound field. In: Signal Processing in Auditory Neuroscience, pp. 67–79 (2019)

5. Wickelmaier, F., Ellermeier, W.: Deriving auditory features from triadic comparisons. Percept. Psychophys. **69**(2), 287–297 (2007)

6. Borg, E.: Noise level, inner hair cell damage, audiometric features, and equal-energy hypothesis. J. Acoust. Soc. Am. **86**(5), 1776 (1989)

7. Leek, M.R., Brown, M.E., Dorman, M.F.: Informational masking and auditory attention. Percept. Psychophys. **50**(3), 205–214 (1991). https://doi.org/10.3758/BF03206743

8. Berglund, B., Hassmen, P., Preis, A.: Loudness and sharpness as determinants of noise similarity and preference. In: INTER-NOISE and NOISE-CON Congress and Conference Proceedings (1996)

9. Jankovic, D., Stevanov, J.: Similarities in affective processing and aesthetic preference of visual. Audit. Gustatory Stimuli. i-Percept. **2**(8), 951–951 (2011)

10. Francesco, A., Östen, A., Jian, K.: dimensions underlying the perceived similarity of acoustic environments. Front. Psychol. **8**(1162), 1162 (2017)

11. Creasey, P.D.: An exploration of sound timbre using perceptual and time-varying frequency spectrum techniques (1998)

A Preliminary Study on the Effect of Somatosensory Games upon Children's Activity Space and Bodily Movements

Hsuan Lin[1(✉)], Ming-Yu Hsiao[1], Yu-Chen Hsieh[2], Kuo-Liang Huang[3], Chia-Wen Tsai[1], and Wei Lin[4]

[1] Department of Industrial Design, Chaoyang University of Technology, Taichung, Taiwan
t2020021@mail.cyut.edu.tw, my.hsiao@msa.hinet.net
[2] Department of Industrial Design, National Yunlin University of Science and Technology, Yunlin, Taiwan
chester@yuntech.edu.tw
[3] Department of Industrial Design, Sichuan Fine Arts Institute, Chongqing, China
shashi@scfai.edu.cn
[4] School of Architecture, Feng Chia University, Taichung, Taiwan
wlin@fcu.edu.tw

Abstract. With the progress of game technology, the interactions between people and input devices have become more diverse and versatile. Meanwhile, with the introduction of somatosensory games, the mode of interaction is no longer restricted to manual operation. Players can input operational instructions into the game through gestures or bodily movements; in addition, the demand for activity space is different from that in the past. When operating game consoles, players not only pay attention to the distance between their eyes and monitors but also need a proper space to move around. Therefore, besides exploring the correlation between bodily activity space and virtual space, this study probes the influence of somatosensory dance games on children's bodily movements. The findings show that by performing dances with the help of somatosensory games, children can understand the dance movements in the games and improve their spatial presence. Moreover, the ages of children have a close relationship with the execution of basic bodily movements. Lastly, children in the youngest group mainly employ the upper body while dancing. The findings can be used as a reference for game development.

Keywords: Dance · Virtual space · Spatial presence · Natural mapping

1 Introduction

As game technology advances constantly, the modes of interaction between people and input devices have become more and more diverse. In the early days, only gamepads and joysticks were available to operate the game and give instructions. With somatosensory devices introduced, the interactive modes of games are no longer limited to manual

© Springer Nature Switzerland AG 2021
P.-L. P. Rau (Ed.): HCII 2021, LNCS 12771, pp. 127–140, 2021.
https://doi.org/10.1007/978-3-030-77074-7_10

operation; that is, players can input operating instructions into games through gestures or bodily movements. In the meantime, the demand for activity space differs from that in the past, for the space is no longer just where a player sits. After the release of Pokémon Go on mobile phones in 2016, its activity space not only allows players to leave their homes but also compels them to move around. Combined with augmented reality (AR), the game transforms a real-life environment into a kind of activity space. Players cannot take their own safety lightly while enjoying their games. Specifically, the scope of activity for modern game consoles is not the same as that in former times. When a game is equipped with a different device, its activity space also changes. Consequently, the player's operational modes turn out to be more versatile; also, the mapping of gameplay as well as the demand for real space becomes an issue.

Launched in November 2006, the somatosensory game console Nintendo Wii has sold over 1,016,300 units, while Wii game discs have sold more than 9,218,500 copies. Nintendo released the Switch game console in March, 2017, selling not less than 61,440,000 units up to June 30, 2020, while over 4,066,700 copies of Switch game discs have sold [1]. The popularity of Wii is related to its unique motion controller. A player operates the controller through realistic bodily movements; then, the player's detected movements are input into the game environment synchronically [2]. The somatosensory game has grown into the fastest-growing product in the digital game industry. In addition to its own technological advancement, which is the decisive factor in the game result [3], its diversified operational modes and somatosensory games are the main reasons for its being loved [4]. Since Nintendo debuted a fitness-themed game called Ring Fit Adventure for the Switch console in October, 2019, the product has been well-received in Taiwan. Its players need to wear Joy-con controllers, producing bodily movements to interact with the game content. Apart from the adventure fun in fighting monsters and beating levels, the game can exercise different muscle groups and provide aerobic exercise. During the game, calories and heartbeats are calculated to measure the player's health levels, or health points. Various combinations of games enable different players to achieve their specific goals of fitness.

The family computer (FC) launched by Nintendo in the 1980s as the first generation of the home video game console (home console or console), also known as the TV game console or TV game device, provided games for home players through an electronic game device [5]. In the beginning, home game consoles output images solely from televisions or computers. However, with the recent advancement of science and technology, images can be viewed on mobile devices. The visual images are upgraded from 2D planar to 3D stereoscopic ones; furthermore, with virtual reality (VR) and augmented reality (AR) developed, those images are continuously improved. Previously, the input device was simply a gamepad or joystick operated by fingers. Nowadays, by using somatosensory devices to detect gestures and bodily movements, players can interact with virtual objects on screens. Thus, not only are the diverse operating modes of games realized, but the fun deriving from them is also enhanced.

The interactive modes of home game consoles are becoming more and more diversified. Nevertheless, aside from the accidents caused, the space required to interact with game consoles at home or indoors has been ignored. Although the government promulgated the "regulatory rules about the safety of children's amusement facilities in various industries" in 2003 [6], the above-mentioned rules were mainly aimed at the safety connected with public facilities. Actually, the issue of space safety associated with children who interacted with game consoles at home was neglected. Therefore, to prevent fun from decreasing due to poor space during the game, this study focuses on the activity space where children interact with game consoles at home. After Wii first distributed Madden in 2007, EA Canada added different game features to its follow-up programs, called family play and all play. These features were mainly meant to make it easier for casual gamers to play the game [7]. Wii developers reduced the complexity of motion controls to ensure that the game would feel natural and easy-to-use [7].

With somatosensory games getting more and more heterogeneous, the way to manipulate game consoles has changed greatly. Formerly, gameplay relied merely on gamepads or joysticks with direction keys and buttons, but now it requires bodily movements to play games. To control a game console, you must not only pay attention to the distance between your eyes and the display but also have a proper space in which to move around. In view of the above, the purposes of this study are set as follows: 1) to explore the mapping relationship between bodily activity space and virtual space, 2) to identify the space requirements of somatosensory games and real-life activities, and 3) to probe the influence of somatosensory dance games on children's movements.

2　Literature Review

2.1　Sports Games and Parent-Child Interaction

Sports games contain the components essential to some aspects of children's growth. After children have grown up, both their physical activity characteristics and personality characteristics are deeply affected by the living and learning environment in their childhood [8, 9]. Parents' participation in children's games and interaction with children amount to an important factor. Hellstedt [10] pointed out that parents should be a positive role model when participating in children's sports games and that parents should savour the sports they are engaged in as well as set their own goals in order to build a healthy lifestyle. The game should concentrate on the process rather than the result, emphasize the activity or game behavior itself without any goal pursued, and have a positive impact [11, 12].

The values and features of sports games themselves are positive and meaningful to young children, but there are other influential factors, including adults, environment, and equipment which will all have an impact on sports games [9]. Besides, as children and adolescents employ interactive games combined with fitness-training equipment, which has a stimulating effect, they can achieve better training effects [13]. If parents and their children play augmented reality (AR) games side by side, they will have more common topics, deepening the parent-child relationship [14]. Therefore, somatosensory sports games integrated with parent-child interaction are worthy of being promoted and will bring positive benefits. However, when children play virtual somatosensory games without their parents' company for lack of an interactive activity space, they will inevitably be disappointed.

Griffing [15] believes that adults must, through observation, comprehend and designate various conditions for children's games, provide sufficient amounts of time and space to enrich the contents of the games, stimulate learning by means of play materials, and make good use of children's preparatory experiences to help children quickly blend into various games. Parents' participation in children's games can render support to children, focus children's attention, lengthen the durations as well as refine the qualities of such games, and establish a harmonious parent-child relationship because of the above interactions [9]. Parents' involvement in games exerts a positive effect on children's development; hence, parents should not underestimate the importance of their assistance in children's sports games.

2.2 Game Controllers and Operation

Game controllers produce a great influence on user interfaces (UI). For instance, while using a gamepad to control the cursor, the user will feel clumsy, desiring a good interface to operate the keyboard or gamepad more smoothly. For that reason, in addition to choosing the controller, the interface design needs to be considered. Moreover, to help the user get close to the operational method of a real object, similar behaviors are adopted as the basic operation of a game controller. Nonetheless, the user may still have unnatural actions while operating the joystick and buttons on the controller. For example, some shooting games with guns provide natural mapping for the behavior-related motor skills in real life [16]. Likewise, when playing dance machine games, you can record the correspondence between steps and rhythms through the control device of the dance mat, without the need to design other controllers.

Related studies indicated that to facilitate the gameplay, a player must be allowed to practice the game's operational techniques. The player should, in accordance with his/her own intentions, be able to control the characters in the game appropriately through the game controller [17]. Also, the player should be able to move game characters effectively and easily in a complex game world, easily manipulating the situations and devices that help him/her achieve his/her goals [18]. The player can feel the sense of control over the game interface together with the game controller. Proficiency in operating the control system is an important part in the process of playing most games. The game controller as well as the core buttons should be simple enough to learn quickly and to improve the sense of control [19], which means the extensibility of advanced operations [20]. The player should be able to freely set his/her controller to suit his/her learning and playing styles [21, 22]; otherwise, the game should be designed to support different learning and playing styles [18].

2.3 Natural Mapping

In order to control video games, users must pay much attention to creating the mental maps of game environments, record objects and landmarks for future reference, and coordinate visual attention with motor behaviors [23]. Only by doing so can users control video games effectively. Mapping is an important design concept in arranging the control interface and the display; furthermore, natural mapping means, through similarity between spaces, guiding the user to directly understand the relationship between the controller and the device action [24]. Natural mapping can effectively help users predict the manipulation of interactive devices, strengthening their understanding of how and why they conduct an operation [4]. When the executable operation of a device is obvious and its controller together with its function corresponds to natural mapping, it is easy-to-use. Though very simple, this principle is rarely incorporated in a design [24]. Operational presence in video games can lead players to four types of natural mapping in the process of operation: 1) directional natural mapping, 2) kinesic natural mapping, 3) incomplete tangible natural mapping and (4) realistic tangible natural mapping [4]. Directional natural mapping is the most basic form which provides directional control for a player to interact with virtual objects on the screen when the player manipulates the control device. In video games, directional natural mapping is the simplest and most commonly-used control mode. Through the device, the player inputs up, down, left, or right, any of which corresponds to the directional action in the game environment, to control such devices as joysticks, keyboards, and gamepads smoothly. Kinesic natural mapping refers to controlling the characters or movements in the virtual game environment through the player's bodily movements in real life, such as Kbox Kinect, instead of providing realistic tangible controllers. Incomplete tangible natural mapping means that the player is provided with a certain controller which partially simulates the feel of its equivalent in the virtual world or game environment, such as the Switch gamepad. Realistic tangible natural mapping means that an input device is the same as or similar to a tangible one in real life. As the player uses the tangible device to perform a certain physical action in real life, the same thing will happen in the virtual world, such as steering wheels and pistols. Skalski et al. [4] compared two controllers, one based on realistic tangible natural mapping (for example, a steering wheel controller for driving games or a bongo drum controller for rhythm games) and the other based on directional natural mapping (for example, joysticks and buttons). According to the above researchers, the former controller aroused and brought more fun to the player. It is also generally thought that fun is the result of more natural exercise and interaction with games [25]. Additionally, better graphics and more realistic controls may create a lot more fun [7].

2.4 Spatial Presence

Video game technology can be applied to natural mapping and interaction, and it may produce gaming's presence-inducing capacity [26]. Game technology and development not only enhance graphic and audio recognition but also stimulate other senses, such as using vibration feedback controllers to reproduce physical tactile sensations. Of course, compared with the dark and quiet movie theater, video games provide limited insulation, which will still be affected by the surroundings [16]. In terms of vision, using virtual reality (VR) technology to present game images greatly reduces the interference coming from the surrounding environment. In the VR game environment, it is necessary to wear head-mounted displays, gloves, and headsets to provide adequate insulation. In that way, the sense of immersion can be strengthened, far surpassing that resulting from current standard video games, and there will be a long forward leap in experience presentation [16]. Although past games had a first-person perspective, most video game simulations not only failed to outline the natural mapping of aggressive behavior but also failed to prevent players from being disturbed by the surrounding environment, which affected their vision and sound [16]. Providing a complete mental model is not merely the key to understanding the presence experiences in the video game but also the key to understanding these presence-inducing experiences. Both spatial presence and fun affect natural mapping in the somatosensory controller. Sklski et al. [4] adopted different control devices to conduct driving experiments in video racing games, discovering that devices based on realistic tangible natural mapping, such as steering wheels, could produce a stronger sense of presence and lead to more fun.

2.5 Summary

Besides enriching the game content, the advancement of game technology renders the visual, auditory, and tactile environments more realistic, which means enhancing the realistic effect of the game environment. With the introduction of somatosensory games, not only do many video games combine somatosensory interaction, but more and more games are also aimed at supporting multi-player interaction. The number of players is increased from the original 2 to 4 or even 8, so relatives and friends can enjoy multi-player interactive games to take more pleasure. This study introduces somatosensory games into children's interactive games to analyze children's dance movements and then explores the demand of somatosensory games for space in the virtual and real worlds. This study centers on the activity space of children aged 5 to 8, exploring the vertical and horizontal activity spaces, the spatial requirements of the virtual activity versus the real activity, and the influence of interactive somatosensory movements.

3   Research Methods

With somatosensory interactive games becoming more and more common, having fun together at home is a new trend in science and technology. Playing games with parents or friends can increase family affection or friendship. Based on the research purposes and problems, this study adopted a multi-player dance game, named Just Dance 2020, for bodily sports. As a kind of game software for research, this dance-driven somatosensory game served to explore the correlation between bodily activity space and virtual space, analyze the bodily movements in children's virtual dance games, understand the demand for usable space in the current indoor environment, and identify the corresponding effects of somatosensory games on children's dance movements. The main focuses of this study are explained below:

3.1   Analysis of Children's Dance Movements

This study selected some pieces of dance music suitable for children as the experimental dance music and invited three teachers, who taught children or dance courses, to participate in expert interviews, as shown in Table 1. Meant to analyze dance movements, the pieces of dance music evaluated were mainly taken from Just Dance 2020, the game software and somatosensory game supported by Nintendo Switch. In the above game, there is an exclusive child mode, which provides eight songs specially designed for children. In addition, there are four songs in the simple mode, so twelve dances in total were chosen to analyze movements. As shown by the analytic results shown in Table 2, the dances in the child mode are divided into two categories, i.e., single dance and pas de deux. The footsteps need to move around or just move in the same place. In terms of dance movements, the eight dances in the child mode are all suitable for children. Of the four dances in the simple mode, two songs, Bady Shark and Always look on the bright side of life, are highly popular to and suitable for children. The other two dances, Bad Boy and Bangarang, are too complicated in movement, too swift in rhythm, too wild in swing, and too fast in change; in consequence, they were excluded from the list.

Table 1. Experts' working experience

Expert	Gender	Occupation	Age	Working years
Expert 1	Male	Dance teacher	58	28
Expert 2	Female	Child-caring teacher	52	30
Expert 3	Male	PE teacher	48	13

Table 2. Analysis of dances

Dance song	Single dance/Pas de deux	Duration (seconds)	Stepping in place/Changing positions (left-right swap)	Game mode	Suitable/Unsuitable
Freeze Please	Single	98	Displacement	Child	Suitable
Happy Birthday	Single	94	Step in place	Child	Suitable
Jungle Dances	Single	136	Changing positions	Child	Suitable
Kitchen Kittens	Single	105	Changing positions	Child	Suitable
Magical Morning	Single	92	Changing positions	Child	Suitable
Mini Yo School	Pas de deux	125	Changing positions	Child	Suitable
My Friend The Dragon	Pas de deux	107	Changing positions	Child	Suitable
The Frog Concert (as a warm-up)	Single	125	Stepping in place	Child	Suitable
Baby Shark	Single	86	Stepping in place	Simple	Suitable
Bad Boy	Single	187	Changing positions	Simple	Unsuitable
Always look on the bright side of life	Pas de deux	198	Changing positions	Simple	Suitable
Bangarang	Single	213	Stepping in place	Simple	Unsuitable

3.2 Experimental Design

Children aged 3–8 were recruited as the experimental participants in this study, which complies with the standard classification attributes of toys [27]. Children under 6 are preschool ones, while those aged 6 to 8 belong to elementary school belong to elementary school pupils. As a kind of digital entertainment software, the game software in this study is classified as general [28], meaning that children over age 0 can participate. In addition, after referring to Wu Zhengzhong's classification of children [29], this study divided children into three groups, namely, those aged 3–5 being the youngest group, those aged 6–7 being the middle group, and those aged 8–9 being the oldest group. Originally, eighteen children (six boys and twelve girls) were enrolled, but five girls did not participate due to shyness. Finally, thirteen children participated in the dance game, with their parents' approval conferred in advance. The children participants were put into three age groups, i.e., 3 in the youngest group, 5 in the middle group, and 5 in the oldest group, with an average age of 7 (SD: 1.41).

3.3 Experimental Tools

Employed to measure and record the range of children's dance movements, the experiment tools consist of a Nintendo Switch host, Just Dance 2020 game software, a 55-inch screen with 4K high resolution 3840×2160 (model: 55PUH6193), and two video recorders. Accompanied by the expert teachers, the children participants entered the sports environment and stood in front of the screen at a distance of 200 cm. Before the start of the experiment, the children were required to practice The Frog Concert, which lasted 125 s. Warming-up was conducted through the etude so that the participants might become familiar with the game environment to reduce the learning effect. After practicing, the children formally presented the first dance, Bady Shark, which took 86 s. After finishing, the children had a 5-min rest and then performed the second dance, Happy birthday, which lasted 94 s. The entire experiment took about 25 min. A number of yellow tapes were pasted on the floor to mark and measure the length of a certain displacement. After the participants finished the two dances, their displacements were recorded through the video recorders, with the front, back, left, and right displacements measured respectively. After the lengths as well as the number of displacements were recorded, the results were stored in the video and then analyzed.

3.4 Experimental Procedure

In this study, the dance game software Just Dance 2020 on the Switch console was used to analyze bodily movements. The children participants performed an etude, The Frog Concert, as a warm-up before the formal experiment so that they might be familiar with the dance images and execute some simple movements. After practicing, they performed the two formal dances. The first was Bady Shark, a highly popular children's song, whose dance movements are simple. At the 52nd second of the dance, there was a short trot on the spot for 7 s. The second was Happy Birthday, lasting 94 s, and there were eight displacements in total, which happened at the 14th, 20th, 26th, 32nd, 63rd, 71st, 77th, and 83rd second. To mark the distance away from the concentric center, yellow tapes were pasted on the floor at a distance of 75 cm, 100 cm, 125 cm, 150 cm, 175 cm, 200 cm, 225 cm, 250 cm, and 275 cm in a rectangular space. The experimental procedures are as follows: 1) The purpose, method, and steps of the experiment were explained to the children participants. 2) The participants filled in their personal information, including age, gender, height, and weight. 3) The participants were requested to perform The Frog Concert as a warm-up so as to get adapted to the game environment. They started dancing under the leadership of the dancers on the screen, which took 125 s. 4) The first formal dance Bady Shark was performed and measured, lasting 84 s. 5) The participants had a 5-min break. 6) The second dance Happy Birthday was performed and measured, lasting 96 s. The whole experiment finished, taking 20–25 min.

4 Results

The participants' displacements in the first dance, Bady Shark, are shown in Table 3. However, the virtual dancers in the game did not move left or right but trotted in place for seven seconds. During the dance, the participants moved 50 cm to the left four times, 75 cm to the left only once, and 50 cm to the right two times; the farthest displacement of 75 cm happened to one boy merely once; their front and back displacements did not exceed 50 cm.

In the second dance, Happy Birthday, the virtual dancers displaced themselves eight times. As shown in Table 3, the participants moved 50 cm to the left eight times, 75 cm 3 times, and 100 cm once. The number of displacements to the right was obviously the greatest. The participants moved 50 cm to the right sixteen times, 75 cm fourteen times, 100 cm seven times, 125 cm three times, and 150 cm once to the far right. There was no forward displacement, but the participants moved backwards for 50 cm nine times, for 75 cm four times, and for 100 cm only once.

Table 3. Displacement

Direction	Left			Right					Back		
Displacement (unit: cm)	50	75	100	50	75	100	125	150	50	75	100
Bady Shark	4	1		2							
Happy Birthday	8	3	1	16	14	7	3	1	9	4	1

Unit: number of displacements

The comparison between boys' and girls' displacements is shown in Table 4. In the first dance, Baby Shark, boys moved 50 cm to the left 3 times, and girls only once; in addition, a single boy moved left for 75 cm only once. In the second dance, Happy Birthday, boys and girls had the same number of displacements to the left at the length

Table 4. The number of displacements for different genders

Direction		Left			Right					Back		
Displacement (unit: cm)		50	75	100	50	75	100	125	150	50	75	100
Bady Shark	Boys	3	1		2							
	Girls	1										
Happy Birthday	Boys	4	2	1	10	8	2	2	1	2	0	0
	Girls	4	1	0	6	6	5	1	0	7	4	1

Unit: number of displacements

of 50cm, i.e., four times. Boys moved 75 cm to the left twice and 100 cm only once. Girls moved 75 cm to the left only once. There were more displacements to the right than to the left. Boys moved 50 cm to the right ten times while girls six times. Boys moved 75 cm to the right eight times while girls six times. Boys moved 100 cm to the right twice while girls five times. Boys moved 125 cm to the right twice while girls only once. Boys moved 150 cm to the far right once. Additionally, girls moved backward more often than boys, for boys moved 50 cm backward twice, while girls moved 50 cm backward seven times, 75 cm four times, and 100 cm only once.

The displacements of different age groups are compared and shown in Table 5. In the first dance, the youngest group did not move farther than 50 cm; the middle age group moved 50 cm to the left and to the right only once each; the oldest group moved 50 cm to the left three times, 75 cm to the left only once, and 50 cm to the right twice.

Table 5. Displacements for different age groups

Direction		Left			Right					Back		
Displacement (unit: cm)		50	75	100	50	75	100	125	150	50	75	100
Bady Shark	Youngest											
	Middle	1								1		
	Oldest	3	1	0	2							
Happy Birthday	Youngest	1			2	2	1					
	Middle	4	1		6	6	5	2		5	3	1
	Oldest	3	2	1	8	6	1	1	1	4	1	

Unit: number of displacements

In the second dance, the youngest group moved 50 cm to the left once; besides, it moved 50 cm and 75 cm to the right twice each, and 100 cm only once. As for the middle age group, it moved 50 cm to the left four times and 75 cm only once. It moved 50 cm and 75 cm to the right six times each, 100 cm five times, and 125 cm twice. The number of backward displacements for the middle age group is the largest of three, for it moved 50 cm back five times, 75 cm three times, and 100 cm once. The oldest group got the largest number of left displacements, which is significantly larger than that of the other two groups, for it moved 50 cm to the left three times, 75 cm twice, and 100 cm once. Its number of right displacements for 50 cm is also the largest of all, up to 8 times. In addition, it moved 75 cm to the right six times, but 100 cm, 125 cm, and 125 cm once each. It moved 50 cm backward four times and 75 cm only once.

5 Discussion

This study mainly explores the participants' displacements, with virtual movements taken as a benchmark. In video games, natural mapping is commonly thought of as how closely actions represented in the game environment match the natural actions used to bring about change in a real environment [30]. This study matches a somatosensory game to children's physical movements in the dance. As the results show, children can quickly understand the gameplay and dance along with the virtual characters on the screen. Through natural mapping, the dance movements in the virtual game and children's bodily movements in the real environment are closely matched, with fun in the game obtained. Natural mapping is related to the technology, environment, or subjective perception of interactive games [31]. Natural mapping is interpreted as the ability of the system to match its control to changes in the environment in a natural and predictable manner [30].

Presence is the feeling that a person experiences in the environment. However, it does not refer to the environment that exists in the real world but the medium that enables the person to feel the surrounding environment through interactive and controlled psychological processes [30, 32]. The game Just dance 2020 belongs to kinesic natural mapping [4], allowing players to employ the same physical movements in the real world to control the game environment. Humans' perceptive system has been optimized for interactions in real life. Consequently, when humans control their limbs to generate movements, they should have a high-level sense of presence [33]. Although the participants in this study had not performed or been familiar with the dance movements in the game before, they carried out the same movements as the virtual characters under the leadership of the latter, interacted with the game, and achieved much success. Tamborini et al. [16] mentioned that realistic tangible natural mapping allows users to execute behavioral mental models easily through a controller which has the same shape as that in real life. In addition, it can help users quickly eliminate their unfamiliarity with controllers and game actions during virtual games. Because of obvious similarity between game actions and real actions, the executability of these mental models will be elevated, with spatial presence further enhanced. By means of Joy-con, Just dance 2020 monitors the player's actions and determines whether a dance movement has reached the designated position. The Joy-con controller is just attached to the player's hand to help the game grasp the position of the player's hand, instead of executing an action through the buttons on the controller. Therefore, even if the player is a child, he/she can enhance his/her spatial presence easily and take much delight in the game.

In this study, the participants were requested to perform two dances, both of which are extremely familiar. The first is Bady Shark, lasting 86 s, and the second is Happy birthday with a duration of 94 s. The former's dance movements are mostly upper limb movements. Only when the dance progressed to the 52nd second, was there a 7-s trotting in place. After that, the children turned around on the spot, continuing to do the trotting without leaving their original positions. In the latter dance, the participants displaced themselves eight times and then returned to their original positions. There were two ways for the participants to move. The first was to hop right and then left to return to the original position; the second was to trot along a circle to return to the original position.

As the results show, the dance movements in the first dance, Bady Shark, depend mainly on hands. The virtual dancers on the screen are all dancing in place without moving left or right. As a result, the number of displacements in the first dance is smaller than that in the second dance, with the farthest displacement for the first is only 75 cm to the left.

In the second dance, Happy Birthday, the participants moved right eight times and then returned to their original positions. As shown by the findings, the participants followed the virtual dancers on the screen, moving to the right and then moving left. Their farthest displacement to the right is as long as 150 cm. Meanwhile, the number of left displacements increased after the participants left their original positions and then returned. In addition, after returning to the original positions, girls tended to move backwards more often than boys. Girls moved 50 cm backwards seven times and 75 cm four times, while boys moved 50 cm backwards only twice.

This study compares the displacements for three groups of children, discovering that children in the youngest group had significantly fewer foot movements when dancing. Their upper-body movements kept up well with the virtual dancers' movements, but their lower-body movements were rarely found. It is suggested that the youngest group noticed only the gestures and manual movements of the upper body. In addition, when the second dance reached the time to move left, the manual movements of the youngest group followed those of the virtual dancers on the screen, but their feet still stayed in place. Only one of the children moved the feet and left the original position when the dance was almost over. The above results are closely connected with the three stages of basic movements proposed by Gallahue, Ozmun & Goodway [34]. Basic movements can be divided into three types: stability, locomotion, and manipulation based on their respective properties. Mobility mainly includes walking, running, jumping, and hopping. Five-year-olds are mostly unskilled in mobility-related movements. In the second dance, they hopped right and then hopped left to their original positions. Another movement was for the participants to trot to the right along a circle and return to their original positions. The youngest group might be unskilled in mobility-related movements, so they displayed only hand movements of the upper body without taking steps or performing the movements of the lower body.

6 Conclusion

This study explores the influence of somatosensory dance games on children's bodily movement. The results show that somatosensory games can help children understand dance movements so that they may easily improve their spatial presence and take more pleasure in the games. However, as more footsteps need to move around, the children's displacements increase in number and length. Children's ages have a close relationship with the execution of basic bodily movements. Children in the youngest group depend mainly on the movements of the upper body when dancing, so they seldom take their footsteps to move away from their original positions. It is proposed that follow-up research may aim at improving parent-child interaction through virtual games.

Video gamers should pay serious attention to the following warnings: not spending too much time playing, keeping alert to dangers, not becoming indulged in games, doing some stretching before sports games to warm up, not taking over strong exercise for fear of getting hurt, and not performing overlong movements beyond the limitation of muscles.

Acknowledgements. This study was subsidized under National Sci-tech Programs supervised by the Ministry of Science and Technology, the Executive Yuan, with it coded MOST 108-2221-E-165-001. At the same time, the authors hereby extend our sincere gratitude to the reviewers who gave many valuable suggestions for revision of this paper.

References

1. Co., N.: Dedicated video game sales units (2020)
2. Hartley, M.: Why is the Nintendo Wii So Successful? Smarthouse—The Lifestyle Technology Guide Website, 12 September 2007
3. Ivory, J.D., Kalyanaraman, S.: The effects of technological advancement and violent content in video games on players' feelings of presence, involvement, physiological arousal, and aggression. J. Commun. **57**, 532–555 (2007)
4. Skalski, P., Tamborini, R., Shelton, A., Buncher, M., Lindmark, P.: Mapping the road to fun: Natural video game controllers, presence, and game enjoyment. New Media Soc. **13**, 224–242 (2011)
5. Wikipedia: Family Computer (2020)
6. T.M.O.T. Interior: Regulatory rules about the safety of children's amusement facilities in various industries (2003)
7. Limperos, A.M., Schmierbach, M.G., Kegerise, A.D., Dardis, F.E.: Gaming across different consoles: exploring the influence of control scheme on game-player enjoyment. Cyberpsychol. Behav. Soc. Netw. **14**, 345–350 (2011)
8. Johnson, J.E., Christie, J.F., Yawkey, T.D.: Play and Early Childhood Development, 2nd edn. (1992), (Wu Xingling, Guo Jinghuang, Trans.). Taiwan: Yang-Chih (1987)
9. Johnson, J.E., Christie, J.F., Yawkey, T.D., Wardle, F.P.: Play and Early Childhood Development, 2nd edn. Scott Foresman & Co, Glenview (2003)
10. Hellstedt, J.C.: Kids, parents, and sports: Some questions and answers. Phys. Sportsmed. **16**, 59–71 (1988)
11. Garvey, C., Rubin, K.: Play: The Crucible of Learning. Harvard University Press, Cambridge (1997)
12. Damon, W., Lerner, R.M., Eisenberg, N.: Handbook of Child Psychology, Social, Emotional, and Personality Development. Wiley, Hoboken (2006)
13. Martin-Niedecken, A.L., Götz, U.: Design and evaluation of a dynamically adaptive fitness game environment for children and young adolescents. In: Proceedings of the 2016 Annual Symposium on Computer-Human Interaction in Play Companion Extended Abstracts, pp. 205–212 (2016)
14. Lin, H., Huang, Kuo-Liang., Lin, W.: A preliminary study on the game design of pokémon GO and its effect on parent-child interaction. In: Rau, P-L Patrick. (ed.) HCII 2020. LNCS, vol. 12192, pp. 115–127. Springer, Cham (2020). https://doi.org/10.1007/978-3-030-49788-0_9
15. Griffing, P.: Encouraging dramatic play in early childhood. Young Child. **38**, 13–22 (1983)
16. Tamborini, R., Eastin, M.S., Skalski, P., Lachlan, K.: Violent virtual video games and hostile thoughts. J. Broad. Elec. Media **48**, 335 (2004)
17. Pagulayan, R.J., Keeker, K., Wixon, D., Romero, R.L., Fuller, T.: User-centered design in games (2003)
18. Gee, J.P.: Learning by design: good video games as learning machines. E-learn. Digit. Media **2**, 5–16 (2005)
19. Johnson, D., Wiles, J.: Effective affective user interface design in games. Ergonomics **46**, 1332–1345 (2003)

20. Desurvire, H., Caplan, M., Toth, J.A.: Using heuristics to evaluate the playability of games. In: CHI 2004 Extended Abstracts on Human Factors in Computing Systems, pp. 1509–1512 (2004)
21. Federoff, M.A.: Heuristics and Usability Guidelines for the Creation and Evaluation of Fun in Video Games. Citeseer, Princeton (2002)
22. Adams, E.: The Designer's Notebook: Bad Game Designer. No Twinkie (2004)
23. Grodal, T.: Video games and the pleasures of control. Media entertainment: The psychology of its appeal, pp. 197–213 (2000)
24. Norman, D.: The Design of Everyday Things, Revised and Expanded edn. Basic Books, New York (2013)
25. Lindley, S.E., Le Couteur, J., Berthouze, N.L.: Stirring up experience through movement in game play: effects on engagement and social behaviour. In: Proceedings of the SIGCHI Conference on Human Factors in Computing Systems, pp. 511–514 (2008)
26. Lachlan, K., Krcmar, M.: Experiencing presence in video games: the role of presence tendencies, game experience, gender, and time spent in play. Commun. Res. Rep. **28**, 27–31 (2011)
27. CN Standards: CNS general number: 797, category number: 7066 (2007)
28. Taiwan Entertainment Software Rating Information. https://www.gamerating.org.tw/Grade.aspx
29. Woo, J.-C.: A study on the application of emotional design on the anxiety and pain alleviation in pediatric outpatients. J. Design **17**, 69–92 (2012)
30. Steuer, J.: Defining virtual reality: dimensions determining telepresence. J. Commun. **42**, 73–93 (1992)
31. Lee, K.M., Park, N., Jin, S.-A.: Narrative and Interactivity in Computer Games (2006)
32. Gibson, J.J.: The Ecological Approach to Visual Perception, Classic Psychology Press, Hove (2014)
33. Biocca, F.: Virtual reality technology: a tutorial. J. Commun. **42**, 23–72 (1992)
34. Gallahue, D.L., Ozmun, J.C., Goodway, J.: Understanding Motor Development: Infants, Children, Adolescents, Adults. Mcgraw-hill, Boston (2006)

Motion Illusion on Form with Different Types of Line Graphic

Chih-Wei Lin[1,2]([✉]), Lan-Ling Huang[1,2], Chi-Meng Liao[1,2], and Hsiwen Fan[3]

[1] Fujian University of Technology, Fuzhou 350118, Fujian, China
copy1.copy2@msa.hinet.net
[2] Design Innovation Research Center of Humanities and Social Sciences Research Base of Colleges and Universities in Fujian Province, Fuzhou 350118, Fujian, China
[3] Chiba University, Chiba 263-8522, Chiba, Japan

Abstract. This study observed the presence of rotational speed thresholds to determine the effects of different types of line graphics on the results and perceptions of three motion illusions (i.e., apparent movement, induced movement, and movement afterimage). Moreover, this study explored interactive and causal relationships among line graphics, rotational speed, and the aforementioned three phenomena. Several experiments were conducted by adopting the method of adjustment derived from psychophysical methods. The findings revealed that different line graphics and rotational speed settings resulted in visual disturbances that differed in terms of emergence, strength, and conversion. In addition, different line graphics and rotational speeds had causal relationships with the upper–lower absolute threshold values of said motion movements. Different types of line graphics on forms led to different phenomena of line afterimage (PLAs) between each interstimulus interval (ISI) in apparent movement, created the PLA between each ISI in induced movement when the rotational speed decreased, caused the phenomenon of line mixing when the rotational speed increased, and produced the phenomenon of line curving and the PLA between each ISI in motion afterimage. The line graphics with low curving degrees produced an effect known as visual association.

Keywords: Motion form · Motion perception · Motion illusion

1 Introduction

Visual illusions, perceptions, and patterns of motion forms are derived from combinations of form determinants relevant to motion forms and rotation of the forms concerned. Different combinations of motion form determinants result in different types and degrees of visual illusions, perceptions, and patterns. This study examined diverse visual illusions and motion patterns in different line graphics with varying rotational speed to explore motion perceptions and understand the relationship between motion form and motion perception.

© Springer Nature Switzerland AG 2021
P.-L. P. Rau (Ed.): HCII 2021, LNCS 12771, pp. 141–155, 2021.
https://doi.org/10.1007/978-3-030-77074-7_11

2 Literature Review

Apparent movement occurs when the conditions of visual masking and a continuous stimulus are fulfilled. Under such circumstances, the stimulus that triggers motion perception does not move but real motion objects create the illusion that the stimulus is moving. In this study, different spirals and continuous line graphics on a form were treated as stimuli for apparent movement. Next, the angles caused by rotation of the form (visual disturbances) were utilized to create visual masking effects, the forming elements of which operated according to the principle of apparent movement. When the form continued to rotate, all continuous line graphics formed a continuous movement and created continuous motion perceptions and patterns.

Induced movement mainly exhibits two patterns: induced movement between objects and that between forms [1, 2]. The present study focused on patterns caused by induced movement between forms, which are directly linked with figure–ground differentiation. When a continuous form (pattern) rotates, the induced movement produces the illusion of a rotating figure, whereas the illusion of stationary ground occurs because of assimilation, which separates the figure from the ground. In this study, different spirals and continuous line graphics on forms were treated as stimuli for induced movement. When the form concerned was in continuous rotation, a motion perception and pattern were triggered to create the illusion of a continuously rising line graphic.

Movement afterimage refers to the perception of a motion aftereffect or persistence of vision occurring for a short period after fixation of an object moving quickly in a specific direction. This phenomenon occurs because when the eye perceives that an object is moving quickly or between locations, it tends to separately and temporarily retain object images perceived at different moments on the retina. The image then continues to appear or is retained on the retina for .04–.06 s before being transmitted through the optic nerve to the brain, where the image of object motion or displacement is perceived. If the object continues to move or suddenly stops, the nerve impulses stimulated by the original image do not instantly disappear, but rather remain for a further .04–.06 s, thereby creating the illusion that what was in the original image still exists. Different spirals and continuous line graphics on forms were treated as stimuli for movement afterimage in this study, which found that when a form was in continuous rotation, the line graphic concerned created afterimages and obscure motion perceptions and patterns.

3 Experimental Design and Samples

This study adopted the method of adjustment derived from psychophysical methods to determine the absolute threshold (AT) and implemented laboratory experimentation with a within-subject design. A total of 20 participants were recruited through a nonprobability sampling technique known as judgement sampling. Five line graphics on forms were designed, namely straight, ladder, arc line graphics, curve with short wavelength, and curve with long wavelength. Each line graphic was 10 mm wide and tilted by 15° [3]. Under a 1-M observation distance, each measured cylinder (triangular pyramid) [4] was 25 cm high [5] and 11 cm (12 cm) wide [6] (Table 1).

Table 1. Five lines graphic and cylinder (triangular pyramid)

Type of line graphic	Straight	Ladder	Curve with long wave-length	Curve with short wave-length	Arc
Triangular pyramid sample					
Cylinder sample					
Graphical approach					

4   Results and Discussion

4.1   Experiment of Perceptions of Apparent Movement Stimulated by Different Types of Line Graphics

Experimental Results. The lower AT of the straight line graphic (65.65 rpm) was the lowest, indicating that this line graphic triggered an apparent movement perception earlier than did the other line graphics. The upper AT of the curve with long wavelength (103.9 rpm) was the highest, suggesting that this curve led to longer perceptions of

Table 2. Rotating speed thresholds of apparent movement for the five types of line graphics (rpm)

Type of line graphic	Perception threshold		Mean
Straight	Upper AT	Speed threshold 33.75	99.40
	Lower AT		65.65
Ladder	Upper AT	Speed threshold 23.10	100.30
	Lower AT		77.20
Curve with long wave-length	Upper AT	Speed threshold 21.20	103.90
	Lower AT		82.70
Curve with short wave-length	Upper AT	Speed threshold 18.20	102.55
	Lower AT		84.35
Arc	Upper AT	Speed threshold 21.55	93.80
	Lower AT		72.25

apparent movement compared with the other line graphics. Nevertheless, the gaps in rotational speed among the measured five line graphics were small. The straight line graphic (33.75 rpm) had the highest speed threshold, and thus had the longest duration of motion illusion and the best performance. Overall, the straight line graphic produced the op-timal effect of motion illusion and optimal perception of apparent movement (Table 2) (Fig. 1).

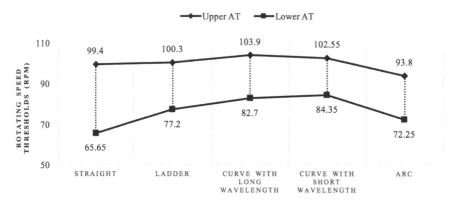

Fig. 1. Run chart in the thresholds of apparent movement in five line graphics (rpm)

Emergence of Line Afterimage Phenomenon. The experimental results demonstrated that different line graphics and different rotational speed settings directly affected the results of motion illusions and apparent movement and the perceptions and discrimination of apparent movement among the participants. Thus, this study identified factors that affect motion illusions and determined interactive and causal relationships among types of line graphics, rotation speed, and apparent movement.

Between each interstimulus interval (ISI), each different type of line graphic can create a different phenomenon of line afterimage (PLA). In other words, when a pyramid rotated, some sections of the line graphics curve exhibited afterimages that fluctuated and shook unevenly, resulting varying degrees of visual disturbances. When the curving degree increased, so did the number of afterimages, consequently exhibiting a more noticeable disturbance (Table 3). When the pyramids with line graphics rotated, visual disturbances known as the phenomenon of line curving (PLC) and phenomenon of line ghosting (PLG) occurred because of the unique shape of triangular pyramids. When the participants perceived and recognized apparent movement, they were visually disturbed by the PLA caused by line graphic rotation and PLC and PLG caused by pyramid rotation. Specifically, disturbances created by straight and arc line graphics were the most minor because these two line graphics were nearly straight. Except for the PLC and PLG, no PLAs appeared. Fluctuation was unnoticeable in these two line graphics, and thus less disturbance was created. Because the fluctuation levels of the ladder line graphic and curves of long or short wavelength were high, the number of afterimages increased. In addition to PLC and PLG, PLA occurred; therefore, the perceived line graphics fluctuated and trembled, creating more visual disturbances. Among the line graphics that exhibited the PLC and PLG, those that were straight caused fewer afterimages, and thus appeared more static with fewer disturbances. By contrast, the number of afterimages increased among the graphics that were highly curved. The perceived images fluctuated and shook, thereby creating stronger disturbances. The aforementioned results demonstrated that the curving degree of line graphics had a causal relationship with the emergence and degree of PLA.

Table 3. Visual disturbances of the five types of line graphics

Type of line graphic	Straight	Ladder	Curve with long wave-length	Curve with short wave-length	Arc
Pattern of line graphic afterimages					
Threshold of rotational speed	65.65-99.4rpm	77.2-100.3rpm	82.7-103.9rpm	84.35-102.55rpm	72.25-93.8rpm

Interactive and Causal Relationships Among Line Graphics, Rotational Speed, and Apparent Movement. The experimental results revealed that different combinations of line graphics and rotational speed could determine the emergence, degrees, and transduction of the PLA, PLC, and PLG, which directly affected the upper and lower AT values of the various line graphics in terms of apparent movement. The curving degrees and rotational speeds of the line graphics affected the presence of apparent movement and the participants' perceptions and discrimination of such movement. Thus, the curving degree and rotational speed had causal relationships with apparent movement. When the line graphic became straight with a lower rotational speed, the number of afterimages decreased and the originally fluctuating and trembling line graphic became stationary. Furthermore, the vision that was originally disturbed by the PLA and PLC was disturbed by only the PLC, indicating a decrease in visual disturbances. Under such circumstances, the lower AT also decreased, suggesting that when the line graphic was straight, the lower AT in apparent movement was identified at an earlier stage. Therefore, the participants first perceived the apparent movement created by the pyramid with a straight line graphic and that with an arc line graphic because their lower AT values (straight = 65.65 rpm; arc = 72.25 rpm) were discovered earlier than those of the other line graphics. Similarly, when the line became straight, the rotational speed increased, the number of afterimages decreased, and the originally fluctuating and trembling line graphic became static. Furthermore, the vision that was originally disturbed by the PLA, PLG, and PLC was disturbed by only the PLG and PLC, indicating a decrease in visual

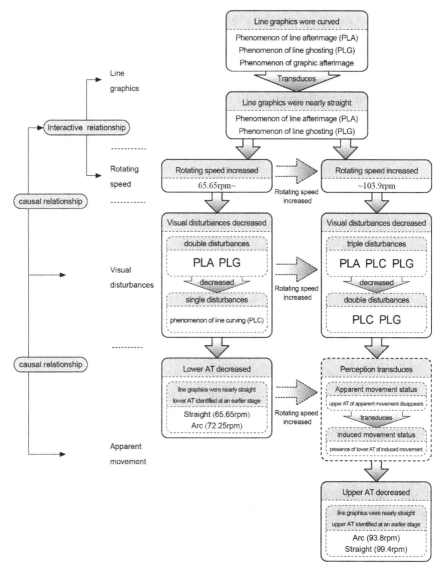

Fig. 2. Interactive and causal relationships among line graphics, rotational speed, and apparent movement

disturbances. However, the upper AT did not increase because of fewer disturbances but instead decreased with the lower AT. This result indicated that the upper AT in apparent movement was identified earlier when the line graphic became straight, possibly because when rotational speed increases, apparent movement, induced movement, and movement afterimage reportedly occur in succession [7]. When apparent movement is perceived and rotational speed increases almost to the transduction threshold, the participant no longer perceives apparent movement (i.e., the upper AT of apparent movement disappears); the perception of apparent movement then transduces into induced movement (i.e., presence of the lower AT of induced movement). In the present study, when the line graphic became straight and the rotational speed increased, the disturbance caused by the PLA disappeared. After the number of afterimages had decreased, the line graphic had stabilized, and the number of visual disturbances decreased, the upper AT appeared earlier. This result proved that the participants more quickly perceived apparent movement created by straight line graphics. Thus, the upper AT values of the arc (93.8 rpm) and straight (99.4 rpm) line graphics were discovered earlier than those of the other line graphics (Fig. 2).

4.2 Experiment of Perceptions of Induced Movement Triggered by Different Types of Line Graphics

Experimental Results. The lower AT of the straight line graphic (44.2 rpm) was the lowest, indicating that the straight line graphic provoked the perception of induced movement earlier than did the other line graphics. In addition, the straight line graphic had the highest higher AT (299.55 rpm), demonstrating that induced movement triggered by this line graphic lasted longer compared with the other line graphics. Furthermore, the straight line graphic had the highest speed threshold (255.35 rpm), and thus had the longest mo-tion illusion with optimal performance. In brief, the straight line graphic produced the optimal effect of motion illusion and optimal perception of induced movement (Table 4) (Fig. 3).

Table 4. Rotation speed thresholds of induced movement for the five line graphics (rpm)

Type of line graphic	Perception threshold			Mean
Straight	Upper AT	Speed threshold	255.35	299.55
	Lower AT			44.20
Ladder	Upper AT	Speed threshold	224.80	283.95
	Lower AT			59.15
Curve with long wavelength	Upper AT	Speed threshold	200.20	267.80
	Lower AT			67.60
Curve with short wavelength	Upper AT	Speed threshold	184.25	257.00
	Lower AT			72.75
Arc	Upper AT	Speed threshold	154.15	228.50
	Lower AT			74.35

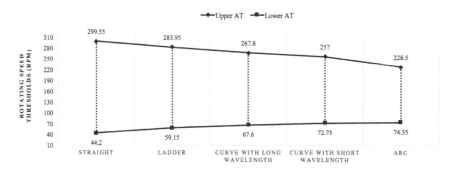

Fig. 3. Run chart in the thresholds of induced movement in five line graphics (rpm)

Emergence of the PLA and Phenomenon of Mixing Afterimages. The experimental results demonstrated that different line graphics and distinct rotational speed directly affected the results of motion illusions and induced movement and the perceptions and discrimination of such movement among the participants. This study determined factors that affected motion illusions and determined interactive and causal relationships among types of line graphics, rotation speed, and induced movement.

Between each ISI, different line graphics and rotational speed settings can create different forms of the PLA and phenomenon of mixing afterimages (PMA; Table 5). Similar to apparent movement, the PLA refers to the fluctuating, trembling, and nonstationary afterimages that appear between each ISI. Such phenomena cause visual disturbances to varying degrees. When the curving and tilting degrees of the line graphic increased, so did the number of afterimages, consequently exhibiting more noticeable disturbances. The PMA occurs when the line graphic concerned exhibits rotary color mixture. When a cone rotated increased, the phenomenon of temporary color mixing was identified on the line graphic [8], thereby creating continual grayscale ramp and a mixing visual effect of grayscale ramp. According to the theory of rotary color mixture, color-mixing images appear because lights of varying wavelengths are equally reflected and absorbed when the in vision in question is stimulated only by black and white [9]. The PMA can cause disturbances to varying degrees. When the curving and tilting degrees of the line graphic concerned were high, the area of continual grayscale ramp expanded, causing stronger disturbances. The PLA disturbance shifted to PMA disturbance as the rotation speed increased. Furthermore, the cylinders adopted in the experiment of induced movement did not have sides or angles as did the aforementioned pyramids because of their curvilinear shapes, and thus the PLC and PLG did not occur when the cylinders rotated. Thus, the participants were only disturbed by the PLA and PMA when perceiving and recognizing induced movement. Specifically, the straight line graphic had the most minor disturbances because even at varying rotation speeds, its straightness did not induce the PLA and PMA. Thus, the straight line graphic created fewer disturbances than did the other line graphics. By contrast, the curving and tilting degrees of the other four line graphics were high, and thus the number of afterimages and the area of continual grayscale ramp increased and induced the PLA and PMA at various rotational speeds. When the cylinders with these line graphics rotated, the participants perceived fluctuated and trembling line graphics and perceived considerable visual disturbances. In summary, at different rotational speeds, the straight line graphic presented few afterimages and the area of continual gray scale ramp was small; this line graphic was relatively static, leading to minor visual disturbances. By contrast, the number of afterimages and area of continual grayscale ramp increased when the curving and tilting levels of the line graphics increased; the line graphic images fluctuated and trembled, creating considerable disturbances. The curving and tilting levels of the line graphics at different rotational speeds had causal relationships with the PLA and PMA.

Table 5. Visual disturbances of the five types of line graphics

Type of line graphic	Straight	Ladder	Curve with long wave-length	Curve with short wave-length	Arc
PLA					
Threshold of rotation-al speed	44.2rpm-	59.15rpm-	67.6rpm-	72.75rpm-	74.35rpm-
PMA					
Threshold of rotation-al speed	-299.55rpm	-283.95rpm	-267.8rpm	-257rpm	-228.5rpm

Interactive and Causal Relationships Among Line Graphics, Rotational Speed, and Induced Movement. The experimental results revealed that different line graphics and rotational speeds can decide the emergence, degrees, and transduction of the PLA and PMA, thereby directly affecting the upper and lower AT values among the measured line graphics during induced movement. The curving levels of line graphics and the rotational speed affected the formation of apparent movement and the participants' percep-tions and discrimination of such movement. Thus, the curving level and rota-tional speed had causal relationships with apparent movement. When the line graphic straightened at a lower rotational speed, the number of afterimages decreased; the origi-nally fluctuating line graphic then turned stationary and PLA disturbances grad-ually diminished. Under these circumstances, the lower AT also decreased, suggesting that when the line graphic was straight, the lower AT of induced movement was iden-tified earlier. Therefore, the participants first perceived the induced movement creat-ed by the cylinder with straight line graphic because its lower AT value (44.2 rpm) presented ear-lier than those of the other line graphics. By contrast, the cylinders with curved and tilted line graphics exhibited their lower ATs later. The lower AT of the arc line graphic (74.35 rpm) appeared last compared with those of the other line graphics, indicating that the participants perceived the induced movement caused by the arc line graphic last. When the line graphic straightened at a higher rotational speed, the area of continual grayscale ramp shrank; the originally fluctuating line graphic then turned stationary and PMA disturbances was alleviated. In this situation, the upper AT value increased when the line graphic was straight, demonstrating that the upper AT of the straight line graphic (299.55 rpm) appeared later than did those of the other line graphics. The participants contin-ued to perceive that the straight line graphic caused induced movement. Conversely, the cylinders with curved and tilted line graphics exhibited their upper ATs earlier; the upper AT of the arc line graphic (228.5 rpm) appeared earliest. The participants quickly ceased to perceive that the arc line graphic caused induced movement (Fig. 4).

4.3   Experiment of Perceptions of Movement Afterimage Triggered by Different Line Graphics

Experimental Results. After the lower AT of the movement afterimage had appeared, perception of such movement never disappears as the rotational speed increases to an infi-nite degree. Thus, upper AT values of movement afterimage were excluded in this study. The lower AT of the straight line graphic (169.8 rpm) was the lowest, indicating that the participants perceived movement afterimage triggered by the straight line graphic earlier than that triggered by the other line graphics. Overall, the straight line graphic induced the optimal effect of motion illusion and optimal perception of movement afterimage (Table 6) (Fig. 5).

Visual Association Effect Derived from Interactions Among Three Visual Distur-bances. The experimental results revealed that different line graphics and rotational speed settings directly affected the results of motion illusions and movement afterim-age and perceptions and discrimination of such movement among the participants. This study attempted to confirm the factors that affected motion illusions and analyzed the interactive and causal relationships among types of line graphics, rotational speed, and movement afterimage.

Between each ISI of movement afterimage, the participants experienced visual dis-turbances to varying degrees, namely the PLA resulting from line graphic rotation and the PLC and PLG resulting from pyramid rotation. At a high rotational speed, perceptions of these three disturbances among the participants began to change. Specifically, when the line graphic straightened, the PLA did not occur; the participants were affected by only the PLC and PLG. However, when the line graphic became curved, the participants iden-tified interactions among the PLA, PLC, and PLG. Originally, the curved sections created illusions of fluctuating and nonstationary afterimages. After the interaction effect had

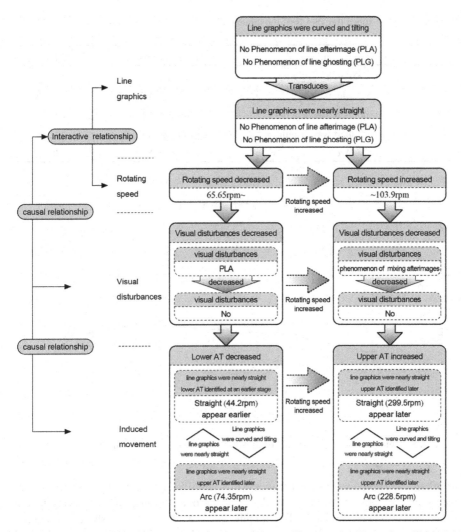

Fig. 4. Interactive and causal relationships among line graphics, rotational speed, and induced movement

become apparent, these three disturbances formed the illusion of motion bounce. This motion illusion produced a visually attractive effect known as visual association effect, which inhibited the participants from perceiving or recognizing movement afterimage (Table 7). Because of the straight line graphic nature, the pyramids with straight and arc line graphics did not exhibit the PLA. Consequently, the visual association effect derived from interactions among the PLA, PLC, and PLG was not identified. These pyramids only caused disturbances in the PLC and PLG. However, the spaces between each line graphic segment of the straight line graphic pyramid were narrower at the top, whereas those of the arc line graphic pyramid were comparable. The straight line graphic pyramid created a strong effect at its top when rotating and strengthening the

Table 6. Rotating speed thresholds of movement afterimage among the five line graphics (rpm)

Type of line graphic	Perception threshold	Mean
Straight	Lower AT	169.80
Ladder	Lower AT	188.70
Curve with long wavelength	Lower AT	195.55
Curve with short wavelength	Lower AT	202.50
Arc	Lower AT	188.15

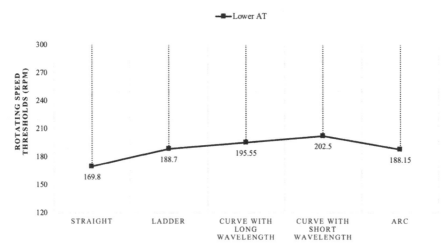

Fig. 5. Run chart in the thresholds of movement afterimage in five line graphics (rpm)

disturbances. Therefore, the disturbances (PLC and PLG) were stronger in the straight line graphic pyramid than in the arc line graphic one. Because of the curving nature of the pyramids with the ladder line graphic and the curves of long/short wavelength, these three pyramids provoked interaction among the PLA, PLC, and PLG and created the effect of visual association at the line graphic curving sections. In conclusion, a line graphic with a lower curving degree creates lower degrees of the PLC and PLA. Without the visual association effect, such line graphics create more noticeable PLC–PLA disturbances. By contrast, a line graphic with a higher curving degree creates a strong illusion of motion bounce as well as the visual association effect, thereby preventing the participants from perceiving or recognizing movement afterimage. The aforementioned results demonstrated that the curving degree of a line graphic had a causal relationship with the emergence and degree of the visual association effect.

Interactive and Causal Relationships Among Line Graphics, Rotational Speed, and Movement Afterimage. The aforementioned results and analyses confirmed that different line graphics and rotational speeds affected the emergence, degree, and conversion of the PLA, PLC, and PLG. These effects in turn directly influenced the lower AT

Table 7. Visual disturbances of the five line graphics

Type of line graphic	Straight	Ladder	Curve with long wave-length	Curve with short wave-length	Arc
	PLA–PLC disturbances	Visual asso-ciation effect	Visual asso-ciation effect	Visual asso-ciation effect	PLA–PLC disturbances
Interaction among the PLA, PLC, and PLG					
Threshold of rotation-al speed	169.8rpm-	188.7rpm-	195.55rpm-	202.5rpm-	188. 15rpm-

of movement after-image under various conditions. Precisely, the interactive relation-ships between the line graphic curving level and rotational speed affected the emergence of movement afterimage and the participants' perceptions and discrimination of such movement. Thus, these results demonstrated that the line graphic curving level and rota-tional speed have causal relationships with movement afterimage. When the line graphic became straight with a lower rotational speed, the number of afterimages decreased and the originally fluctuating and trembling line graphic became static. Furthermore, the vision that was originally disturbed by the PLA, PLC, and PLG was disturbed by only the PLC and PLG, indicating an increase in visual disturbances. The AT also decreased, exhibiting earlier emergence of movement afterimage under disturbances from the PLC and PLG. Therefore, the lower AT values of movement afterimage among the line graphics with lower curving levels appeared at an earlier stage com-pared with those with higher curving levels; specifically, the lower AT of the straight line graphic (169.8 rpm) and that of the arc line graphic (188.15 rpm) presented at an early stage, thereby enabling the participants to first perceive the illusions of move-ment afterimage. When the line graphic became curved with a higher rotational speed, the number of afterimages increased and the originally static line graphic be-gan to fluctuate. The dis-turbances turned from noticeable PLC and PLG to a stronger effect of visual association; in addition, the effect of motion bounce increased. The resultant lower AT then increased, suggesting that a strong visual association effect effectively distracted the participants from perceiving and recognizing movement afterimage. Thus, the movement afterimage was identified late. The line graphics with high curving levels were late in exhibiting the lower AT of movement afterimage compared with the shoe with low curving levels. Specifically, the lower AT of the ladder line graphic (188.7 rpm), that of the curve with long wavelength (195.55 rpm), and that of the curve with short wavelength (202.5 rpm) were recognized late, indicat-ing that the participants perceived the movement afterim-age caused by the ladder line graphic and curves with long/short wavelength after they perceived movement caused by the straight and arc line graphics (Fig. 6).

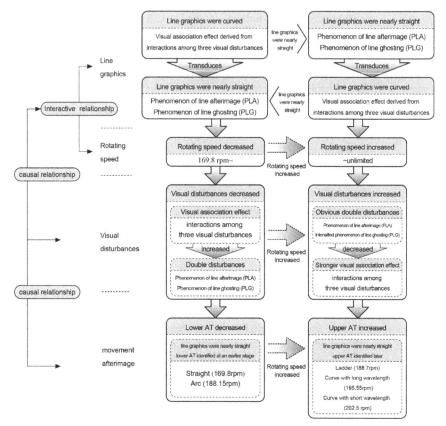

Fig. 6. Interactive and causal relationships among line graphics, rotational speed, and movement afterimage

5   Conclusion

This study observed the presence of rotational speed thresholds to determine the effects of different types of line graphics on the results and perceptions of three motion illusions (apparent movement, induced movement and movement afterimage), Moreover, this study explored interactive and causal relationships among line graphics, rotational speed, and the aforementioned three phenomena. The findings revealed: (1) Different line graphics and rotational speed settings resulted in visual disturbances that differed in terms of emergence, strength, and conversion. In addition, different line graphics and rotational speeds had causal relationships with the upper–lower absolute threshold values of said motion movements. (2) Different types of line graphics on forms led to different phenomena of line afterimage (PLAs) between each interstimulus interval (ISI) in apparent movement, created the PLA between each ISI in induced movement when the rotational speed decreased, caused the phenomenon of line mixing when the rotational speed increased, and produced the phenomenon of line curving and the PLA between each ISI in motion afterimage. The line graphics with low curving degrees produced an effect known as visual association. The straight line graphic produced the optimal effect of motion illusion and optimal perception of three motion illusions.

References

1. Chen, G.D., Chang, C.C.: A study about the induced motion of the rotative speed and the caliber of strand. In: Proceedings of Conference on Asia Society of Basic Design and Art in Tsukuba 2007, pp. 295–300. Japan Society of Basic Design and Art, Japan (2007)
2. Chen, G.D., Chang, C.C., Lin, P.C.: A study on the induced motion of the rotative speed and the width of line. Issue Basic Des. Art **16**, 19–22 (2008)
3. Chen, G.D.: The Study of Kinetic Art Dynamic Optical Illusion on Column with Spiral Pattern. National Taiwan University of Science and Technology, Taipei (2008)
4. Chen, G.D., Lin, C.W., Fan. H.W.: Motion perception on column of rotational dynamic illusion in kinetic art. J. Des. **20**(3), 1–19 (2015).
5. Oyama, T., Imai, S., Wake, T.: New Sensory Perception Psychology Handbook. Seishin Shobo Ltd., Tokyo (2000)
6. Liu, Y.H.: The Research of Preference for Ratio of Geometry Shapes. National Yunlin University of Science and Technology, Yunlin (2010)
7. Chen, G.D., Lin, C.W., Fan. H.W.: The study of motion perception on the rotational motion illusion of cone. J. Sci. Technol. **24**(1), 85–101 (2015)
8. Hung, C.Y., Guan, X.S.: A Study of Subtractive Color Mixture and Rotatory Color Mixture. National Science Council, Taipei (1999)
9. Lee, C.F.: Effects of optical processing on the subjective color. J. Sci. Technol. **11**(4), 279–284 (2002)

Rethinking the Body as a Humanistic Intervention – Teaching HCI with a Recognition of Multiplicity

Hanwei Shi[✉]

School of Art and Design, Guangdong University of Technology, Guangdong, China

Abstract. Over the past decade, the proliferation of embodied interaction based on biosensing technologies and the internet of things (IoT) has increasingly brought academic attention to the humanistic concerns about the usage, the accessibility, and the interpretation of the human body in the form of data. The human body, which has been a rich domain for the exploration of biological mechanics and cultural significance, now is entering a new era when lived bodily experiences can be captured and represented in the abstraction of data. For an educator and practitioner who works with the intersection of body and technology, two pedagogical challenges that one might constantly face are: (1) how to keep a critical position while embracing the novel technologies; (2) how to deal with the tension between the multiplicity of the body discussed in humanities and the universalizing effect of electronic technologies. Particularly, in a country like China proud of her unique body tradition it is impossible to ignore the wider social, cultural and ethical context where the body is embedded.

This article is a response to such challenges. By offering thoughts on the importance of introducing the discourse of the body in the humanities into the ever-expanding realm of human-computer interaction (HCI) studies and embodied interaction design in the context of classrooms at a technical university, I would like to argue that bringing the humanistic conversation of the cultural and philosophical diversity of the body into discussion can be a great complement to the HCI curriculum. Students will benefit a lot from the plural perspectives on the body discussed in humanities and get a valuable opportunity to rethink the role that technology plays in relation to the body. I situate this article within the larger picture of Humanistic HCI approaches, aiming at offering a reflexive examination of the renewed and renewing body-technology relations.

Keywords: Embodied interaction · The discourse of the body · Teaching HCI · Humanistic HCI

1 Introduction

In the ever-expanding fields of human–computer interaction (HCI) studies and interaction design, there has been a proliferation of scholarly work on the technology in relation to the body, highlighted by the exploration of wearable data-tracking devices based

© Springer Nature Switzerland AG 2021
P.-L. P. Rau (Ed.): HCII 2021, LNCS 12771, pp. 156–163, 2021.
https://doi.org/10.1007/978-3-030-77074-7_12

embodied interaction, ranging from medicine to entertainment. A considerable number of the work has put the academic weight on overcoming technological difficulties or optimizing models of analysis, while it seems to be accompanied by an underrepresentation of the discussion on the ethical, social, and cultural significance of the role that novel technologies play in relation to the body. As the intervention of humanistic approaches has entered the scholarship of HCI, exemplified by the manifesto of humanistic HCI proposed by Jeffrey Bardzell and Shaowen Bardzell in 2016, [1] the well-established humanistic discourses, for instance, gender studies, race studies and critical theories, have initiated new trends in HCI with a growing interest in the renewed and renewing relation between body and technology.

This article is along with the effort to reopen the ancient subject of the body in the midst of technological saturation, examining both the cooperation and the tension between the human body and the latest embodied technologies. In particular, this article focuses on the challenges of teaching the design of the wearable devices in the rapidly self-refreshing fields of HCI and interaction design with humanistic concerns. When intensive pedagogical attention has been paid to overcome the technological difficulties, for example, the making of intelligent fabrics, the design of innovative tangible interfaces, the use of software, or data analytics, as an educator and practitioner specializing in new media with a formal educational background of humanities, I believe that there remains room for humanistic concerns, and the study of the wearable in HCI can remarkably benefit from a recognition of the body discussed in the context of art, philosophy, and culture.

This article begins with a review on the successful marriage of humanities and HCI in an attempt to reveal humanistic concerns about datalization of the human body, particularly exemplified by two cases: one is the academic discussion on the quantified self (QS) in media studies, and the other is the recent growth of research on the wearable in relation to the gendered body in HCI. The article then moves to the specific cultural context of China, where the practice and the interpretation of the body can be radically different from the Western tradition. In that sense, taking the body's cultural specificity into account is crucial in terms of teaching HCI related to embodied technologies. The final part is about thoughts on the curriculum design when bringing the body discourse in humanities to the HCI classroom.

2   The Intersection of Humanities and Embodied Interaction HCI

2.1   The Role of Humanities in the Age of Technological Saturation

"The most profound technologies are those that disappear." [2]. This famous claim about the future of computing made by Mark Weiser in the early 1990s has become perhaps the most cited statement that advertises the magical invisibility of the innovative technology under the umbrella of the internet of things (IoT). What is less known is the following part in Weiser's original text, he then continues, "Such a disappearance is a fundamental consequence not of technology, but of human psychology." It seems that what Weiser concerns most is not the celebration of the novelty of technology; instead, he tries to remind us of the fact that the whole process of computing is invisible, yet the effect of how it affects human is not. The ubiquitous spread of technology that intervenes in every corner of our daily lives is accompanied by a psychological effect, which tends to take the measurement and the circulation of personal data for granted. Such taken-for-granted-ness raises a myriad of questions concerning issues around ethics, culture, and politics.

To a certain degree, humanities thrive on revealing such taken-for-granted-ness, or to say, making the invisible visible. As the famous quote from the pioneering media theorist Marshal McLuhan points out, "One thing about which fish know exactly nothing is water, since they have no anti-environment which would enable them to perceive the element they live in." [3]. The fish-water relation in McLuhan's words is a perfect metaphor for the disappearing effect of technologies as used in the wearable. What we need to ask at the moment is the question: can humanities serve as the "anti-environment" that helps make fish aware of the existence of water? By mentioning the prefix "anti-", I suggest not to interpret it as the negative refutation; rather, it is a teaser that seduces a second look at what has been excluded, which is to say, at what goes beyond the uncritical embrace of innovative technologies. Moreover, what makes a good teacher in HCI, to the best of my belief, is not only the skill to swim, but also the ability to detect the boundary of the water.

2.2 Bodies that Matter in the Field of Embodied Interaction

The notion of embodiment is not something new in humanities. Maurice Merleau-Ponty, who offers a phenomenological treatment of the body as the flesh interconnects and exchanges with its surroundings in the 1960s [4] now has been considered as the key figure who has initiated the focus on the embodiment as well as the recalcitrance that challenges the Cartesian divide, which can be described as a series of dichotomy such as body versus mind, culture versus nature, or subject versus object. Despites its philosophical root, the discussion on embodiment in HCI has remained a minor topic for decades because it more or less stands on the opposite side of the dominant scientific method derived from the Cartesian rationalism. It is the innovation of wearable devices based on embodied interaction, arguably, signals the increasing academic recognition of how the latest development of embodied interaction reshapes the landscape of the body. Since the first wearable device, FitBit Classic, launched in 2008, it has never failed to attract scholars' attention in humanities, especially in the fields more inclined to technology such as media studies.

For instance, in her study on the quantified self (QS), digital culture scholar Btihaj Ajana discusses both the benefits and risks of the quantification of the body [5]. Intensively referring to Foucault, she carefully examines how the development of measurement strategies, from charts to self-tracking devices, serves as the biometric power that regulates individuals' behavior, formulates new form of the sense of the self, namely, QS. In addition, the latest biosensing technology is able to separate the deviant from the normative. Ajana's work represents a number of scholars who follow the classic Foucauldian analysis. They tend to view the contemporary digital measuring invention as instrumental in expanding the discussion on embodied technology at the social, political, and legal level, triggering meaningful reflections on the power relation embedded in data. Diverse topics on how the wearable is used as a self-disciplinary practice in courtrooms, workplaces, gyms and hospitals [6–8] have proven that the merging of humanities and embodied interaction can be fruitful.

In addition to media studies, feminist scholars' long-term effort in problematizing the body has provided a myriad of scholarship in order to make way for refreshing the value of the gendered body, which has been historically undermined. For example, Donna Haraway, one of the most prominent feminist scholars working with the intersection of feminism, technology, and science, proposes the body ideal of the cyborg in the high-tech culture [9]. Along with the emphasis on the interconnected nature of the body crystalized in the notion of cyborg, she also makes great contribution in terms of uncovering the taken-for-granted-ness of science's neutral and objective position. For Haraway, the production of knowledge is always and already situated because the way in which one's point of the view is constructed has to do with one's specific position. The trick behind is that the point of view of the privileged, be it the white, the upper class, or the male, has been normalized as the universal vision. By proposing the notion of *situated knowledge*, Haraway reminds us of the interconnected nature at the heart of the body, which makes our perspective unique and never ceases to produce knowledge through the various interactions with the lively world.

Today the mystery of cyborg is no longer a sci-fi fantasy but has become our daily reality due to the rapid development of the wearable. With regard to the fields of HCI and embodied interactions in particular, one of the most valuable discussions comes from the reflection on the embodied technologies in relation to women's health, such as the menstrual tracking devices designed to support the datalization of period cycle [10, 11]. In the article entitled "Multiples Over Models" [12], Sarah E. Fox, Amanda Menking, Jordan Eschler, and Uba Backonja offers a thorough research on current quantification modes utilized in the wearable menstrual tracking devices, applications, and platforms, revealing the striking fact that even a device designed particularly for women can intentionally or unintentionally enforce new forms of oppression when the notions like "the user", "the normal", or "the fertile" are defined by the algorithmic model universally and uncritically. As implied in the witty title, what has been missing from the existing models of the quantification of menstruation is the lived body with its unique situated knowledge, or to say, the multiplicity. The data-driven model risks the danger of universalizing different situations within a rigid formula, and therefore serves as a new form of exclusion.

After a brief review of research on embodied interaction that successfully combine the humanistic concern, I would like to step back to McLuhan's metaphor of fish-water in the Introduction. If humanities can be the "anti-environment" that make a fish aware of the existence of water, it only because it calls for a closer look at what has been excluded from the normal, especially in the age of the intensification of datalization. The regulatory power of quantification can reach to an enormous scale because the latest development of embodied technologies is fully able to "disappear", using Weiser's words, yet keeps affecting both of the physical and psychological status of the human body.

3 Bringing the Body into the HCI Classroom in a Non-western Context

As a new media scholar and practitioner, I have been frequently asked the question by my students: what is exactly the newness of the new media? For me, the newness is neither about the perfection of algorithms nor the progressive development of the technology. By saying this, I do not intend to downplay the role of technology. Rather, as mentioned earlier in this article, my call for more rooms for humanistic concerns when teaching HCI is an effort to raise the students' awareness of the underrepresented realm of "the excluded". The newness shall not be merely treated as a rhetoric strategy often related to an optimistic or progressive implication. Instead, it refers to the desire for questioning the established norms that technically "disappear" like water for fish. For the ever-changing fields of HCI, it is impossible to merely focus on a singular discourse. An interwoven model of thinking that invites diverse discourses is what the article truly embraces.

When taking the course of "Wearable Technologies" as a PhD student, I was also encouraged by my chair to take the seminar of "History of the Body" in the department of History in the same semester. Discussing the body across different academic realms with people from varied educational background was a precious experience. After a close examination of the gendered body, the disabled body, the raced body, the non-human body, and the body in various medical traditions, I actually made an embodied interactive project inspired by the ideas of the non-human assemblage as the thing power discussed in the department of History, and further developed it into a installation work.

The way in which the body is perceived, interpreted, and experienced is deeply rooted in a culturally specific context, resulting in unique body practices, ethics, and aesthetics. This is one of the most valuable insights that I got from the experience of working with the intersection of HCI and humanities. When I started my teaching career in China, one thing struggling me most was how to keep the balance between the necessary datalization of the body and the urgency to maintain the cultural specificity. In addition, most of the classic readings in the current discourse of the body in humanities come from the Western writers, as shown in the previous section such as the Foucauldian mode of the biopolitical power and the feminist problematization of the body. Most of the students attending the section of "Innovative Interface Design" that I teach will graduate with a degree in engineering. They might be unfamiliar with the terminology and the theoretical thoughts, especially considering the fact that the quality of the Chinese translation varies dramatically.

Therefore, I was very cautious about selecting the reading materials. The introduction of *The Expressiveness of the Body and the Divergence of Greek and Chinese Medicine* by Shigehisa Kuriyama is one of the pieces preferred by the students. Kuriyama is a historian of medicine at Harvard University. The reading opens with an interesting visual contract between two figures of the bodies. Side by side, the left is the depiction of the body in traditional Chinese medicine dating back to 1341; the right is the picture of the anatomical musculature from classical Greek. With the sharp contrast between different representations of the human body in the medical traditions of China and Greece, Kuriyama then raises a profound question concerning the cultural specificity of the body: "How can perceptions of something as basic and intimate as the body differ so?" [13] (Fig. 1).

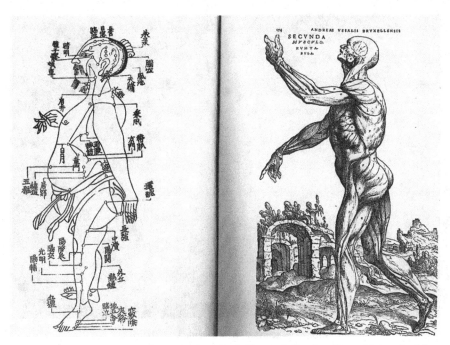

Fig. 1. The left: Hua Shou, Shisijing fahui, 1341, Fujikawa Collection, Kyoto University Library

The right: Fabrica Veslius, 1543, Wellcome Institute, London.

Scanned from *The Expressiveness of the Body and the Divergence of Greek and Chinese Medicine* pp. 10–11.

This become the guiding question that inspired students to rethink the body as a territory for multiple imaginary. In the following week, I asked them to find body practices that illustrated both the cultural specificity of China and the unique means of embodied interactions other than electronic devices. It was more than pleasant to watch what the students had found. Among a myriad of body practices, I was impressed by a student who recorded the process of the traditional beautifying method using cotton thread to erase the wrinkles ("开脸" in Chinese). Another student brought a small stick made by oxhorn, which was the instrument of his mother's daily routine of scraping. Scraping, "刮痧" in Chinese, is a traditional therapy that rubs one's body with a tiny massage tool (see Fig. 2). When asked to quickly brainstorm a sketch of an innovative tangible interface inspired by the body practice, this student came up with an interesting idea: creating an interactive scarping board that can help the user find the acupuncture point correctly.

Though the students still needed to take more courses on data analysis and programming in order to complete their proposals, the outcome of bringing the humanistic conversation of the body into the HCI classroom is promising. Instead of viewing the body merely as a passive object for abstracting data, a close look at the body helps open the students' imagination with a refreshing understanding of the body as a lively territory rich in cultural significance.

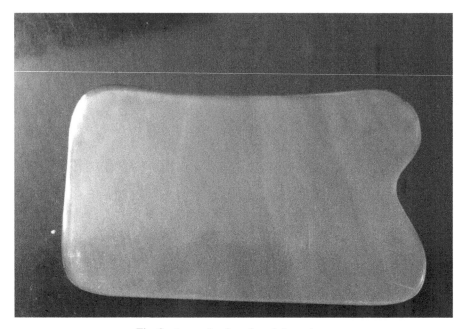

Fig. 2. A scarping board made by oxhorn

4 Conclusion

The body discussed in humanities serves as a useful complement for HCI training. A reflexive examination of the renewed and renewing body-technology relation in the age of the technological saturation is part of the larger picture of Humanistic HCI approaches. For an educator in HCI who wants to improve his or her pedagogical skills, a refreshing look at the body can be helpful.

At the end of the article, I summarize several thoughts on how to successfully make use of the body with humanistic concern in class.

First, the introduction part or the sections that allow students to do some creative exploration will be the proper time to bring the topic of the body into discussion.

Second, understanding the body in terms of multiplicity helps students expand their exploration of forms of interaction of way of knowing and experiencing. Encourage students to observe, record, and make use of the body practices in a specific cultural context.

Third, in order to help students make use of the cultural message inherent with specific body practices, ask them to describe their observations in detail.

Fourth, be careful with the choice of readings. Although the academic fields like philosophy or feminist studies have provided a remarkable number of cannon works concerning the body, students from a non-Western culture can be confused by the unfamiliar terms.

Last but not the least, use multiple vivid examples borrowed from gender studies, race studies, and disability studies to illustrate the notion: there is always the room for designing when we look at what has been excluded from the norms.

References

1. Bardzell, J., Bardzell, S.: Humanistic HCI. Interactions **23**(2), 20–29 (2016). https://doi.org/10.1145/2888576
2. Mark, W.: The Computer for the 21st Century, Scientific American, pp. 94–104, September 1991
3. Marshall, M., Fiore, Q.: War and Peace in the Global Village: An Inventory of Some of the Current Spastic Situations that Could be Eliminated by More Feedforward. McGraw-Hill, New York (1968)
4. Merleau-Ponty, M., Landes, D.: Phenomenology of Perception. Routledge, London and New York (2012)
5. Btihaj, A.: Governing Through Biometrics: The Biopolitics of Identity. Palgrave Macmillan (2013)
6. Crawford, K., Lingel, J., Karppi, T.: Our metrics, ourselves: a hundred years of self-tracking from the weight scale to the wrist wearable device. Eur. J. Cult. Stud. **18**, 479–496 (2015)
7. Rapp, A., Tirabeni, L.: Personal Informatics for sport: meaning, body, and social relations in amateur and elite athletes. ACM Trans. Comput.-Hum. Interact. **25**(3), 1–30 (2018)
8. Moore, P.: Tracking affective labour for agility in the quantified workplace. Body Soc. **24**(3), 39–67 (2018)
9. Haraway, D.: Simians, Cyborgs, and Women: The Reinvention of Nature. Routledge, New York (1991)
10. Almeida, T., Wood, G., Comber, R., Balaam, M.: Interactivity: looking at the vagina through Labella. In: Proceedings of the 2016 CHI Conference Extended Abstracts on Human Factors in Computing Systems (CHI EA 2016), pp. 3635–3638. ACM, New York (2016). https://doi.org/10.1145/2851581.2890261
11. Søndergaard, M.L.J., Hansen, L.K.: Intimate futures: staying with the trouble of digital personal assistants through design fiction. In: Proceedings of the 2018 Designing Interactive Systems Conference (DIS 2018), pp. 869–880. Association for Computing Machinery, New York (2018). https://doi.org/10.1145/3196709.3196766
12. Fox, S.E., Menking, A., Eschler, J., Backonja, U.: Multiples over models: interrogating the past and collectively reimagining the future of menstrual sensemaking. ACM Trans. Comput.-Hum. Interact **27**(4), 24 (2020). Article 22, https://doi.org/10.1145/3397178
13. Kuriyama, S.: The Expressiveness of the Body and the Divergence of Greek and Chinese Medicine. Zone Books, New York (1999)

A Novel Approach Combined with Therbligs and VACP Model to Evaluate the Workload During Simulated Maintenance Task

Bo Wang[1], Zhen Zhang[2], Changhua Jiang[1], Yan Zhao[1], Shaowen Ding[1], Fenggang Xu[1], and Jianwei Niu[2(✉)]

[1] Astronaut Center of China, Beijing 100094, China
[2] School of Mechanical Engineering, University of Science and Technology Beijing, Beijing 100083, China
niujw@ustb.edu.cn

Abstract. Automobile maintenance is a primary necessary means to prolong the service life of the automobile, to improve the reliability of automobile use, and to reduce the operating costs. The cumbersome and bulky automobile equipment usually requires maintenance, which need multiple servicemen collaborate with each other. Workload is an important factor affecting maintenance efficiency, when the workload of servicemen is too high, it will affect the physical and mental health of servicemen, work efficiency, maintenance reliability. Therefore, through the study of multi-person maintenance workload, the performance of maintenance can be evaluated.

We designed a comprehensive evaluation software for maintenance procedures based on C# programming and SQL server database, combined with objective and subjective metrics. Based on the VACP theory, this software is suitable for serial parallel multiplayer task analysis, that really combines analysis of therbligs with workload analysis, and displays the results dynamically. The software has passed usability test that is established through focus group discussion and heuristic evaluation. At present, the proposed software package has a friendly interface design, strong guidance, and good usability of data analysis. It's anticipated to expand this convenient workload evaluation package combined with therbligs and VACP model to evaluate the maintenance work in automobile maintenance in future.

Keywords: Automobile maintenance · Video segmentation · Therbligs · VACP model · Workload assessment

1 Background

Automobile maintenance is one of the main means and necessary method to extend the service life, improve the reliability of use and reduce the operating cost [1]. When the Automobile is repaired, it usually requires the collaboration of multiple servicemen. In order to improve maintenance efficiency, the researchers and engineers are using the continuous upgrading of the automobile hardware equipment availability, while ignoring the

© Springer Nature Switzerland AG 2021
P.-L. P. Rau (Ed.): HCII 2021, LNCS 12771, pp. 164–173, 2021.
https://doi.org/10.1007/978-3-030-77074-7_13

maintenance performance of individual servicemen, work load assessment and design of reasonable maintenance process tasks. Therefore, it is inevitable to cause some servicemen to have heavy tasks, high workload, while other servicemen are idle, therefore the overall maintenance efficiency is reduced. From ergonomics viewpoint, the servicemen are an important factor affecting the maintenance of the Automobile. On the one hand, servicemen can extend the life of the automobile, on the other hand, improving the reliability of the Automobile can protect the lives of servicemen. Therefore, it is of great significance to analyze the automobile maintenance process and assign the maintenance tasks rationally.

Servicemen workload refers to the cost paid by the operator to achieve a certain level of performance [2]. Workload is an important factor affecting maintenance efficiency. When the servicemen workload is too high, it will affect the physical and mental health, work efficiency, and maintenance reliability of the servicemen. Therefore, by studying the multi-person maintenance workload, the maintenance performance can be evaluated. In addition, based on the evaluation results of the multi-person maintenance workload, the maintenance coordination process can be optimized, the maintenance task distribution among the servicemen can be balanced, the maintenance pressure of the servicemen can be reduced, the reliability of the maintenance and the stability of the system can be improved, and the success of the maintenance can be increased rate.

Motivation analysis was invented by the Gilbreths of the United States. An action is composed of one or several basic therbligs [3]. The therbligs are used to study the motion, delete unnecessary or invalid therbligs, simplify the operation process, and formulate standard operating procedures. This research has been widely used in the scientific research and engineering work of traditional basic industrial engineering, but so far most of the combination of dynamic element analysis and task load analysis has been realized at the theoretical level, and the results of the combination of the two have not been displayed.

At present, the workload evaluation model of the existing workload evaluation method, for example, Miller proposed a dynamic evaluation timeline analysis method of serial tasks [4], and Hamilton and Bierbaum proposed a parallel multi-task workload task analysis (Task Analysis Workload, TAWL) evaluation model, this are single-dimensional workload evaluation indicators that cannot accurately evaluate the workload and describe the characteristics of the workload [5]. Later Tian ShuJie et al. proposed a workload evaluation model suitable for serial and parallel tasks and based on a dynamic time window based on time-multiple resource occupancy [6]. However, the model sets the same weight for the four VACP channels, which does not conform to the actual situation. To this end, this paper proposes a workload evaluation index that combines subjective and objective dimensions of resource demand occupancy rate and occupancy time, and establishes a workload evaluation model suitable for serial parallel tasks and four channels with different weights.

This paper has developed a complete set of tools and methods. This software is based on VACP theory, which truly combines dynamic element analysis with workload analysis and displays the results dynamically. First, it supports users to disassemble the video into therbligs and perform analysis of therbligs, and then directly use the GUI-VACP grade score to facilitate the user to perform serial and parallel multi-person task analysis.

The later software automatically generates the workload trend chart, which is of great significance for helping to analyze complex tasks such as automobile maintenance.

2 Method

Therbligs is the basic movement required to complete a job. This software adopts the analysis of therbligs principle of Gilbreth and his wife. The Gilbreths analyzed the movements that require the use of both hands, and the hand movements are divided into 17 basic therbligs. Later, the American Society of Mechanical Engineers added the agent "Find", denoted by F, so that there are 18 basic therbligs of agent analysis. These therbligs can be divided into the following three types.

(1) Necessary therbligs, including: reach grasp, move, assemble, disassemble, use, release, inspect.
(2) Auxiliary therbligs, including: search, find, select, plan, pre-position.
(3) Invalid therbligs, including: hold, unavoidable delay, rest and avoidable delay.

Gilbreth believes that work is composed of certain basic action elements combined in different ways and in different orders. Analysis of therbligs, on the one hand, can help find invalid or wasteful actions of workers, reduce the waste of power and time, simplify operation methods, reduce fatigue, and establish standard operation methods. On the other hand, it can improve work efficiency, delete invalid therbligs, and reduce assistance therbligs. Thereby, Therbligs method can simplify the operation actions and help to find the most reasonable sequence of therbligs for a certain operation.

The work load of the servicemen in the automobile is composed of mental load and physical load. The mental load is related to human perception and cognitive activities. It is composed of visual load, auditory load and cognitive load. Physical load includes psychomotor load related. In this paper, occupancy time and VACP occupancy rate are used as workload evaluation indicators. VACP occupancy rate refers to the degree of resource occupancy on each channel of VACP per unit time, and occupancy time refers to the occupancy time of each agent on VACP channel [7].

The workload is expressed as follows:

$$W = t\xi = t\lambda / T$$

Where: W is the workload; t is the occupancy time; ξ is the VACP occupancy rate; λ is the resource occupancy degree; T is the duration of the dynamic time window. The degree of resource occupancy is observed by the user combined with the attribute description of each therbligs, and all the workloads of the current therbligs are manually evaluated, which can be represented by 0(evaluation values:0), 1–2(evaluation values:0.42), 3–5(evaluation values:0.83), and 6–7(evaluation values:1.25). The specific evaluation is shown in the Table 1 below [8].

According to the "red line" theory in the workload, when the workload reaches 80%, human performance begins to decline, and the actual workload has reached the highest value of 100%. Therefore, after deduction, the 0-point rating is assigned a value of 0, 1–2 points are assigned a value of 0.42, 3–5 points are assigned a value of 0.83, and 6–7 points are assigned a value of 1.25.

Table. 1. Description of VACP model

Load	Rating	Description
V (Visual load)	0	No visual load
	1–2	Visual inspection
	3–5	Visual positioning, identification
	6–7	Visual tracking, reading
A (Auditory load)	0	No auditory load
	1–2	Determine the direction of the sound
	3–5	Determine the sound position
	6–7	Interpret the information in the sound
C (Cognitive load)	0	No cognitive load
	1–2	Unconscious, choose
	3–5	Unilateral assessment
	6–7	Multiple judgments, calculations
P (Physical load)	0	No physical load
	1–2	Discrete behavior
	3–5	Manual control
	6–7	Sequence manual manipulation

3   Results

3.1   Analysis of Therbligs

This function is to analyze the difference in the data of the elements of the video. The user imports the video files into the system, and then manually mark each element video according to the played video, and edit the name of each element. The software back-end enter the class attributes into the database. In the later stage, descriptive statistics are used for the acquisition of the video data according to the sample data volume, and the normal test and the median test are carried out independently. Finally, the measured values of time of each element are obtained, the software transfer statistics results and relevant parameters to the corresponding therbligs database.

The software function enables that the user imports the maintenance video file into the software, cuts out the video of each therbligs according to the played video, and then calculates the mean, median, and normal test of the background video data of the software, and finally transmits the data result to the database.

This is the main interface of the system as shown in the Fig. 1.

Fig. 1. System main interface

Users can click "video analysis" button to open the file and import the video into the system, as shown in the Fig. 2.

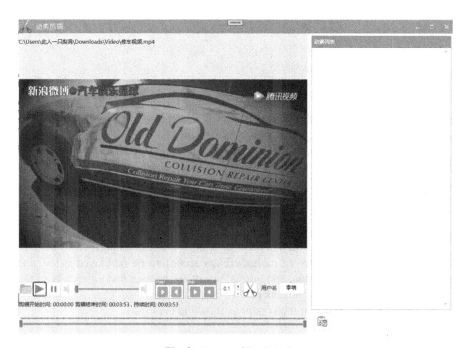

Fig. 2. Import video interface

The user clicks ✂ to cut the video clip of the therbligs, and clicks ✎ , 🗑 to modify or delete the therbligs information, as shown in the Fig. 3.

Fig. 3. Motile cut interface

The user clicks 📋 to analyze, and the system will perform statistics on the mean, median, and data interval of the data later, as shown in the Fig. 4.

动素名称	均值（秒）	中位数（秒）	数据区间（秒）	P-VALUE	置信区间
插	4.000	4.000	4.000 -4.000	0	[0, 0]
定位	3.000	3.000	3.000 -3.000	0	[0, 0]
拧	3.000	3.000	3.000 -3.000	0	[0, 0]

Fig. 4. Motive analysis interface

3.2 VACP Model

Based on the digital therbligs video, the user manually scores the current therbligs visual load, auditory load, cognitive load, and physical load in four levels: no, low, medium, and high. The software will first build a single load model based on the mathematical model constructed in this article, and create a multi-dimensional workload evaluation model. The system generates dynamic charts later. In order to measure the dynamic changes of the servicemen's workload during the maintenance process, it can later help design the maintenance process. The later software automatically generates a single-person comprehensive workload line chart to vividly display the workload change process in the maintenance process, or generates a multi-person maintenance comprehensive workload line chart to facilitate multi-person load trend analysis.

The function is that the user manually scores the visual load, auditory load, cognitive load, and physical load of the current motility element based on digital maintenance tasks. The software backend outputs a single-person load chart model and a multi-person load chart based on the VACP theory to measure the dynamic changes of the workload during the maintenance process.

The user edits the digital workflow, as shown in Fig. 5.

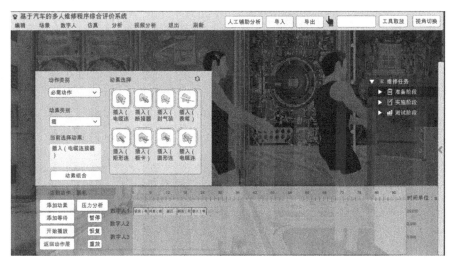

Fig. 5. Action add interface

The user clicks "pressure analysis" to pop up the VACP scoring interface, which will display all the motifs contained in the current actions of the digital human. Users can score "visual load", "auditory load", "cognitive load" and "physical load" respectively. The grading interface is shown in Fig. 6.

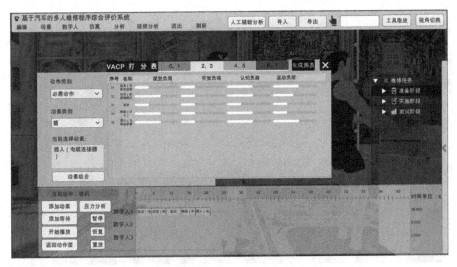

Fig. 6. VACP scoring interface

On this basis, the user clicks "Generate Chart" to generate load evaluation results based on the current action such as "visual load", "auditory load", "cognitive load" and "physical load". The results are shown in Fig. 7.

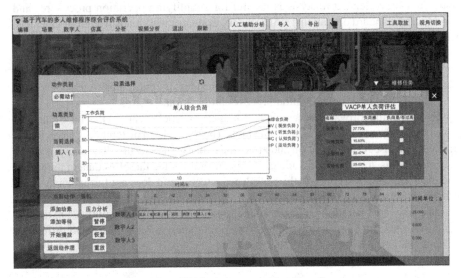

Fig. 7. VACP single integrated load interface

Based on the analysis of the comprehensive load of a single person, it supports the evaluation of the comprehensive load of multiple persons. The user can edit the maintenance process of multiple maintenance personnel according to the above process. The result is shown in Fig. 8.

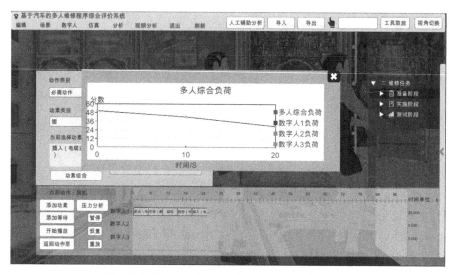

Fig. 8. VACP multi-person integrated load interface

4   Discussion and Conclusion

Workload research is helpful for the optimization of multi-person maintenance procedures; dynamic element analysis is helpful for simplifying operation procedures and formulating standard operations; VACP theory evaluates workloads from four channels, which is more scientific and comprehensive. This article proposes a method to combine the two organically. This method has been successfully applied in complex tasks such as automobile maintenance, and is expected to be extended to load analysis of other complex operations, greatly improving efficiency. Besides, it contributes to team selection and training and team collaboration process optimization.

However, this software has two shortcomings: this article assumes that the 4 channels have four levels: none, low, medium, and high. The model level difference is large, which may easily cause result deviation and affect the experimental results. The analysis of the therbligs in this article only calculates the mean value of the therbligs data in the video, and the normal analysis obtains the confidence interval, and does not further analyze the difference between the therbligs.

The software operation is simple and clear, and the results are intuitive. On the one hand, the software successfully helped realize the analysis of world video difference through statistical analysis. On the other hand, the VACP was scored manually, and dynamic charts were generated, which successfully realized multi-dimensional workload assessment.

Acknowledgements. This research is supported by Equipment Advanced Research Project (Grant NO. 61400040103), the Open Funding Project of National Key Laboratory of Human Factors Engineering (Grant NO. 6142222190309), the Foundation of National Key Laboratory of Human Factors Engineering (SYFD1700F1801, SYFD180051801), China.

References

1. Keiichi, W., Masakazu, N.: Estimation of skill degree for casting worker by using Therblig analysis. In: The Proceedings of Conference of Kanto Branch , vol. 2003, no. 9 (2003)
2. Salas, E., Dickinson, T.L., Converse, S.A., et al.: Toward an understanding of team performance and training. In: Swezey, R.W., Salas, E.: Teams: their training and performance. Westport, CT, no. 3, pp. 3–29 Ablex Publishing, US (1992)
3. Chen, S.G., Wang, C.H., Chen, X.P., et al.: Study on changes of human performance capabilities in long-duration spaceflight. Space Med. Med. Eng. **28**(1), 1–10 (2015). (in Chinese)
4. Miller, K.M.: Timeline analysis program (TLA-1), final report: NASA-CR-1 44942. National Aeronautics and Space Administration, Langley Research Center, Hampton (1976)
5. Hamilton, D., Bierbaum, C.: Task analysis/workload (TAWL): a methodology for predicting operator workload. Proc. Hum. Factors Ergon. Soc. Ann. Meeting **34**(16), 1117–1121 (1990)
6. Tian, S.J., Wang, B., Wang, L., et al.: Workload assessment model based on time-multiple resource occupation. J. Beijing Univ. Aeronaut. Astronaut. **12**, 2497–2504 (2017). https://doi.org/10.13700/j.bh.1001-5965.2016.0896
7. Nguyen l,Y., Green, P., et al.: Visual, auditory, cognitive, and psychomotor demands of real in-vehicle tasks: UMT R I2006–20. University of Michigan Transportation Research Institute, Ann Arbor, pp. 40–45 (2006)
8. Hart, S.G., Staveland, L.E.: Development of NASA-TLX (Task Load Index): results of empirical and theoretical research. Adv. Psychol. **52**(6), 139–183 (1988)

Shift in Computation – Tangible to Intangible

Yufan Xie[✉]

University of Southern California, Los Angeles, CA 90007, USA
yufanxie@usc.edu

Abstract. The design industry is dominated by ocularcentric culture – visualization and formal generation are generally regarded as the goal of production, focusing on transform intangible message into tangible visual forms. When computational tools intervene the process, we habitually use them as shortcut to complex forms, ignoring other potentials. There should be a shift in computation, realizing multi-sense in space for human. The shift is a transition from ocularcentric to all-sense culture. It challenges the singular process of "visualization" and "generative form", by integrating two-way transformation between tangible message and intangible message. We should re-consider the process and result of computation – how do we transform tangible visual system into a source of intangible messages serving all senses? Meanwhile, we are in an era of interactivity. As emerging interactive mediums such as game engine, VR/AR devices extend the border of interaction, allowing real-time transmission of complex behavior and spatial information. The shift requires reflection on design thinking and our roles. When computation relates to space, the shift of computation is also the shift of space, in which architect will be a key role of constructing complex spatial system and human-space interaction. The shift addresses interdisciplinary and multi-sensory integration, aiming at diversity of spatial experiences and value of individuals, as thus inclusive and humanistic space can be built, to serve everyone.

Keywords: Architecture · Ocularcentrism · Multi-sensory · Interdisciplinary

1 Introduction

In recent years, computation and digital tools have played an important role in architecture and design, being widely applied in parametric design. How intangible factors (load, heat etc.) can be retrieved/mapped/visualized, then transformed into a solution, and finally being constructed in an efficient way - are keys of the workflow. In the past decade, many researches have shown the power of algorithms in formal generation (using algorithm to generate ideal forms) and optimization (using algorithm to achieve ideal performance). Since 2017, I've been researching on generative systems and explore aesthetics of visual form. While exploring related topics, an idea of "non-visual design" emerged in my mind, which was defined as the opposite process of "visualization". As a mixed-background researcher involving art, architecture and electronic music, it's been on my mind for years that – is it enough to produce "looking good" but generally repetitive visual form? Is visual form or visualization the only result of computation? Can computation and generative design serve other senses?

© Springer Nature Switzerland AG 2021
P.-L. P. Rau (Ed.): HCII 2021, LNCS 12771, pp. 174–184, 2021.
https://doi.org/10.1007/978-3-030-77074-7_14

The article will introduce my early exploration on generative form and interactive acoustic space, to present a transition from tangible to intangible and challenge the predominant ocularcentrism [1] in design. In this shift, "Tangible" refers to visual forms. "Intangible" refers to messages of non-visual and untouchable medium - for instance, sound. Such shift focuses on both visual and non-visual fields, as defining space as a bridge between different senses and groups. I believe it is a decisive and inevitable trend of being "intangible", as interactive media and augmented reality will allow dynamic and intangible mediums to come into play. The shift is exactly what Marshall McLuhan argued as "a change over from the eye to the ear" [2]. It will re-define spatial system as a future interface, in which architect will be a key role in constructing mechanism to bridge behavior and multiple messages.

2   Tangible Computation

Algorithms, such as Differential Growth [3], Reaction Diffusion [4], and other complex systems are generally discussed in the field of parametric design and digital fabrication, and used as prototype of spatial generation. Related researches in generative algorithms, such as *"Cellular Growth"* by Andy Lomas [5] and *"Growth Forms"* by George W. Hart [6] have shown us how complex forms of visual aesthetics can be generated through computer algorithms, which is capable for complex form and simulations. They have taken impact on design culture and architecture industry for a long period of time. Their ideas also triggered my earlier studies. At this stage, my researches focused on visualization and generative forms. Intangible factors are decisive for outcome – different parameters will lead to drastically different visual forms.

2.1   Experiments

Inspired by precedents, my project *"Self-Organizing System of Basic Behaviors"* in 2017 [7], explored generative form as a particle-based system which combined 4 basic behavior on each single particle. The system is built in Grasshopper [8], utilizing Anemone component [9] for iteration. From a single system, thousands of results can be generated by changing factors of 4 basic behaviors (see Fig. 1) - birth and death of child is based on Voronoi [10] density of mother (the area of each single Voronoi cell), whereas attracting and repelling functions are based on vector field [11] generated by the least and most dense particles. The system is self-adaptive and environment-responsive, different factors and environments will produce drastically different results (see Fig. 2). Prompted by its visual compelling results and formal potentials, I started experiments on 3d printing and urban forms with knowledges on such generative systems.

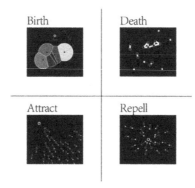

Fig. 1. Behaviors – birth and death, attract and repell.

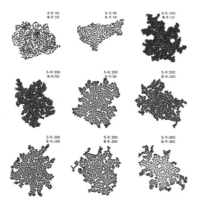

Fig. 2. Self-organizing results. By changing key factors – Search Range (S-R) and Birth Range (B-R), various forms can be generated.

In 2019, the collaborative project "*The Unknown Cities*" by Yan Wu, Yu Chen and me [12] (exhibited at 2019 Bi-City Biennale of Urbanism/Architecture [13]) is an experiment on generation of future urban form. It experimented on computational form as urban system and 3d printing. The project consists of 4 3d printed objects, which were generated with grasshopper definition (see Fig. 3). 4 cities were designed based on sci-fi narrative by Yan Wu, proposing 4 imaginary form of future urban system under extreme climate change. Computation functions as a simulation of urban system.

It's clear to see the visual priority in these early researches. Meanwhile I began to rethink the visual culture. Though it's common to see that one single algorithm can leads to various visual forms, actually most designers are doing similar and predictable researches on computational forms. For instance, the Voronoi system has been produced as various types of products such as housing, façade, furniture and so on...... which have visual similarity. Basically because 1. algorithms are used to generate a specific category of shape, and 2. algorithm were considered as a basis of computational design. Then if we put concepts aside, projects using the same algorithm prototype will present a highly similar visual appearance in most cases. Just in speaking of results, I think it's worth

Fig. 3. 3D Printing of the Unknown Cities. Vector Field is used in "Floating City" (top-left), algorithm "Diffusion Limited Aggregation" [14] is used in "Pipeline City" (bottom-left), Differential Growth is used in "Hyperplastic City" (bottom-right) and "Underwater City" (top-right).

thinking about what other form of result can we produce, even using the same algorithm? Looking for other potentials of these geometries and computational workflow became one of my concerns.

Due to my personal focus on interactive media and music narratives, I started researches on topics of acoustic space. I define the later researches, as "intangible computation".

3   Intangible Computation

The intangible computation argues the goal of computational design, while in most conditions, it serves mere visual form. The "intangible" refers to non-visual medium and senses, for instance – sound, which is intangible and plastic. Such computation challenges the subjectivity of visual thinking, different from ocularcentric culture which regard all other senses as secondary to "visualization". For example, generally when designer deal with music related narratives, they regard music as a background track (such as film production), or a source of "visualization" (audio-visual performance etc.) It's common to see that, in architecture, the translation between music or dance and a static shape through a notation or drawing. Even when new technology came into play, most designers still focus on how music signals or feelings can be visualized as an image, a sculpture or a literal architecture. In other words, designers tend to "see the music" more than "hearing the space".

By contrast, intangible computation addresses the multi-sense process and result of design by, rethinking spatial system and integrating concepts from other industries, since our industry hardly reflect senses and design is seen as visual production by far. It is a shift on design culture, of how we use technologies, who we serve, and how we define interdisciplinarity.

3.1 Experiments

In my later researches, sound narrative and interactive acoustic space are core topics. I setup several rules to link spatial interaction with sound synthesis and real-time mixing in electronic music. Space and interaction are placed as the core mechanism, to link computation with sound. There's been several precedents on space-driven acoustic system, such as Theremin (1920) by Leon Theremin [15] which use motion of hands to modulate parameters of sound and Photophore Synth (2015) developed by Taika System [16], using spatial information of multi-agent system [17] to modulate parameters of synthesizer [18]. Such systems both utilize spatial behavior as key factors to control sound, whereas they mainly focus on synthesis of a single sound, which didn't involve the architecture of music, or macro mixing mechanism of sound. In my experiments, to be more precise on human scale and 3d space, the interaction and experience through space become essential sources of music control and mixing.

Spatial Scene Structure. As an early experiment on acoustic space, *"GAP+ Architecture of Sound"* in 2019 [19] reversed the traditional message-to-space workflow (visualization) into space-to-message generation. Inspired by non-linear electronic music performance of Live Set [20], the project fully developed a sound mixing mechanism as an interactive acoustic space, transforming architectural definitions into a framework of non-linear acoustic experience. As user moves through architectural grids, different groups of sound are triggered and mixed in real-time (see Fig. 4). The project argues that - is solid, or visual form the only goal of computational system? Can music narrative be spatialized interaction?

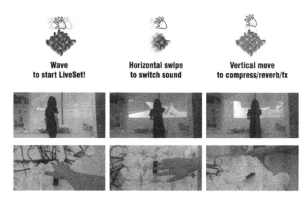

Fig. 4. GAP+ interaction with Leap Motion [21], Grasshopper and MAX/MSP [22] definition. The project constructed an interactive system for live performance based on location and gesture.

On sound narrative side, the project is an experiment using tangible form as definition of music scene structure. It utilizes space zones generated by binary tree (see Fig. 5), as a real-time structure for live-set control. Each zone equals to a scene group of sound in live set, containing several synced sound loops. Based on spatial locations, music scenes can be switched and effect of soundscape can be changed. With the flexibility of virtual space and sound processing, the mechanism is similar to a walkthrough with various control

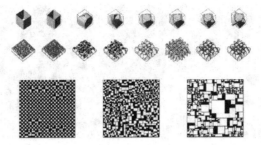

Fig. 5. Binary Tree system and GAP+ generation.

over music scenes and effects. As an architectural project, it is imagined as a digital museum and shared community in a solid form in reality. The mechanism functions as daily walking experiences in acoustic space by sound attenuation and filtering between solid walls. During a walk, sounds are played interior and passed through openings on exterior walls, finally mixed and heard by a specific exterior location. Furthermore, for music industry, it reveals a spatial narrative controlled by audiences.

This project is a primary exploration on intangible computation, using sound to show great potential of spatial narrative for non-visual media. It also addressed the key role of time – it is real-time media that allows other senses to challenge visual culture. Then, in the second stage of intangible computation, I decided to utilize game engine to explore acoustic space as a complete virtual system of new music narrative.

Spatial Modulation. As an extension, a work-in-progress project of mine has been launched since 2020 September. This project argues how architectural definition of fields can be used as source of modulation [23], in which human behavior and location can take effect on our surroundings in virtual space. As shown in Fig. 6, parameters mapped upon space are read through location of users, meanwhile through u, v coordinates, parameters can be edited and retrieved in real time. As a result, the repetitive pattern of field, or in another word, spatial modulation will generate patterns of experience. The pattern in time becomes a pattern of subjective experience, which inherently links space and sound, when user walk through and sample from the field. How fast we move, and how we move or behave, is decisive to the pattern of modulation and experience. In speaking of music, such variation on patterns of modulation is exactly where electronic music is built from. Generally in an electronic track or performance, loops of clips were arranged on timeline, and shaped by modulation from multiple sources - such as LFO (Low Frequency Oscillator) [24] and Mod Wheel [25]. The shaped sounds, present a varied pattern over loops. Such feature is commonly used to build energy and narrative in various genres and lives. The project first proposes spatialized and interactive modulation with complex mapped control in an architectural way, reflecting music narrative in the era of augmented reality.

Meanwhile, the pattern presents a variation in time, depending on the direction and speed of movement. With repetitive, predictable pattern of field, people can understand the scale of space and locate themselves. Such mechanism reaveals an augmented reality serving acoustic world, people can experience the space even without visual content.

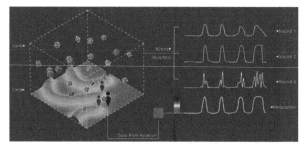

Fig. 6. User behavior on field - spatial interaction can be recorded as via a map, then output as modulation source. Modulation is sampled based on user's location in space.

Fig. 7. Early test on field reading and writing. Data in field being sampled according to *x, y* of user, to modulate sound.

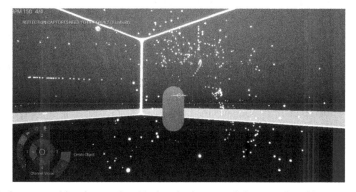

Fig. 8. Early test on object interaction. Each point is a sound element placed in space, which can be triggered and moved by user. Their sound attributes are also mapped along x, y, z coordinates of space, as user move them around, the sound changes and also indicates a specific location and scale.

Unreal Engine [26] is used as research tool (see Fig. 7). The gaming environment inherently provide human scale interaction and spatialized sound. In this system, space become a key factor of intangible computation. There are some other features testing on the generation aspect of sound, such as collision (see Fig. 8). The collision event simulate the reality but take more control over the virtual sound synthesis. It also becomes a random factor in sound narrative. At the same time, as sounds are spatialized by attenuation in Unreal Engine, the virtual space can be more detailed and precise.

4   The Shift

Shifting from tangible to intangible in computation is a decisive and inevitable transition, which will be largely enhanced by various interactivity in the coming era of augmented reality. The shift will expand the border of interaction from visual to other senses. For example, sound is regarded as minor sense for visual people, but it is also an important way of understanding the world of visual impairment, which is generally ignored in this visually designed world. In fact, the urban environment we are living in, which seems to be convenient, are assumed to be used by sighted people, and designed in a sighted mode of thinking. I scheduled an interview with human echolocation expert, Daniel Kish who "lost his sight in infancy and taught himself to echolocate, reveals what it's like to see with sound" [27]. Our conversation also argued the effect of visual culture, in which the ignorance of acoustic environment will lead to inconvenience for blind community.

"......buildings are visually marked up, you got arrows, you got signs, they are created for sighted people in mind. They are created for sighted people to find their way around, and they are never ever created to make blind people to find their way around......those kinds of places will be very tricky for a blind person to sort out, they are very difficult to navigate......" said Daniel Kish in interview "*Listener*" [28].

Under such condition, making our environment acoustic-friendly (e.g. sound navigation) and serve for non-visual people will become important topics. Especially when we discuss interactive architecture and smart city, the intangible space should be taken as. Essentially, virtual technologies will allow this shift, though we have to admit limitations of physical rules, such as load, aging, assembly are hindering the way towards a highly-intelligent environment in material world. In a virtual environment, the intelligence will be largely enhanced, and intangible computation will have influence on everybody.

Precisely, as technologies are developing, the shift can be realized through three conceptual aspects (see Fig. 9):

1. First, the shift redefines the role of designer and user. Architects should be redefined as a framework producer, which is more or less like a game designer. Meanwhile user get involved, as a more powerful role in the final result. How user can participate to be more autonomous in given environment is much more important.
2. Then the shift happens on production and experience. In the future, we produce interactive framework and as designer we release control over the experience. Then user will intervene the process as a variable, by which the content requires, and will no longer be static.
3. Finally, the shift happens on medium and content. It's based on the rise of virtual world, then dynamic and intangible mediums can be much more operable in space.

With diversified medium, the content can serve all senses, equally considering our sensory system as diversified ways of experiencing space, rather than just visual aesthetic.

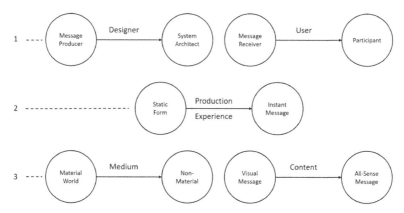

Fig. 9. Three aspects of how the shift can be realized through conceptual changes.

The first and the second shift have been accepted as a common sense in interaction related industries, but have not been widely recognized by architects when architecture field is still dominated by static visual culture. The fact is, only with realizing two shifts above, then third shift can happen. The third shift is a goal for all design and art related industries. I think it is valuable to jump out of our scope of visual culture and re-consider our practice as multi-sense experience. Especially the singular process of "visualization" is no longer enough or suitable in the context of interactivity. Besides, I would presume architect as a dualistic role in spatial computation to intermediate user and space, on scale, behavior and framework. In speaking of interactive, or intelligent architecture, the richness and diversity of message will be a concern of mental life and humanity.

As a researcher in electronic music and acoustic space, I believe in this shift, architect will be involved as innovator of spatial interface for other industries. In the history of music, innovations of interface – from notation to live looping [29], then to DAW (Digital Audio Workstation) [30] – brought drastic changes on music narrative, performance and experience. As virtual technologies are fast developing, we should ask ourselves - what is the next generation of interface? Or in another word – when everything is becoming spatial, what is a spatial narrative for ear, nose or body? What should spatial interface be like? How do we synthesis a sound? How do we perform? How do we construct music?

5   Conclusion

After experiments on interactive acoustic space, it's been validated that it is feasible to utilize computation for intangible computation such as acoustic space. The next plan is developing the spatial modulation project. This project will develop architectural space as an interface and acoustic system, challenging the inherent ocularcentrism in field of design, architecture and art. The research will give ideas in the forthcoming augmented age, though it has shortages based on architectural-defined discourse and it hasn't figured out a complete framework for the narrative yet. In later plans, through collaboration with professional sound designer and musicians for advice, I will further examine this framework.

The shift in computation is reflecting our monolithic visual culture. Visual is not the only destination, it is just one choice of our senses. Criticizing the ocularcentric culture doesn't refer to anti-visual, it basically reveals multi-sensory and diversified future in a broader sense. It reflects on my own visual researches, and on our overall mindset.

The shift in computation is inevitable, when interactivity enhance experiences, and the diversity of senses will sooner or later, become a topic concerning all user groups. Comparing to limited interaction in material world, the virtual space is more intelligent and flexible. Bypassing physical restrictions, computation will serve better for human sense in virtual world. "Intangible" world will connect different fields and groups in the coming years.

The shift in computation is moreover a shift in mind. Most people are born visually, growth visually and trained visually, which lead to our ocularcentric mind and its products. Inherently, we think visually, and construct our world in such a mindset, though space is just a data, which can be understood through different messages, by different user groups. Maybe visual people would not care the acoustic world, but it doesn't mean nobody cares. The subconscious ignorance over "minor senses" exactly embodies our "blindness". We should re-consider, to make a shift in our visual mind, a shift in our ocularcentric culture, a shift in industrial standards. We need to ask ourselves - Who do we serve? What do we design? What should we consider?

Acknowledgements. This article special thanks to Professor Zhiyong Fu for invitation. Special thanks to the mentor of project "GAP+ Architecture of Sound" - Professor Yufang Zhou in Central Academy of Fine Arts in China. Thanks to Professor Yan Wu and artist Yu Chen as collaborator of "The Unknown Cities". Thanks to Professor Lisa Little in University of Southern California as mentor of the in-progress "Spatial Modulation" research. Special thanks to Daniel Kish from World Access for the Blind organization for the interview "Listeners".

References

1. Ocularcentrism. https://www.oxfordreference.com/view/10.1093/oi/authority.201108031 00245338. Accessed 28 Jan 2021
2. Marshall McLuhan Speak l Living in an Acoustic World. https://marshallmcluhanspeaks.com/lecture/1970-living-in-an-acoustic-world/index.html. Accessed 28 Jan 2021
3. Differential Growth. https://dictionary.apa.org/differential-growth. Accessed 28 Jan 2021
4. Reaction-Diffusion System. https://en.wikipedia.org/wiki/Reaction%E2%80%93diffusion_system. Accessed 28 Jan 2021
5. Lomas, A.: Cellular forms: an artistic exploration of morphogenesis. In: ACM SIGGRAPH 2014 Studio. ACM, New York (2014)
6. Hart, W.G.: Growth forms. In: Proceedings of Bridges 2009 (2009)
7. Cellular Growth: Self-Organizing of Basic Behaviours. https://www.youtube.com/watch?v=r3US4T3gfgw. Accessed 28 Jan 2021

8. Grasshopper-Algorithmic Modelling for Rhino. https://www.grasshopper3d.com/. Accessed 28 Jan 2021
9. Anemone. https://www.food4rhino.com/app/anemone. Accessed 28 Jan 2021
10. Voronoi Diagram. https://en.wikipedia.org/wiki/Voronoi_diagram. Accessed 28 Jan 2021
11. Vector Field. https://en.wikipedia.org/wiki/Vector_field. Accessed 28 Jan 2021
12. The Unknown Cities. https://uvnlab.com/the-unknown-cities-en/. Accessed 28 Jan 2021
13. Shenzhen-Hongkong Bi-City Biennale of Urbanism/Architecture. https://www.szhkbiennale.org.cn/. Accessed 28 Jan 2021
14. Diffusion Limited Aggregation. https://en.wikipedia.org/wiki/Diffusion-limited_aggregation. Accessed 28 Jan 2021
15. Theremin. https://en.wikipedia.org/wiki/Theremin. Accessed 28 Jan 2021
16. Photophore. https://www.taikasystems.com/photophore. Accessed 28 Jan 2021
17. Multi-Agent System. https://en.wikipedia.org/wiki/Multi-agent_system. Accessed 28 Jan 2021
18. Synthesizer. https://en.wikipedia.org/wiki/Synthesizer. Accessed 28 Jan 2021
19. GAP+. https://uvnlab.com/gap-en/. Accessed 28 Jan 2021
20. Live Set. https://en.wikipedia.org/wiki/Live-set. Accessed 28 Jan 2021
21. Leap Motion Controller. https://www.ultraleap.com/product/leap-motion-controller/. Accessed 28 Jan 2021
22. MAX/MSP. https://cycling74.com/products/max. Accessed 28 Jan 2021
23. Modulation. https://en.wikipedia.org/wiki/Modulation. Accessed 28 Jan 2021
24. Low-Frequency Oscillation. https://en.wikipedia.org/wiki/Low-frequency_oscillation. Accessed 28 Jan 2021
25. Mod Wheel. https://electronicmusic.fandom.com/wiki/Mod_wheel. Accessed 28 Jan 2021
26. Unreal Engine. https://www.unrealengine.com/. Accessed 28 Jan 2021
27. Kish, D.: Human echolocation: How to "see" like a bat, vol. 202, no. 2703, pp. 31–33, 8 April 2009. RELX Group, U.S. (2009)
28. Listener - We Are Heading Towards an Acoustic Future. https://youtu.be/q-8fFQoapvk. Accessed 28 Jan 2021
29. Live Looping. https://en.wikipedia.org/wiki/Live_looping. Accessed 28 Jan 2021
30. Digital Audio Workstation. https://en.wikipedia.org/wiki/Digital_audio_workstation. Accessed 28 Jan 2021

Future Footprint: A Future Signal-Driven Design Ideation Tool

Lin Zhu and Zhiyong Fu[✉]

Tsinghua University, Beijing, China
zhu-l20@mails.tsinghua.edu.cn, fuzhiyong@tsinghua.edu.cn

Abstract. The processes of design thinking usually begin with the divergence of thinking to seek inspiration and opportunity points. For the designers, they focus on the design theme, combine their existing experience, and capture creative signals through data collection, brainstorming, and other design methods. Traditional design tools rarely adapt to the multiplicity and uncertainty of design objects as they move into the long-term future. There arises a need in the teaching of design, that is to accurately identify opportunities for innovation amid a multitude of potential signals. And guide designers to reply the uncertain future, conceive the future in different ways, stimulate their implicit memory to generate design ideas from the inside out, and further influence their current choices and actions. To meet this challenge, this study proposes a design thinking tool that inspire future thinking: Future Footprint (FF). Which can assist users envision the future of products and services through the integration of design and futurology tools. Based on the signal-oriented preliminary data, broaden the thinking of users through the way of path thinking. Inspire users to generate breakthrough creative ideas driven by technology and social humanity, and produce the conception of the future. This tool has been applied in the research and teaching of the course "Product and Service Design" to guide students to propose design ideas. The specific use methods and results of the tool will be further described in this paper.

Keywords: Design thinking · Future vision · Future footprint (FF) · Ideation tool

1 Introduction

Nik Baerten promotes that "Exploring the future is a complex process. Without a different perspective, we can walk into a very uncertain or frightening environment. If you lose synchronicity or asymmetry of information, it creates a lot of big problems." [1] Future-oriented designers could conduct trend research and develop scenarios, and future-oriented design quickly approaches science fiction when the time scale approaches the distant future. In addition to the decline in the validity of analytical data, the further away the "truth" in the future, the change is greater in its value. Simon describes this phenomenon as "discounting the future" [2]. Compared with the way of imagination in science fiction that from "0 to 1", explore the emerging technologies could be regarded as exploration from "1" to infinity. The future is unpredictable and disobedient, considering the future variability may support us better determine the direction of the present

© Springer Nature Switzerland AG 2021
P.-L. P. Rau (Ed.): HCII 2021, LNCS 12771, pp. 185–194, 2021.
https://doi.org/10.1007/978-3-030-77074-7_15

and increase the possibility of a better future. Moreover, factors that may lead to an undesirable future can be identified and prevented previously. Realistic future-oriented research on design tools responds to concerns about the future and translates them into training designers for future sensitivity.

1.1 Design for the Long-Term Future

In the era of integration and crossover, design discipline has emerged increasingly branches that are interpenetrated with science and technology, society and humanities, and shows the trend of multi-disciplinary cooperation to deal with complex problems. Microsoft announced the "City next" smart city plan at The WPC 2013 conference. Sony Group proposed "Hidden Senses" in 2018, exploring how people interact with the technology inhabiting the lifestyles of tomorrow [4]. Matthieu Cherubini created self-driving cars in different algorithms according to the variety of environments, each one follow certain moral principles/behavior settings, and embedded in the simulation environment of unmanned virtual cars [5]. We see a series of new designs based on the future development trends of technology, showing future life scenarios and new modes of human-computer interaction. This kind of development inspires the thinking of human- ity, the change of lifestyle of human society and values. The relationship between society, technology and culture has become the concern of researchers.

The term Futurology was first proposed and used by the OssiP Flechtheim in 1943. Futurology is increasingly introduced into the theoretical development of design as it seeks a systematic, pattern-based understanding of past and present, and the possibil- ity of identifying future events and trends [6]. Design methods such as Critical Design (Anthony Dunne, 1999), Speculative Design (Dunne & Raby, 2013), and Design Fic- tion (Julian Bleecker, 2009) present to re-guide the design trend and direction. Jeffrey Bardzell and Shaowen Bardzell propose strategies for applying a humanistic orientation to the future of design. Dunne and Raby conclude a design method for design artifacts that are oriented towards design and idealistic future artifacts [7] William R. Huss pro- poses that the development of the situation is more important than the farsightedness [8]. Rather than predicting the uncertainty, design is used to open up possibilities which can be discussed, debated, and define a better future for a particular group of people [9].

1.2 From Divergency to Constriction

Norman proposes seven principles to enable Designers to adopt a user-centered approach, one of the seven is "Use both knowledges in the world and knowledge in the head." (Norman 1988, 189–201) Design development can be considered as the integration of external and internal information. Design Thinking, which was formally provided by Peter G. Rowe [10] refers to the design process that using the Thinking mode and method, is a set of the people-oriented design concept, an innovative methodology to solve the problem, and could be simply interpreted as an innovative method or tool. David Kelley proposes five steps of design thinking: Empathy, Define, Ideation, Prototype, and Test. "Double drill Design model" raised by the British Design Council. All of them reflect a commonness of design thinking, namely the divergent-focus mode. Expand the scope of the problem, further examine the underlying causes of the problem separately, and then sight on the description of the problems. The whole process experiences the stages of "divergence" together with "summary". In the "divergence" stage, we try to open our minds tremendously; In the "summary" phase, we condense and refine our insights and ideas.

As a tool to inspire the depth and breadth of design, the majority of design thinking tools manifesting in the form of filling-in tools, which are endowed with rules to a certain extent. They rely on the existing cognitive basis, put effort in the short-term benefits, neglect the meaning and value of design outputs and their programme in the long-term development. Dig deeper to find that design tools are structured expressions of a particular way of thinking. Therefore, in the study of design tools, new thinking patterns can be proposed as a response to the humanistic perspective in the conception of future scenarios. We hope that through the design of this kind of tools to assist designers to design the future with a holistic view and the future development thinking.

1.3 Signals Acquisition

Saiya Berlin proposed that taking "the sense of reality" as the premise to solve problems is one of the qualities that practitioners acquired to obtain, which requires individuals to reflect on the real existence and practical activities [11]. Since emerging technologies are usually based on the application precedents of existing one. Various technological systems not only develop, but also form a new intersecting environment among each other. As the dependence of Internet technology on electricity shows that the development of science and technology is exactly as the coevolution of biology, facing a constantly changing situation [12]. "Reality" exists as the opposite of fantasy, it determines that people should understand history, life, and the world in the context of the times.

Here, we regarded the invisible future opportunities in the real world as signals, which not be made up out of thin air, but be discovered by creative people. Finding and accurately capturing the direction and path of development will not only help designers get out of the current thinking, but also give them the support of planning long-term goals under ideal conditions. Roberto Viganti proposed that the creative process is from the inside out, and that the way to construct possible hypotheses is based on reflection and self-criticism rather than on the creation of ideas by using personal insights [13].

Differ from the conventional self-access data, we advocate the path of discovery through the organic combination of external information and internal knowledge. Cognitive psychologists have identified three key steps in learning: 1. Coding, 2. Consolidation, and 3. Retrieval [14]. It is also pointed out that complex knowledge is three-dimensional, which requires multiple kinds of explicit and implicit memory. Implicit memory is the idea that we automatically retrieve past experiences when we understand new ones (Doug Larson, 2011). We take this opportunity to report on our method innovation. This paper is organized as follows: First, we explicate the concepts of Design Thinking, Design Futures. We then describe an established Design ideation tool: Future Footprint, and introduce the 3 part of it, the "99 News", the "What will be" and the "Future territory map". Next, we report findings from a case study that we conduct in the class.

2   Future Footprint (FF)

This study provides a tool that integrates futurology and design thinking, to inspire users to think about realistic design driven by technological and socio-humanistic background, promote thinking about user-responsive design and design-reflective design. To address this challenge, we introduce a new design method that we call "Future Footprint (FF)". FF helps users contextualize multiple applications as well as critical aspects of individual applications. Quickly foregrounding potential issues before any implementation. By structuring comparison, FF also injects time to reflect upon issues, practices, and opportunity areas, helping teams to reform their hypotheses and produce a more adept understanding both of the technique growth and social trends. We have used FF to explore over 100 future signals, prototyping 4 application variations over the course of six weeks.

2.1   Research Stance

Don't try to speculate about the future. What we can do is to understand the forces that are likely to influence future change. In the process, develop your strategy and constantly look for weak links, forces for change, and trajectories (Arnold Wasserman, 2020). View of the "three horizons" (Curry et al. 2008; Sharpe 2015; Sharpe et al. 2016) as a way of thinking about the world and exploring the possibilities of taking practical action yourself. A variety of future technologies used to enrich the dialogue and thus provide concrete action towards the foreseeable future. [15] Research Through Design (Jonas, 2007:190), design practice is transformed into design knowledge. We came up with the "Future Footprint" ideation tool.

2.2   Structure of Future Footprint (FF)

"Future Footprint" is a design thinking tool to inspire future thinking, designed to guide users to build future thinking. Giving assistance to the users to reflect on future visions of products and services by integrating design and futurology tools. The specific problems are placed in the future world to explore solutions. In the shaping of the future world, the attitude and character of future thinking are endowed to the design through the design of macro patterns such as world outlook and values. Future Footprint (FF) combines science and design thinking, such as second-hand research [16], HMW (How Might We) [17], Territory Map [18], Brainstorming [19]. Based on the way of thinking at the same time to join in future tools such as "Future Signs", "Future triangle", "Future Signals", "CLA" [20] and Episodic Future Thinking [21]. Future Footprint (FF) consists of 1. "99News", 2. "What will be" and 3. "Future Territory", which are interrelated with each other and progressively layer by layer. The structure and operation mode of this Tool were introduced below (see Fig. 1) formed in 3 levels, they are Internal Motivation, Tool Body, and External Form. Each step of the tool corresponds to the transformation of signals in the brain.

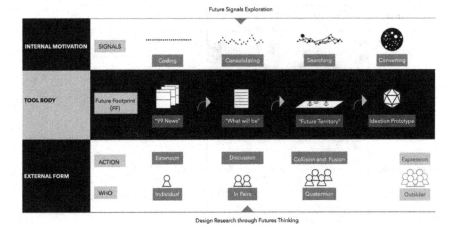

Fig. 1. Future Footprint (FF) operation mode

2.3 Procedures of Future Footprint (FF)

Step 1. "99News". The first part is called 99News (see Fig. 2). Which composed of 99 current news items. Use a combination of pictures and text introduction. Each card contain one news, includes tittle, image recognition, and text introduction as the figure showed. During the preparation step, we sincerely suggest the organizers to browse and read the latest news information from various aspects. Specifically, we recommend to collect from the following aspect: emerging foods, materials and processing, behavior and relationships, new nature, and urban transportation, but not limited to these. Then collect and capture the development trends of science and technology and hot topics before making this.

In the first of three step, a set of 99 news (It should be noted that the 99 cards here are based on the number of participants and should be adjusted accordingly if the number of participants is different in your usage scenario. We recommend 3–10 cards per participant, which can be adjusted according to the schedule). News is randomly assigned to 16 users. Each user receives about 5 cards for independent reading (see Fig. 3). On account of the inside-out process, as an individual, of understanding and absorbing what each card offers. Ideally, users can obtain: What- stands for the latest information, Who- stands for stakeholders/users, When- stands for application scenarios/opportunities, Why- stands for deep reasons such as ethics, human and nature, and How- stands for application possibilities of technology.

Step 2. "What will be". The second step is about how the future will be (see Fig. 4). In the second stage, participants worked in groups of two and wrote down predictions for the future based on the news content in the previous step. According to the content of the previous step, filling in the blank below "In the future, will _____ _____, because…" retained and processing information of the repetition in the head. The three blanks correspond to the subject, the action/change, and the technical means respectively. The process of stimulating the related memory information and expressing their views

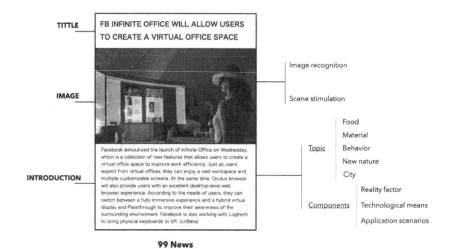

99 News

Fig. 2. "99News" structure diagram

Fig. 3. Future Footprint -99 News

on the news obtained through discussion can achieve information sharing in a short time with other participant groups. Differ from the direct output of keywords such as "brainstorm" and put up a variety of colorful sticky notes. Thinking in a network rather than a dot way of thinking, a complete sentence description will lead users to think about the relationship between subject and technology, as well as the trends that technology affects automatically.

Step 3. Future Territory. In the third step, we asked participants to brainstorm ideas about the future signs (see Fig. 5). Groups of two members are disrupted to the groups of four. Each group of four people integrated the acquired information according to the three categories of signal, question, and explanation. They expanded the content to form multiple three-dimensional triangles of future signs. The three-dimensional triangle in the figure has three faces, representing the signal, the problem, and the interpretation. Users are requested to fill in the contents of these three parts in order and then fold them into three-dimensional triangles. It should be noted that each three-dimensional triangle

Fig. 4. Future Footprint -What Will Be

representing a complete signal thinking. The "signals" are the messages from newspapers or news, and with a certain number and visibility. The "problem" in this situation is that what we see and what we feel is happening around us, it's an objective fact. The "interpretation" is the person who receives the signal and the problem's understanding of the future, biased toward subjective inferences.

Fig. 5. Future Footprint -future territory map

Record the information obtained in the thinking process in written form, extract the possible future development directly on the existing information and express their ideas. During this step, each group arrange around 20 future sign triangles according to their relevance to the topic, and then discusses them in the physical space (a piece of paper as a territory map). By adjusting the positions of different future signs, the group explore how the possibilities of technology would affect our lives, and reached an agreement with their team members on the design scheme.

3   Results and Findings

The tool has been used in the "Product and Service Design" course of Tsinghua University Academy of Fine Arts in the fall of 2020. The course is an elective course open to the entire university, with 6 undergraduates and 10 graduate students from engineering, liberal arts, and design majors. Between September-November 2020, sixteen participants studied in 4 co-design groups. It had greatly stimulated students' enthusiasm for thinking about the future. We collected 99 current news items, participants made more than 100 "what the future holds" predictions and captured 80 signs of the future. During the course, all hard paper filled in by participants were recorded periodically and recycled after the course. The students who participated in this course conducted a semi-structured interview about their experience of using the tool, and then we introduced the structure and the usage of the tool to several teachers in the design field.

By sorting out students' and teachers' feedback, we had gained inspiration for improving the use and Settings of this tool. Some interviewees gave a positive evaluation of the use of the tool, and further suggested that the content of "99News" should be adjusted according to the theme. Some interviewees had shown their haze: "This tool has opened up my mind to the use of technology, but I have a hard time stepping out of the scene to think of other possibilities in a short time." The instructor should give more detailed guidance in filling out each step of the tool so that the user can understand each step more accurately. We appreciated the cherish comments from the interviewees and concluded that Future Footprint (FF) as a teaching tool requires to be further improved as follows. 1. Some interviewees agreed on the connectivity of the tool between each step but also pointed out that the perception was not clear in the use process and we would add some guidance accordingly in the following use. 2. By sorting out the text content filled in by users, we find that users have multiple preconceptions for the same subject. Based on the second step, we will expand it again to stimulate their thinking to the greatest extent.

4   Limitations

Although we recorded the students' output electronically and analyzed the text content, the content was more diverse than the design ideas proposed by students who participated in the course in previous years. We know that this course is the first validation of the tool as the creator of the tool has participated in the testing process of the tool throughout the whole process, and the collected data and feedback are subjective in a certain extent. At the content level, since the tool contains data packages, the preparation of the tool is required to sort out and collect current news in advance. The preparation of the tool requires to spend a certain amount of time and to find the appropriate information platform, the content should be updated before each use. At the method level, the verification method is relatively simple, and demands to be verified from multiple perspectives and multiple environments, to make a more accurate evaluation of the applicability of the tool and promote more in-depth suggestions for improvement. At the formal level, this tool is used in the hard paper version, the collection of using procedures and records requires manual input. In the following research, we decide to make targeted adjustments to the above problems. At present, we have used the combination of Web3D [22] and WebAR [23] to conduct visual design for the first step "99News" of the tool, and will gradually realize the electronic process of the entire tool. We are planning to further fine-tune the content of the tool by introducing the tool to more design educators through the open-source platform, tracking the process and results of its use.

5   Conclusion and Discussion

A tool not only contains its intrinsic value of usage, but also reflects the needs, perceptions, and values of the people who create, own, and use it. Future Footprint (FF) is a design tool integrating futurology and design thinking, guiding users to capture external signals. Its inspiring design thinking about the future driven by technology and social and cultural background, and forming future signals through internal processing. We live in a world with numerous problems, and solving those requires us to be able to imagine the future in diversity ways with economic and human behavior. The signals of the future are hidden everywhere in the present. Future is untouchable, the way we design anticipated a revolution to safeguard us accurately capture and realize those futures. 'Now' is the overlap of 'Past' and 'Future'. We will continue to update and iterate on the tool with the expectation of inspiring more design stakeholders to sight on design for the long-term future.

Acknowledgments. This paper is supported by Tsinghua University Teaching Reform Project (2020 autumn DX05_01 Creativity, innovation and entrepreneurship education). The figures used in this paper are from the works of students of Tsinghua University, and the feedback on the use of tool is from the students who participated in the Product and Service Design course in 2020. We would like particularly to acknowledge all the members, along with Deli Peng, Guoyue Liu, Ke Wang, Duo Li, Yawen Zheng, Ke Chen, Xinyi Guo, Zhuoqi Jia, shiqi Gao, Taotao Tang, Xiaohan Zhang, Jingjing Wu, Xuemei Yu, Yidan Wu, Yanting Chen, and Anji Li. Expressing our sincere thanks to their works.

References

1. Baerten, N.: SpeculativeEdu l Interview: Nik Baerten (2021). https://speculativeedu.eu/interv iew-nik-baerten/
2. Simon, H.A.: The Sciences of the Artificial, 3rd edn. vol. 157. MIT Press, Cambridge (1996)
3. Microsoft.Smart cities: Enhance citizen experiences, increase sustainability and resilience, and promote innovation for your city services. [EB/OL]. https://www.microsoft.com/en-us/ industry/government/smart-cities. Accessed 11 June 2013
4. Hidden Senses Concept/Stories/Sony Design (2021). https://www.sony.net/SonyInfo/design/ stories/hidden_senses/
5. Cherubini, M.: Ethical Autonomous Vehicles 2013–2017[EB/OL] (2017). https://mchrbn.net/ ethical-autonomous-vehicles/

6. Darity Jr., W.A.: International Encyclopedia of the Social Sciences. MI: Macmillan Reference USA, 2008, vol. 4000, p. 9 (2008)
7. Dunne, A., Raby, F.: SPECULATIVE EVERYTHING: Design, Fiction, and Social Dreaming, vol. 12. The MIT Press Cambridge, Massachusetts, London, England (2013)
8. Huss, W.: A move toward scenario analysis. Int. J. Forecast. **4**(3), 377–88 (1988). https://doi.org/10.1016/0169-2070(88)90105-7
9. Dunne, A., Raby, F.: SPECULATIVE EVERYTHING: Design, Fiction, and Social Dreaming, vol. 15. The MIT Press, Cambridge (2013)
10. Rowe, P.: Design Thinking, p. 20. MIT Press, Cambridge (1998)
11. Berlin, I., Gardiner, P., Hardy, H.: The Sense of Reality, pp. 23–30. Pimlico, London (1996)
12. Kelly, K.: What Technology Wants, pp. 20–40. Penguin Books, New York (2014)
13. Verganti., R. : Overcrowded: Designing Meaningful Products in a World Awash with Ideas, pp. 234–238. The MIT Press, London (2016)
14. Brown, P.: Make it Stick: The Science of Successful Learning, pp. 79–83. Harvard University Press, New York (2014)
15. Sharpe, B., Hodgson, A.: International Futures Forum, Aberdour, Fife, UK (2017)
16. Muhleman, A. T.: The internet, secondary research and adhesives (strategic solutions). Adhesives & Sealants Industry, October 2014
17. Crawford, K.: How Might We? – A Design Thinking Exercise for Problem Solving (2017). https://spin.atomicobject.com/2018/12/12/how-might-we-design-thinking/. Accessed 3 Jan 2021
18. Richard, P., Mavor, A. (eds.): Human-System Integration in the System Development Process: A New Look. Committee on Human-System Design Support for Changing Technology. National Academies Press (2007)
19. David, H.: Visual Tools for Constructing Knowledge. VA: ASCK, Alexandria (1996)
20. Inayatullah, S.: Six pillars: futures thinking for transforming. Foresight **10**(1), 8–12 (2008). https://doi.org/10.1108/14636680810855991
21. Suddendorf, T., Corballis, M.: The evolution of foresight: what is mental time travel, and is it unique to humans? Behav. Brain Sci. **30**(3), 299–313 (2007). https://doi.org/10.1017/s0140525x07001975
22. Pavel, H., Emiliyan, P.: Study of 3D technologies for web. In: Digital Presentation and Preservation of Cultural and Scientific Heritage, pp. 309–314. Emiliyan Petkov (2019). https://www.researchgate.net/publication/336374062_Study_of_3D_Technologies_for_Web
23. Harding, C.: WebAR is Almost Here—And About to Change Everything (2018). https://cortneyharding-72342.medium.com/webar-is-almost-here-and-about-to-change-everything-711c5069a8c0

Cross-Cultural Product Design

Cross-Cultural Design: A Set of Design Heuristics for Concept Generation of Sustainable Packagings

Xin Cao[1,2]([⊠]) [iD], Yen Hsu[1] [iD], and Weilong Wu[1] [iD]

[1] The Graduate Institute of Design Science, Tatung University, No.40, Sec. 3, Zhongshan N. Rd. 104, Taipei City 10461, Taiwan
[2] Fuzhou University of International Studies and Trade, 28, Yuhuan Road, Shouzhan New District, Changle District 350202, Fujian, China

Abstract. The design teams responsible for developing sustainable packagings often lack sufficient knowledge to generate different solution concepts. This lack of knowledge is an obstacle to the effective development of sustainable packagings. In order to facilitate the generation of sustainable packagings concepts, this study proposes design heuristics for sustainable packagings concept generation (DHFSP), a collection of design heuristics that can provide sustainable design teams with the information they need for sustainable packagings design concept generation. A step-by-step process for constructing a DHFSP for sustainable packagings is presented, an example of a sustainable packaging is developed, and the usefulness of the DHFSP in aiding concept generation for sustainable packagings is validated through an empirical evaluation.

Keywords: Sustainable design · Design heuristic · Design knowledge

1 Introduction

The increasing internationalisation and globalisation of communications, commerce and industry is leading to a wide cultural diversity of users of information, services and products. If generic products are usable and attractive to such a wide range of users, culture becomes an important product design issue. Designers of products are therefore faced with the challenge of designing across cultures and need to elaborate and adopt design methods that take into account cultural patterns, factors, expectations and preferences, and to develop product concept design methods that are adapted to global users.

In the field of industrial products, the high volume and frequency of transporting products has led to the over-consumption and waste of product packaging, which is a concern for companies and designers worldwide. A number of scholars in academia have proposed definitions of sustainability [1]. The Brundtland report defines sustainable development as "development that meets the needs of the present without compromising the needs of future generations [2]". While this definition is explicitly people-centred and

© Springer Nature Switzerland AG 2021
P.-L. P. Rau (Ed.): HCII 2021, LNCS 12771, pp. 197–209, 2021.
https://doi.org/10.1007/978-3-030-77074-7_16

focuses on social justice and human needs, the goal of sustainability has been explicitly focused on the environment throughout the decades of the environmental movement.

Ceschin and Gaziulusoy (2016) categorise sustainable design methods over the past decade into four innovation categories in chronological order: product innovation, product-service system innovation, socio-spatial innovation and socio-technical system innovation [3].

The scope of this study is in the area of sustainable product innovation (cross-cultural sustainable product innovation design methods), where most of the research has been conducted on design methods that focus on improving existing products or developing new products, such as sustainable behavioural design methods, bionic design methods and Base of the Pyramid design methods. These design methods can provide designers with a set of design strategies to reduce the environmental impact of a product throughout its life cycle [4]. Ceschin and Gaziulusoy (2016) state that all three of these methods contribute to the development of sustainable products.

Despite the continuous refinement of design methods, design teams developing sustainable packaging products today still face many problems, one of which is the lack of design heuristics needed to generate problem-solving concepts.

The ideal sustainable packaging product design team should include product designers and users [5]. In most cases, product designers lack expertise in the type of sustainable packaging product under consideration and experience with the target users. The use of a user-centred human-centred design method can be an effective remedy to this deficiency, helping designers to understand and define design problems [6]. However, once a design problem has been defined and entered into the concept generation phase of the solution, the design team is likely to encounter a lack of design knowledge [7].

Currently, design teams rely heavily on the knowledge of team members to generate and explore solution concepts, and the brainstorm method appears to be the most commonly used design approach [8]. The problem is that the design knowledge within the design team may not be sufficient to explore the design domain effectively. Despite their design expertise, designers often do not have much practical experience in solving specific sustainable product design problems. The variety of materials and processes used in sustainable packaging products is so great that designers with extensive design experience often find it difficult to solve specific sustainable product design problems in real life [9]. Some users have designed their own sustainable packagings when project budgets were insufficient. However, they often do not have the knowledge to generate solution concepts. Even if they understand the problem context and user needs, they are not design experts who specialise in creating problem-solving concepts. The lack of knowledge required by the design team to generate problem solution concepts is a certain barrier to sustainable product development.

In order to solve the above problem of insufficient design knowledge within the design team, this study integrates excellent examples of sustainable packagings by searching keywords and databases, extracts and summarises their design heuristics, and proposes a DHFSP (design heuristic for sustainable products) for sustainable packagings. The study also proposes design heuristics for sustainable products (DHFSP). Previous design heuristics related to DHFSP have been proposed by several academics [10]. However, the previous design heuristics were all generic problem solving strategies and did not address the sustainable packaging product category. As a result, designers have had to experimentally test each design in the absence of knowing the success rate of each design in solving sustainable packagings design problems.

The DHFSP are design heuristics that brings together design knowledge that may inspire designers and help them to conceptualise solutions for specific sustainable packagings, based on the experience of analysing past examples of design solutions that meet specific sustainable packagings design requirements. DHFSP can create design heuristics for different types of sustainable packagings, and making it available to sustainable design teams can go a long way to alleviate this knowledge gap. This study presents a step-by-step process for constructing DHFSP for sustainable packagings and how it can be used in the brainstorming process of conceptualising sustainable packagings. To demonstrate the usefulness of DHFSP in the design of sustainable packagings, this study presents a case study of sustainable products design for packagings.

2 Literature Review

2.1 Sustainable Behavioural Design Methods

The sustainable behavioural design methods can provide designers with a set of design strategies to reduce the environmental impact of a product throughout its life cycle [4]. However, this method does not focus on the impact that the user's behaviour may have on the product as a whole. The way a user interacts with a product can have a significant impact on the environment [11]. As a result, design researchers have begun to explore the role of design in influencing user behaviour and have developed various design research methods that build on various behaviour change theories. Because there are many different models of behaviour change in the social sciences, the behaviour change approach incorporates many different design models. For example, the Sustainable Behaviour Design Model, developed at Loughborough University, is based on behavioural economics and proposes a set of design intervention strategies based on informing, empowering, providing feedback, rewarding and using [12]. The intentional design method draws inspiration from various domains and proposes eight lenses through which to understand and influence aspects of individual behaviour and situations [13]. Examples of sustainable behavioural design applications that can currently be found in the literature include the environmental dimension (encouraging users to adopt more environmentally sustainable uses) [14] and the social dimension (enabling users to adopt healthier lifestyles) [15].

2.2 Bionic Design Methods

The bionic design methods are premised on the use of nature as models, metrics and guides [16]. Using nature as models require studying nature's models and processes and adapting them to human problems and using ecological criteria to judge the 'correctness' of innovations. The rationale for using nature as an ecological criterion is that over 3.8 billion years of evolution, nature has learned what works and what is appropriate. Using nature as a guide, the emphasis is on learning from nature rather than using it. The bionic design methods define three theoretical and practical levels of bionicism: firstly, imitating natural forms, secondly, imitating natural processes, and thirdly, imitating ecosystems [16]. The bionic design methods promote the use of waste as a resource and the closing of cycles in production and consumption. There are various methods and tools available for integrating bionic design into the product design process, such as the Chakrabarti system [17] and the release card set [18]. Although imitating nature is an old and effective method in design and innovation, the claim that innovation generated by imitating nature is sustainable is somewhat misleading, as isolating principles, structures or processes from nature and imitating them does not necessarily fall under the category of sustainability [18].

2.3 Base of the Pyramid Design Methods

In addition to environmental issues, in the last decade design researchers have also started to address social issues, with a particular focus on the Base of the Pyramid design methods (BoP). From a methodological point of view, over the last few years researchers have come up with various design manuals and tools that offer different and complementary methods. The most iconic are: Design for Sustainability [19], D4S [20] and Design for Business Development. The human-centred design toolkit provides guidelines and tools for user-centred design. the BoP protocol [21] and the market creation toolkit [22], which provide methods and tools for business modelling. Gomez Castillo (2012) proposes a methodological framework that integrates the tools provided by these methods.

3 Presenting the DHFSP Development Process

In this study, a three-step process based on empirical construction was developed for a DHFSP for sustainable packagings design, assuming that the design objective was specified as "product packagings are over-consumed and wasteful, and that sustainable goals such as recycling and effective recovery of packaging are to be achieved". The following are the development process of the DHFSP:

3.1 Select Search Keywords and Database

The DHFSP development process starts with a search for the corresponding keywords in order to find the corresponding products and patents. This keyword selection is very important, as it determines the search results and thus the quality of the design.

The first category includes terms such as 'product', 'packaging' and 'invention', as well as their synonyms and singular-plural forms. The lexicon used in this study includes Roget's thesaurus [23] and Wordnet [24], which can be used to identify synonyms and singular plural forms. The second category of keywords includes adjectives such as 'sustainable', 'recyclable' and 'low carbon', which describe categories of sustainable design. The third keyword category includes verbs such as 'conserve', 'reuse' and 'reduce', which describe how sustainable design works. Synonyms and related terms were identified by checking the SCDict academic dictionary and encyclopaedia. The search formula combines the keywords in each category with the Boolean operator OR.

In addition to searching for keywords, it is also necessary to prepare corresponding databases or computer-aided tools, such as ProCAPD [25], ENVOPExpert [26], Sustain-Pro [27], etc. The use of internet search engines: Google search engine and the United States Patent and Trademark Office (USPTO) database can be useful for searching for relevant examples of sustainably relevant products.

3.2 Data Collection and Pre-processing

After selecting keywords and databases, relevant examples of sustainable packagings and patents are retrieved. The selection is then carried out using pre-defined criteria. The criteria include: 1. relevant packagings or patents that meet the design requirements; 2. three criteria that meet the economic, environmental and social requirements of sustainable design.

The format used to describe products and patents varies. It is therefore important to standardise their format: 1. the problem that the product or patent addresses; 2. how the product or patent solves that problem. The process of format standardisation helps designers to understand the product or patent of an invention, facilitates the subsequent integration of data and the discovery of design huristics. In addition to filtering, duplicate examples (examples that are very similar in terms of function, behaviour and structure) are also combined.

The format used to describe products and patents varies. Therefore, it is important to standardise their format: 1. the problem that the product or patent addresses; and 2. how the product or patent solves that problem. The process of format standardisation helps designers to understand the product or patent of an invention [28], facilitates the subsequent integration of data and identifies design heuristics.

In the above standard format, the 'base design idea' of a product corresponds to the functional behavioural structure of the product in terms of the functional behavioural structure (FBS) ontology of the design object proposed by the designer. The FBS model, proposed by Gero (1990), describes the component relationships, behaviours and object structures of design objects. In addition, the functions in the FBS model can be tailored to the purpose of the product in the DHFSP construction process.

3.3 Discovering Design Heuristics Using the Kawakita Jiro (KJ) Method

After the design heuristics have been extracted from the packagings and patents collected in the previous stage, the designers or a team of experts in the field work together to integrate the analysis using the Kawakita Jiro (KJ) method, which helps designers to collate large amounts of unstructured data and find hidden meanings in the summarised data. The analysis process are as follows.

1. Prepare a small card for each product or patent, describing how it solves the design problem it addresses through the standard format identified in the previous stage, and requiring each card to be simple and easy to understand. These cards are then placed together on a larger work surface so that all participating members can see them all.

2. The participants work individually and in silence. They find cards that are similar to each other according to a specific sustainable design question and place them close to each other on the table. Repeat this until there is no possibility of re-grouping. Each participant can adjust the cards of the others after they have been grouped until there is no disagreement. If a card can belong to more than one group, it can be copied in different groups.

3. Participants discuss the card grouping patterns on the work surface. If necessary, minor changes can be made to the grouping results. Participants check the similarities between each card according to its design function. The participants then described the commonalities expressed in each group's card design in the form of design heuristics. Figure 3 provides a set of examples to illustrate how design heuristics can be extracted from a group of products or patents.

4. Participants grouped the design heuristics from each group into larger groups. The participants check the similarity of the design heuristics in each group and explain the commonalities with higher level design heuristics. The similar groups are then brought closer together to form supergroups, until no further grouping is possible. This grouping creates a tree-like hierarchy of design heuristics, in which the lowest to highest level of design heuristics is shown. The development process of the DHFSP is shown in Fig. 1, providing knowledge of the different levels of design enlightenment.

The DHFSP method proposed in this study presents the final set of design heuristics in the form of a booklet that describes each design heuristic in the form of illustrations with text. During the conceptual design phase, designers can quickly generate design ideas and generate further ideas from that design idea [29]. The standard brainstorming method usually does not require the use of any design aids but relies on the designer's own design knowledge, and the DHFSP method can be used at this stage to facilitate the generation of design ideas.

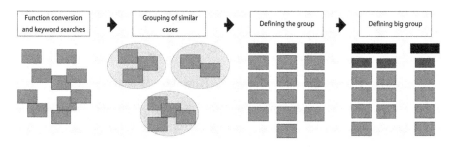

Fig. 1. The DHFSP development process

4   DHFSP Method Applied to Product Packagings Development

Based on the development process described in Chapter 3, DHFSP was applied to the design of product packagings. The case search was conducted through three categories of keywords combining functional, behavioural and structural aspects: the first category includes "packaging" and its plural forms; the second category includes "sustainable" and its synonymous adjectives; the third category includes The third category includes verbs such as "save" and "reuse". The keywords are then grouped together using the operator OR, and the keywords are then linked together using the operator AND to generate a search formula. The search formulas are imported into the Google search engine database and the USPTO database to search for relevant products and patents.

A total of 158 packaging examples (121 existing products and 37 patents) were searched from the keywords. The 158 examples were then filtered and 120 examples that met the above criteria were selected and retained, with each example converted to a standard format. Finally, the KJ method as shown in Fig. 1 was used for the grouping analysis.

As shown in Table 1, a total of 15 design heuristics were identified with their detailed descriptions, and the number of examples for the 15 design heuristics in the DHSFP methodology ranged from 3 to 14 examples, with a significant difference in the number of examples. As shown in Fig. 2, a total of 3 levels of design enlightenment were identified by further grouping.

Figure 3 provides an example of how design heuristic can be extracted from a group of similar packagings and patents, showing five packagings in the same group using the KJ method. These inventions aim to achieve the design objective of 'sustainable product packagings' and are similar in that they can be converted for use as accessories to their packaged products in addition to their function as packagings.

The design heuristics in Table 1 can help designers or design teams to generate conceptual solutions for sustainable packagings, and this study has created a handbook as an aid to design, which contains examples of products for each design heuristic and their textual descriptions. Figure 4 and Fig. 5 illustrate some of the content of the manual to illustrate the presentation of design heuristics in the DHSFP.

This study also provides a preliminary evaluation of the DHFSP, not with the aim of providing a comprehensive validation of the proposed method, but only preliminary information on its potential and shortcomings. Furthermore, the novelty of the DHFSP can only be assessed through comprehensive feedback on the concept product, which is limited in scope to conceptual design aspects. However, innovative concepts are closely related to novelty. Sarkar and Chakrabarti (2011) suggest that differences in the level of product innovation depend in part on differences in their degree of novelty. Therefore, this study provides an initial assessment of the DHFSP through the use of three assessment criteria (novelty, quantity and quality) by experts. These criteria have been used extensively in previous studies to assess design ideas. The results of the preliminary expert assessment indicate a positive impact of the DHFSP in supporting the performance of sustainable packagings concept generation (novelty, quantity and quality).

Table 1. Fifteen design heuristics for sustainable packagings

No.	Design heuristics	Description	No. of examples
1	Combine several products into one	Sustainable packagings can be created by combining the functionality of several different products	5
2	Reduce the amount of packaging material	Sustainable packagings can be created by reducing the amount of packaging material	13
3	Use recyclable packaging materials	Sustainable packagings can be created by using recyclable packaging materials	12
4	Use reusable packaging materials	Sustainable packagings can be created by using of reusable packaging materials	9
5	Use waste packaging materials	Sustainable packagings can be created by using of waste packaging materials	10
6	Use naturally degradable packaging materials	Sustainable packagings can be created by using of naturally degradable packaging materials	8
7	Use low-cost packaging materials	Sustainable packagings can be created by using of low-cost packaging materials	12
8	Integrate the packaging with the product	Sustainable packagings can be created by integrating the packaging with the product	4
9	Convert into accessories for the product	Sustainable packagings can be created by converting into accessories for the product	5
10	Convert into secondary packaging for products	Sustainable packagings can be created by converting into secondary packaging for products	8
11	Be available for different sizes of products	Sustainable packagings can be created by adapting to different sizes of products	5
12	Reduce extra packaging between product packages	Sustainable packagings can be created by reducing of extra packaging between product packages	4

(*continued*)

Table 1. (*continued*)

No.	Design heuristics	Description	No. of examples
13	Use green patterns and colors	Sustainable packagings can be created by using green patterns and colors	14
14	Use modular architecture	Sustainable packagings can be created by using modular modular architecture	5
15	Convert packaging functions	Sustainable packagings can be created by converting product packaging functions	6

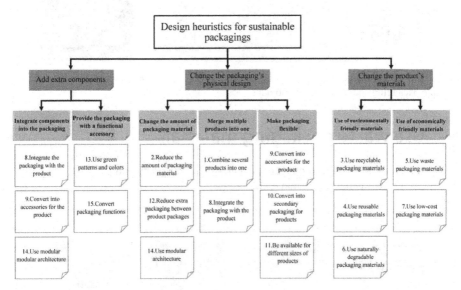

Fig. 2. A hierarchical structure of the design heuristics for sustainable packagings

Identify similarities between example packagings

A group of example packagings (a)~(e) has been collected. They share a similarity that can be converted into accessories for the product.

(a)　　　　　　　(b)　　　　　　　(c)　　　　　　　(d)　　　　　　　(e)

Fig. 3. An example illustrating the extraction of a design heuristic from a group of similar example packagings

Reduce the amount of packaging material

- A sustainable product packaging can be created by reducing the amount of packaging material.

Fig. 4. The design heuristic, 'Reduce the amount of packaging material', and corresponding example packagings described in the booklet

Use waste packaging materials

- A sustainable product packaging can be created by using of waste packaging materials.

Fig. 5. The design heuristic, 'Use waste packaging materials', and corresponding example packagings described in the booklet

5 Summary

This study proposes design heuristics DHFSP for sustainable packagings that takes into account the wide cultural diversity of global users and is a globally adaptable method for product concept design that can support the concept generation of sustainable packagings. The DHFSP is the result of experience gained through the analysis of past design solution examples. As shown in Fig. 1, this study proposes a step-by-step process for the DHFSP. The DHFSP development process uses the KJ method to extract the corresponding design heuristics from the example data in an inductive manner, and then presents the multi-layered design heuristics knowledge through a hierarchical organisation.

To further describe the practicality of the DHFSP, sustainable packagings is used as an illustrative example. As shown in Table 1, 15 design heuristics were derived from the analysis of 158 sustainable products found in the keyword search. Three levels of design heuristics were identified through a grouping analysis using the KJ method, as shown in Fig. 2.

The study also compared the DHFSP (for sustainable packagings) with three previous design heuristics (SCAMPER, TRIZ and 77 Design Heuristics). The 15 design heuristics in the DHFSP were found to be different from the three previous design heuristics in terms of content, with only a few design heuristics from SCAMPER, TRIZ and 77 Design Heuristics having similar meanings to the DHFSP. SCAMPER, TRIZ and 77 Design Heuristics are all generic problem solving strategies that are not specific to the sustainable product category. As a result, designers have to experimentally test each design heuristic without knowing the success rate of each design heuristic in solving sustainable design problems. DHFSP is independent from other generic design heuristics and can make an additional contribution to the concept generation of sustainable packagings by inspiring designers worldwide and helping them to conceptualise sustainable packaging solutions.

DHFSP can be used as a solution for the current lack of specialist conceptual tools and low R&D investment in sustainable packagings development, and can play an important role in supporting sustainable packagings design activities. For the DHFSP method to have a real impact, there is a need to develop DHFSP that cover a wide range of sustainable product types, and this may require a large team effort. Currently, this research project is developing several DHFSP for sustainable products design.

References

1. Author, F.: Article title. Journal 2(5), 99–110 (2016)
2. Holden, E., Linnerud, K., Banister, D.: Sustainable development: our common future revisited. Glob. Environ. Change **26**, 130–139 (2014)
3. Brundtland, G.H.: Our common future—call for action. Environ. Conserv. **14**(4), 291–294 (1987)
4. Ceschin, F., Gaziulusoy, I.: Evolution of design for sustainability: from product design to design for system innovations and transitions. Design Stud. **47**, 118–163 (2016)
5. Pigosso, D.C.A., McAloone, T., Rozenfeld, H.: Characterization of the state-of-the-art and identification of main trends for ecodesign tools and methods: classifying three decades of research and implementation. J. Indian Inst. Sci. **95**(4), 405–428 (2015)
6. Allen, M., et al.: Involving domain experts in assistive technology research. Univ. Access Inf. Soc. **7**(3), 145–154 (2008)

7. Carmien, S.P., Fischer, G.: Design, adoption, and assessment of a socio-technical environment supporting independence for persons with cognitive disabilities. In: Proceedings of the SIGCHI Conference on Human Factors in Computing Systems (2008)
8. Hwang, D., Park, W.: Design heuristics set for X: a design aid for assistive product concept generation. Design Stud. **58**, 89–126 (2018)
9. Gulliksen, J., Lantz, A.: Design versus design-from the shaping of products to the creation of user experiences. Int. J. Hum.-Comput. Interact. **15**(1), 5–20 (2003)
10. Hersh, M.A.: The design and evaluation of assistive technology products and devices part 1: design (2010)
11. Hwang, D., Park, W.: Development of portability design heuristics. In: DS 80–4 Proceedings of the 20th International Conference on Engineering Design (ICED 15) Vol 4: Design for X, Design to X, Milan, Italy, 27–30 July 2015 (2015)
12. Sherwin, C., Bhamra, T.: Ecodesign innovation: present concepts, current practice and future directions for design and the environment. In: Design History Society Conference. University of Huddersfield, Huddersfield (1998)
13. Bhamra, T., Lilley, D., Tang, T.: Design for sustainable behaviour: using products to change consumer behaviour. Design J. **14**(4), 427–445 (2011)
14. Lockton, D.: Design with intent: a design pattern toolkit for environmental and social behaviour change. Brunel University School of Engineering and Design Ph.D. theses (2013)
15. Tang, T., Bhamra, T.: Putting consumers first in design for sustainable behaviour: a case study of reducing environmental impacts of cold appliance use. Int. J. Sustain. Eng. **5**(4), 288–303 (2012)
16. Ludden, G.D.S., Offringa, M.: Triggers in the environment. Increasing reach of behavior change support systems by connecting to the offline world. In: 3rd International Workshop on Behavior Change Support Systems, BCSS 2015. CEUR (2015)
17. Benyus, J.M.: Biomimicry: innovation inspired by nature. Morrow, New York (1997)
18. Chakrabarti, A., et al.: A functional representation for aiding biomimetic and artificial inspiration of new ideas. Artif. Intell. Eng. Design Anal. Manuf. **19**(2), 113–132 (2005)
19. Volstad, N.L., Boks, C.: On the use of biomimicry as a useful tool for the industrial designer. Sustain. Dev. **20**(3), 189–199 (2012)
20. Castillo, L.G., Diehl, J.C., Brezet, J.: Design considerations for base of the pyramid (BoP) projects. In: Proceedings of the Nothern World Mandate: Culumus Helsinki Conference (2012)
21. Crul, M.: Design for sustainability: a practical approach for developing economies. UNEP/Earthprint (2006)
22. Simanis, E., Hart, S.: The base of the pyramid protocol: toward next generation BoP strategy. Cornell Univ. **2**, 1–57 (2008)
23. Larsen, M., Flensborg, A.: Market creation toolbox: your guide to entering developing markets. DI International Business Development (2011). https://www.access2innovation.com/download/market_creation_toolbox.pdf. Accessed April 2018
24. Jarmasz, M., Szpakowicz, S.: Roget's thesaurus and semantic similarity. In: Recent Advances in Natural Language Processing III: Selected Papers from RANLP, vol. 2003, p. 111 (2004)
25. Budanitsky, A., Hirst, G.: Evaluating wordnet-based measures of lexical semantic relatedness. Comput. Linguist. **32**(1), 13–47 (2006)
26. Kalakul, S., et al.: Computer aided chemical product design–ProCAPD and tailor-made blended products. Comput. Chem. Eng. **116**, 37–55 (2018)
27. Halim, I., et al.: A combined heuristic and indicator-based methodology for design of sustainable chemical process plants. Comput. Chem. Eng. **35**(8), 1343–1358 (2011)

28. Carvalho, A., Matos, H.A., Gani, R.: SustainPro—a tool for systematic process analysis, generation and evaluation of sustainable design alternatives. Comput. Chem. Eng. **50**, 8–27 (2013)
29. Ramani, K., et al.: Integrated sustainable life cycle design: a review. J. Mech. Design **132**(9) (2010)

Exploring the Integration of Emotion and Technology to Create Product Value—A Case Study on QisDesign Lighting

Jen-Feng Chen[(⊠)], Po-Hsien Lin, and Rungtai Lin

Graduate School of Creative Industry Design, National Taiwan University of Art,
New Taipei City, Taiwan
t0131@ntua.edu.tw, rtlin@mail.ntua.edu.tw

Abstract. The Strategic Roadmap 2025 released by Lighting Europe described the key focuses in three development stages for the lighting market over the next 10 years. During the first stage, the focus of development lies in technology—specifically, in the application and popularization of high-efficiency LED light sources. At the second stage, manufacturers advance to the smart lighting control system, thereby achieving sustainable development through energy saving and environmental designs. At the third stage, the strategic focus lies in the mental aspect, the intention of which is to develop human-oriented lighting designs. Accordingly, well-being and health can be promoted through improvement of lighting quality. These three stages of strategy will provide new momentum for the growth of the lighting market and improve people's quality of life. This study adopted a literature review, expert discussion, and survey analysis to identify human-oriented luminary design factors that integrate sensitivity and technology, with the aim of integrating Taiwan's lighting technology with consumers' emotional need and introducing emotional value in products.

Keywords: Human-centered · Perceived quality · Well-being · Emotional value

1 Introduction

The "experience economy" is trending worldwide, where consumers are seeking meaning and feeling when buying products. Corporations have started using technology related to big data analysis, eye tracking, and neuroscience to identify consumer preferences and market trends. The goal is to create a unique user experience and deliver pleasurable sensations and emotional satisfaction. Understanding the emotional needs of consumers is the key for a corporation to triumph. Nokia has a famous phrase, "human technology", which implies that only by satisfying human needs can the value of a technology be foregrounded. Rungtai Lin [24] once said that product design has shifted from the past focus on design for function to design for feeling. The core value of design has changed from the rational value of emphasizing functions and biological needs to the emotional value of emphasizing aesthetics and psychological needs. The lighting industry too must stop stressing functions and costs and shift to a human-centered design mentality.

© Springer Nature Switzerland AG 2021
P.-L. P. Rau (Ed.): HCII 2021, LNCS 12771, pp. 210–223, 2021.
https://doi.org/10.1007/978-3-030-77074-7_17

Since 2012, the manufacturing and sale of conventional incandescent light bulbs have been banned. Because of its price and production advantages, LED has become the major light source of the 21st century. Developments in LED luminaries have evolved from a center of technology breakthroughs, such as the pursuit of brightness, efficiency, and smart system control, toward human-centered designs that emphasize pleasing sensual organs and offering comfort and a sense of well-being [22]. Therefore, luminary designs must feature favorable illumination functions as well as satisfy psychological needs for pleasure and comfort—and thus increase needs to wants and desires in consumer hearts—to become attractive products with high market competitiveness and brand value.

The trend of product design has gradually leaned toward designs that elicit emotions of happiness, pleasure, and satisfaction [15]. An LED light source is rich in color, small in size, low in temperature, and high in directivity. Numerous innovative thoughts have emerged in LED lamp designs and applications. For example, the Philips Living Colors lamps are rich in color and shift between colors of light to rapidly alter the atmosphere of a space, satisfying people's varying situational needs. The Taiwan luminary brand QisDesign, which has received numerous awards worldwide, utilizes the characteristics of novel light sources to integrate design creativity with technology and aesthetics. This enables the company to create products with high added value that are sold worldwide and win multiple design awards, shedding light on LED luminaries in the early stage of LED light source popularization. The designs from QisDesign have received positive reviews from consumers, and the company has established a unique brand image in the market for luxurious luminary products.

As the technology continue to improve, LED light sources must both provide favorable lighting functions and satisfy people's emotional needs. The application realm of LED light sources is expanding. In addition to indoor and outdoor lighting, LED lighting can be used in every aspect of life, such as in medical lighting, automobile lighting, plant lighting, and display backlighting. This development trend signifies that lighting and life have an increasingly intimate relation. While relevant technology continues to mature, the barrier to industry entry decreases. Consequently, Taiwan's luminary industry cannot solely focus on function, cost, and mass production orientations when developing products. To outmaneuver the competition created by price cuts, product design must be human-orientated by deeply investigating the effect of lighting on the mind and the body such that with the emergent trend of smart technology, emotional design can bring lighting closer to human nature and thus create product value.

2   Literature Review

2.1   Principles for Designing Superior Light Sources in LED Luminaries

Sunlight, air, and water are three indispensable elements of life. The US Department of Energy National Laboratories, in the 2001 National Human Activity Pattern Survey [14], reported that American people spent 86.9% of their time in various types of buildings. This accentuates the importance of indoor lighting. Lighting quality affects human physical and mental health. According to the International Commission on Illumination (CIE), adequate illumination can create an environment in which people can see clearly, move safely, and complete visual tasks effectively and precisely without causing inappropriate visual fatigue or discomfort. A light source design must meet relevant international standards in creating superior luminary products. According to the National Standards of the Republic of China (CNS) CNS12112, the lighting environment defined for lighting design in a workplace is such that lighting facilities must satisfy visual comfort, effectiveness, and safety needs. To meet these needs, designers must pay attention to lighting environment parameters such as illuminance, luminance, uniformity, glare, color render index, and color temperature. The European EN 12464-1 standard stipulate specifications for illuminance, uniformity, glare control, and contrast sensitivity. The US Department of Energy has also established standards for the color render index, blink, glare, and network lighting for solid state lighting [7].

The current study reviewed international indoor lamination standards and identified the following luminary light source assessment principles.

(a) Illuminance: The unit is lux. It refers to the degree of brightness of an object or surface shone by the light source. The intensity of illumination is the basic condition for people to see clearly. A low mean illuminance tires the eyes and potentially causes nearsightedness, whereas discomfort is induced if illuminance is excessively high. According to the CNS, indoor spaces have different standards for illuminance. For example, the illuminance of a general office should be 500–750 lx, whereas that of a corridor or tea and water room should be 150–200 lx [11, 13].

(b) Uniformity: In a work environment, people see various surfaces. If the illuminance varies greatly, then time is needed to adjust when the eyes move from one surface to another. Such adjustment causes visual fatigue. Therefore, adequate illumination requires uniform illuminance. The ratio of the minimum illuminance to mean illuminance should be as close to 1.0 as possible. Generally, it should be at least 0.5 [2, 27].

(c) Glare: Glare refers to the discomfort a bright light source or object causes the eyes. For example, if an LED light source or bulb directly shining into the eyes or reflecting on desktops or objects into the eyes causes discomfort and the inability to see clearly in the visual range, this strong light is glare. The Glare index (UGR) is a glare calculation method proposed by the CIE. The UGR value of most lighting systems is 10–30, with a high index value indicates greater discomfort. A UGR value lower than 10 is considered not discomforting. A UGR value of <10 is a global standard for glare [4, 5].

(d) Flicker index: According to researchers, flicker may trigger disease symptoms. Under more severe scenarios, it may trigger an epileptic episode. A power supply output voltage ripple may cause a luminary to flicker, and when the eyes are exposed to a 70–160 Hz high-frequency flicker for a prolonged period, physical discomfort, headache, and vision damage can be caused. Therefore, the CIE requires a flicker index or FI ≤ 0.02 to reduce the impact of flicker on the human body [12, 17].

(e) Color rendering index: Color rendering refers to the degree of which the color of an object is revealed after a light source shines on the object. Proximity to the color under natural light indicates high color rendering. The CIE have proposed 14 colors for testing the color rendering index (CRI). When the CRI value is close to 100, color rendering is high. General indoor illumination requires CRI >80. In response to consumers placing greater emphasis on illumination quality, many manufacturers have increased their standard to a CRI value of >90 [23].

2.2 LED Luminary Design for Inducing Senses of Pleasure and Comfort

Multiple international standards regulate basic requirements for indoor lighting, and they are critical bases for lighting design. However, these regulations are not in line with the current lighting design trend that is human-centered and values perceived quality. Therefore, numerous major international illumination manufacturers in recent years have proposed two major design directions that focus on (1) the correlation between light source comfort and work efficiency and (2) the improvement of light quality to increase senses of pleasure and well-being.

Regarding improvement of light sources and work efficiency, PHILIPS Lighting noted that according to research, improving lighting conditions can increase the degree of comfort, with a positive impact on employee performance. For example, employee mental and memory functions improved by 25%, their efficiency in handling phone calls increased by 12%, and their productivity increased by 23%. A smart interconnection system lighting company, Interact, stated that light controls life functions and determines the internal biological clock of humans, profoundly influencing their mood. Imitating the changing sequence of natural light sources can help adjust the biological clock of employees and increase their vitality. For older workers, adjusting light and providing higher illuminance help them see greater detail [9, 10].

Regarding light quality and well-being, the Dutch physicist Kruithof depicted a curve based on psychological physics data and proposed the principle of illumination and color temperature combination. When high illumination is paired with high color temperature and low illumination paired with low color temperature, optimal atmospheres that are comfortable and pleasurable are created [3, 18]. Osram proposed several suggestions in this regard. For example, color temperature affects mood. White warm light has a calming effect, whereas white natural light has a strong encouraging effect. Therefore, a lighting system can use dynamic lighting in its design to adjust biological rhythms. Moreover, PHILIPS Lighting and Osram have both noted the positive influences of allowing people to feel a blue sky and white clouds indoors. Their studies have verified that sunshine can elicit a subjective sense of well-being, and that people enjoy sitting by the window or in natural lighting while working [16, 19]. On the basis of aforesaid studies, superior artificial lighting technology can be used for users working indoors to experience a blue sky and white clouds and thus feel pleasure and happiness.

2.3 Discussion on Human-Centered Designs

Maslow's Hierarchy of Needs and the Levels of Need for Luminary Design. Maslow's hierarchy of needs is well known and provides a critical basis in the design field for exploring people's needs. Its main concept is that needs are in a hierarchy, and the satisfaction of needs starts from the bottom level of this hierarchy. Only when one layer of needs is satisfied do people pursue needs in the next, higher level of needs [6]. In the book Designing for Emotion, Aarron Walter describes a hierarchical structure of user needs that is highly similar to that of Maslow. Walter maintained that only after satisfying basic needs, such as functions, reliability, and usability can a product meet higher, pleasure-oriented needs [1]. In a similar manner, design of luminary products must first meet the basic illumination needs at the bottom hierarchical level and then advance to superior illumination needs, and finally to human-centered illumination needs regarding pleasure. Basic illumination needs focus on the effectiveness, safety,, and cost of a luminary and its ability to be mass-produced. Superior illumination needs emphasize smart control, energy saving, appearance and style, color, and texture. In human-centered illumination needs, the senses of pleasure and comfort brought by a luminary and personalized uniqueness are emphasized.

Norman's Three Levels of Emotional Design. In the book Emotional Design, Donald A. Norman proposes three major layers of emotional designs. (1) The visceral level denotes the first impression of a product. It refers to the initial influence of a product, including its appearance, texture, and feel. The visceral response is earlier than conscious thought. It is a rapid decision made regarding good or bad, beautiful or ugly, and safe or dangerous. It involves viewers' direct response to and first visual impression of a product. (2) The behavioral level is related to the use and experience of a product. Experience involves various aspects, including functionality, understanding, and usability. Whether a product has effective functions and whether the functions can be easily identified, understood, and operated determine the experience of using the product. (3) In the reflective level, product characteristics reflect a consumer's self-image, personal satisfaction, and memory. This level emphasizes the underlying message, cultural elements of a product or its meaning and effectiveness. Its most crucial value is to satisfy emotional needs, and among them, the most critical is to foster self-image and social status [20, 28].

A Design Framework Integrating Emotional Need and Technology. Norman suggested that in the success of launching a product, the emotional factors of its design might be more critical than practical factors [21]. Demirbilek and Sener (2003) stated that a designer should understand emotions, users, and product design because these three aspects affect choice in purchase and use [8]. Product design changes over time. The core value of design has shifted from emphasizing rational value, such as functions and physiological need, to emotional values that emphasize aesthetics and psychological needs [26]. In the 21st century where digital technology continues to advance, designers must integrate emotional design factors with rational factors to create values for brands.

On the basis of to the five hierarchical levels of need proposed by of Maslow and the visceral, behavioral, and reflective design levels established by Norman, the relationships between technology, humanity, and cultural creativity described by Lin (2005) indicate that design styles are defined through function, friendliness, fun, uniqueness, and pleasure. Norman's three design characteristic levels denote differing but clear design focuses [25]. Based on this hierarchical structure, luminary design needs were compiled to propose the current study's research framework (Fig. 1), which clearly reflects the luminary design factors of different levels and their corresponded relationships.

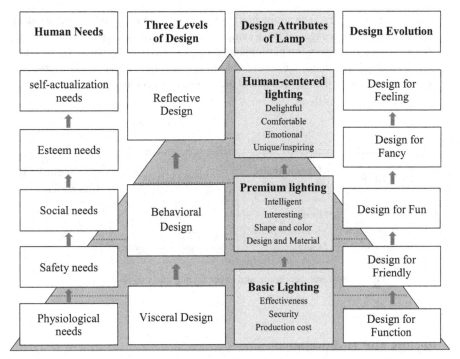

Fig. 1. Lighting design, between human needs and technology (Adopted: Rung-Tai Lin, 2005)

3   Research Methodology

We devised a questionnaire based on Norman's three levels of design to explore the performance of QisDesign luminaries at the visceral, behavioral, and reflective levels, to gauge participant preferences, and to further understand luminary design factors that integrate emotional need with technology.

(1) Luminary sampling. After discussion with experts, we adopted the three factors of color, style, and function and arrived at eight combinations. Each combination was represented by one light product. Thus, eight luminary samples were used.

(2) The experts discussed and selected suitable semantic adjectives. From relevant academic research and magazines, we collected adjectives related to the appearance of luminaries and feelings about light sources. A total of 15 senior designers and professors in design discussed and voted to select the adjectives.

(3) Questionnaire design. For the visceral level, the adjectives selected by experts and questions about appearance and lighting effects were used to explore participants' visceral feelings. Regarding the behavioral level, the participants were asked to try out the luminary samples and provide feedback on the ease of understanding and ease of operation of the product functions. As for the reflective level, we used a video to display situational images of the luminary samples for the participants to perceive the imagery the luminary attempted to deliver.

(4) Questionnaire survey. The eight luminary samples were placed in independent spaces. Situational images of these samples were repeatedly projected on the wall. After viewing the video, the participant completed the questionnaire. While completing the questionnaire, they were able to operate the switches of the luminary samples and change their angles to feel the products in person. In all, 72 participants were recruited (Fig. 2).

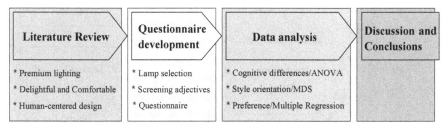

Fig. 2. Research procedure and experimental analysis flowchart.

4 Results and Discussions

In total, 72 valid questionnaires were retrieved. Reliability analysis was conducted to explore the internal consistency of each questionnaire construct. By removing a single item and observing the degree of reduction of the Cronbach α for each construct, we assessed item selection and reliability. The analysis revealed that Cronbach α was .949, indicating high reliability. After individual items were removed, α was reduced, indicating that the instrument had favorable item selection and reliability.

4.1 Perceptual Differences by Sex, Educational Attainment, and Age

The sex and educational attainment of the participants was used as the independent variable. Seven factors of the luminary samples and scores from two comprehensive assessments were used as dependent variables. We conducted independent samples t tests and discovered that neither sex nor educational attainment exhibited significant differences in the assessment scores of the product characteristics in any dimension.

The age of the participants was then used as the independent variable to examine the same dependent variables. Analysis of variance was adopted to test whether age exerted a significant effect on product characteristic assessment. The results indicated that except pleasurable feelings and uniqueness, the remaining factors did not exhibit significant differences (Table 1). Post hoc testing revealed that the mean score of participants aged 40–49 years was significantly higher than that of participants aged 19–29 years. Comparing the mean scores of the different age groups revealed descending order of those 40–49 years old > 30–39 years old > 19–29 years old. The uniqueness factor also exhibited significant differences. A post hoc test indicated that the mean score of participants aged 40–49 years was significantly higher than that of those aged 30–39 years. Comparing the mean scores of the different age groups revealed descending order of those 40–49 years old >19–29 years old >30–39 years old. Thus, the participants aged 40–49 years scored pleasurable feelings and uniqueness significantly higher than did the other age groups. Although this age group identified with these two factors the most, the participants of other age groups also exhibited a high degree of identification.

Table 1. Cognitive differences for product factors by age group

Variables	Works	F	M	Scheffe's post hoc
Pleasurable	19–29 years old	3.696*	4.03	3 > 1
	30–39 years old		4.16	
	40–49 years old		4.72	
Uniqueness	19–29 years old	4.746*	4.44	3 > 2
	30–39 years old		4.08	
	40–49 years old		4.72	

*$p < .05$

4.2 Spatial Cognition Distribution and Six Basic Relationships Among the Eight Luminary Samples

The participants' cognition was analyzed spatially using multidimensional scaling analysis, in which the appearance and illumination effect of the eight luminary samples were examined along with six product styles (Fig. 3). Of the six product styles, modernity, geometry, simplicity were associated with the appearance dimension, and gentleness, warmth, and calm were associated with the illumination effect dimension. The stress index and RSQ were .10266 and .94636, respectively, indicating the dimensional space exhibited favorable goodness of fit.

In Fig. 3, Luminary P1 is located at the far right of the dimensional space. It exhibited the most consistency with the two imagery directions of appearance and illumination effect; therefore, P1 most accentuated the characteristics of modernity, geometry, simplicity, gentleness, warmth, and calm. P5 is located at the top of the dimensional space, closest to the appearance dimension, and it had the characteristics of modernity, geometry, and simplicity. P7 is located at the bottom of the dimensional space, closest to the illumination effect dimension, and it had the characteristics of gentleness, warmth, and calm. P2 was at the opposite direction from P1, and thus, it was considered having a traditional, organic, and complex appearance. The top three luminary samples in terms of overall preference were P1, P3, and P4 (Table 3), all of which are situated at the right side of the dimensional space. Their styles were associated with senses of modernity, geometry, simplicity, gentleness, warmth, and relaxation (Table 2).

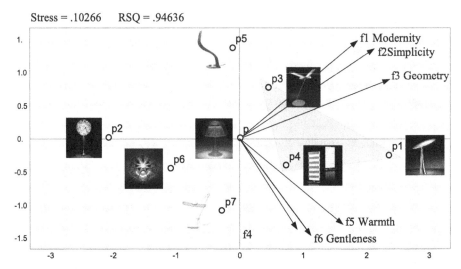

Fig. 3. Two dimensional cognitive space of the eight luminary samples

Table 2. Mean scores of the eight luminary samples with regard to the six product styles.

	P1	P2	P3	P	P5	P6	P7	P8
f1 Modernity	4.49	3.72	4.25	4.25	4.07	3.78	3.51	3.86
f2 Simplicity	4.54	2.31	3.94	3.51	4.06	2.69	3.32	3.49
f3 Geometry	3.68	3.19	3.60	3.67	2.86	3.10	2.57	3.38
f4 Relaxation	3.68	3.28	3.50	3.86	3.08	3.43	3.60	3.49
f5 Warmth	4.13	2.65	3.17	3.72	3.18	3.11	3.68	3.44
f6 Gentleness	4.38	2.96	3.33	3.68	2.88	3.50	3.69	3.46

4.3   Analysis of the Three Levels of Design and Sense of Pleasure

To explore the effect of the three design levels (i.e., the visceral, behavioral, and reflective levels) on sense of pleasure, we took the three luminary samples with the highest overall preference to conduct multiple regression analysis. In Tables 4, 5 and 6, sense of pleasure was the dependent variable. Appearance preference, illumination effect preference, and illumination effect comfort at the visceral level; functional ease of understanding and ease of operation at the behavioral level; and uniqueness at the reflective level were adopted as independent variables to examine the correlation between each characteristic and sense of pleasure. The F value of the overall regression model of P1, P3, and P4 reached a strongly significant level, indicating that the six independent variables are suitable for predicting the sense of pleasure score. The R value of the multiple correlation

Table 3. Overall preference ranking.

Rank	1	2	3	4	5	6	7	8
No.	P1	P3	P4	P2	P7	P8	P6	P5
Product								
Mean Scores	2.36	3.82	3.96	4.71	4.96	5.26	5.39	5.54

coefficients and the coefficient of determination R^2 value respectively indicated that the independent variables explained 58.3% and 76% of the total variation of the dependent variable. Of the standard regression coefficients of the independent variables, the β values of the functional ease of operation and appearance preference of P1 were both significant, suggesting that they were two major factors affecting pleasurable feelings toward the product. All R values for P3 and P4 were significant, verifying that the six independent variables were significantly correlated with the dependent variable. The β value of P4's illumination effect preference, illumination effect comfort, and functional ease of operation were significant, showing that illumination effect preference, illumination effect comfort, functional ease of operation, and appearance preference were major factors affecting pleasurable feelings toward a product.

Table 4. Multiple regression analysis of the variables in the visceral, behavioral, and reflective levels and the pleasurable experience of P1

Independent variable	Predictor variable	B	r	β	t
Pleasurable	Appearance preference	.289	.403***	.293	2.790*
	Illumination preference	−.106	.213	−.080	−.668
	Illumination comfort	.264	.268	.184	1.781
	Ease of understanding	.124	.409	.144	1.501
	Ease of operation	.480	.646***	.533	5.602***
	Uniqueness	−.062	.303**	−.050	−.477
	R = .734[a]	$R^2 = .538$		F = 12.626***	

*$p < 0.05$ **$p < 0.01$ ***$p < 0.001$

Table 5. Multiple regression analysis of the variables in the visceral, behavioral, and reflective levels and the pleasurable experience of P3

Independent variable	Predictor variable	B	r	β	t
Pleasurable	Appearance preference	.159	.563***	.194	1.778
	Illumination preference	.142	.656***	.158	1.061
	Illumination comfort	.330	.654***	.317	2.328
	Ease of understanding	.085	.320**	.099	1.041
	Ease of operation	.140	.475***	.158	1.524
	Uniqueness	.119	.490***	.115	1.148
	R = .764[a]		R² = .583		F = 15.144***

$^{**}p < 0.01$ $^{***}p < 0.001$

Table 6. Multiple regression analysis of the variables in the visceral, behavioral, and reflective levels and the pleasurable experience of P4

Independent variable	Predictor variable	B	r	β	t
Pleasurable	Appearance preference	.020	.554***	.027	.346
	Illumination preference	.270	.760***	.359	3.800***
	Illumination comfort	.261	.766***	.328	3.310**
	Ease of understanding	.059	.425***	.038	.526
	Ease of operation	.356	.625***	.291	3.599**
	Uniqueness	.103	.394***	.061	.884
	R = .872[a]		R² = .760		F = 34.291***

$^{**}p < 0.01$ $^{***}p < 0.001$

5 Conclusions

This study employed a literature review, expert discussion, and survey analysis to gain a further insight into which LED luminary design factors are preferred by users and bring a sense of pleasure. The study results may serve as a critical reference for future designers integrating LED technology and emotional needs of users in order to create emotional value in luminary products. Through the aforementioned research and analysis, we arrived at the following conclusions.

(1) Through the literature review, we learned that adequate illumination must meet superior light source design principles addressing comfort and pleasurable sensations. Regarding these principles, we reviewed international organization regulations and deduced five critical indices that must be met: illuminance, uniformity, glare, flicker, and color rendering. Regarding users' emotional needs, we explored the design trend of major international luminary brands and examined relevant research to conclude that next-stage illumination design will focus on satisfying users' sense of pleasure and comfort.

(2) The survey results revealed that cognitive differences for the characteristics in the three design levels, namely visceral, behavioral, and reflective, did not vary by sex or educational attainment. The different age groups exhibited significant differences in sense of pleasure and uniqueness. Regarding sense of pleasure, participants aged 40–49 years rated it significantly higher than did participants aged 19–29 years. Regarding uniqueness, participants aged 40–49 years rated it significantly higher than participants aged 30–39 years did. Although other design characteristics differed nonsignificantly with age, the mean scores of these characteristics were high, suggesting that the three design levels of the luminary products merit reference and in-depth research.

(3) The analysis indicated that the luminary sample that ranked first in preference also scored highest in style imagery in terms of appearance and illumination effect. The luminary samples ranked second and third also scored high in these two aspects. Accordingly, the geometry, simplicity, and modern feel of appearance and a gentle, warm, and relaxed illumination effect affected participant preference for luminaries.

(4) The luminary ranked first had β values for functional ease of operation and appearance attaining significance. The luminary samples ranked second and third were significantly for all R values. The six dependent variables, namely appearance preference, illumination effect preference, illumination effect comfort, functional ease of understanding, functional ease of operation, and uniqueness, were significantly correlated with sense of pleasure. The luminary sample ranked third in preference had significant β values for functional ease of operation and appearance. Analysis of these three luminaries indicated that the QisDesign luminaries featured design factors at the visceral and behavioral levels that elicited users' sense of pleasure, whereas factors at the reflective level did not exhibit a significant effect.

(5) For future studies on luminaries and user perception, researchers should consider light source factors along with appearance factors at the visceral level to comprehensively examine product characteristics. Second, at the reflective level, researchers can convey more critical product information, such as a product's underlying cultural significance and other meanings such that survey respondents can have higher-level experiences. Finally, we suggest that this research framework be used in the emotional value of other prominent luminary brands to acquire more profound referential data for future design.

References

1. Walter, A.: Designing for Emotion. A Book Apart (2011)
2. Slater, A.I., Boyce, P.R.: Illuminance uniformity on desks: where is the limit? Light. Res. Technol. **22**(4), 165–174 (1990)
3. Chang, C.-Y., Shie, M.-Y., Feng, C.-C., Chang, J.-Y., Lai, S.-J.: Psychological responses toward light and heat in interior lighting. J. Design Sci. **12**(1), 103–127 (2009)
4. Chang, C.-F., Wang, T.-Y., Liu, C.-H.: Research and analysis of the unified glare rating (UGR) of indoor lighting. J. Technol. **28**(4), 243–249 (2013)
5. Sun, C.-C., Chen, C.-H.: Development of LED and its application in lighting. J. Adv. Technol. Manage. **2011**, 1–23 (2011)
6. Chien, C.-W., Lin, C.-L., Lin, R.-T.: The study of period style in products. J. Natl. Taiwan Coll. Arts **99**, 1–27 (2016)
7. Martinsons, C., Zissis, G.: Potential health issues of solid-state lighting. International Energy Agency Solid State Lighting Annex, Report number: Energy Efficient End-Use Equipment (4E) (2014)
8. Demirbilek, O., Sener, B.: Product design, semantics and emotional response. Ergonomics **46**(13–14), 1346–1360 (2003)
9. Juslén, H., Tenner, A.: Mechanisms involved in enhancing human performance by changing the lighting in the industrial workplace. Int. J. Ind. Ergon. **35**(9), 843–855 (2005)
10. Heschong Mahone Group: Daylighting in Schools. An investigation into the relationship between daylight and human performance. Detailed Report. Fair Oaks, CA (1999)
11. Hsiao, H.-C.: Classroom lighting status quo and improvement plan. Educat. Bimonthly **333**, P13-17 (1995)
12. Bullough, J.D., Sweater Hickcox, K., Klein, T.R., Narendran, N.: Effects of flicker characteristics from solid-state lighting on detection, acceptability and comfort. Light. Res. Technol. **43**(3), 269 (2011)
13. Smolders, K.C.H.J., de Kort, Y.A.W., Cluitmans, P.J.M.: A higher illuminance induces alertness even during office hours: findings on subjective measures, task performance and heart rate measures. Physiol. Behav. **107**(1), 7–16 (2012)
14. Klepeis, N., Nelson, W., Ott, W., et al.: The national human activity pattern survey (NHAPS): a resource for assessing exposure to environmental pollutants. J. Expo. Sci. Environ. Epidemiol. **11**, 231–252 (2001)
15. Hsiao, K.-A., Chen, P.-Y.: Cognition and shape features of pleasure images. J. Design **15**(2), 1–17 (2010)
16. Leather, P., Pyrgas, M., Beale, D., Lawrence, C.: Windows in the workplace: sunlight, view, and occupational stress. Environ. Behav. **30**(6), 739–762 (1998)
17. Lu, S., Liu, L., Yu, A.: Exploring the Stroboscopic analysis of lighting products and its impact on functional lighting. Light Light. **2014**(4), P22-27 (2014)
18. Hsieh, M., Li, G.: Effects of illumination types and color temperatures on office worker's moods and task performance from the viewpoint of eco-efficiency -a case of subjects at the age of 20–28. J. Archit. **102**, P1-18 (2017)
19. Wang, N., Boubekri, M.: Investigation of declared seating preference and measured cognitive performance in a sunlit room. J. Environ. Psychol. **30**(2), 226–238 (2010)
20. Norman, D.A.: The Design of Everyday Things. Basic Books, New York (1990)
21. Norman, D.A.: Emotional Design: Why We Love (or Hate) Everyday Things. Basic, New York (2004)
22. Ogando-Martínez, A., López-Gómez, J., Febrero-Garrido, L.: Maintenance factor identification in outdoor lighting installations using simulation and optimization techniques. Energies **11**(8), 2169 (2018)

23. Pimputkar, S., Speck, J., DenBaars, S., et al.: Prospects for LED lighting. Nat. Photon **3**, 180–182 (2009)
24. Lin, R.-T.: The integration of Human needs and technology - the Cultural Innovation. Sci. Dev. **396**, 68–75 (2005)
25. Lin, R., Su, C.-H., Chang, S.-H. From cultural creativity to qualia. In: Conference of Taiwan Institute of Kansei. Tunghai University, Taiwan (2010)
26. Lin, R.-T., Lin, P.-H.: A study of integrating culture and aesthetics to promote cultural and creative industries. J. Natl. Taiwan Coll. Arts **85**, 81–105 (2009)
27. Lin, S.-S., Ke, J.-T., Wu, C.-F.: Exploring Key Factors of LED Lamps for Writing Scenarios. J. Design, 1249–1254 (2015)
28. Su, W.-Z., Lin, P.-H., Han, F.-N.: The study of cultural and creative products' emotional design and customer purchase intention. J. CAGST 213–224 (2018)

Identification of Product Functional Images Among Older Adults

Li-Hao Chen[1]($\boxtimes$), Yi-Chien Liu[2], and Chun Wang[1]

[1] Fu Jen Catholic University, No. 510, Zhongzheng Rd., Xinzhuang District,
New Taipei City 24205, Taiwan
[2] Cardinal Tien Hospital, No. 362, Zhongzheng Rd., Xindian District,
New Taipei City 23148, Taiwan

Abstract. Older users require substantial consideration in product design. Functional images play a crucial role in the interaction users have with products. This study investigated the characteristics of functional images that are identifiable for older users. Six sets of images representing different product functions were employed in the test, and each set contained images represented based on five levels of stylization from concrete to abstract. A total of 30 older users participated in the image identifiability sequencing test and interview. The results indicated that excessively simplified and abstract images were difficult for participants to identify, whereas those depicted through pictorial illustration or graphic rendering were the easiest to recognize. The results are expected to provide a reference in product design for older users.

Keywords: Older adults · User interface · Intuitive interaction

1 Introduction

Following technological development, product appearance is no longer the sole concern for product designers. User interface planning and information presentation substantially affect the interactivity between users and products. Enabling users to operate products smoothly is paramount. Courage and Baxter [1] contended that products should be user-friendly rather than requiring users to adapt to product design. In Taiwan, where population ageing has reached a critical level, older users must be prioritized when considering product design. Regression in physical and cognitive functions affects older users' product use performance [2]. However, images, labels, and symbols on products are components of the user interface [3] and can be utilized as a means of communication between users and products [4]. In other words, these images represent the intended functions of products and play a crucial role in the interaction between users and products. Wang and Chen [5] applied six sets of functional images, which included both concretized and abstract images, and explored their identifiability among older users. The test function images used in the study were divided into six groups: (1) reservation, (2) cooking, (3) adjustment, (4) switch, (5) timing, and (6) menu functions. Each group consisted of five sample images (realistic or abstract). The test images were redrawn in

© Springer Nature Switzerland AG 2021
P.-L. P. Rau (Ed.): HCII 2021, LNCS 12771, pp. 224–233, 2021.
https://doi.org/10.1007/978-3-030-77074-7_18

black and white. The size of each image (2 × 2 cm) was identical to the typical size of an icon in a user interface (see Fig. 1). Twenty-five elderly people were invited to participate the test as the subjects, and they were asked to list the five images based on the level of recognition.

The results revealed that these users identified the concretized and familiar images the most easily; variations were observed in their ability to recognize moderately simplified images, and associating abstract images with real-life objects or events was difficult for these participants. This finding indicated that the abstract images did not effectively convey the intended functions of the products and thus confirms that users' prior product usage experience and knowledge critically affects their cognition of functional images. Images intended for older users must describe product functions comprehensively. However, most functional images for products currently in use have undergone certain levels of simplification. In-depth research on the effect of different levels of image simplification on the ability of older users to understand functional images is paramount.

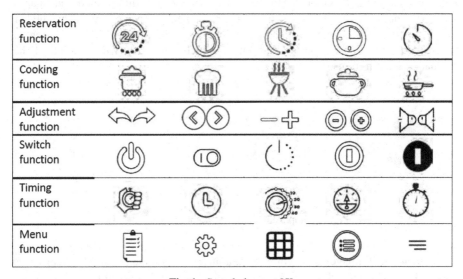

Fig. 1. Sample images [5].

Designers typically adopt a variety of approaches to stylize objects as simplified images [6]. Meyer and Lavenson [7] categorized functional images, from the most concretized to the most abstract, into the following five levels of stylization (Fig. 2):

(1) Natural photography: presents a real object in a planar space using concrete photography. This is the highest of the five levels of stylisation.
(2) Pictorial illustration: presents a 3D image through illustration. Visual details are supplemented to make the image closely resemble natural photography.
(3) Graphic rendering: shape and profile of an object are maintained, whereas its colours and 3D regions are planarized.
(4) Graphic symbology: details of the shape, colour, and spatial attributes of an object are omitted for planarized and cubic presentation.

(5) Abstract symbology: represent an object as an abstract graphic symbol, and express the concept of the image using extremely simple geometric lines.

| Natural photography | Pictorial illustration | Graphic rendering | Graphic symbology | Abstract symbology |

Fig. 2. Five levels of stylization [7].

Ryan and Schwartz [8] stated the levels of recognition in various presentation forms, simplified and complex graphics (Fig. 3). The results of the study stated that images with higher definition are not necessarily more recognizable (such as Fig. 3D had higher recognition than Fig. 3A and B; the recognition of Fig. 2C was the lowest). Currently, most product function images are simplified, which might affect recognition of the products' function for users.

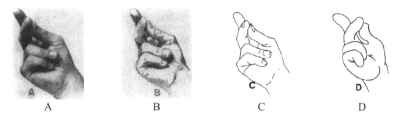

A   B   C   D

Fig. 3. Graphical simplification of the hand [8].

As mentioned, images and symbols are a means of communication between users and products. Understanding older users' identification of functional images is crucial for designing user interfaces and intuitive operation designs that fulfil the needs of these users. Following Wang and Chen s' study [5] mentioned above, this study further explores the different levels of image simplification on the ability of older users to understand functional images.

2   Method

2.1   Participants

The participants of this study were 30 adults aged 65 years or higher (26 women and 4 men), who exhibited normal vision, language communication, and behaviors and consented to participate in the image identification test. The average age of the participants was 71.4 years (SD = ± 5.7).

2.2 Test Procedure

The researchers explained the goal and required tasks of this study before commencing the test. After the participants understood the details of the tasks, they were asked to practice sequencing the demonstration images; the sequencing results were not included in the data analysis. The test employed six sets of images, each containing images of all the aforementioned five levels of stylization; all the images were printed on 13 × 13 cm paper cards (Fig. 4). The test sites were the participants' home or isolated, undisturbed locations. The test was conducted on each participant individually. The participants were provided with an unlimited amount of time to view each set of images and sequence them according to their functional identification level. The easiest images to identify were listed as Sequence 1, the second easiest images to identify were listed as Sequence 2, and so on. The most difficult images to identify were listed as Sequence 5. The participants indicated the sequences of the images verbally, and the researchers recorded the sequencing results. After the image s process was completed, the researchers conducted a short interview with the participants to ascertain the frequency with which the participants used home appliances and the reasons for their sequencing results.

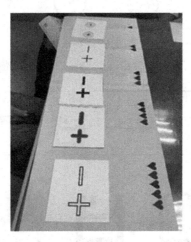

Fig. 4. Image cards used in the test.

2.3 Test Samples

The six sets of functional images concerned the following functions: delay start, cooking, adjustment, timer, power, and archive. Each set consisted of images designed based on the aforementioned five levels of stylization proposed by Meyer and Lavenson [7], as indicated in Table 1. To prevent the influence of participants' perception by colour, all of the images were printed in greyscale. Furthermore, to prevent the learning effect in participants, the images in each set were displayed in a random order rather than according to their level of stylization (Fig. 5).

Table 1. Test images.

Name	Images and codes				
	A_1	A_2	A_3	A_4	A_5
Delay start					
	B_1	B_2	B_3	B_4	B_5
Cooking					
	C_1	C_2	C_3	C_4	C_5
Adjustment					
	D_1	D_2	D_3	D_4	D_5
Timer					
	E_1	E_2	E_3	E_4	E_5
Power					
	F_1	F_2	F_3	F_4	F_5
Archive					

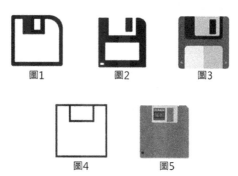

Fig. 5. Random arrangement of the test image.

3 Results

3.1 Image Sequencing Results

Table 2 lists the sequencing results of the six sets of images by the participants. The respective sequencing results of the delay start, cooking, adjustment, timer, power, and archive sets were $A_3 \rightarrow A_2 \rightarrow A_1 \rightarrow A_4 \rightarrow A_5$, $B_3 \rightarrow B_1 \rightarrow B_4 \rightarrow B_2 \rightarrow B_5$, $C_2 \rightarrow C_3 \rightarrow C_4 \rightarrow C_5 \rightarrow C_1$, $D_2 \rightarrow D_1 \rightarrow D_3 \rightarrow D_4 \rightarrow D_5$, $E_4 \rightarrow E_3 \rightarrow E_2 \rightarrow E_1 \rightarrow E_5$, and $F_2 \rightarrow F_4 \rightarrow F_3 \rightarrow F_1 \rightarrow F_5$, respectively. According to a Kendall's W analysis, with the exception of the archive set, all of the sets reached the level of significance in

Table 2. Statistics for test images.

Name	Image	Mean	SD	Kendall's W	Significance (p)
Delay start	A_1	2.90	1.21	0.546	0.000
	A_2	1.93	0.69		
	A_3	1.80	0.85		
	A_4	4.00	0.74		
	A_5	4.37	1.22		
Cooking	B_1	2.57	1.36	0.202	0.000
	B_2	3.33	0.84		
	B_3	2.43	1.25		
	B_4	2.57	1.38		
	B_5	4.10	1.49		
Adjustment	C_1	4.40	1.28	0.588	0.000
	C_2	1.50	0.82		
	C_3	2.00	0.79		
	C_4	3.37	0.93		
	C_5	3.73	0.69		
Timer	D_1	2.50	1.76	0.234	0.000
	D_2	2.37	1.13		
	D_3	2.67	0.76		
	D_4	3.23	1.17		
	D_5	4.23	1.28		
Power	E_1	3.07	1.70	0.080	0.048
	E_2	2.90	1.16		
	E_3	2.67	1.06		
	E_4	2.63	1.59		
	E_5	3.73	1.29		
Archive	F_1	3.03	1.69	0.031	0.451
	F_2	2.70	1.24		
	F_3	3.00	1.49		
	F_4	2.83	1.15		
	F_5	3.43	1.46		

their image sequence consistency ($p < .05$). However, in contrast to the relatively high sequence consistency in the delay start and adjustment sets, that in the cooking, timer, and power sets was low.

The results indicated that the participants did not identify the natural photography images the most satisfactorily; rather, they most easily identified the pictorial illustration and graphic rendering images. With the exception of the images in the adjustment set, the abstract symbology images were the least identifiable for the participants. This was because the images in the adjustment set were difficult to distinguish among the five levels of stylization, rendering the examination of their identifiability trend difficult.

3.2   Interview Results

The interview results were divided into two parts, namely the frequency with which participants used home appliances and high-technology products and their opinions regarding the functional images on product user interfaces. Table 3 lists the frequencies with which participants used home appliances and high-technology products. Most of these participants had used home appliances frequently prior to this study. However, only half of the participants had used high-technology products prior to this study, indicating that the participants varied regarding their need and acceptance for these products.

Table 3. Interview with participants regarding home appliance and high-technology product usage frequency.

Question	Responses		
Used home appliances?	Yes: 30 participants (100%)	No: 0 participants	
Frequency of using home appliances	Always: 25 participants (83%) (Every day)	Often: 5 participants (17%) (3–5 times per week)	
Used high-technology products?	Yes: 14 participants (47%)	No: 16 participants (53%)	
Frequency of using high-technology products	Always: 9 participants (64%) (Every day)	Often: 4 participants (29%) (3–5 times per week)	Rarely: 1 participant (7%) (1–2 times per week)

Table 4 presents the participants' opinions regarding functional images in product user interfaces, which reveal that most of the participants considered concretised images to be the best style for representing products. Many of the participants expressed that applying both text and images improved their understanding of product functions, but some contended that text used in user interfaces was small and required additional effort to read. The participants in favor of simplified images maintained that interface images should be illustrated through a simple and clear style.

Table 4. Interview on the image styles.

Age	Image style						
	Concretized	Abstract/Simplified	2D	3D	Image only	Text only	Image and text
65–70 (18)	11 (61%)	2 (11%)	2 (11%)	3 (17%)	7 (39%)	2 (11%)	9 (50%)
71–75 (6)	3 (50%)	1 (16%)	2 (34%)	0	3 (50%)	1 (17%)	2 (33%)
76–80 (4)	3 (75%)	0	0	1 (25%)	2 (50%)	0	2 (50%)
81up (2)	0	0	1 (50%)	1 (50%)	0	0	2 (100%)

4 Discussions

The results of this study indicated that oversimplified images (abstract symbology) were the least identifiable and those depicted by pictorial illustrations and graphic rendering were the most identifiable for the participants. For example, as shown in Fig. 6, for the delay start set, the participants indicated that A_3 and A_2 did not differ significantly but that A_3 was more identifiable than A_2 because it included a larger black area. On the other hand, A_5 was considered to be difficult to recognize as a representation of time by most of the participants because only four lines were used to express time intervals.

A2

A3

A5

Fig. 6. Test images.

With the exception of the archive set (F), the image sequencing results in all of the sets reached a level of significance. Because most of the participants had no experience using computer disks, they were unable to sequence the archive set images directly. When the participants understood the product function conveyed by a set of images, they were able to sequence the images according to their identifiability. Most studies have focused on users with normal cognitive and behavioral capabilities, and those examining older users have primarily focused on users aged 65 years or older. The lifestyles of participants varied considerably regarding their needs for product design [2] because of usage experience, motor functions, and cognitive abilities.

Research has indicated that older users' familiarity with technology must be considered when examining their intuitive interactions [9, 10]. Wang and Chen [5] also contended that older users could most easily identify concretized and familiar images. Substantial differences exist among older users regarding education level, technological familiarity, and cognitive ability. Therefore, functional images intended for use by older users should have designs that maximize concretization and avoid excessive simplification. Furthermore, the experience and knowledge users have of product usage critically affect their intuitive interactions with products. Understanding target users' knowledge concerning the product functions conveyed by images is paramount when investigating the design of intuitively interactive functional images.

5   Conclusions

This study investigated older users' cognitive understanding of product functional images. The results indicated that images depicted through pictorial illustration or graphic rendering were the easiest to identify for older users, whereas images depicted through abstract symbology were difficult for these users to apprehend. All of the test images applied in this study were printed without colors. Therefore, future studies may employ identical images printed in different colors to explore the effect of colors on the identifiability of functional images for older users. Additional image styles can also be employed in further research. Comparing and generalizing these research findings may further clarify the image characteristics that enable older users to intuitively recognize product functions, thereby facilitating intuitive interactions between users and products.

Acknowledgements. This research was supported by the grant from the Ministry of Science and Technology, Taiwan, Grant MOST 109-2410-H-030-014.

References

1. Courage, C., Baxter, K.: Understanding Your Users: A Practical Guide to User Requirements Methods, Tools, and Techniques. Morgan Kaufmann Publishers, Boston (2005)
2. Lee, C.F.: Approaches to product design for the elderly. J. Design 11(3), 65–79 (2006). (in Chinese)
3. Hackos, J.T., Redish, J.C.: User and Task Analysis for Interface Design. Wiley, New York (1998)
4. Thimbleby, H.: User Interface Design. ACM Press, New York (1990)
5. Wang, C., Chen, L.H.: Cognitive study of products' user interfaces for use by elderly people. In: Bohemia, E., de Bont, C., Holm, L.S. (eds.) The Academy for Design Innovation Management. LNCS, vol. 4, pp. 179–187. Design Management Academy, London (2017). https://doi.org/10.21606/dma.2017.66
6. Hsu, C.C., Wang, R.W.Y.: The Relationship between shape features and degrees of graphic simplification. J. Design 15(3), 87–105 (2010). (in Chinese)
7. Meyer, R.P., Laveson, J.I.: An experience judgment approach to tactical flight training. In: Sugarman, R.C. (ed.) the Human Factors Society 25th Annual Meeting, pp. 657–660. Human Factors Society, Santa Monica (1981)

8. Ryan, T.A., Schwartz, C.B.: Speed of perception as a function of mode of representation. Am. J. Psychol. **69**, 60–69 (1956)
9. Blacker, A., Hurtienne, J.: Towards a unified view of intuitive interaction: definitions, models and tools across the world. MMI-Interaktiv **13**, 36–54 (2007)
10. Blackler, A., et al.: Researching intuitive interaction. In: Roozenburg, N., Chen, L.-L., Stappers, P.J. (eds.) IASDR 2011 Diversity and Unity, vol. 1, pp. 1–12. TU Delft, Delft (2011)

Universal Design: Auxiliary Chopsticks Design for the Elderly

Chien-Chih Chen[1,2]([✉]) and Chiwu Huang[1]

[1] College of Design, National Taipei University of Technology, Taipei, Taiwan
chen@mail.mcut.edu.tw, chiwu@ntut.edu.tw
[2] Industrial Design Department, Mingchi University of Technology, New Taipei City, Taiwan

Abstract. The concept of "Universal Design" is to design products that are easy to use for everyone. With the rapid increase of the elderly population in the world, the principle of "Universal Design" has become one of the essential knowledges and design considerations for designers. The physical and mental decline of the elderly makes it difficult for them to use traditional chopsticks. The purpose of this research is to use the principle of "universal design" to design a set of auxiliary chopsticks for the elderly. First of all, this study found out the problems of using chopsticks for the elderly. Then, through expert focus group interviews and morphological analysis, the characteristics and functional requirements of chopsticks were summarized. The authors conceived creative ideas and made evaluation prototypes. The new chopsticks have passed two-stage assessment of elderly subjects. In all 7 universal design principles, the evaluation of auxiliary chopsticks is superior to that of traditional chopsticks. Further research may be needed to make improvements.

Keywords: Universal design · Senior living aids · Auxiliary chopsticks

1 Introduction

Healthy adults are the default user base for most products. However, in real life, the characteristics of the elderly, children, pregnant women, the disabled and other population groups are very different from normal adults. Many studies have shown that muscle strength decreases significantly with age and affects people's ability [2, 9, 12].

Due to the physical and mental decline, the elderly cannot use normal methods or appliances in their daily lives. For example, in most Asian countries, chopsticks are the main tableware. However, as people get older, it becomes more and more difficult to use chopsticks. Based on the principle of "Universal Design", a set of auxiliary chopsticks for the aged is designed, which reduces the requirement of muscle strength and accuracy, and increases the practicability of chopsticks. It is hoped that this study can improve the difficulty of using chopsticks for the elderly and provide reference for the research and development of the auxiliary chopsticks for the elderly.

© Springer Nature Switzerland AG 2021
P.-L. P. Rau (Ed.): HCII 2021, LNCS 12771, pp. 234–243, 2021.
https://doi.org/10.1007/978-3-030-77074-7_19

2 Literature Review

2.1 Universal Design Principles

Universal design's philosophy is to design products and environments that are suitable for a wider range of users, without the need for adaptation or special design [10, 11]. Table 1 shows the philosophy of universal design and traditional designs. The goal of universal design is to obtain the maximum intersection of traditional product design and special design, as shown in Fig. 1.

Table 1. Philosophy of universal design and traditional designs

Universal design	Traditional design	Special design
1: Equitable Use The design is useful and marketable to any group of users	• For ordinary adults under normal use needs • Not always taking into account the needs of the disadvantaged.	• For specific individuals with specific needs, case by case • Not suitable for ordinary people, such as tailor-made wheelchairs
2: Flexibility in Use The design accommodates a wide range of individual preferences and abilities		
3: Simple and Intuitive Use Use of the design is easy to understand, regardless of the user's experience, knowledge, language skills, or current concentration level		
4: Perceptible Information The design communicates necessary information effectively to the user, regardless of ambient conditions or the user's sensory abilities		
5: Tolerance for Error The design minimizes hazards and the adverse consequences of accidental or unintended actions		
6: Low Physical Effort The design can be used efficiently, comfortably, and with a minimum of fatigue		
7: Size and Space for Approach and Use Appropriate size and space is provided for approach, reach, manipulation, and use regardless of user's body size, posture, or mobility		

Fig. 1. Universal design VS traditional product design

2.2 Aging and the Need for Auxiliary Chopsticks

The physiological structure of the human body can be roughly divided into the central nervous system, peripheral nervous system, skeletal muscle system and cardiovascular system [7]. The decline in physiological structure is the most important factor that affects the elderly when handling products. The most affected by aging is the ability of vision, hearing and muscle control. These three are also the most important capabilities when using the product. Table 2 lists the changes in related physiological structures, effects and some design suggestions [1].

It has been found from many studies that the elderly often encounters many difficulties in daily activities [1, 3]. Although the physical and mental functions of the elderly cannot be improved, they can use assistive devices to improve their independence, efficiency and safety.

The use of assistive devices can have the following functions: 1. Make work more efficient, including labor saving, time saving or making work more comfortable; 2. Promote independence of life and improve quality of life. The use of assistive devices can not only reduce the burden of caregivers, but also allow people with physical and mental disabilities to be free from the limitations of their physical functions. 3. Allow people with dysfunction to start using the functions of the body as early as possible, not only to keep their ability from degrading, but also to promote the development of functions. 4. Increase the safety of activities [8].

Taiwan entered the aged society in 2018 (more than 14% of the population is elderly). In 2026, Taiwan will become an over-aged society (more than 20% of the population is elderly). The transition time will be faster than in Europe, the US and Japan. The aging rate is second only to that of Japan. Due to the increase in the elderly population, there are no suitable products to meet the needs of the elderly in the existing market, especially the tableware used in meals. Therefore, this study was carried out to study the auxiliary chopsticks, hoping to help the elderly to eat and their ability to take care of themselves.

Table 2. The design suggestions of the human-machine interface for the elderly

Physiological structure changes	Effects of changes in physiological structure	Design suggestions
Central nervous system changes • Brain cell reduction • Brain weight reduction • Reduced blood flow in the brain • Autonomic nervous aging	• Memory decline • Poor learning ability • Weakened logical thinking ability • Slow response speed • Reduced sleep time • Poor temperature regulation	• Simplified operation procedures • Promoting cognitive function • Group keys according to importance • Avoid possible confusion, • Avoid irrelevant information • Using one-handed operation
Peripheral nervous system – visual changes • The crystal of the crystal turns yellow, turbid, loss of flexibility • The pupil becomes smaller • Control the muscle decline of the crystal	• Absorb more short wave (blue) • Increased scattering, difficult to focus • Reduced field of view • Less transmitted light • The closest distance one can see clearly increases • Reduce light regulation • Increased discomfort to glare	• Provide adequate lighting • Avoid glare • Use a non-reflective surface • Use patterns and text as much as possible • Isolate important information • Use appropriate text size and stroke thickness • Strengthen the contrast between text and background • Use contrasting colors
Peripheral nervous system – auditory changes • The tympanic membrane becomes thin and atrophic • Ossicular degeneration in the middle ear • Loss of neuro elements in the cochlea	• Reduced sensitivity to sound (10 decibels higher than young people at age 70 to be heard) • Reduced range of audible frequencies, especially higher frequencies above 2000 Hz	• Providing volume control to suit the hearing level of different users • Providing multiple sensory means for alarms, signals and control keys, such as a combination of sound and visual display • Reduction of noise from products • Use low-frequency sounds as a warning and urgent message
Skeletal muscle system changes • Bone loss • Spine atrophy, bending • Joint and ligament sclerosis • Articular cartilage wear • Muscle volume reduction	• Osteoporosis, easy to fracture • Bent body, height becomes shorter • Reduced limb mobility and a sense of stiffness • Arthritis • Weakened muscle strength • Decreased movement agility • Decreased endurance	• The size and shape of the controller can be easily held with one hand • Surface texture can help hold • The required force is multiplied by 75% of the data found • Provide multiple operation modes, and use larger levels of movement and joints • Avoid accidental touches of important keys • Provide an adjustable user interface

2.3 Related Research

Lee and Chen (2012) conducted a study on the performance of the chopstick auxiliary device and its position on the chopsticks. The results showed that the 15 mm wide auxiliary device performs best in terms of clamping force, accuracy, stability and subjective evaluation [5]. Wu (2009) pointed out that increasing the holding position of the chopsticks handle can effectively reduce the force applied when operating the chopsticks, and the design of the slanting front end and face of the chopsticks tip can reduce the hand bending and increase the contact area with food [13]. According to Lee (2007), the chopstick auxiliary device can successfully guide or correct the scissors operator to switch to the correct chopstick [4]. Studies by Lin (2002) show that chopsticks with textured tips are more likely to pick up larger objects. Table 3 summarizes the relevant research [6]. On the basis of the above-mentioned literature, some design specifications of chopsticks were established in this study.

Table 3. Summary of studies on chopsticks design

Authors (year)	Research method	Conclusion
Yu-Chi Lee, Yi-Lang Chen, 2012	24 males with shearing experience use 3 size (15, 25 and 35 mm) chopsticks and 6 auxiliary combinations at 2 positions (on top of chopsticks and 1/3 stem) to perform 3 simulation tasks (Force, accuracy and stability)	The size of the auxiliary chopsticks affects the performance and subjective evaluation. The auxiliary equipment with a width of 15 mm shows the best performance. The position of the auxiliary equipment only affects the squeezing force
Sui-Chi Wu, 2009	Use myoelectric signal to record the muscle activity of four sets of chopsticks, the subject's subjective preference scale	Larger chopstick stem could decrease the operation force effectively; rake and flat tip of chopsticks could decrease hand bending and increase the area of food contact
Yu-Chi Lee, 2007	40 subjects, through the use of non-dominant hand combined with auxiliary tools to simulate novice chopsticks with no experience to evaluate the learning effect of auxiliary tools	The new auxiliary device can guide the scissor operator using the chopsticks correctly. The assist device provides a quick, easy and effective method for beginners to learn how to hold chopsticks correctly
Tzu-Fen Lin, 2001	Used 8 different combinations of experimental chopsticks (4 kinds of materials, pulled and pulled), the effects of job performance, subjective evaluation and subjective preference	The experimental results prove that the pliers-style gripping method and the chopsticks with textured chopstick tips are easier to pick up larger objects (longan)

3   Method

3.1   Problems and Requirements

Chopsticks are shaped pairs of equal-length sticks used as eating utensils in most of East Asia for over millennia. Because the use of traditional chopsticks requires fine finger muscle control, which is more difficult for the elderly. Therefore, in order to find out the inconvenience and shortcomings of traditional chopsticks, this study observed the user's operating procedures. Then, focus group interviews were conducted with 2 human factors engineering experts and 2 product design experts to synthesize chopsticks that meet human factors, usability, and structural requirements. Finally, the problems of the traditional chopsticks and the functional requirements of the new auxiliary chopsticks are listed, as shown in Table 4.

Table 4. Traditional chopsticks problems and new chopsticks requirements

Traditional chopsticks problems	New chopsticks requirements
• For ordinary adults • Not flexible • Not easy to learn • Hard to Use • Fall off easily • Laborious • Not easy to pick up food	• Suitable for more groups • Flexible • Facilitate learning structure • Easy to use • One-piece chopsticks • Labor saving design • Easy to pick up food

3.2   Prototyping and Pilot Experiment

After several developments and reviews, the author produced a detailed evaluation of the prototype through a 3D printing method. The final design is shown in Fig. 2. In the course of development, the author recruited 10 elderly consultants as design reference in order to seek the opinions of the elderly users. When the final prototype was completed, the consultants assessed the use of chopsticks.

According to the technology acceptance model (TAM), Likert scales were used to investigate perceived usefulness (users' awareness of improving product efficiency), perceived ease of use (users' perception of product ease of use), use attitude (users' preference for products), and use behavior intention (users' willingness to use products in the future).

All the 10 elderly consultants who participated in the questionnaire were volunteers. Among them, 3 males and 7 females were healthy and could live independently with an average age of 80 years. They generally admire new chopsticks and are willing to use them. The results are shown in Table 5.

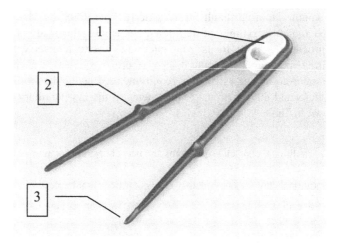

1. Detachable silicone elastic for labor saving
2. Enlarged finger grip area for comfort
3. Enlarged tip using an arc shape to increase the efficiency of clamping
4. One-piece design, the chopsticks will not separate during use
5. Using the gripping method to replace pinsers-pinching mode

Fig. 2. Features of new auxiliary chopsticks

Table 5. TAM statistical results

	Age			Education
	65–75	75–85	85–95	College
Male 3	0	1	2	3
Female 7	3	3	1	7
Item	Cognitive usefulness	Cognitive ease of use	Willingness to use	Use attitude
Score	3.9	4.0	4.1	4.0

3.3 Assessment of New Auxiliary Chopsticks

In order to further verify whether the new auxiliary chopsticks comply with the seven principles of universal design. A Likert scale questionnaire was designed to evaluate the performance of traditional wooden chopsticks and new auxiliary chopsticks under seven universal design criteria.

Subjects
In the study, 12 elderly people aged 65 to 75 were surveyed by questionnaires and the actual use of traditional wooden chopsticks and new auxiliary chopsticks.

Experimental Chopsticks
The wooden chopsticks used in the experiment were 220 mm in length. The chopsticks ranged in width from 6 to 3 mm diameters from handle to tip, respectively.

In contrast to traditional chopsticks, the new auxiliary chopsticks are 220 mm in length, ranged in width from 7 to 3 mm diameters from handle to tip, respectively. The chopstick has an elastic plastic between its two long sticks so it stays open at maximum, which also helps save effort, as shown in Fig. 2. In the middle of the chopsticks there is an enlarged finger rest for greater comfort. In order to improve the clamping efficiency, an enlarged circular arc surface is added to the clamping tip.

4   Results and Discussions

The statistical results of the questionnaire are shown in Table 6. Results of ANOVA showed the auxiliary chopsticks were significantly better than that of the traditional chopstick in all seven assessments of universal design principle ($p < 0.05$). The most highly rated item was "Low Physical Effort", followed by "Simple and Intuitive Use". On the contrary, the worst item for traditional chopsticks was "Low Physical Effort", and the second worst item was "Simple and Intuitive Use". Figures 3 shows the results of the evaluation using radar maps.

Table 6. Assessment of new auxiliary chopsticks and traditional chopsticks (n = 12)

UD principle	Auxiliary chopsticks		Traditional chopsticks		Differences
	Mean	SD	Mean	SD	
1. Equitable Use	4.08	0.67	3.58	0.51	0.50*
2. Flexibility in Use	4.00	0.74	3.25	0.75	0.75*
3. Simple and Intuitive Use	4.25	0.45	3.00	0.60	1.25*
4. Perceptible Information	3.92	0.67	3.08	0.51	0.84*
5. Tolerance for Error	4.00	0.60	3.42	0.67	0.58*
6. Low Physical Effort	4.42	0.51	2.92	0.51	1.50*
7. Size and Space for Approach and Use	3.83	0.72	3.17	0.72	0.66*

*$p < 0.05$

Fig. 3. Assessment results of new auxiliary chopsticks

5   Conclusion

The elderly often encounters problems when using traditional chopsticks, such as the difficulty of precise operation of chopsticks, laborious, hard to pick up food and so on. Using the concept of "Universal Design", a new type of chopsticks device was developed to assist the elderly to eat more effectively. This study improves these problems through expert interviews, morphological analysis and creative research. The new chopsticks went through two stages of evaluation, proving that the usability of auxiliary chopsticks is superior to traditional chopsticks in all assessments of universal design principle. The advantages of the new design are easy to use, labor saving and easy to pick food. The development of the chopsticks auxiliary device can not only help the elderly or the disabled, but also benefit others.

References

1. Chen, C.-C.: Human-machine interface design beyond age. Mingchi Inst. Technol. J. **28**, 169–176 (1996). (in Chinese)
2. Hurley, B.F.: Age, gender, and muscle strength. J. Gerontol. Ser. A **50**, 41–44 (1995)
3. Kroemer, K., Kroemer, H., Kroemer-Elbert, K.: Chapter 12: designing for special populations in ergonomics, pp. 601–645. Prentice Hall, NJ (1994)
4. Lee, Y.-C.: An auxiliary device for chopsticks operation to improve the food-serving performance. Mingchi University of Technology, Master thesis, New Taipei City (2008). (in Chinese)

5. Lee, Y.-C., Chen, Y.-L.: Optimal combination of auxiliary device size and its location on chopsticks for food-gripping performances. Percept. Motor Skills: Motor Skills Ergon. **115**(1), 187–196 (2012)
6. Lin, T.-F.: The study of the effects of pinching operation, material of chopsticks, and carved groove on food-serving performance. Huafan University, Master thesis, New Taipei City (2001). (in Chinese)
7. Liu, S.: Gerontology and Gerontology. Hochi Publisher, Taipei (1991)
8. Luo, J.: Assistive devices, friends of assistive devices newsletter, no. 5, pp. 3–5. Resource services for first rehabilitation assistive devices, Taipei (1998). (in Chinese)
9. Melzer, I., Benjuva, N., Kaplanski, J.: Age related changes in muscle strength and fatigue. Isokinetics Exerc. Sci. **8**, 73–83 (2000)
10. Pirkl, J.J.: Transgenerational Design—Products for an Aging Population. Van Nostrand Reinholk, New York (1994)
11. The Center for Universal Design, The Universal Design File: Designing for People of All Ages and Abilities NC State University (1998)
12. Ustinova, K.I., Ioffe, M., Chernikova, L.: Age-related features of the voluntary control of the upright posture. Hum. Physiol. **29**, 724–728 (2004)
13. Wu, S.-C.: Ergonomic evaluation and redesign of chopsticks for the elderly. Kaohsiung First University of Science and Technology, Master thesis, Kaohsiung (2009). (in Chinese)

Placemaking with Creation: A Case Study in Cultural Product Design

I-Ying Chiang[1,2(✉)], Rungtai Lin[1], and Po-Hsien Lin[1]

[1] Graduate School of Creative Industry, National Taiwan University of Arts,
New Taipei City, Taiwan
iychiang@mx.nthu.edu.tw, rtlin@mail.ntua.edu.tw,
t0131@ntua.edu.tw
[2] Department of Arts and Design, National Tsing Hua University, Hsinchu, Taiwan

Abstract. Urban development is often a priority policy measure for national governments, leading to shocks such as demographic dissimilarity and imbalances in urban-rural development. Humanity is now facing undeniable challenges that extend to localities, societies, and ways of life. Taking a small mountain village as its case study, this study investigates how to discover and mobilize local cultural resources while using cultural added-value design innovation to create qualia products and captivating experiences. Our results show that the design, development, and promotion of local cultural products has huge potential in stimulating visits from non-locals, re-attracting former residents, and turning intermediary towns into flourishing examples of the aesthetic economy. In this light, the questions of how to integrate the lifestyle of local residents, how to conduct fieldwork surveys and data collection, how to investigate models for the recognition and communication of cultural codes, and how to develop methods of extraction and frameworks of analysis for cultural factors, thereby providing modes of translation for cultural product design, are all well worthy of closer examination. This study is of important practical value for increasingly marginalized communities that are promoting regional revitalization in terms of placemaking. By developing unique local design and shaping the aesthetics and individuality of local brands, we can facilitate both cultural communication and sustainable innovation in the face of neoliberalistic global competition, thereby achieving the shared ideal of innovative transformation.

Keywords: Placemaking · Cultural product · Cultural innovation · Translation · Qualia · Hengshan

1 Introduction

In 2017, Taiwan's Executive Yuan, faced with problems such as declining birth rates, an aging population, and imbalances in urban-rural development, announced three major policy initiatives, namely, "Living and Working in Peace", "Growth in Nature", and "Balanced Taiwan". "Balanced Taiwan" envisions using local strengths to develop local industries. In doing so, it hopes to reverse the brain-drain, encourage young people to

© Springer Nature Switzerland AG 2021
P.-L. P. Rau (Ed.): HCII 2021, LNCS 12771, pp. 244–261, 2021.
https://doi.org/10.1007/978-3-030-77074-7_20

return to their hometowns, and solve demographic dissimilarities. In 2018, the National Development Council proposed its "National Strategy for Regional Revitalization", proactively promoting "placemaking" policies and fully engaging in related work. The council clearly specified that it would prioritize revitalization in agricultural, mountain and fishing villages, intermediary towns, and aborigine villages [11]. In 2019, the National Development Council ushered in the "Inaugural Year of Taiwan's Placemaking". Ever since, all kinds of issues and visions for regional revitalization have been the source of enthusiastic discussion and aspirational targets in the spheres of industry, government, education, and research. However, the questions of how to prepare the ground for regional revitalization, how to enable cultural communication and innovative industry to both take root, and how to achieve the ideal of sustainable development in rural areas driven by an aesthetic economy are intimately connected to the cultural consciousness of communities and their capacities for cultural translation. The uniqueness of regional cultures and the innovative construction of regional knowledge have already become core factors in the competitiveness of nation states. Provisions for the health of economies and the wealth of societies, while partially established on the foundations of technical proficiency and technological development, should also make efforts to discover innovative advantages and potentials in local culture, using cultural innovation for the communication of cultural wealth, shaping lifestyle choices, increasing industry competitiveness, and driving regional development [2].

Regional societies harbor rich cultural resources that derive from the texture of local everyday life. Regional character leads to a unique sense of beauty that is charged with cultural significance. Nourishing the diverse activities and products of regional cultures can lead to the development of a local aesthetic that is an important asset. It can also attract the participation and support of visitors (i.e., consumers). The changes in Taiwan's current demographics, its regional economic development and past regional policies all bear striking similarities to Japan. For example, the implementation of community empowerment, guidance for regional industries, and rural renewal policies are all similar to the spirit and approach of Japan's regional revitalization. As a result, this study argues that in the investigation of relevant regional revitalization research, we cannot neglect the experience and results of Japan's placemaking. Japan's experience in experiential marketing using local people, events, and attractions to reshape the value of remote regions is extraordinarily rich. For example, Niigata Prefecture's Echigo-Tsumari "Art Triennial" and "Satoyama Jujo" are both renowned instances of regional revitalization. Furthermore, since 2005, Niigata Prefecture has promoted a "Hyakunen-monogatari" (centennial values) project, using innovative design to transform and preserve traditional arts, showcasing the full scope of local arts and the charm of innovative design to the world, thereby paving the way for regional revitalization. These measures, which demonstrate the added value of cultural innovative design, have greatly inspired this study.

This study focuses on regional communities, more specifically on the Hengshan region. In our fieldwork we collected precious regional cultural resources, discovered unique regional aesthetics, extracted exclusive cultural elements, and investigated how "cultural literacy", "design coding", and "cognition and dissemination processes" can turn cultural resources, such as local cultural attractions, cultural and historical heritage, and natural scenery into "regional cultural products". This study also focuses on how "innovation" derived from regional culture is "transformed" with "value added" design, stimulating possibilities for regional revitalization. We hope that this study will be of practical use for mountain villages and intermediary towns that are promoting placemaking.

2 Literature Review

2.1 Cultural Resources and Design Translation

Cultural Spaces and Cultural Products. Leong and Clark have proposed a stratified framework for the study of cultural product design that divides culture spatially into outer, mid, and inner strata. Leong and Clark discuss the cultural carriers in each of these strata: the outer stratum includes tangible, physical mediums; the mid stratum covers rituals and customs of social behavior; and the inner stratum points toward intangible spiritual concepts [4]. Rungtai Lin has integrated this concept with Norman's three aspects of "Emotional Design" (visceral level, behavioral level, and reflective level) to analyze the cultural product attributes and design characteristics of each stratum [8]. This research serves as the theoretical basis and design standpoint of this paper's study of cultural integration (Fig. 1).

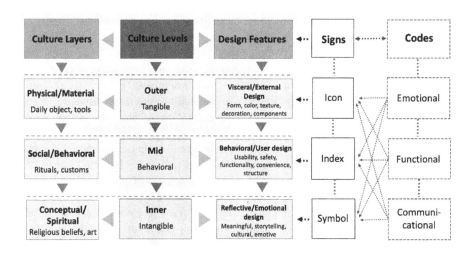

Fig. 1. Stratified framework for cultural spaces and sign coding [4, 7, 8, 13]

Signs, Codes, and Cultural Systems. Our understanding of signs is based on Ferdinand de Saussure's (1857–1913) research in the field of semiotics. Saussure proposed a key concept of semiotics, namely, that signs can be divided into the "signifier" and the "signified". The signifier is a form of expression, such as the shape of words, or sounds.

The signified is a concept or an object, such as the interpretation of words or their content. Saussure made a further distinction: there is not necessarily a relation between the signifier and signified, but rather an arbitrary, evolving method of use [13, 15]. Charles Sanders Peirce (1839–1914) proposed a further theory of types of signs, dividing them into icons, indexes, and symbols. In icons there is an obvious similarity between the signifier and the signified. In indexes there is an integral relation between the signifier and the signified. In symbols, there is an arbitrary and conventional relation between the signifier and the signified [13].

Some signs refer to broader systems of meaning. J. E. Williamson (1978) calls these "referent systems". S. Hall (1980) labels them code. G. Rose (2006) argues that code is a conventional form of meaning production that certain groups or communities share in common. She argues that the code itself influences the sign's content. We can use code as a means to gain entry to referent systems [13]. In regard to product design, Ming-Huang Lin views code as the smallest design unit, dividing it into three aspects: function code, emotion code, and communication code. This approach is a response to modernism, post-modernism, and product semantics in the history of industrial design. Lin argues that these three types of code can coexist within one carrier. Moreover, the effect of such coexistence and the interaction between code and signs influence our understanding of the world in step with differences in our personal cognition, senses, imagination, and feelings [7].

Solomon has explored the cultural production systems of intertextual concepts: The process of cultural production includes symbolic and semiotic cultural roots. The two guiding systems of innovative design and market management have also been introduced into the cultural production system, thereby influencing the consumer (audience). Subsequent consumer activities (social interaction) slowly form contemporary culture, establishing a symbiotic circle [14].

The cultural products discussed in this paper refer to the cultural content of cultural artifacts. Considering and examining these concepts anew, we use design philosophy to find innovative forms for cultural elements that accord with the present times. Designers use signs and their applications to allow consumers (the audience) to understand cultural products. By exploring the semiotic uses of cultural product design, they reflectively extend the meaning of cultural code to a spiritual plane. This study argues that the three types of signs proposed by Peirce, together with Ming-Huang Lin's three types of code, echo the three different translated uses of cultural space proposed above and the design coding of different product attributes (Fig. 1).

Design Translation of Regional Culture. Culture often manifests itself through cultural carriers. But how should we select appropriate regional cultural carriers? How can we extract unique cultural elements? "Unique insight" may appear to be a key talent, but in fact "cultural literacy" can be easily fostered; we need to personally visit a place, and get to know its people, events, and attractions first-hand through sincere exchange. After selecting a regional cultural carrier, which is to say after "investigation/setting the context", we must consecutively engage in "interaction/story telling" and "development/script writing" in order to begin the process of "implementation/product design". Then, according to the strata of the cultural carrier and the attributes of the product, we "transfer", "transit", or "transform" in order to extend the meaning and translate the cultural code. Finally, we complete the regional cultural product, offering a delightful experience of qualia products. This study has touched on the key points of culturally integrated design and the value added model of cultural product design [6, 8], which form the key framework for our "design translation model for regional culture" (Fig. 2).

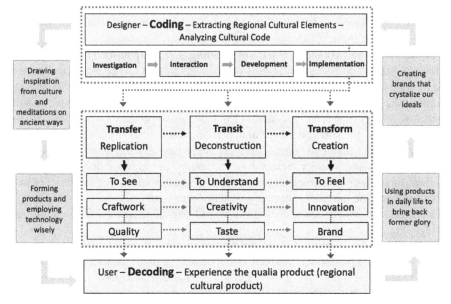

Fig. 2. Design and translation model for regional culture [6, 8]

2.2 Cognition and Dissemination Models of Cultural Products

Building on cognitive psychology, Donald A. Norman proposed the three main aspects of emotional design: visceral level, behavioral level, and reflective level. In the course of designing products, in addition to outer "aesthetic" elements, designers also need to consider the "functional" behavior that facilitates cognition and the "emotional" effect that concerns users' inner "feelings". This discourse has become an important yardstick that continues to influence contemporary product design. Norman showed that any product has three different psychological phenomena: 1. The designer's model, i.e., the designer's vision of his or her product during the moment of creation; 2. The user's model, i.e., the user's cognition of how a product or appliance should be used or operated; 3. The system form, i.e., the written explanation or manual that accompanies a product, conveying information about the system design and attending product impressions [12]. Often, designers expect that the designer model and user model will match. However, designers' expectations and users' understanding do not always meet perfectly. As a result, this study argues that it is of vital importance to investigate the mental cognition and information communication models associated with regional cultural product design (Fig. 3).

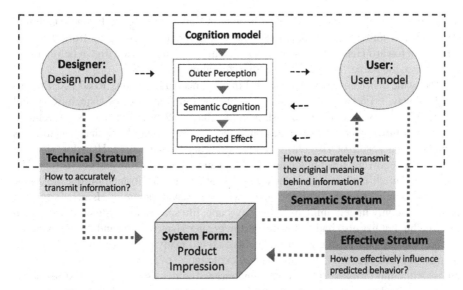

Fig. 3. Cognition and dissemination models of cultural products [9, 12]

2.3 Introduction to the Research Area

The research area of the present study is Hengshan Township in Hsinchu County. We investigate the cultural resources of the "Small (Shallow-dish) Mountain Town", its cultural literacy, cultural translation, added-value design, and cultural innovation to explore local wisdom, the aesthetics of daily life, as well as how life products invested with cultural meaning are created and innovated. As a result, understanding the historical background and contemporary issues is an important preliminary step before diving into the body of our research.

For nearly 300 years, from the T'ien-chi era (1624) during the Ming dynasty to the Kuang-hsu reign period under the Ch'ing dynasty, immigrants from the Minnan and Eastern Canton regions of mainland China crossed the straits by a number of means to reclaim and cultivate land in Taiwan. The immigrants formed a diverse society of land reclaimers. Before the Japanese Colonial Period, this process of reclamation progressed from south to north, initially driven by ethnic Chinese migrants from the prefectures of Ch'uanchou and Changchou and then by migrants from the eastern Cantonese prefectures of Ch'aochou, Huichou, and Chiayingchou. To distinguish the latter group from the migrants from Ch'uanchou and Changchou, these Cantonese reclaimers were referred to as "Hakka" people. Hengshan Township is located in northern Taiwan, in the upper reaches of the T'ouch'ien river, in the Youluo river catchment basin. From the Ch'ing times onward, the reclaimed land in the protected area and strategic pass area to the east of the T'uniu boundary has been inhabited mainly by Hakka communities [10]. The breadth of Hengshan Township from east to west is roughly 13.8 km. Its length from north to south is roughly 14.2 km. Its total ground area is 66.35 km^2. Approximately 90% of this area is hilly or mountainous. What remains is alluvial plain land. The whole area is below an altitude of 1,000 meters above sea level. The terrain rises from the northwest to the southeast, approaching the Snow Mountain Range. During the Japanese Colonial Period, it was known as "T'ien-pei (Sky-back) Mountain". Looking up from the Chu-tung basin toward Hengshan, it is as if a horizontal screen has been drawn before one's eyes, hence the name "Hengshan" (i.e., horizontal mountain) [3].

Today the district of Hengshan Township is connected by County Road 120, Provincial Highway 3, Freeway 68, and a railroad to the municipalities of Hsinchu City and Zhubei City as well as other townships and villages. While transport is convenient, freedom of movement has also accelerated the flow of people. Hengshan Township is located in the center of Hsinchu County, bordering Hsinchu City. Hsinchu City is a historically renowned town in the north of Taiwan with nearly 300 years of history. The town of "Chu-ch'ien", which formed in the 1830s, was transformed during the 1980s into a nationally renowned "technology town". The development of Hsinchu's science park has not only hugely influenced the cityscape, life spaces, and industrial ecosystem of Hsinchu, but has also had an impact on the greater Hsinchu area and the surrounding villages [1]. In the past, the government directed too many resources toward the development of cities, leading to the unremarked exploitation of resources in villages close to cities. Although farming villages were encouraged to develop leisure industries, these were conceived to meet the demands of city dwellers. As a result, efforts still need to be made to find a balance mechanism between the demands of urban and rural communities that both preserves the agency of villages and is environmentally sustainable. In 2017 the EU, facing changes in urban-rural relations in villages surrounding cities, together with the complications of an aging population, proposed a "smart village" project, arguing that "smart" denotes people, not technology. The agency of villages should not be limited to their immediate confines but should extend across districts, regions, and even national boundaries to more effectively integrate external resources and opportunities, search for innovative solutions, stimulate local development, and find venture opportunities for cooperation between urban and rural areas [5]. The Hengshan area is currently facing the shocks and difficulties of demographic dissimilarity (declining birth rate and aging population) and an imbalance in urban-rural development (prioritizing the city, neglecting villages). In this challenging context, what are the developmental restrictions and revitalization prospects of small mountain towns? These important issues are of vital interest to this study and the local residents.

3   Methodology

3.1   Case Study: Regional Cultural Product Design in Hengshan

This study focuses on Hengshan Township, conducting design development across three strata, namely: cultural space, cultural strata, and the attributes of cultural products. In terms of Product A. *"Landscape on Table"*, its attribute is outer. The external design was executed in reference to the visceral level, to achieve a sense-based design goal. Industrial remains were translated into concrete plant pots (Fig. 4). In regard to product B. *"Hengshan Tea Set"*, its attribute is middle. A functional design was executed in reference to the behavioral level to achieve a cognition-based design goal, culminating in a tea set design for leisure purposes (Fig. 5). For product C. *"Tranquility in Hengshan"*, its attribute is inner. An emotional design was executed in reference to the reflective level, to achieve a design goal that conveyed inner spirit and arts. The finished product is a quietly graceful piece of brooch (Fig. 6). While the three groups of cultural product attributes in this study each have varying designs and goals based on their orientations to cultural strata, in terms of cultural coding they are not confined to a unitary stratum of design development.

Fig. 4, 5, 6. Product A: *"Landscape on Table"*, Product B: *"Hengshan Tea Set"*, Product C: *"Tranquility in Hengshan"* (from left to right)

Fig. 7, 8, 9. The Asian concrete factory, Hengshan orange orchard, Organic tea farm (from left to right)

(1) Outer Level (Material)

- *"Landscape on Table":* A concrete plant pot translated from industrial remains. (Materials: Concrete, Brass) (Product A) (Fig. 4)
- **Selection of cultural carrier:** Asian Concrete Factory (Fig. 7)
- **Product Attribute Orientation:** Visceral level/External design
- **Cognitive Communication Model:** Technical stratum/Outer perception
- **Design Concept:** When diving deep into the investigation of the Hengshan area, one discovers the love-hate relationship that residents have with the Asian Concrete Company. On the one hand, they are thankful for the job opportunities that it has brought to Hengshan; on the other, they are bothered by the environmental pollution it has caused. No matter one's assessment, the Asian Concrete Park encompassing Jiouzantou station has stood witness to the highs and lows of Hengshan's development. This study drew inspiration from the form of the Asian Concrete Factory to design miniature landscapes using succulent plants.

(2) **Middle Level (Behavioral)**

- *"Hengshan Tea Set":* A tea set design for leisure purposes. (Materials: Glass, Maple Wood, Brass) (Product B) (Fig. 5)
- **Selection of cultural carrier:** Hengshan orange orchard (Fig. 8)
- **Product Attribute Orientation:** Behavioral level/Functional design
- **Cognitive Communication Model:** Semantic stratum/Significant cognition
- **Design Concept:** Early on, Hakka communities, known for their thrift and economy, cultivated the tradition of making dried, and pickled food products. On the sun facing, wind protected slopes of Hengshan, they planted acres of orange orchards. For this case study, we cooperated with local farmers and processed food workers to distill the sweet and sour flavor of local oranges into citrus conserve. Accompanied by fresh fruit, the conserve is further concocted into a delectable marmalade snack that can be enjoyed with tea. The designer used processed products from Hengshan's orange orchards (dried orange peel, citrus conserve, and orange marmalade), matched with an artistically crafted tea set, to open a window onto an innovative tea-drinking culture. We hope that the use of these superior products will provide a qualia experience that will drive the development of local industry. The explanation of the cultural code implementation and design coding for this case study is as follows (Fig. 10).

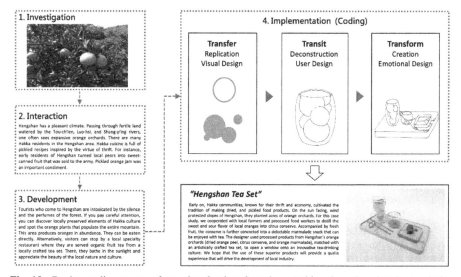

Fig. 10. Design coding process for regional cultural products (taking the cultural code translation of *"Hengshan Tea Set"* as an example)

(3) **Inner Level (spiritual)**

- *"Tranquility in Hengshan":* A quietly elegant metallic accessory. (Materials: Silver, Enamel, Mussel-Pearl) (Product C) (Fig. 6)

- **Selection of Cultural Carrier:** Shak'eng organic tea farm (Fig. 9)
- **Product Attribute Orientation:** Reflective level/Emotional design
- **Cognitive Communication Model:** Effective stratum/Inner spirit
- **Design concept:** Drawing on her recent experiences exploring the Hengshan area, the designer endeavored to embody the core values of "peace, sense impressions, and connection", to interpret an aesthetic that is peculiar to Hengshan. The works in this series, situated within the familiar botanical cultural carrier of "tea leaves", employ the subtle modulations of enamel coloring, matched with a metallic structure, and dotted with pristine pearls, to express the artist's longing for the quiet nature of Hengshan, articulating a shared ideal of local life and an aesthetic full of elegant taste.

3.2 Questionnaire Design

This study conducted design experiments for three case studies from different strata, employing the psychological models of mental cognition and information communication (Fig. 3) to design assessment questions corresponding to each stratum. The results obtained from the questionnaires were analyzed to ascertain the evaluations of potential consumers about regional cultural products, thereby testing the effectiveness of the "design and translation model for regional culture" (Fig. 2) and the "design coding process for regional cultural products" (Fig. 10) in this study.

We used a Google form to conduct an online survey with a total of 32 questions. The questionnaire was split into three sections: the first section included the image, reference picture (cultural carrier), and the design concept of each group of products. Participants were asked to assess the strength of their impressions across three cognitive strata (technical, semantic, and effective) for each group (Table 1) to test the design orientation and cognitive communication effect of the product's attributes. In the second section, participants were queried about their views of local cultural products and asked to select their favorite product (Table 2) to test the correlation between local cultural characteristics and preferences. The final section of the survey asked for participants' basic information, including the regional attribute information of their place of residence and work.

3.3 Interviewees

A total of 190 people participated in the questionnaire evaluation. After inspection, we determined that their contributions were all valid. Interviewees were all potential consumers from the Chinese speaking world. Table 3 shows the distribution of different participants.

4 Results

4.1 Overall Assessment

The three case studies of this design experiment largely garnered a high degree of recognition for "representing specialty products of the Hsinchu Hengshan region," scoring 4

Table 1. Questionnaire content related to "product attribute cognition and information communication results" (taking Product A as an example)

Questionnaire Goal	Questions
Technical Stratum (Outer Perception)	a1. What do you think of the similarity and associations between the appearance (shape, color, or texture) of this product and the design reference image (concrete factory)? Low ← □1 □2 □3 □4 □5 → High a2. What do you think of the functionality of this product when you use or operate it (can it be placed on a table or used as a micro potted plant)? Low ← □1 □2 □3 □4 □5 → High
Semantic Stratum (Significant Cognition)	a3. How much does this product have an "imaginative storytelling" quality? Low ← □1 □2 □3 □4 □5 → High a4. How much does this product make you "reflect on the history of industrial development"? Low ← □1 □2 □3 □4 □5 → High
Effective Stratum (Inner Spirit)	a5. How much does this product make you feel the "special character of local culture"? Low ← □1 □2 □3 □4 □5 → High a6. How much do you feel that this product represents the "specialty products of the Hsinchu Hengshan region"? Low ← □1 □2 □3 □4 □5 → High
Regression Analysis	a7. How much do you like the product *"Landscape on Table"* (concrete decorative item)? Low ← □1 □2 □3 □4 □5 → High

points or above in the 5-point scale: Product A = 48.2%, Product B = 62.2%, Product C = 50.8%. The three case studies also won a high degree of recognition for "the degree of preference for this product": Product A = 61.1%, Product B = 69.9%, Product C = 65.9%. In both "representing specialty products of the Hsinchu Hengshan region" and "the degree of preference for this product", Product B was most highly assessed.

Table 4 shows that there is a significant difference in the preference of different genders for Product A ($p < 0.05$). Females like Product A more than males.

Table 5 shows that there is a difference in the preference of different age groups for Product A. The preference among the under-25 s is significantly more favorable than among the 46–55 s. 46–55 s like it significantly more than above-65 s. That is to say the preference for Product A is highest among the under-25 s.

Table 2. Questionnaire content related to "common views of regional cultural products"

Goal	Questions
To test the relation between the specialties of regional culture and preference; to attract purchases, the connection to visiting the area	2-1. Which characteristics of the "regional cultural product" design are the most important? ☐Appearance ☐Usability ☐Storytelling ☐Conveying the content of regional culture 2-2. Money permitting, would you be attracted to buying these products because of their "regional cultural features"? Low ← ☐1 ☐2 ☐3 ☐4 ☐5 → High 2-3. Choose a product that you believe has the most convincing "regional cultural features." ☐A ☐B ☐C 2-4. Choose your favorite product. ☐A ☐B ☐C 2-5. Does a product with "regional cultural features" make you want to visit the region? Low ← ☐1 ☐2 ☐3 ☐4 ☐5 → High

Table 3. The distribution of different interviewees

Gender		Education		
Male	Female	High School	University	Graduate School
81	109	10	98	82
Age				
Under 25	26 – 35	36 – 45	46 – 55	56 – 65	Above 65
40	25	22	61	27	15
Background				
Design	Arts	Humanities	Sciences	Other
46	24	45	55	20
Current Place of Residence				
City	Suburb	Remote Village	Other
121	58	11	0
Place of Work				
City	Suburb	Remote Village	Other
134	41	7	8

Table 6 shows that there is a significant difference in the preference of interviewees with different professional backgrounds for Product A. Although "Scheffe's post hoc" does not show a significant difference, a review of the averages reveals that interviewees with a background in design-related spheres preferred Product A. The discrepancy is

Table 4. A T-test of the preference of different genders for regional cultural products

Items	Variable	n	M	SD	t
A	Male	81	3.52	1.195	−2.215*
	Female	109	3.88	1.052	
B	Male	81	4.04	0.872	−0.909
	Female	109	3.92	0.914	
C	Male	81	3.78	0.949	−1.276
	Female	109	3.95	0.937	

*p < 0.05

Table 5. An analysis of variance in preferences for regional cultural products among different age groups

Works	F	Variables	M	Scheffe's post hoc
A	5.464***	1. Under 25	4.33	1 > 4, 1 > 6
		2. 26–35	3.92	
		3. 36–45	4.00	
		4. 46–55	3.39	
		5. 56–65	3.56	
		6. Above 65	3.07	

***p < 0.001

largest between those with design and science backgrounds (Product A, 1 > 4). There is also a significant difference in the preference of interviewees with different professional backgrounds for Product B. Here, the discrepancy is largest between those with humanities and arts backgrounds (Product B, 3 > 2).

Table 7 shows that there is a significant difference in the preference of interviewees residing in different places for Product A. "Scheffe's post hoc" reveals that those living in the city-bordering suburbs were more favorable in their preferences than those from agriculture, fishing, or mountain villages. That is to say, suburban residents were most favorable in their preference for Product A.

Table 8 shows that there is a significant difference in the preference of interviewees working in different places for Product A. "Scheffe's post hoc" reveals that those working in the suburbs were more favorable in their preferences than those who were either retired or working in agriculture, fishing, or mountain villages. That is to say, suburban workers were most favorable in their preference for Product A.

Table 6. An analysis of variance in preferences for regional cultural products among different professional backgrounds

Works	F	Variables	M	Scheffe's post hoc
A	2.715*	1. Design	4.04	1 > 4
		2. Arts	3.92	
		3. Humanities	3.82	
		4. Sciences	3.40	
		5. Other	3.45	
B	3.374*	1. Design	4.00	3 > 2
		2. Arts	3.54	
		3. Humanities	4.27	
		4. Sciences	3.82	
		5. Other	4.15	

$*p < 0.05$

Table 7. An analysis of variance in preferences for regional cultural products among different places of residence

Works	F	Variables	M	Scheffe's post hoc
A	7.571***	1. City	3.61	2 > 3
		2. Suburb	4.12	
		3. Remote village (Agriculture, Fishing, Mountain)	2.91	

$***p < 0.001$

Table 8. An analysis of variance in preferences for regional cultural products among different places of work

Works	F	Variables	M	Scheffe's post hoc
A	3.174*	1. City	3.69	2 > 3
		2. Suburb	4.07	
		3. Remote village	3.14	
		4. Other	3.00	

$*p < 0.05$

Table 9 shows that there is a high degree of statistical significance in the regression model for **Product A**. F value = 82.719 ($p < 0.001$); the correlation coefficient R = 0.756, the coefficient of determination R squared = 0.572. In terms of the regression coefficient, the ß value was highest for the technical stratum ($t = 6.815$, $p < 0.001$). Next highest was the effective stratum ($t = 3.406$, $p < 0.01$). There is a high degree of statistical significance in the regression model for **Product B**. F value = 92.378 ($p < 0.001$); the correlation coefficient R = 0.774, the coefficient of determination R squared = 0.598. In terms of the regression coefficient, the ß value was highest for the technical stratum ($t = 5.187$, $p < 0.001$). Next highest was the semantic stratum ($t = 4.776$, $p < 0.001$). There is a high degree of statistical significance in the regression model for **Product C**. F value = 96.206 ($p < 0.001$); the correlation coefficient R = 0.780, the coefficient of determination R squared = 0.608. In terms of the regression coefficient, the ß value was highest for the technical stratum ($t = 5.351$, $p < 0.001$). Next highest was the semantic stratum ($t = 3.631$, $p < 0.001$).

Table 9. Multiple regression analysis for the product design's technical, semantic, and effective strata on product preference

Independent variable	Predictor variable	B	SE	ß	t
A	Technical stratum (Outer perception)	0.573	0.084	0.485	6.815***
	Semantic stratum (Significant cognition)	0.128	0.087	0.116	1.465
	Effective stratum (Inner spirit)	0.236	0.069	0.239	3.406**
	R = 0.756	$R^2 = 0.572$	F = 82.719***		
B	Technical stratum (Outer perception)	0.408	0.079	0.356	5.187***
	Semantic stratum (Significant cognition)	0.435	0.091	0.354	4.776***
	Effective stratum (Inner spirit)	0.139	0.071	0.149	1.959
	R = 0.774	$R^2 = 0.598$	F = 92.378***		
C	Technical stratum (Outer perception)	0.507	0.095	0.451	5.351***
	Semantic stratum (Significant cognition)	0.365	0.100	0.326	3.631***
	Effective stratum (Inner spirit)	0.044	0.069	0.047	0.645
	R = 0.780	$R^2 = 0.608$	F = 96.206***		

**p < 0.01 ***p < 0.001

From the results of multiple regression analysis, we were able to learn that intervie-wees' preferences can be predicted through performance across technical, semantic, and effective strata. In particular, the technical stratum (Outer perception) is a key element in determining preference.

4.2 Analysis of Placemaking

Through the completion and analysis of questionnaires, this study revealed the evalua-tions of 190 potential consumers on regional cultural products. These may be summarized as follows: Most importantly 85.5% of interviewees believe that regional cultural prod-ucts "can convey the content of regional culture". 83.9% of interviewees (selecting 4 or above on the 5-point scale) believe that products "with regional cultural features" will strongly attract purchasers. 71% of interviewees (selecting 4 or above) believe that prod-ucts "with regional cultural features" will inspire a strong desire to visit the place of pro-duction. The multiple-choice assessment for products "with regional cultural features" was: Product A = 33.7%. Product B = 45.1%. Product C = 21.2%. The multiple-choice assessment for "my favorite product" was: Product A = 36.8%. Product B = 39.4%. Product C = 23.8%. This shows that the more interviewees judged a product to have "regional cultural features", the more they liked it. On this basis, developing "products with regional cultural features" will attract visitors from other areas, stimulating the experience economy and opening up opportunities for regional revitalization.

On the basis of the results above, we conducted a reliability estimate of the three levels (visceral, behavioral, and reflective) of cultural product attributes in relation to the three strata (technical, semantic, and effective) of the "cognitive and communication model". Our results show that the Cronbach's α coefficients were Product A = 0.908, Product B = 0.904, Product C = 0.928 (Cronbach $\alpha > 0.9$), proving that the inner consistency of the scale is sufficient to support this research results.

5 Conclusions

This paper has been written from the perspective of placemaking, investigating how "cultural literacy", "design coding", and "cognitive and communication processes" can turn regional cultural resources into "regional cultural products". As stated in the fore-word, the focus of our research has been on how "innovation" in regional culture" can "transform" and "add value" to design, thereby assisting increasingly marginalized "intermediary towns" in sparking regional development. Our conclusions are as follows.

(1) **The participation and cooperation of local residents is needed for the selection of cultural carriers:** Although the scope of this study was a marginal mountain vil-lage with an aging population, local residents still exhibited a great deal of commu-nity spirit, characterized by unity and local pride. They were very happy to engage with our research team, helping us to select cultural carriers that local residents were clearly able to recognize and agree with. Their support was crucial in avoiding the danger of cultural misappropriation and developing the design experiments of this paper.

(2) **The flexible use of cultural code can help fruitfully analyze cultural products.** We conducted design experiments on the "stratified cultural space and sign coding framework" and "design translation model for regional culture" that we formulated based on previous scholarship articulating the three levels of cultural space and the three attributes of product orientation. When engaging in the actual process of design, we made full use of cultural sign coding. While each product had its own design goals, they are not confined to a unitary stratum of design development. That is to say, different types of code can coexist within one cultural carrier, furnishing richer possibilities for the interpretation of cultural products.

(3) **The functionality of cultural products should not be restricted to the niche tastes of small groups:** Most importantly, the results of our questionnaire show that 85.5% of interviewees believe that regional cultural products can "convey the content of local culture". All of the products in our study share this quality. They were all, furthermore, recognized for "representing the specialty products of the Hsinchu Hengshan region". In the final multiple-choice questions on "the products with the strongest regional features" and "your favorite product", Product C, which had a high degree of preference in individual assessments, fell short. We hypothesize that products with a "specific function" are less likely to be liked. For instance, Product C is a brooch accessory for females and is therefore not suitable for the daily needs and lives of the target audience.

(4) **An accurate vision of the target audience can help in the design and promotion of cultural products:** The results of our questionnaire show that different ages and professional backgrounds lead to discrepancies in the cognition of and preferences for cultural products. The degree of preference will influence the motivation to purchase products and visit the region. We argue that before embarking on cultural product design, one should first determine the audience to which the area under revitalization should appeal. From there, one can investigate the cognition and reception of that audience to the different cultural strata and consider the use of cultural coding translation. This will aid the design, development, and promotion of cultural products.

An emphasis on urban development has led to imbalances in urban-rural development, leading many villages to face the difficulties of marginalization. This study has carried out experiments in the Hengshan area of northern Taiwan, developing an innovative model for designing regional cultural products, furnishing a method of translation for regional cultural resources, establishing a system of knowledge for regional design, and using design innovation to unfold the rich prospects of regional culture. We hope that in the future, the manufacturing of cultural products will marry the local wisdom of regional craftsmen, exploring the integrated realization of innovative design, local arts, and inter-regional marketing to achieve the shared ideal of placemaking.

Acknowledgments. This research has benefited from Taiwan's Ministry of Science and Technology's humanities innovation and social enterprise project bursary (MOST 108-2420-H-007-010-HS1). We would also like to thank National Tsing Hua University's Hsinchu smart urban-rural revitalization project team, participants in the Hengshan area, and interviewees for their support.

Finally, we would like to thank the senior faculty from the Graduate School of Creative Industry at the National Taiwan University of Arts for their valuable suggestions.

References

1. Chiu, H.-W.: Hardship of pioneer - social innovation along the Taiwan route 3. Taitung J. Hum. **5**(2), 1–36 (2015). (in Chinese)
2. Hsu, C.-H.: Transformation model for cultural creative product design (doctoral dissertation). National Taiwan University of Arts, New Taipei (2014). (in Chinese)
3. Huang, W.-C.: Hsinchu County Annals of Taiwan Province. Cheng Wen Publishing, Taipei (1983). (in Chinese)
4. Leong, D., Clark, H.: Culture-based knowledge towards new design thinking and practice– a dialogue. Des. Issues **19**(3), 48–58 (2003)
5. Lin, F.-J.: The revitalization plan for smart urban-rural hsinchu: local vitality and capability establishment (midterm report: MOST 108-2420- H-007-010-HS1). Ministry of Science and Technology, Taipei (2019). (in Chinese)
6. Lin, C.-L., Chen, S.-J., Hsiao, W.-H., Lin, R.: Cultural ergonomics in interactional and experiential design: conceptual framework and case study of the Taiwanese twin cup. Appl. Ergon. **52**, 242–252 (2016)
7. Lin, M.-H.: Sign and Code in product design. J. Design **5**(2), 73–82 (2000). https://doi.org/10.6381/JD.200012.0073. (in Chinese)
8. Lin, R.-T.: Transforming Taiwan aboriginal cultural features into modern product design: a case study of a cross - cultural product design model. Int. J. Design **1**(2), 47–55 (2007)
9. Lin, R.-T.: From service innovation to qualia product design. J. Design Sci. **14**(S), 13–31 (2011). (in Chinese)
10. Lu, C.-Y.: Explore the shaping of Hakka region around Qiunglin township and Hengshan township in Qing Dynasty (Master's thesis). National Taiwan Normal University, Taipei (2013). (in Chinese)
11. National Development Council: National Strategy for Regional Revitalization. Executive Yuan of Taiwan, Taipei (2018). (in Chinese). https://www.ndc.gov.tw/Content_List.aspx?n=78EEEFC1D5A43877. Accessed 28 Jan 2020
12. Norman, D.A.: Emotional Design: Why We Love (or Hate) Everyday Things. Basic, New York (2004)
13. Rose, G.: Visual Methodologies: An Introduction to the Interpretation of Visual Materials (K.-C. Wang, Trans.). Socio Publishing, Taipei (2006). (Sage Publications of London, 2001). (in Chinese)
14. Solomon, M.R.: Consumer Behavior: Buying, Having and Being. Prentice Hall, New Jersey (2003)
15. Su, P.-H.: Viewers and Codes of Images: A Series of Semiotic Study and the Cross-Media Interpretation. The Liberal Arts Press, Taipei (2013). (in Chinese)

Usability of Self-service Beverage and Snack Vending Machines

Zi-Hao Ding, Lan-Ling Huang$^{(\boxtimes)}$, and Shing-Sheng Guan

FuJian University of Technology/Design Innovation Research Center of Humanities and Social Sciences Research Base of Colleges and Universities in Fujian Province, No. 3 Xueyuan Road, University Town, Minhou, Fuzhou 350118, China

Abstract. The purpose of this research was to evaluate the usability of self-service beverage and snack vending machines. The sizes and positions of the three functional configurations (i.e. product selection, payment, and product collection) were evaluated by users. The results of this study can be summarized as follows: (1) Users preferred larger-sizes on payments and product collections. (2) For the positions of three functional configurations, users like their positions concentrated, and with the shortest movement displacement for the upper extremities, which may have better motion efficiency. (3) About the Comprehensive score for three projects, a total of 202 subjects participated in the evaluation. Project A is better than existing vending ma-chines (Project C). Project A is more suitably for user expectations and needs. There-fore, this research proposes improvement design suggestions for the sizes and positions of three functional configurations. The results of this study will provide a design reference for the vending machine industries.

Keywords: Usability · Functional configurations · Self-service vending machines

1 Introduction

Self-service vending machines are commonly used devices for commercial sales and can be automatically delivered based on consumer choice and payment. It is not limited by time and place, easy to manage. It is a brand new form of commercial retail. In 2017, the "new retail" concept emerged in China, the Internet and traditional retail giants entered the unmanned retail model, and the vending machine industry also entered the era of intelligence [1]. People and commerce are shifting further from cash to digital payments, and many are now adopting mobile wallets such as Alipay and WeChat Pay. Nearly 45% of the Chinese population used mobile wallets in 2018, one of the highest adoption rates in the world [2].

Affected by the COVID-19 in 2020, the development of unmanned retail industry has become faster and more diverse. Consumers' consumption patterns have also changed. Many unmanned retail companies have turned to selling self-service retail equipment [3], such as self-service delivery, self-service take-out, contactless mask vending machines,

© Springer Nature Switzerland AG 2021
P.-L. P. Rau (Ed.): HCII 2021, LNCS 12771, pp. 262–272, 2021.
https://doi.org/10.1007/978-3-030-77074-7_21

self-service vending machines, automatic vending cabinets, self-pickup cabinets and other shopping methods. This will become a new direction for the vending machine industry, and it may also be one of the important equipment in life.

Studies about automated vending machines focused on the type and quality of food offered [1], on the engineering methods applied to vending machine systems [4, 8], on how to improve interaction by connecting vending machines to Internet-based services [5, 6], and accessibility evaluation [7, 9]. However, human factors received less attention, and there are few studies that investigate the usability of automated vending machines by user experiences.

At present, China's vending machines have a limited variety of products. The products are mainly concentrated in beverages and snacks. Compared with foreign vending machines, China's vending machine industry still need to be developed [1]. Data from iResearch show that in 2017 [12], 29.5% Chinese retail users who have used vending machines. Beverages and snacks are the most purchased items in unmanned stores in Chian. 63.4% of unmanned store users have purchased drinks and 57.7% of users purchased Snacks [12]. Studies also shown that more than 50% users in China do not use self-service vending machines because they are afraid of making mistakes and unwilling to learn the operation steps [5]. Naddeoa et al. (2019) also shown that use problems of the vending machines, such as the height of the payment is inappropriate, the position of product collection is low, and poor motion efficiency [10]. These problems may affect the consumer experience.

The purpose of this research was to evaluate the usability of self-service beverage and snack vending machines. It is hoped that the results of this research could be used to improve design existing vending machines to meet user needs, and design reference for vending machine developers.

2   Method

This research included three parts: (1) to survey the appearance sizes of self-service beverage and snack vending machines frequently used on the market. (2) to survey the sizes and positions of the three functional configurations (i.e. product selection, payment, and product collection) on the self-service beverage and snack vending machines that users prefer. (3) to evaluate the satisfaction of self-service beverage and snack vending machines in different configurations. The contents of each part are described as follows:

First Part. Types of self-service beverage and snack vending machines that have largest Market share was survey. The appearance sizes and three functional configurations sizes of self-service beverage and snack vending machines were collected or measured by researchers.

Second Part. Researchers conducted user interviews to survey the sizes and positions of three functional configurations (i.e. product selection, payment, and product collection) that users preferred. Researchers recruited users with experience in use vending machines, and users with no experience. Each user is required to complete two tasks according to their operating requirements: (1) to select the appropriate size model of the three functional configurations, (2) to place the appropriate operating positions of the three functional configurations on an appearance model of the vending machine (see in Fig. 1). Before the interview, the researcher explained to the function and usage of the vending machine to each user, so that the user understand three functional configurations of the vending machine. The researcher records the size of the three functional configurations selected and the coordinates of the three functional configurations by each user. The position of the three functional configurations is based on the lower left corner of the model board as the coordinate origin.

Third Part. A questionnaire was designed to investigate the relative importance of each evalua-tion index, and comprehensive evaluation of three self-service beverage and snack vending machines (from the results of second part) by users. The questionnaire in-cludes two parts: (a) characteristics, and (b) relational matrix analysis (RMA) was applied to evaluate the self-service beverage and snack vending machines. The steps of the relational matrix analysis are: (1) Establishing an evaluation index system; (2) Establishing a weight system; (3) Establishing individual index evaluation standards; and (4) Comprehensive evaluation. The evaluation index items and individual index evaluation standards are proposed according to interview experts (ergonomics ex-perts or product design experts).

The weight of each evaluation index item and comprehensive evaluation were evalu-ated by users. The method of paired comparisons was used to decide the weight of each evaluation index item.

Data Analyses. All data were analyzed with SPSS for Windows version 13.0. The characteristics of the two groups and individual index evaluation of three projects were analyzed with descriptive statistics. The Fisher's LSD Exact Test was used for evaluation index items (product selection, payment, product collection, product collection, and Overall coordination). Differences were considered significant when $p < 0.05$.

3   Result and Discussion

3.1   The Appearance Sizes of Self-service Beverage and Snack Vending Machines Frequently Used on the Market

In China, the vending machine manufacturer, Dalian Fuji Bingshan Vending Machine Co., Ltd. (Fuji) market share reached 53%. For vending machine operators, Beijing Ubox Science & Trading Co., Ltd. (Ubox) self-service vending machine market share exceeds 40%, which is a leading company in the industry; Shanghai Miyuan Beverage Co. Ltd. (Miyuan) has a market share of about 19%, ranking second; Guangzhou Fuhong's market The share is 15%, ranking third; the market share of other companies is less than 10% [11].

Based on the above literature survey results, this research selected the vending machine manufacturer or operator with the largest market share in China's vending machine market as the survey objects: Fuji (suppliers), Ubox (operators) and Miyuan (operators). A total of eight self-service beverage and snack vending machines were selected that frequently used on the market in China. The appearance size and three main functional areas of each selected model have been proposed (see Table 1).

The eight beverage and snack vending machines have an average height of 1854 mm (SD 36) and a width of 1165 mm (SD 55). Researchers made a 1:1 appearance model of vending machine according to the average height and wide sizes of eight beverage and snack vending machine models for user interview. Also, we selected the three functional configurations (i.e. payment, Product selection and Product collection) of the vending machine with large differences as the experimental model (see Table 2).

Table 1. Eight models of vending machines were frequently used on the market in China.

Suppliers or operators	Vending machines models	Appearance size	Size of payment	Size of Product selection	Size of Product collection	Pictures
		Height (H) and Wide (W), mm				
Dalian Fuji ingshan	1. FVM-CP23-B2PIT (screen size:7")	H 1830 W 1165	H 850 W 850	H 200 W 120	H 170 W 750	
	2. FVM-CP23-B2PIT (screen size: 10.1")	H 1830 W 1165	H 850 W 850	H 340 W 190	H 170 W 750	
	3. FVM-CP23-B2PIT (screen size: 23.6")	H 1830 W 1165	H 870 W 620	H 510 W 250	H 180 W 750	
	4. FVM-Multi vending machine	H 1830 W 1055	H 1100 W 650	H 480 W 250	H 200 W 410	
Beijing Ubox	5. Ubox Beverage Vending Machine	H 1894 W 1165	H 580 W 630	H 310 W 280	H 110 W 750	
	6. Ubox Food Vending Machine	H 1920 W 1180	H 1120 W 700	H 300 W 170	H 120 W 450	
Shanghai Miyuan	7. Miyuan Beverage Vending Machine	H 1830 W 1170	H 720 W 750	H 150 W 100	H 170 W 790	
	8. Miyuan Food Vending Machine	H 1870 W 1261	H 1110 W 550	H 570 W 310	H 210 W 520	

3.2 The Sizes and Positions of Three Functional Configurations that Users Preferences

A total of twenty users were interviewed and field tested (see Fig. 1). Ten users who have experience in using vending machines are arranged in group A. The group A consisted of four males and six females, with an average age of 31.5 years (SD 9.4) and height of 168.3 cm (SD 8.3). The other ten users who have no experience are arranged in group B, consisted of seven males and three females, with an average age of 55.5 years (SD 6.9) and height of 166.7 cm (SD 5.2). Analysis of the age of two groups showed significant differences, $t(18) = 6.458$, $p = 0.000$). There were no statistically significant differences among the two groups with regard to height.

Table 2. Nine types and sizes of three functional configurations utilized for the user interview.

No	Product selection	No	Payment	No	Product collection
A	H 630 W 630	D	H 380 W 190	G	H 160 W 300mm
B	H 830 W 480	E	H 260 W 150	H	H 130 W 560
C	H 440 W 470	F	H 160 W 90	I	H 130 W 330

Height (H) and Wide (W)
Unit: mm

The Sizes of Three Functional Configurations in Group a (Have Use Experience).
For the Product selection, nine users preferred A (see Table 2), the other one user selected B, and no user selected C. Nine users indicated that the size of A is appropriate. When they stand in front of the vending machine, they are in the field of vision.

For payment, seven users preferred D, the other three users selected E, and no users selected F. Seven users said that the larger screen size for the payment can display more information to operate easily.

For the Product collection, seven users preferred H, the other three users selected G, and no user selected I. Seven users said that a wider product collection size can avoid the need to move their bodies left and right when picking up goods. Three users who selected G said that the product collection size is moderate and can be placed on a high place without bending over to pick up the goods.

The Sizes of Three Functional Configurations in Group B (no Use Experience). For the Product selection, seven users preferred A (see Table 2), the other three user selected B, and no user selected C. Five of seven users showed that the size of A is appropriate to select goods. The opinion is same to group A.

For payment, six users preferred E, the other three users selected F, and only one user selected D. Four of the six users said that the payment screen (E) is suitable for them to operate, while the larger screen (D) makes them feel stressed.

For product collection, eight users chose G, the other two users chose H, and no user chose I. Six of the eight users indicated that the size of H is appropriate, and the exact location where the product falls is clear, so that the product can be obtained quickly.

Overall, the sizes and positions of the three functional configurations in group A are compared with group B found that the three functional configurations in group B are more concentrated. The possible reason for this result is that users in group B prefer the position with the shortest displacement during the purchase process. Compared with existing vending machines, the product collection positions in groups A and B are higher. This means that the product collection of existing vending machines needs to consider ergonomics in vending machine design (Fig. 2).

Fig. 1. The 1:1 model set-up for user interview.

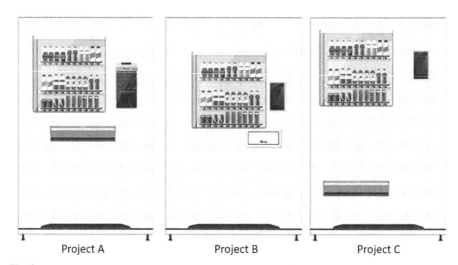

Project A   Project B   Project C

Fig. 2. The results of three main functional areas that users preferences. Project A is proposed by group A, and Project B is proposed by group B. Project C is existing vending machine used on the market.

3.3 Relational Matrix Analysis Was Applied to Evaluate the Self-service Beverage and Snack Vending Machines

Establishing Evaluation Index System. Three experts (one ergonomics expert and two product design experts) were interviewed to propose the evaluation index of self-service beverage and snack vending machines. The evaluation index were as follows:

A. The position suitability of product selection (whether the position is convenient to select the product).
B. The position suitability of payment (whether the position is convenient to operate).
C. The position suitability of product collection (whether the position is easy to pick-up product).
D. Motions efficiency (the number of actions is less, the movement distances of the hand is shorter, and natural).
E. Overall coordination (whether the position of three functional configurations are appropriate).

Establishing the Weight of Each Evaluation Index Item. A total of 202 subjects filled in the online questionnaire in china, 142 females and 64 males. Their age ranged from 15 to 26 years old (92%), 27 to 38 years old (8%). Each subject evaluated the relative importance of the five evaluation index items. Using the pairwise comparison to calculate the weight in each evaluation index item.

The result (see in Table 3) found that the weight of overall coordination is 0.3, followed by movement efficiency is weight 0.2, the position suitability of product selection is weight 0.2, the position suitability of payment is weight 0.2, and the position of product collection is 0.1. Users thinned that overall coordination is an important factor for the three functional configurations of self-service vending machine.

Establishing Individual Index Evaluation Standards. Three experts (one ergonomics expert and two product design experts) were inter-viewed to propose the individual index evaluation standards. The items were listed on a 5-point scale with 1 signifying "Very bad", 2 signifying "poor", 3 signifying "nor-mal", 4 signifying "better" and 5 being "excellent".

Table 3. Weights of evaluation index items

Evaluation index items	Score of Paired comparison										Cumulative score	Weights
	1	2	3	4	5	6	7	8	9	10		
A. The position suitability of product selection	1	1	0	0							2	0.2
B. The position suitability of payment	0				0	1	1				2	0.2
C. The position of product collection			0		1			0	0		1	0.1
D. Motions efficiency			1		0		1		0		2	0.2
E. Overall coordination				1		0		1	1		3	0.3
Count	1	1	1	1	1	1	1	1	1	1	10	1

Comprehensive Evaluation. A total of 202 subjects evaluated five evaluation index items for three projects. The result (see Table 4) of comprehensive score shown that project A is significantly high-er than project B and project C. Also, both Project A and Project B also are better than Project C (existing vending machine used on the market).

For the product selection, Project A is significantly higher than project B and project C, $F(2, 603) = 8.723, p = 0.000$.

For the payment, Project A is significantly higher than project B and project C, $F(2, 603) = 10.348, p = 0.000$. It can be explained as the users preference for a larger payment size, it has a better user experience and convenient to use.

For the product collection, Project A is significantly higher than Project B and Project C, $F(2, 603) = 10.926, p = 0.000$. Also, Project A is significantly higher than Project B and Project C on motions efficiency, $F(2, 603) = 12.536, p = 0.000$.

For the overall coordination, Project A is significantly higher than Project B and Project C, $F(2, 603) = 10.046, p = 0.000$. This result can be explained that the three functional configurations preferred by users are more concentrated. The possible reason for this result is that during the purchase process, the user likes the position with the shortest movement displacement of the upper limbs.

Table 4. Comprehensive scores

Projects	The evaluation index					Comprehensive score
	The position suitability of product selection	The position suitability of payment	The position of product collection	Motions efficiency	Overall coordination	
Weights	0.2	0.2	0.1	0.2	0.3	
Project A (mean, SD)	3.7 (0.9)	3.9 (0.9)	3.8 (1.0)	3.8 (0.9)	3.8 (0.9)	3.8
Project B	3.5 (0.9)	3.6 (0.9)	3.6 (0.8)	3.6 (0.8)	3.6 (0.9)	3.6
Project C	3.3 (0.9)	3.5 (0.9)	3.3 (0.9)	3.4 (1.0)	3.4 (0.9)	3.4

Project A is proposed by group A, and Project B is proposed by group B. Project C is existing used on the market.

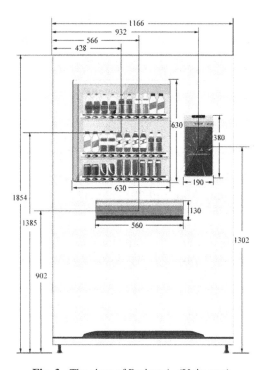

Fig. 3. The sizes of Project A. (Unit: mm)

Overall, the results of this research show that the sizes and locations of the three functional configurations of Project A are suitably for user expectations and needs. The size and location of the three functional configurations of Project A is proposed for design reference (see Fig. 3).

4 Conclusion

The purpose of this research was to evaluate the usability of self-service beverage and snack vending machines. The results of this study can be summarized as follows:

(1) Users (with use experiences on self-service Beverage and Snack Vending Machines) preferred larger-sizes on payments and product collections.
(2) For the positions of three functional configurations, users like their positions concentrated, and with the shortest movement displacement for the upper extremities, which may have better motion efficiency.
(3) About the Comprehensive score for three projects, a total of 202 subjects participated in the evaluation. Project A is better than existing vending ma-chines (Project C). Project A is more suitably for user expectations and needs. Therefore, this research proposes improvement design suggestions for the sizes and positions of three functional configurations. The results of this study will provide a design reference for the vending machine industries.

Based on the results of the study, this research will extend to investigate the position of the facial recognition area (payment) on the vending machine.

Acknowledgment. Thanks to the Education department of Fujian Province Project (JAS170296), and Fujian University of Technology Project (GY18093).

References

1. Frost & Sullivan. https://www.frostchina.com/?p=9168. Accessed 21 Jan 2021
2. Craig, G.: Vending Machines Get Smart With AI. https://www.insight.tech/content/vending-machines-get-smart-with-ai#main-content. Accessed 21 Jan 2021
3. Retail Customer Experience. https://www.retailcustomerexperience.com/news/covid-19-fosters-self-service-innovation-in-china/. Accessed 21 Nov 2016
4. Guan, S., Yu, F.: Human-machine interface design of vending machines based on eye trackin. Packaging Eng. **40**(08), 230–236 (2019)
5. Miao, Y.F., Huang, Y.Q., Gao, Y.G.: User-centric self-service vending machine interface design. J. Mach. Des. **1**, 126–128 (2017)
6. Gao, Y.: Research on human-computer interaction of intelligent vending system based on service design. Design **32**(07), 40–43 (2019)
7. Caporusso, N., Udenze, K., Imaji, A., Cui, Y., Li, Y., Romeiser, S.: Accessibility evaluation of automated vending machines. In: Di , Giuseppe (ed.) AHFE 2019. AISC, vol. 954, pp. 47–56. Springer, Cham (2020). https://doi.org/10.1007/978-3-030-20444-0_5
8. Naeini, H.S., Mostowfi, S.: Using QUIS as a measurement tool for user satisfaction evaluation (case study: vending machine). Int. J. Inf. Sci. **5**(1), 14–23 (2015)
9. Yu, N., Mei, J., Chen, Z.: Research on the improvement of vending machine design based on human factors engineering. J. Chongqing Univ. Technol. **31**(3), 43–51 (2017)

10. Naddeoa, A., Califanoa, R.: The effect of spine discomfort on the overall postural (dis)comfort. Applied Ergonomics **74**, 194–205 (2019)
11. Forward the economist. https://www.qianzhan.com/analyst/detail/220/190307-397298e6. html#comment. Accessed 21 Jan 2021
12. iResearch. https://report.iresearch.cn/report/201709/3064.shtml. Accessed 21 Jan 2021

Smart Product Design for Food Waste Problem in the Canteen of Chinese University

Xinrong Han, Bingjian Liu$^{(\boxtimes)}$, Xu Sun, and Jiang Wu

University of Nottingham Ningbo China, Ningbo, China
bingjian.liu@nottingham.edu.cn

Abstract. The food waste problem has become increasingly severe, especially in China's developed areas. In the south-east, the food waste problem increased dramatically, especially among young people from 18–29. In this study, 12 university students were interviewed to find out the reasons for food waste. The root causes were found to be the lack of awareness and bad eating habits. At the same time, there is a strong intention that they try to eat healthily and keep in 'good' 'figure'. A smart design solution was proposed based on the calories burnt and stomach capacity to precisely suggest the volume of food people need in one meal, thus reducing the chance of producing food waste. User evaluations were carried out, and the results were mostly positive.

Keywords: Smart product design · Food waste · Diet culture · Canteen · China

1 Introduction

Food waste is a serious problem in modern society for it not only waste resources on the earth, but also cause a serious environmental issue when dealing with the landfill. According to sustainable service company RTS [1], the US wasted 80 billion pounds of food in 2020, and it occupies 30–40% of all food supply in the US. Food waste has an apparent negative impact on the environment, society, and economics [2]. Food is the single most frequently occurred material in our landfills. When food decomposes, it produces methane, a greenhouse gas 21 times more potent than $CO2$. Landfills generate 20% of all methane emissions, which causes serious environmental damage [3]. As for the social impact, food may be no longer sufficient, facing the dramatically increasing population [4]. In the aspect of the economy, FAO (2014) estimated there is 2.6 trillion USD every year associated with food waste, equivalent to 3.3% of the global GDP [5].

With the biggest population, the food waste problem in China has become increasingly serious and is expected to increase by 278–416 million tones as the rapid economic growth [6]. Studies have shown that it is important to deal with campus food waste in China for two reasons. The first one is that in China, the numbers of students on campus have increased a lot, and college students have accounted for more than 10% of the local population. Second, campus food waste solutions can be a typical solution and offer "good practice" to other situations [7].

© Springer Nature Switzerland AG 2021
P.-L. P. Rau (Ed.): HCII 2021, LNCS 12771, pp. 273–287, 2021.
https://doi.org/10.1007/978-3-030-77074-7_22

Studies show that there is a huge generation difference in food-wasting behaviors. Consumers in China have experienced starvation in the 1960s hardly waste food [8]. According to the research from Food and Culture, the young generation contributes to 7.3 million tons of food waste each year in UK [9]. As young people (20–29 years old) in developed areas act worse in wasting food, solving the food waste problem of this generation is urgent. The root cause of their wasting behavior should be found out as well as the motivation factor of food saving behavior. The specific investigating group is university students since they are the typical group of young people who are easily getting access to.

Since food waste is a severe problem, as shown in the data above, reducing food waste is urgent. The food waste hierarchy in Fig. 1 shows the priority of treating food waste: food waste prevention, re-use, recycle, recovery and disposal, respectively [1]. According to the hierarchy, the prevention of food waste should be given priority when considering solutions, and in the prevention stage, dieting behavior is an important aspect.

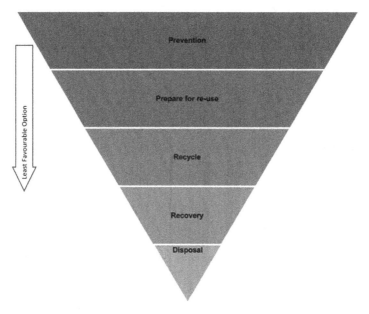

Fig. 1. Food waste dealing hierarchy [2]

Dieting behavior is frequently mentioned in the publications. Studies have shown that dieting knowledge has a significant influence on household food waste [10]. For example, young adults (18–25 years) are generally vulnerable to weight gain who are also the group of people that have improper diet behavior [11]. This study aims to further explore the relationship between eating habits, the motivation of dieting behavior, and food saving actions and, a design solution was proposed the food waste problem in the context of the university canteen. In the design process, problems are firstly identified, and related programs are analyzed, then the target was set to be the young generation, and detailed user interviews were carried out to find out the root cause of food wasting behavior. Next, design solutions were proposed, and experiments were done to test the feasibility of the plan. In the end, conclusions were made, and further research plans were stated.

2 Related Studies

In this section, the related studies on food waste are investigated to understand food production and consumption.

2.1 Food Waste from the Production Line

Many governments and organizations are currently carrying out campaigns to deal with the food waste problem. The food waste problem can either be blamed on the producer or customer. Food waste generated from the production line, and some programs to prevent customer food waste were listed and analyzed.

The reason for the food waste problem from the production line can be categorized into the following four points.

- Production line inefficiencies
- Packaging failure
- Waste from restarting the production line
- No financial benefit if redistribute food [12]

Although these defects can be eased by refining the production line, like shortening the time for the machine to reach the production standard, the liftable span is really limited. Most of these issues are related to technical bottlenecks and do not contribute to a large amount of food waste, focus should be turned to the customer side instead.

2.2 Customer Side Food Interventions

There are many different interventions against customer food waste, including informational interventions which are the most frequently used. This includes education and training on food waste problems, prompting methods such as signs and stickers, or video portraying certain practices [13]. Although these kinds of methods cost the least, the level of involvement is limited, so they are not always effective. Some people may even not notice the information notification attaching to the table when they are having a meal.

Other programs focus on the specific situation only. For instance, Harvard Business School carried out a program of diverting food from events and conferences. These foods are often untouched [14]. This program has its drawbacks. First, it does not consider how to avoid food waste but instead focus on dealing with excess food. Second, the food rediverted after the conferences may decrease quality and may affect people's health. Third, it may be effective in specific situations, but cannot solve the problem in the broader setting.

Practitioners are also paying attention to HCI in solving the issue of food waste, some small-scale design products for food producers using some interactive technologies, buy doing these, they want to promote sustainable customer buying behavior. Others focus on where food waste goes. For instance, they install BinCam on the rubbish bin and post the picture of wasted food on Facebook for reflection [15]. There is a trend to solve social problems using technology, and the usage of the mini camera mentioned above can effectively document daily food waste for further investigation, it can also raise people's awareness of how much food they waste and the content of the food they waste, thus reduce the chance they buy extra food in the future.

However, there is still a gap to increase the practicality of these methods. Several aspects can still be improved. First, a higher level of involvement is needed. Some methods (especially information intervention) only provide the customer with some messages without getting feedback. Second, using technology to influence people's behavior. Although HCI has been applied in some practices, they are mainly used as documentation and raising people's awareness, the level of interactivity should be increased. Third, detailed user research should be carried out to find the motivation behind customer food saving behavior. Dieting behavior is to be considered since there is a study stating that dieting behavior is closely related to food saving [10].

2.3   How to Decide How Much Food People Need?

Calories Need. The amount of food people need depends on many factors, including height, age, sex. The general state of health, job, leisure time activities, physical activities, genetics, body size, environmental factors, composition, and medications are taken [16]. Although many factors are influencing the food needed, the main measuring standards are calorie burnt and basic body parameters (Table 1).

According to the data from MedicalNewsToday, calorie requirement difference on gender and age is plotted, and there is an increase in trend from age 2 to 18, while the decrease from 18 years old, male generally require more calorie intake than female as shown in Fig. 2 [16].

Stomach Capacity. Experts have suggested that using calories to evaluate the number of food people should eat is not enough. Another important factor to evaluate how much food people can eat is stomach capacity. The stomach capacity can alter with age and body size, and people may feel discomfort if the stomach is stretched beyond its normal volume [17]. If the stomach is full, the body will send a satiety signal to the brain, and one will no longer willing to eat more. At the same time of eating meals, the stomach is also digesting the food, different food has different digestion rate, so the signal of feeling full may also be delayed. The phases of feeling from hangry to full are shown in Fig. 3 below.

From the study done by JC. Lynegaard et al. (2020) about piglet stomach capacity, it is shown that the capacity of the stomach has a linear relationship with body weight (Fig. 4). From the limited data, it shows that the higher the body weight, the bigger the capacity of the stomach [18] (Table 1).

Table 1. Factors that influence calorie consumption (Yvette Brazier 2018) [16]

Age	Sedentary level		Low active level		Active level	
	Male	Female	Male	Female	Male	Female
2–3 years	1,100	1100	1,350	1250	1,500	1400
4–5 years	1,250	1200	1,450	1250	1,650	1400
6–7 years	1,400	1300	1,600	1500	1,800	1700
8–9 years	1,500	1400	1,750	1600	2,000	1850
10–11 years	1,700	1500	2,000	1800	2,300	2050
12–13 years	1,900	1700	2,250	2000	2,600	22501
14–16 years	2,300	1750	2,700	2100	3,100	2400
17–18 years	2,450	1750	2,900	2100	3,300	2400
19–30 years	2,500	1900	2,700	2100	3,000	2350
31–50 years	2,350	1800	2,600	2000	2,900	2250
51–70 years	2,150	1650	2,350	1850	2,650	2100
71+ years	2,000	1550	2,200	1750	2,500	2000

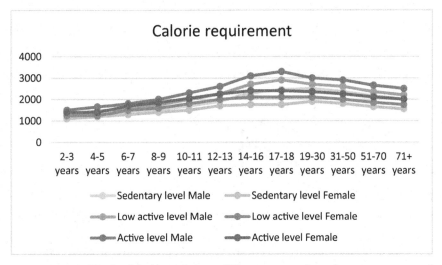

Fig. 2. Calorie requirement difference on gender and age

Calorie burnt and stomach capacity is two important factors influencing a person's need for food. They can be indicated and measured when suggesting the amount of food needed.

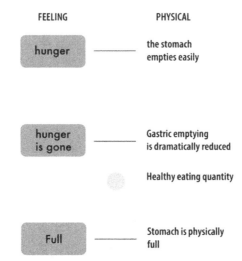

Fig. 3. Stomach capacity and feelings of fullness in different stages of eating

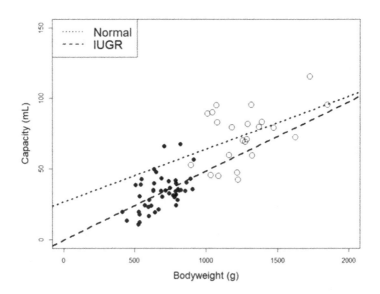

Fig. 4. Stomach capacity versus body weight (JC. Lynegaard et al. 2020) [18]

2.4   Food Consumption in the Canteen

According to the research, food waste in the canteen (meal catering system) is generated through four aspects: spoilage, meal preparation, unserved food, and plate waste. The latter two points which are defined as untouched food, are estimated to occupy approximately 14% per capita of waste in EU27 [19].

A study conducted in the school of UL canteen shows that most users left at least 1/3 of the food, indicating a lack of awareness, the performance of student's food saving behavior was measured using the waste consumption index (WI), the lower the WI, the higher the consumption.

$$WI = \frac{RMW \times 100}{AMW \times numMD} \tag{1}$$

RMW = rejected meal weight
AMW = Average meal weight
numMD = number of meals served per day

WI measurement: Great-0.0 to 3.0%; Good-3.1 to 7.5%; Bad-7.6 to 10.0%; unacceptable-over 10.0%

The result from the UL canteen shows that the WI was classified mainly as Bad and Unacceptable, representing a serious problem in canteen food waste. The average food waste per student is about 220g [19].

3   Interview of University Students

To further understand the food waste situation in a Chinese university, the interview was conducted with 12 university students to investigate their eating habits and the reasons behind the action of wasting food.

3.1   Questions for the Interview

Consumer behavior and foodservice mode research regarding food waste by Jessica, et al. (2015) was used as a reference and two questions were prepared for the interview [20]:

- Do you waste food? / Why do you waste food? (Consumer behavior)
- Under which circumstances do you most likely to waste food? (food service mode)

In addition, according to the research by Quested et al. (2013), that consciousness is also an important factor in food waste and the following two questions were asked [8].

- If the food is too much for you, do you aware of that if you ordered too much before you made the order?
- If you realized, do you still want to order that much food?

3.2   Analysis of the Interview

The interview content is guided by these questions but not limited to, aiming to create an open chat and finding out what people think. For the first question 'Do you waste food?/Why do you waste food?', only 1 out of 12 people said he never wasted food, the rest of the people said they wasted food in different situations. The reasons that they mentioned are (1) Part of the meal they don't like so they only pick out the part they like and eat them (2) The meal is too much for them to eat (3) When they tried new dishes and found that they did not like it (4) They overbought or overorder the meal because they felt hungry at the beginning.

The second question 'Under which circumstances do you most likely to waste food?' The result is that 5 out of 20 people mentioned that they tend to waste food when the serving mode is self-serving and since they care about the variety of the food, they tend to pick up different food, which is more than what they need. What's interesting is that almost half of the students interviewed replied they cannot eat up the rice in one meal.

When asking the two questions about the consciousness aspect of wasting food, (1) 10 out of 12 people said they only decide how much food they need to eat based on common sense, and they are quite confident about their standard. (2) None of them know the exact amount of food they need to eat. (3) 5 out of 12 even said they never thought of eating up all the food ordered, but only finishing those parts that they like to eat. Table 2 shows the frequency of mentioned statements in the interview.

Table 2. Frequency of mentioned statement.

Statements	Frequency of mention
Do not know how much they need to eat (only based on common sense)	10/12
Rice is often too much	8/12
Only like to eat certain kind of food	6/12
Not want to eat up all the food/ do not care if they have to throw them away or not	5/12
Tend to waste food at the buffet	5/12
Would like to get a scientific guide on healthy eating	11/12

3.3 Findings from the Interview

Reasons. From the FoodPrint organization (2020), the reasons for wasting food can be oversized portions, the inflexibility of meal selection size, and extensive menu choices. It can also be the result of some over-buying and over-preparing behavior [21].

From the interview, the following problems of the participants were found:

- Not aware of how much food and money wasted
- Overestimate the amount of food that can be eaten.
- Tend to eat only the part that the author likes.
- Waste a lot of rice.

From the literature review and interview, the root cause of young people wasting food can be summarized to be the lack of awareness of food waste and bad eating habits.

Motivation Factor. As mentioned in the interview results, almost all the people interviewed mentioned that they would like to get a scientific guide on healthy eating. According to the study conducted by A.S.poobalan et al. (2018), the characteristics of the young generation can be summarized into four aspects [11].

- They tend to care about the variety of 'food'.
- They eat more food if they love it.
- The main motivation factor for diet behavior was' self-'appearance'.
- They have positive attitudes towards a healthy diet, while do not eat healthily. The self-reported overweight/ obesity trend is about 22%.

Based on the existing issues of non-healthy eating and their strong motivation to have good self-'appearance', having a healthy meal and keep fit can be the motivating factor to have healthy eating habits. Therefore, it is reasonable to connect the controlled food quantity with reducing food waste. In addition, if people get the proper amount of food that they need, it can reduce the chance to have surplus food and thus reducing food waste. Calorie burnt and stomach volume are two factors that limit a person's food intake, certain data need to be collected to precisely calculate the food need.

4 Design Solution

4.1 Design Flow

From the interview and the literature research, it is found that the key thing behind the food waste behavior of young people is the lack of awareness and bad eating habits. There is also a finding of their strong desire to eat properly and healthily to have a 'good' 'figure'.

The research and experiment above indicate that there is a way to inform the user of the detailed calorie and volume they consume every day, and they will be less likely to waste food. Based on the research, a design solution logic was generated as shown in Fig. 5.

Three information sources will indicate the meal selection process of users.

- The user's data, like age, sex, weight, height, etc.
- The user's calorie burnt, which can be measured using the app on the smartphone, as well as the calorie of food taken (through food detecting)
- The data collected by the apps, which includes daily consumption and daily burnt calorie, this data will give the user a better understanding of their eating habit and enable them to eat more healthily.

This system can be applied in a buffet or canteen, where the user can self-serve the food for themselves. It can not only better suggest the amount of food needed for the food products but also help people reduce food waste and get better health.

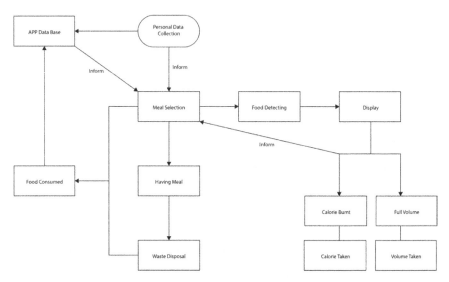

Fig. 5. Design solution logic

4.2 Calculation of Food Intake

After knowing how much food people need, it is also important for the system to calculate the calorie of food people get on their plate, the following two methods were tested for the system.

Based on Resistance Difference of Food. Different food has different resistance range, in this experiment, two different fruit, apple (500–1000) and cucumber (<500) were used as specimens, and the algorithm was set to display the relevant fruit name on the LCD, this easy-to-understand principle works when we want to detect the food in the plate (Fig. 6).

A circuit was built, and the code was uploaded to the Arduino Uno board to simulate the food detecting process (Fig. 6). From the experiment, utilizing resistance difference was proved to be an effective way to detect different kinds of food and may be possible to apply to the system. Although different food may have overlap in their resistance range, it still may work in the canteen setting when there are only limited types of food (Fig. 7).

Spectral Analysis. Spectral analysis is also an effective method to detect the food content; it is a more precise food-scanner than using resistance range [22]. For instance, researchers developed the TellSpec to detect nutrients and calories of the food, whose working principle is similar to the technology used to detect cancer in blood samples. The system of TellSpec has three parts: (1) Spectrometer scanner (2) an algorithm that exists in the cloud (3) an interface on your smartphone, as shown in Fig. 8.

Volume. The food volume calculation should also be considered since volume capacity is another criterion to measure how much food people need. After acquiring the information about food type, relevant density (ρ) values which are pre-stored in the database

```
const int frootSense = 0;
int frootResistance, high = 0, low = 1023;
int frootDetect;
#include <LiquidCrystal.h>              //remove this if u don't have LCD, anyw
LiquidCrystal lcd(12,11,5,4,3,2);       //remove this if u don't have LCD, anyway:
void setup(){
Serial.begin(9600);
lcd.begin(16, 2);                       //remove this if u don't have LCD, anyways it
lcd.clear();                            //remove this if u don't have LCD, anyways
}
void loop()
{
lcd.clear();                            //remove this if u don't have LCD, anyways it wont af
lcd.setCursor(0,0);        //remove this if u don't have LCD, anyways it wont affect.
lcd.print("Food Detect:");              //remove this if u don't have LCD, anyways it wont
lcd.setCursor(0,1);        //remove this if u don't have LCD, anyways it wont a
frootResistance = analogRead(frootSense);
Serial.print("Resistance:");
Serial.print(frootResistance);
Serial.print("\n");
if (frootResistance>1000 & frootResistance<1500){
Serial.print("Apple \n");
lcd.print("Apple");        //remove this if u don't have LCD, anyways it wont affect.
}
else if(frootResistance>500 & frootResistance<1000){
Serial.print("Cucumber");
lcd.print("Cucumber");     //remove this if u don't have LCD, anyways it wont affect.
}
else {
Serial.print("No Food \n");
lcd.print("No Food");      //remove this if u don't have LCD, anyways it wont affect.
}
delay(1000);
}
```

Fig. 6. Programming for the resistance range test

Fig. 7. Food testing with Arduino

Fig. 8. TellSpec for nutrients detection [23]

will be exacted. Another information that needed to know is the gram of food and the volume can be calculated using the equation:

$$V = \frac{m}{\rho} \tag{2}$$

In this experiment, programs were written, and a weight detector using the HX711 Balance module with the load cell was used, which can measure the weight of the food. (Fig. 9). The testing device is shown in Fig. 9.

From the weight test, precise weight result was got in grams. It proved that utilizing a weight sensor is a possible way to detect the volume of food, with the help of other data available.

4.3 Final Concept

The concept from this design flow is called Future Canteen, a new way people interact to select the meal in the canteen. By selecting the type of food and put them on the plate, the relevant information about calorie consumption and volume capacity will show on the right side respectively (Fig. 10). Based on the research, two evaluation standards were used, calorie burnt and stomach capacity.

4.4 User Feedback

To test the effectiveness of this system, five students were interviewed, and feedbacks were analyzed. 3 out of 5 students say that this system will effectively enable them to know how much food they can eat and expect to get good health. One student mentioned that he always overestimates the amount of food he can eat due to the feeling of the

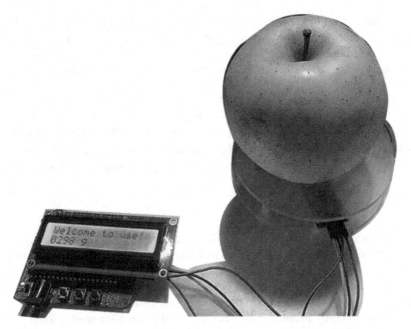

Fig. 9. Apple weight test

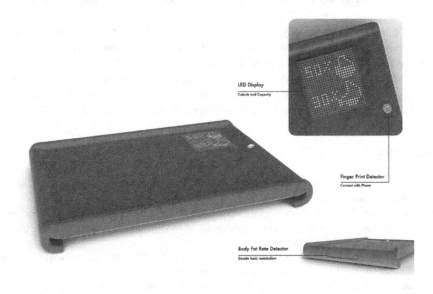

Fig. 10. Plate with measuring capability

hanger and waste a lot of food instead and this system will help him to avoid such a situation. Another student says she has the requirement to control the food intake, and it is always good to know more about herself. There are a few negative feedbacks as well, two students concern that this system may somehow limit their freedom of selecting food, they worry they will not be able to eat whatever they like affected by these recommendations. Although there is both positive and negative feedback, most of them are positive.

Several implications from these feedbacks were found. Firstly, the two standards (calorie burnt and stomach capacity) should not be so strict, and there should be more data analysis to find a proper range of calorie and capacity. By doing this, the system provides more freedom for people to select the meal. Secondly, a customized plan may be preferred for people with a specific requirement for eating (e.g., control sugar intake).

5   Conclusion

Food waste is a serious global issue and in the context of Chinese university canteen, the situation has unique characteristics, which can influence the diet culture and habits. In this study, a design study to quantify 'people's diet scientifically, thus reducing food waste and enabling and enabling young people to live a healthier life. Compared to other programs, this proposal presents a more effective method to solve food waste, since it aims to change young' people's eating behavior and meets their need to keep in good figure. It also utilizes the current technology to collect and provide useful information to the users. The study will reduce food waste and go further to contribute to environmental protection.

Since the target group is young people in developed area of China and this study only focus on university students, the data may not be that representative without other young people in different background involved.

Future studies should further test the effectiveness of this system, for instance, typical user target will be recruited for longitudinal empirical evaluations, and their food waste data will be recorded and evaluated in the long run in the canteen setting.

References

1. Rts, Food waste in America in 2020. https://www.rts.com/resources/guides/food-waste-ame rica/. Accessed 23 Jan 2021
2. Papargyropoulou, E., Lozano, R., Steinberger, J.K., Wright, N., BinUjang, Z.: The food waste hierarchy as a framework for the management of food surplus and food waste. J. Cleaner Production **76**, 106–115 (2014)
3. EPA (2020) Landfill Methane Outreach Program (LMOP). https://www.epa.gov/lmop/basic-information-about-landfill-gas. Accessed 23 Jan 2021
4. Stancu, V., Haugaard, P., Lähteenmäki, F.: Determinants of consumer food waste behaviour: two routes to food waste. Appetite **96**, 7–17 (2016)
5. Philippidis, G., Sartori, M., Ferrari, E., M'Barek, R.: Waste not, want not: A vio-economic impact assessment of household food waste reductions in the EU. Resources, Conservation and Recycling **146**, 514–522 (2019)
6. Negri, C., et al.: Anaerobic digestion of food waste for bio-energy production in China and Southeast Asia: a review. Renewable Sustainable Energy Rev. 133 (2020)

7. Zhang, D., Hao, M., Chen, S., Morse, S.: Solid waste characterization and recycling potential for a university campus in China. Sustainability 12 (2020)

8. FoodPrint (2020). The Problem of Food Waste. https://foodprint.org/issues/the-problem-of-food-waste/. Accessed 24 Jan 2021

9. Steeden, K.: Generation Waste: why are millennials throwing away so much food? (2017). https://sustainablefoodtrust.org/articles/generation-waste-why-are-millennials-throwing-away-so-much-food/. Accessed 23 Jan 2021

10. Min, S., Wang, X., Yu, X.: Does dietary knowledge affect household food waste in the developing economy of China? Food Policy (2020)

11. Poobalan, A.S., Aucott, L.S., Clarke, A., Smith, W.C.: Diet behaviour among young people in transition to adulthood (18–25 year olds): A mixed method study. Health psychology and behavioral medicine 2(1), 909–928 (2014)

12. Blake, M.: Enormous amounts of food are wasted during manufacturing - here's where it occurs (2018). https://theconversation.com/enormous-amounts-of-food-are-wasted-during-manufacturing-heres-where-it-occurs-102310#:~:text=In%202017%2C%20the%20UN%20estimated,gigatonne%20is%20a%20billion%20tonnes).&text=This%20means%20that%20just%20under,during%20manufacturing%2C%20distribution%20and%20retail. Accessed 23 Jan 2021

13. Stockli, S., Niklaus, E., Dorn, M.: Call for testing interventions to prevent consumer food waste. Resources Conservation Recycling 136, 445–462 (2018)

14. Gilboa, Y.: Reducing food waste at Harvard Business School (2017). https://green.harvard.edu/tools-resources/case-study/reducing-food-waste-harvard-business-school. Accessed 23 Jan 2021

15. Ganglbauer, E., Fitzpatrick, G., Comber, R.: Negotiating food waste: using a practice lens to inform design. ACM Trans. Comput.-Hum. Interact. 20 (2013)

16. Brazier, Y.: How much food should I eat each day? (2018). https://www.medicalnewstoday.com/articles/219305#quick_facts. Accessed 24 Jan 2021

17. Fronthingham, S.: How Big is Your Stomach? (2018). https://www.healthline.com/health/how-big-is-your-stomach#takeaway. Accessed 24 Jan 2021

18. Lynegaard, J.C., Hales, J., Nielesen, M.N., Hansen, C.F., Amdi, C.: The stomach capacity is reduced in intrauterine growth restricted piglets compared to normal piglets. Animals 10 (2020)

19. Pinto, R.S., et al.: A simple awareness campaign to promote food waste reduction in a University canteen. Waste Management 76, 28–38 (2018)

20. Aschemann-Witzel, J., De Hooge, I., Amani, P., Bech-Larsen, T., Oostindjer, M.: Consumer-related food waste: causes and potential for action. Sustainability 7, 6457–6477 (2015)

21. Quested, T.E., Marsh, E., Stunell, E., Parry, A.D.: Spaghetti soup: the complex world of food waste behaviours. Resources Conservation Recycling 79, 43–51(2013)

22. Goisser, S., Wittmann, S., Fernandes, M., Mempel, H., Ulrichs, C.: Comparison of colorimeter and different portable food-scanners for non-destructive prediction of lycopene content in tomato fruit. Postharvest Biol. Technol. 167 (2020)

23. Indiegogo: Tellspec: Food Scanner (2014). https://www.indiegogo.com/projects/tellspec-what-s-in-your-food#/. Accessed 26 Jan 2021

24. Cox, A.J.: Variations in size of the human stomach (1945). https://www.ncbi.nlm.nih.gov/pmc/articles/PMC1473711/pdf/calwestmed00012-0041.pdf. Accessed 24 Jan 2021

Cross-Cultural Research on Consumer Decision Making of HNB Product Modeling Based on Eye Tracking

Lizhong Hu[1], Lili Sun[1], Yong Zha[1], Min Chen[2], Lei Wu[2], and Huai Cao[2(✉)]

[1] Anhui Key Laboratory of Tobacco Chemistry, Anhui Tobacco Industrial Co., Ltd., Hefei 230088, People's Republic of China
[2] School of Mechanical Science and Engineering, Huazhong University of Science & Technology, Wuhan 430074, People's Republic of China
caohuai@hust.edu.cn

Abstract. This paper reports on a cross-cultural study on modeling styles in different countries in the field of HNB product, which is used to measure the consumer decision-making. Based on the eye tracking method, we conducted an experiment study. The independent variables were the HNB product in international brand and domestic brand. The dependent variables include the pupil diameter and subjective evaluation. A total of 68 subjects participated in the experiment. The main findings of this study were as follows. (1) Based on the heat map analysis, the logo, decorative ribbon and indicator light are the main of consumer's visual attention areas. (2) The mean pupil diameter of international brand was higher than domestic brands. However, the difference between the maximum and minimum pupil diameter in domestic brands was higher than international brand. (3) From the perspective of subjective evaluation, the international brand has the higher respondents score than domestic brands. In addition, this study provides a method for reference value with consumer decision-making for the design in related consumer electronic product fields.

Keywords: Cross-cultural · Decision making · Eye tracking · Modeling

1 Introduction

HNB product (Heat-not-burn) have become an important replacement product to traditional cigarette, because it can bring the same taste as real cigarettes while reducing the harm to the human body. Compared with traditional cigarettes, HNB product have obvious advantages. It not produces second-hand smoke, so it not affects the surrounding people. In terms of the chemicals released by cigarettes, HNB product can prevent the production of 95% of harmful substances. From the perspective of the impact on the human body, HNB product is only equivalent to 5% of traditional tobacco. Research from the royal college of medicine in the UK has been committed to research on the effects of HNB product on the human body. The impact of HNB product on the human

P.-L. P. Rau (Ed.): HCII 2021, LNCS 12771, pp. 288–299, 2021.
https://doi.org/10.1007/978-3-030-77074-7_23

body is significantly lower than that of traditional cigarettes. Therefore, HNB products are expected to become effective alternative tools to help to protect human health.

In the research field of consumer products, improving users' consumption experience and enhancing users' repurchase of products play an important and positive role in the word-of-mouth dissemination of corporate brands. The reasons behind consumers' purchasing behaviors are diverse and complex, often at the subconscious level. Consumers themselves are difficult to realize the reasons behind their purchasing decisions. Therefore, research on the reasons why consumers make purchase decisions with the help of relevant tools can better help enterprises gain in-depth insights into consumer behavior. Eye-tracking experiments can provide quantified visual attention and convey unique consumer behavior. Plot customers' shopping behavior characteristics and preferences by overlaying and analyzing consumer data. Enterprises are choosing to use eye-tracking data to help the business make relevant decisions scientifically. Eye tracking can effectively reveal these consumer's visual behaviors, provide enterprises with clear information that affects purchasing decisions, and obtain deep consumer behavioral insight. At present, the eye tracking analysis of HNB products is a gap in the academia and industry, through the scientific analysis of eye tracking behavior. It can evaluate a series of important user experience issues such as consumer' cognitive process, visual concerns and decision making of HNB products, and guide enterprises to effectively carry out innovative research and development in HNB product design areas.

The objectives of this study were as follows: (1) Provide objective data insight for design develop by evaluating the decision making of mainstream HNB products in the market. (2) Through the eye tracking experiment research on the cross-cultural of three types of HNB products brands: international brand, domestic brand (China tobacco), domestic brand (others), the distribution factors of users' visual attention and decision-making explored. (3)Through the subjective evaluation and combining with the experimental data of eye movement, the design attention factors of cross-cultural HNB products were analyzed, which can provide reference for relevant consumer electronic design field.

2 Literature Review

There is a close relationship between human eye movement and the psychological mechanism of visual information processing. In the process of product cognition, users will produce a series of rapid eye movement. By studying the users' first effective visual fixation point, gaze plot and heat map, the user's perception and cognitive process could be effective studied, and then the human cognitive mechanism and behavior can be explain. Traditional research methods rely on explicit behavior and subjective evaluation based on users (observation, questionnaire, interview, etc.), which has subjectivity defects. Physiological data represented by eye movements are more objective and can truly reflect people's behavioral intentions. For example, eye tracking can clearly record consumer's eye movement behavior, which is an important indicator to study people's objective thinking process. Eye movement reflects consumer's visual strategies when watching product modeling design, and it is also an objective presentation of the subjective thinking process. Eye tracking is a powerful tool that can objectively measure consumer attention and spontaneous feedback to product and marketing messages. These insights can help marketers effectively design modeling elements to capture consumers' attention. Eye tracking enables decision-makers to know in real time how consumers interact with products in different cultural and understand their level of cognitive engagement. This method can reduce the recall error and social desirability bias effect, which usually ignored in traditional methods, as shown in Fig. 1.

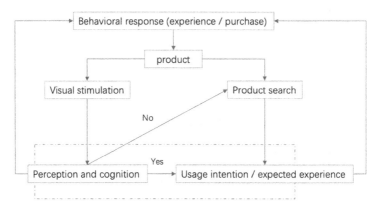

Fig. 1. Principle diagram of eye tracking research for product modeling

In the research of eye tracking method in product modeling design field, Dun (2019) studied consumers' purchase decisions based on eye-tracking experimental results [1]. Hu (2019) studied the preference of young consumers for the appearance design of smartphones through the eye tracking technology [2]. Ou (2018) used eye-tracking experiment to conduct bionic design of product form, and proposed a method to determine bionic morphological characteristics by measuring eye movement data [3]. Su (2017) proposed the method of applying physiological arousal quantity to evaluate product modeling design elements [4]. Wang (2016) used eye tracking as research tool to carry out feature extraction and cognitive analysis of automobile styling from the perspective of feature plane [5]. Hu (2015) evaluated the appearance design of fracturing trucks through eye tracker analysis combined with analytic hierarchy process (AHP) [6]. Tang (2015) adopted the evaluation methods of subjective psychological and objective physiological quantity in the process of user experience as the evaluation indexes of automobile design [7].

In the field of international related research, Gere (2020) conducted an eye tracking study on food images, and used a bottom-up evaluation method based on the size of stimuli, picture background, food display direction, and the number of products under test, to study consumers' visual attention behaviors toward food [8]. Guo (2016) studied pictures of different smartphones and analyzed the capture of attention by product attributes [9]. Liu (2020) studied the design of campus street lamps as an example, used emotion measurement and eye-tracking technology to evaluate the appearance design of products [10]. Nayak (2019) reviewed the scope of application eye movements as an objective visual aesthetic analysis method and systematically presented them after extensive literature retrieval [11]. Khalighy (2015) provided a methodology to quantify the visual aesthetics in product design by applying eye-tracking technology [12]. Guo (2019) simulated an eye- tracking experiment in the aesthetic process of a desk lamp and conducted subjective evaluation to obtain a result of quantifying the visual aesthetic of the product [13].

In the present research, visual behavior trajectories in different areas of product modeling and appearance design tested by using eye-tracking technology experiments. There were also extensive literature searches to study the development path of visual aesthetic evaluation and the development potential of eye movement research. Through the literature study, the current user experience analysis of HNB products, especially the research on quantitative and scientific eye tracking testing has good research potential and need to be future explored.

3 Experiment Method

3.1 Definitions

The research approach of this experimental study was as follows: firstly, the eye- tracking device was adopted to allow users to observe the pictures of HNB product. By capturing the user's eye movement, the quantitative eye-movement data, gaze plot, heat map can be obtained during the test. Through eye movement data, the interest areas of HNB product design were analyzed. Through the above objective eye movement test data, combined with the user's subjective questionnaire, the reasons for the user's interest in the product can be more effectively analyzed, providing relevant theoretical support for the modeling design of HNB products. The independent variables were the characteristic and appearance elements of different HNB products in international brand and domestic brand. The dependent variables include the pupil diameter and subjective evaluation. The experimental theoretical framework, as shown in the Fig. 2.

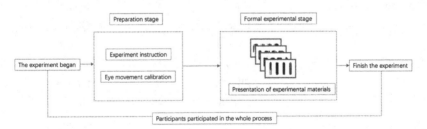

Fig. 2. The theoretical framework of the experiment

3.2 Participants

A total of 68 people were enrolled in this experiment, among which 53 were males accounting for 77.94%, female accounted for 22.06%. The age distribution in 20–30 years old, 30–40 years old and over 40 years old, accounting for 42.65%, 35.29%

and 22.06% respectively. In terms of profession, there are 14 corporate managements, accounting for 20.59%, 42 professional skill workers accounting for 61.77%, 6 administration staffs accounting for 8.82%, and 6 others accounting for 8.82%. The number of participants with education background below bachelor's degree, bachelor's degree and master's degree are 50%, 36.76% and 13.24% respectively.

3.3 Experiment Stimuli and Tasks

According to comprehensive considerations such as product popularity and market distribution, 16 typical HNB products from domestic and international brands were selected. Specifically, there are 5 international brands, 5 domestic brands (China tobacco), 6 domestic brands (others), as shown in Table 1.

Table 1. Mainstream brands of HNB product

Category	The brand name	Product name
International brand	Philip Morris International	Iqos3.0
	British American Tobacco	Glo
	Japan Tobacco	Ploom
	Imperial Tobacco	Pulze
	KT&G	Lil
Domestic brand (China tobacco)	Yunnan Tobacco	Webacco
	Sichuan Tobacco	Kungfu
	Hubei Tobacco	MOK
	Guangdong Tobacco	ING
	Heilongjiang Tobacco	KOKEN
Domestic brand (others)	Buddy	Ibuddy
	Fog&Frog	QOQ
	AVBAD	Avbad
	JOUZ	Jouz
	PlUSCIG	Pluscig
	ISMOD	Ismod

In this study, we used the Tobii X300 desktop eye tracker to conduct experiments. The experimental samples presented in the screen (1920*1080 pixel), allowing users to observe freely, carefully observe the details of the product modeling they are interested in, and collect consumer's eye movement data. Present only one product at the same time and each product picture presented for 10 s. Finally, the Likert scale was used to recording the aesthetics, attractiveness and purchase intention. Using eye-tracking methods, we can analyze participants' eye attention behavior and consumer decision-making during actual tasks, as shown in Fig. 3.

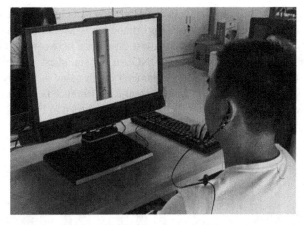

Fig. 3. Participant in the experiment and the environment

4   Result Analysis

4.1   Analysis of Heat Map

From the heat map, the visual hot area of the product mainly focuses on the position of logo, decorative ribbon and indicator light. Each model has its own characteristics, and the heat map focuses on different places with different models, as shown in Fig. 4.

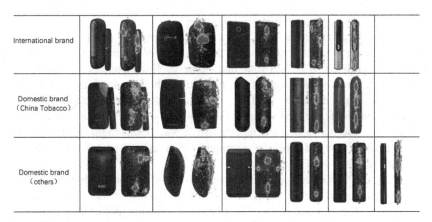

Fig. 4. The heat map visualization in the experiment

4.2   Pupil Diameter

In the statistics of pupil diameter of international brand, the largest mean pupil diameter was found in the Ploom. Pulze had the smallest mean value of pupil diameter, and Iqos 3.0 had the smallest difference between the maximum and minimum value of pupil diameter. In the statistics of pupil diameter of HNB products in domestic brand (China tobacco), the maximum mean value of pupil diameter is MOK, the minimum mean value of pupil diameter is ING, and the smallest difference between the maximum and minimum mean value of pupil diameter is ING. In the statistics of pupil diameter of domestic brand (others), the largest pupil diameter mean value is Ibuddy, the smallest pupil diameter mean value is pluscig, and the smallest difference between the maximum and minimum average pupil diameter is pluscig too, as shown in Fig. 5.

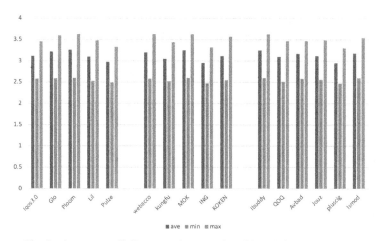

Fig. 5. Average pupil diameter of cross-cultural brands in HNB product

The average pupil diameter of three types of HNB brands in cross-cultural was calculated, and the average pupil diameter of the international brand = 3.14 mm, domestic Brand (China tobacco) = 3.11 mm, and Domestic Brand (others) = 3.13 mm. By comparing the average pupil diameters of international brands, domestic brands (China tobacco) and domestic brands (others), it can be found that the average pupil diameters of international brands are the largest, followed by domestic brands (others) and domestic brands (China tobacco). From the perspective of cross-cultural cognition, the degree of emotional activation to the appearance design of international brands was higher than domestic brands.

However, the maximum pupil diameter of domestic brand (China tobacco) = 3.52 mm, international brand = 3.50 mm, domestic brand (others) = 3.48 mm. The minimum pupil diameter was international brand = 2.56 mm, domestic brand (others) = 2.55 mm, domestic brand (China tobacco) = 2.54 mm. According to the mean value of the maximum pupil diameter, the highest was domestic brands (China tobacco) and the lowest was domestic brands (others), international brands is in the middle level. The difference between the maximum pupil diameter and the minimum pupil diameter of domestic brands (China tobacco) was the most highest. This indicates that among different brand cognition, domestic brands (China tobacco) also has a higher visual stimulus and emotional activation for consumers.

4.3 Subjective Evaluation

In addition to the analysis of objective eye movement data, the experiment also collected subjective evaluations from consumers, giving them the Likert scale of 1–5 based on their intention to buy (1 being "very unlikely" and 5 being "very likely"). Aesthetics score: In the statistics of international brand, the aesthetics score data were as follows: Iqos3.0 (M = 3.67; SD = 0.99), Glo (M = 3.39; SD = 1.00), Ploom (M = 2.71; SD = 1.15), Lil (M = 3.78; SD = 0.85), Pulze (M = 4.02; SD = 0.89). In the domestic brand (China tobacco) statistics, the aesthetic scores was as follows: Webacco (M = 2.46; SD = 1.04), Kungfu (M = 3.34; SD = 1.04), MOK (M = 3.43; SD = 1.06), ING (M = 2.48; SD = 1.18), KOKEN (M = 2.74; SD = 1.02). In domestic brand (others) statistics, the aesthetics score data was as follows: Ibuddy (M = 2.63; SD = 1.11), QOQ (M = 3.17; SD = 0.93), Avbad (M = 3.66; SD = 1.07), Jouz (M = 3.54; SD = 0.91), Pluscig (M = 3.04; SD = 1.18) Ismod (M = 3.54; SD = 1.06). The overall sample of international brand has a higher score. The highest rating of international brand is Pulze, while the lowest rating is Ploom. Among domestic brands (China tobacco), the highest score was MOK and Kungfu. Among Domestic brands (others), the highest rating Avbad and is Jouz. as shown in Fig. 6.

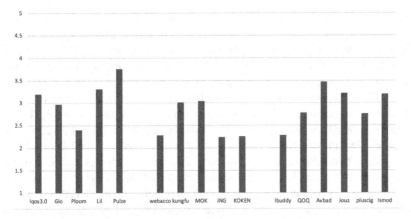

Fig. 6. Average aesthetic score of cross-cultural brands

Attractiveness score: In the statistics of international Brand, the attractiveness score were as follows: Iqos3.0 (M = 3.67; SD = 0.99), Glo (M = 3.39; SD = 1.00), Ploom (M = 2.71; SD = 1.15), Lil (M = 3.78; SD = 0.85), Pulze (M = 4.02; SD = 0.89). In the Domestic brand (China tobacco) statistics, the attractiveness score was Webacco (M = 2.28; SD = 1.03), Kungfu (M = 3.01; SD = 1.02), MOK (M = 3.04; SD = 1.06), ING (M = 2.24; SD = 1.10), KOKEN (M = 2.26; SD = 0.92). In domestic brand (others), the attractiveness rating data were as follows: Ibuddy (M = 2.28; SD = 0.97), QOQ (M = 2.78; SD = 0.98), Avbad (M = 3.47; SD = 1.18), Jouz (M = 3.22; SD = 0.95), Pluscig (M = 2.76; SD = 1.25) Ismod (M = 3.20; SD = 1.04). The highest international brand score was Pulze. The highest score of domestic brands (China tobacco) was MOK, with an average score of 3.04. The domestic brands (China tobacco) with a mean score lower than 2.5 were Webacco, ING and KOKEN respectively. The attractiveness of international brands was generally higher than domestic brands (China tobacco) and domestic brands (others), as shown in Fig. 7.

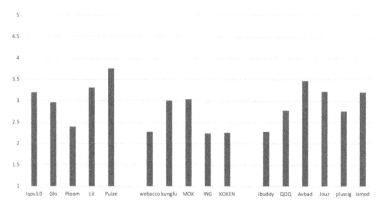

Fig. 7. Average attractiveness score of cross-cultural brands

Purchase intention: In the statistics of international brand, the purchase intention were as follows: Iqos3.0 (M = 2.76; SD = 1.30), Glo (M = 2.63; SD = 1.23), Ploom (M = 2.06; SD = 1.21), Lil (M = 2.90; SD = 1.14), Pulze (M = 3.26; SD = 1.16). In domestic brand (China tobacco), the score data of purchase intention were as follows: Webacco (M = 2.01; SD = 1.13), Kungfu (M = 2.50; SD = 1.18), MOK (M = 2.63; SD = 1.24), ING (M = 1.82; SD = 1.05), KOKEN (M = 2.00; SD = 1.04). In domestic brand (others), the score data of purchase intention were as follows: Ibuddy (M = 2.00; SD = 1.07), QOQ (M = 2.44; SD = 1.22), Avbad (M = 2.96; SD = 1.33), Jouz (M = 2.79; SD = 1.28), Pluscig (M = 2.36; SD = 1.22) Ismod (M = 2.65; SD = 1.15). The score of purchase intention is lower than the evaluation of aesthetics and attractiveness. The highest score was Puzle in international brands and the lowest score was ING for domestic brands. The overall rating of international brands is higher than that of domestic brands. Overall, the modeling design of international brand is higher competitive than domestic brand in the purchase intention aspect, as shown in Fig. 8.

Regardless of the product modeling, only according to the cultural category, from the subjective evaluation, the score from aesthetics to attractiveness to purchase intention is in a descending state. Aesthetic score was generally higher and the purchase intention was generally lower, attractiveness was in the middle level. From the subjective evaluation data, both in domestic and international brands, elliptical shape scoring is generally higher than the slender shape scoring. However, there are extreme differences in the evaluation of thin shape, the highest score and the lowest score both appear in the thin strip shape. This suggests that elliptical modeling is a relatively safe design strategy. If the product designer wants to innovate, it can make thin and long shapes design, but this will have some decision making risk.

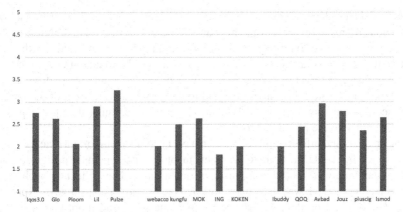

Fig. 8. Average purchase intention score of cross-cultural brands

5 Conclusion

Based on the eye tracking method, this paper conducted an experiment study. The independent variables were the HNB product in international brand and domestic brand. The dependent variables include the pupil diameter and subjective evaluation. Total of 68 subjects participated in the experiment, and the main conclusions were as follows:

(1) Based on the visualization analysis in heat map, the logo, decorative ribbon and indicator light are the main areas of consumer's visual attention. The appearance of HNB products could divided into the following three categories: box, box&pen and pen. Most HNB products designed with the shape of large rounded corners. Cultural differences mainly reflected in the aspect of product modeling.

(2) According to eye-tracking data analysis, different pupil diameter data obtained through experimental tests. The highest mean pupil diameter was international brand, followed by domestic brands. The difference between the maximum pupil diameter and the minimum pupil diameter was 0.98 mm for domestic brands (China tobacco), 0.94 mm for international brands, and 0.93 mm for domestic brands (others). It showed that the consumer had higher visual stimulation and emotional activation in domestic brands (China tobacco) and international brand.

(3) From the perspective of subjective evaluation, the international brand has the higher respondents score and favorability than domestic brands, indicating that domestic brands still have a large expansion capacity for improvement.

In the cross-cultural of HNB product design, whether international brands or domestic brands, the product's aesthetics, attractiveness and purchase intention are complementary to each other. The degree of aesthetics determines the degree of attractiveness and determines the consumer's purchase intention. In the product modeling design, whether international or domestic brand, bigger modeling volume means higher visual stimulation, consumers are more likely to notice and feedback on higher purchase intention. Under the same type of brands, the styling differentiation of international brand is obvious and the product identification is higher. Different models of international brand have obvious representative styling. There are also many similar modeling designs lack of strong recognition of the design language. For example, the appearance of Ibuddy is similar to that of Ismod, and the appearance of Jouz is similar to that of QOQ. However, there are some innovation ideas, such as Avbad's innovative ox-horn shape design. In the future design, more attention should be paid to the brand identification of the product, and the iconic modeling language will be more competitive, rather than just depending on the difference of the product name.

Data analysis and information acquisition have become an important emerging trend in the research field of UX design. Researchers in the data age can make design decisions based on data analysis. The development of quantitative data collection such as eye tracking has changed the way designers think and the research perspective of experience design. The original product experience has become more and more humanized and refined in the data era. UX thinking has shifted from the study of consumer behavior and subjective psychological feelings to the aspects of objective eye tracking data analysis and information integration. In addition, this paper may has some limitations in the experiment. In this experiment, the exposure rate and popularity of international brands may slightly higher than domestic brands, which could lead to the difference in users' familiarity. Future experiments will study the impact of such difference in a more detailed consideration.

Acknowledgments. The research supported by research on quantification analysis technology of eye tracking in user experience of HNB product (project number: 202034000034009).

References

1. Dun, L., Xiong, Y., Yang, S., Ye, C., Zhang, R.: Influence of product types and representation forms on consumers' purchase decisions based on eye-movement technology. Packag. Eng. **40**(18), 214–219 (2019)
2. Hu, Y., Qiao, L., Zhu, J., Yu, M., Han, Q.: Eye movement for appearance design of smart phone. Packag. Eng. **40**(16), 171–176 (2019)
3. Ou, X., Zhou, Z., Liu, B., Zhang, X.: Product form bionic design based on eye-tracking technology. Packag. Eng. **39**(22), 144–150 (2018)
4. Su, J., Qiu, K., Zhang, S., Xiao, L., Zhang, X.: Evaluation method study of product modeling design elements based on eye movement data. J. Mach. Des. **34**(10), 124–128 (2017)
5. Wang, Z., Li, H.: Automotive styling feature extraction and cognition based on the eye tracking technology. Packag. Eng. **37**(20), 54–58 (2016)
6. Hu, W., Zhang, K., Zhang, Z.: Evaluation on modeling design of fracturing truck based on eye-tracking technology. J. Mach. Des. **32**(06), 109–112 (2015)
7. Tang, B., et al.: User experience evaluation and selection of automobile industry design with eye movement and electroencephalogram. Comput. Integr. Manuf. Syst. **21**(06), 1449–1459 (2015)
8. Gere, A., et al.: Structure of presented stimuli influences gazing behavior and choice. Food Quality Preference **83**, 103915 (2020)
9. Guo, F., Ding, Y., Liu, W., Liu, C., Zhang, X.: Can eye-tracking data be measured to assess product design?: Visual attention mechanism should be considered. Int. J. Ind. Ergon. **53**, 229–235 (2016)

10. Liu, P., et al.: An aesthetic measurement approach for evaluating product appearance design. Math. Probl. Eng. **2020**, 1–15 (2020)
11. Nayak, B., Karmakar, S.: Eye tracking based objective evaluation of visual aesthetics: a review. In: Rebelo, F., Soares, M.M. (eds.) AHFE 2018. AISC, vol. 777, pp. 370–381. Springer, Cham (2019). https://doi.org/10.1007/978-3-319-94706-8_41
12. Khalighy, S., Green, G., Scheepers, C., Whittet, C.: Quantifying the qualities of aesthetics in product design using eye-tracking technology. Int. J. Ind. Ergon. **49**, 31–43 (2015)
13. Guo, F., Li, M., Hu, M., Li, F., Lin, B.: Distinguishing and quantifying the visual aesthetics of a product: an integrated approach of eye-tracking and EEG. Int. J. Ind. Ergon. **71**, 47–56 (2019)

On the Zen Connotation in Product Design

Tze-Fei Huang[✉] and Po-Hsien Lin

Graduate School of Creative Industry Design, National Taiwan University of Arts,
Daguan Road, Banqiao District, New Taipei City 22058, Taiwan
t0131@ntua.edu.tw

Abstract. The demands of modern consumers have evolved from the basic physical and psychological to spiritual pursuits and aesthetics experiencing needs. On
the other hand, the style of product design has escaped from simply changing
the outer form, and is more focused on the intangible value starting from the
soul. Design style is not limited to metaphysical expressions but oriented towards
metaphysical interactive experience. Modern people are rich in economic life but
lack of soul. Obtaining spiritual sustenance through religion has become the goal
that modern people pursue, and since simplicity in the Zen image is also close to
the basic concept of Zen, the application of Zen to art design is accepted by the
majority of ethnic groups and has turned a design style.

Therefore, this research will focus on the Zen design of products, which can
give designers a reference for the transmission of Zen imagery, shaping elements
and processing techniques in designing Zen products. The core value of Zen is
explored through literature analysis in the early stage, followed by the questionnaire survey based on the semantic difference to evaluate the feelings of people with religious beliefs in design backgrounds, people with religious beliefs in
non-design backgrounds, and the general public. Finally, the results obtained are
assessed in a multidimensional manner so as to investigate the differences. This
research hopes to give designers interested in Zen a clearer picture of the Zen
image.

Keywords: Product design · Zen · Elements of style

1 Introduction

Subsequent paragraphs, however, are indented. The world is currently in a state of
chaos, and the emptiness, the pursuit of nature, and simplicity represented by Zen can
give modern people who are lost a chance to breathe. Now the role of designers is no
longer simply to provide product functions and the pursuit of beauty but to shoulder the
responsibilities of the era, the society, and the culture. Therefore, it is very important
and far-reaching to understand and recognize the origin of its culture. A comprehensive
survey conducted by the Religious Counseling Division of Taiwan's Ministry of Interior
Affairs in 2005 showed that 35% of Taiwan's population consider themselves Buddhists.
The economic and aesthetic values of Buddhist cultural relics are at the service of the
religion, not only providing an artistic image and imagery for the creation of religious

© Springer Nature Switzerland AG 2021
P.-L. P. Rau (Ed.): HCII 2021, LNCS 12771, pp. 300–310, 2021.
https://doi.org/10.1007/978-3-030-77074-7_24

consciousness but also offering specific material conditions and expression channels for the implementation of religious practices, promoting the development of the Buddhist culture.

The United Nations Educational, Scientific and Cultural Organization (UNESCO) defines cultural and creative goods as consumer goods that convey opinions, symbols and lifestyles. The demand characteristics of modern consumers have transformed from the basic physical and psychological to the spiritual levels to pursuing and experience aesthetics. On the other hand, the style of product design has escaped from simply changing the outer form, and is more focused on the intangible value starting from the soul. It coincides with the insight and emptiness of "mind" in Zen. Therefore, the design style is no longer limited to metaphysical expressions but oriented towards metaphysical interactive experience. As early as the end of the 19th century, the West had begun to notice this mysterious Eastern thought, and this power has quietly spread in the field of art and continued to this day. In recent years, the so-called Zen style has emerged in many design industries including space design, architecture, furniture, and product design. The Zen style referred to herewith is not directly related to the religious Zen but is purely applied to the beauty or aesthetics presented in design through the imagery extension. The Zen-rich oriental aesthetics "Wabi-Sabi Beauty" has become one of the modern significances of Japanese graphic design mainstream.

This research project uses a quantitative analysis of the current market's demand for Zen-related cultural relics, classifies and analyzes Zen's aesthetic characteristics, establishes visual evaluation and symbolic meaning analysis that express Zen, and analyzes whether Zen imagery products sold by Taiwan Buddhist organizations are loved by consumers through a questionnaire survey, subsequently exploring whether there is a significant difference in consumer perceptions of Zen, which can provide a reference for future Zen product design. The purpose of this research is to understand the current market's preference for Zen products and the perception of Zen. The product images used in this research are mainly based on Taiwan Buddhist organizations, commercial Buddhist cultural relics, and works with Zen imagery created by artists. This research will mainly discuss the Zen style design, hoping to give designers a reference to the perception of Zen, shaping elements, and processing techniques when designing Zen style products.

2 Literature Discussion

The section of Literature Discussion helps us understand the definition and main ideas of Zen. Apart from this, it also addresses the expression of Zen in the field of art, which offers better understanding of artists' cognition of Zen as well as the Zen concepts presented through artistic techniques. This section can serve as the basis for this research.

2.1 Definitions and Accounts of Zen

In recent years, the term "Zen" has appeared in various design fields or general social activities. As it is widely used in art and cultural domains, it remains what is Zen and what Zen is in the terrain of design. The definition of Zen has been discussed by both Chinese and foreign scholars. For example, the founder of Taiwan Buddhist Institute Fagushan, who was the fifty-generation heir of the Cao Dong School of Zen, Master Shi Sheng-Yan, explained Zen in the book Zen and Enlightenment [6]. He classified Zen into four types: (1) Meditation-The proper Buddhist noun for this notion is dhyāna, which is translated into Chinese with the meanings of concentration, calmness, and contemplation. (2) Four Meditation Heavens-It is translated in Sanskrit as catvāri dhyānāni. The first meditation refers to being free from birth and happiness; the second meditation refers to the creation of joy; the third meditation refers to being free from joy and happiness; the fourth meditation refers to being pure. (3) Sitting meditation-As the name suggests, sitting meditation is the use of sitting posture to achieve the purpose of meditation. (4) Zen-It originated in India, grew up and matured in China, and then spread to Korea, Japan, and Vietnam. Zen emphasizes subjectivity, and does not pay attention to traditional authority and external revelation. The ultimate goal is to make people see their own minds and cut off all intellectual thoughts and emotional disturbances. Another intellectual, known as the "World Zen Master", who was nominated for the Nobel Peace Prize in 1963, was the Japanese Buddhist scholar- Daisetz Teitaro Suzuki. In his book Zen and Japanese Culture [2], Zen is defined as follows. (1) It refers to being straight on the spiritual entity, ignoring all forms. (2) Zen refers to striving to seek the existence of spiritual entities in any form. (3) Zen believes that incomplete forms and defective facts can express the spirit. (4) Zen denies formalism, traditionalism, and ritualism, resulting in the direct exposure of the spirit, returning to its loneliness and solitariness. (5) This transcendental loneliness and "absolute" solitude are the spirit of poverty and asceticism, which excludes all traces that may be non-essential. (6) This kind of lonely spirit is not clinging to anything in ordinary words. (7) If loneliness is understood in the absolute sense of Buddhists, it is that Zen is silent in the wild, from the humble wild weeds to the so-called highest form in nature. Daisetz Teitaro Suzuki believes that the most effective way for people who learn Zen to achieve spiritual enlightenment is to experience and practice Zen, which also means pondering or contemplation. Based on the above, we can first learn that Zen has the meanings of pondering and ruminating as well as utilizing intuition and instinct to calm the troubled mind.

2.2 Zen Connotation in the Domain of Art

The Zen style in the field of art is the art of insight into the essence of human life, not aesthetic contemplation. The "insight" of Zen and the "inspiration" of art belong to the same spiritual phenomenon. In Chinese art, Zen is more manifested in painting, poetry, and calligraphy. Chinese art adopts a contrasting artistic conception of virtual and real, which perfectly expresses the beauty of obscurity, beauty of nature, and beauty of erratic nature. It is a manifestation of emptiness. Through the depiction of natural landscapes and scenery, it shows the perception of living, the society, and life or the calm and indifferent state of mind. Ink expresses pure consciousness in simple colors, through the black and white of the void and the reality: white is where there is no ink, that is, the void, and black refers to where there is ink, that is, the reality. The void and the reality, one concealing and one appearing, interact to form the beauty of the artistic conception of painting tension. The characteristics of dynamicity and metaphor in calligraphy convey the ethereal and transcendent and the profoundly connotative Zen state. The influence of Zen on Japanese culture is even greater, extended in drama, painting and calligraphy writing, among which the beauty of simplicity is the best representation of the Zen mood, with minimal and simple design forms to achieve both functional and aesthetic effects. The use of unique quality and texture of the material, without special surface treatment, conveys a delicate, calm, and invisible Zen aesthetics, expressing the rich connotation meaning with the least internal representation. The characteristics of Zen art culture, in terms of Chinese and Japanese Zen style arts, mark the unity of the three aesthetics of Confucianism, Taoism, and Buddhism, bringing it to maturity. Art works abounding in the Zen style often employs multi-layered images to arouse the rich artistic associations of viewers. Since Zen focuses on experience and intuition, in order to be enlightened with Zen in the process of daily life, one must experience the true nature of Zen from ordinary times. As Dazhuo Suzuki stressed, "An artist must enter the object, feel it from the inside, and live his life by himself." Finally, by discussing the performance of Zen culture in traditional arts such as poetry, calligraphy, and painting, it can be concluded that Zen culture emphasizes the essence of things in line with natural rhythms, such as "quaintness", "quietness", "emptiness", "faintness", "thoughtlessness", "harmony", and "leisureliness". Suzuki also pointed out that Zen art is characterized with non-balance, asymmetrical, and unilateralism, poverty, simplicity, loneliness, solitude, quietness, among many characteristics [3]. The above-mentioned Zen culture can become the appearance and language image to be conveyed in the product design in terms of artistic characteristics. The characteristics of Zen thinking, such as metaphor, metonymy, epiphany, etc., are used as application in product design and provide guidance in aesthetic spirit.

3   Research Methodology

3.1   Procedure and Framework

The research is divided into four steps as follows:

Phase 1. We conduct a literature review on the definition and interpretation of Zen. In this part, it will focus on the related discussion of Zen and the development of Zen in art and design, identifying the codes and images of Zen from the literature serving as basis of the questionnaire survey at the next phase.

Phase 2. Through the analysis of the connotation of the concrete and abstract symbols of Zen related products and the perceptual image research of many product shapes in the past, the Semantic Differential (SD method) is used at this phase. The questionnaire samples are based on Taiwanese Buddhist organizations. With commercial Buddhist cultural relics and works created by artists with Zen imagery as the main objects, Donald A. Norman's "Emotional Design" [1] is used as the basis for coding.

Phase 3. In the Semantic Difference experiment, 10 participants with design education, 41 ones with art-related backgrounds, and 10 ones with no art or design education are used. It is also tested whether participants have a religious belief so as to help us detect cognitive differences between them.

Phase 4. Based on the data crossing different items and types, we analyze consumers' preferences for Buddhist cultural relics currently on the market and works created by artists with Zen imagery.

3.2 Coding

Through the interaction between people and things, we can feel the message or meaning that things are intended to express, and convert it into the appearance of the product, unveil the sensual inside our hearts, and open up the perceptual level of people's superficial consciousness. The involvement and moving experience generated by the interaction between the user and the product are derived. Therefore, aside from the basic functions, whether the product can provide users with additional feelings has also been a critical issue. Using Donald A. Norman's "Emotional Design" (2005), this research proposes three levels of design concepts for design coding, attempting to find out the involvement and moving experience generated by the interaction between the user and the product.

(1) Instinct level-is the exterior design attractive?
(2) Behavior level-does the Buddhist cultural products bear associations with the religion?
(3) Reflection level-does it have Buddhist cultural connotation or profoundly present the meaning of Zen?

The level of instinct reflects the attractiveness of the product's "appearance" to users, the level of behavior reflects the "function" and "practicality" of the product, and the level of reflection indicates the imagery elements of the product's "meaning behind". Therefore, if the product intends to attract attention, instinctive design is the best way. If you take human beings as the starting point, you can implement the design at the behavioral level. If you want to inspire people to think more deeply, you have to design at a reflective level. Zen, as a branch of the Zen schools, is one of the practices of Buddhism, so it is intriguing whether Zen products can profoundly present the essence of Zen, to transform and translate the form and spiritual connotation of things.

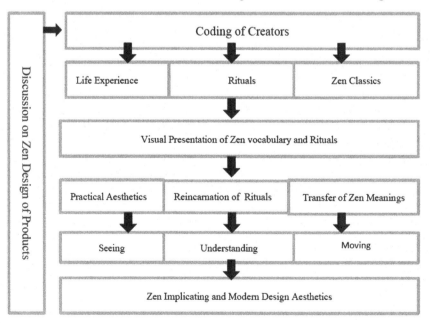

Fig. 1. Three levels of Zen product design (sorted out and drawn in this study)

3.3 Sampling

The product images used in this research are mainly based on Taiwanese Buddhist organizations, commercial Buddhist cultural relics, and works with Zen imagery created by artists. The samples are collected according to the three categories: (1). Religious body; (2). Ritual tools and utensils; (3). Daily necessities. The questionnaire is based on the semantic images and the Zen thinking integrated in the literature discussion.

Table 1. Experimental samples (1. Religious body)

Table 2. Experimental samples (2. Ceremonial tools and utensil)

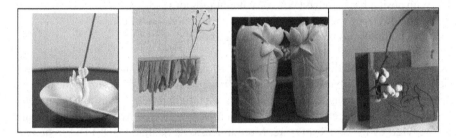

The questionnaire is based on the semantic images and the Zen thinking integrated in the literature discussion valuation Criteria: Two-way Semantic Vocabulary of Zen Imagery:

Table 3. Experimental samples (3. Daily necessities)

1. Rustic or Gorgeous
2. Ordinary or Elegant
3. Brisk or Stable
4. Straightforward or Obscure
5. Secular or Religious
6. Concise or Complicated
7. Friendly or Serious
8. Technical or Artistic
9. Free or Temperate
10. Do you feel Zen from this product?
11. Is this product associated with any religion?
12. How much you like this product?
13. Do you usually like products with a Zen style?
14. Please choose a product that you think is the most Zen-like (single-choice question).
15. Please select one of your favorite products (single- choice question).

According to the Cognitive analysis of the works by the respondents with different backgrounds:

(1) With design background (With or without religious belief)
(2) With art related background (With or without religious belief)
(3) General public (With or without religious belief)

In the Semantic Difference experiment, 10 participants with design education, 41 ones with art-related backgrounds, and 10 ones with no art or design education are used. It is also tested whether participants have a religious belief so as to help us detect cognitive differences between them. The data were collected from the questionnaires and analyzed by the SPSS software, Multidimensional Scaling (MDS) Discussion and analysis of product image and shape, Finally, the characteristic words were generalized and summarized.

3.4 Questionnaire

In this survey, questionnaires with a 5-point scale are executed online, which are divided into two parts: the first part is the sub-item; the second part is the overall response to the tested sample. We also collect the basic information of the subjects (Table 4).

4 Results

4.1 Overall Evaluation

Table 5 lists the samples with different average and standard deviations of the subjects' evaluation of the following aspects. The samples are divided into three categories: 1. Religious body; 2. Ritual tools and utensils; 3. Daily necessities. Therefore, different types of samples will have different numbers of items. The results show that (1) regarding products in the category of "religious body" (P1–P4), they directly correspond to

Table 4. Content of the Questionnaire (P1 as the instance).

	How does the product feel?	
	1. Rustic or Gorgeous	Low □1 □2 □3 □4 □5 High
	2. Ordinary or Elegant	Low □1 □2 □3 □4 □5 High
	3. Brisk or Stable	Low □1 □2 □3 □4 □5 High
	4. Straightforward or Obscure	Low □1 □2 □3 □4 □5 High
	5. Secular or Religious	Low □1 □2 □3 □4 □5 High
	6. Concise or Complicated	Low □1 □2 □3 □4 □5 High
	7. Friendly or Serious	Low □1 □2 □3 □4 □5 High
	8. Technical or Artistic	Low □1 □2 □3 □4 □5 High
	9. Free or Temperate	Low □1 □2 □3 □4 □5 High
	10. Do you feel Zen from this product?	□A □B
	11.Is this product associated with any. religion?	□A □B
	12. How much you like this product?	Low □1 □2 □3 □4 □5 High
	13. Do you usually like products with a Zen style?	□A □B

14. Please choose a product that you think is the most Zen-like (single-choice question).	□1 □2 □3 □4 □5□6 □7 □8 □9 □10□11 □12
15. Please select one of your favorite products (single- choice question).	□1 □2 □3 □4 □5 □6 □7 □8 □9 □10□11 □12

strong religious associations, showing calmness, directness, and high technicality. (2) Regarding products in the "ritual appliance" category (P5 –P8), with religious symbols or religious representative symbols, they have a more direct association with religion, showing simplicity and high artistry. (3) Regarding products in the category of "daily necessities" (P6 –P12), products with higher scores in religious representative symbols show lower artistic performance, while those with drawing patterns related to religious bodies exhibit higher complexity and seriousness.

4.2 Analysis of Product Attributes

The product attributes are classified through multi-similarity and multi-dimensional scaling (MDS), and the relative preferences after classification are found. The stress coefficient of this study is 0.12933, and the RSQ value is 0.92100. Therefore, this model has high consistency. As shown in Table 5, the analysis of the nine attribute factors and the representative products of each factor, the subjects generally report that P6's artistic creation is a Zen implicating product, because there is no direct representation of religious symbols, so free and obscure items score higher. In addition, P1, P2, and P3 are faith bodies (Buddha statues), but P1 has simple colors and presents a simple image compared to the other two. P3 is a Buddha statue created by the artist. It is tested in partial realistic and partial abstract expressions, bringing subjects feelings of gorgeousness, complexity,

Table 5. Average ratings and standard deviations of evaluation.

	P1	P2	P3	P4	P5	P6	P7	P8	P9	P10	P11	P12
f1 Rustic - Gorgeous	2.00	2.30	4.20	2.27	3.68	2.32	3.20	2.22	3.75	3.53	1.67	3.28
f2 Ordinary - Elegant	3.82	3.97	4.08	3.12	3.90	3.27	3.22	3.50	3.23	3.53	2.85	2.83
f3 Brisk - Stable	4.28	3.55	3.20	4.23	2.58	2.97	3.37	3.22	3.22	3.17	3.72	3.68
f4 Straightforward - Obscure	2.65	2.88	2.43	3.12	3.17	3.97	2.73	3.33	2.00	2.67	3.77	2.05
f5 Secular - Religious	4.45	3.82	3.85	3.67	3.23	2.43	2.68	3.30	3.85	3.62	2.27	4.28
f6 Concise - Complicated	1.93	2.18	3.88	2.37	2.50	2.42	3.02	2.15	3.32	3.25	1.98	3.08
f7 Friendly - Serious	3.27	2.63	2.93	3.38	2.32	2.80	2.68	2.47	3.38	3.17	2.40	3.78
f8 Technical - Artistic	3.47	4.05	3.77	3.60	3.48	3.58	2.95	3.82	2.37	2.82	3.77	2.40
f9 Free - Temperate	3.57	2.42	2.43	3.10	2.70	2.37	2.95	2.53	3.47	3.27	2.57	3.87

and briskness. P12 is a religious symbol for daily necessities (accessories) and is highly realistic, which will make subjects feel serious and restrained (Fig. 2, Table 6).

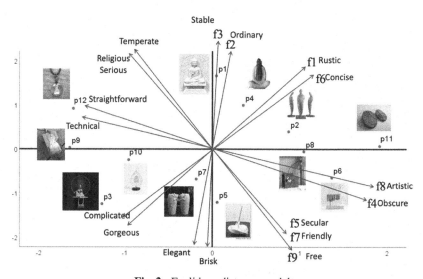

Fig. 2. Euclidean distance model

5 Conclusion

According to the results of this study, the more Zen inspiring they are, the more religious they are. Preferences are not related to Zen connotation. The products which are favored by subjects tend to be more abstract or artistic, which of preferences are not influenced by religion background and Zen implicature. P3 with the strongest preference is the religious body (Buddha statue) as the artist's creation. Among some of the realistic and abstract expressions, they are gorgeous, complex, and highly technical, which is why

Table 6. Nine-attribute factor analysis and their representative items.

Styles and Characteristics		Factor Loading			Representative Item		
		Factor 1	Factor 2	Factor 3	Item 1		Item 2
Secular - Religious	f5	.909	.085	.354		Secular	Religious
Friendly - Serious	f7	.867	.109	-.339		Friendly	Serious
Free - Temperate	f9	.837	.-086	-.446		Free	Temperate
Straightforward - Obscure	f4	-.805	-.523	-.024		Obscure	Straightforward
Rustic - Gorgeous	f1	.184	.963	.112		Gorgeous	Rustic
Concise - Complicated	f6	.238	.891	-.041		Complicated	Concise
Brisk - Stable	f3	.599	-.707	-.112		Brisk	Stable
Ordinary - Elegant	f2	-.002	.196	.957		Elegant	Ordinary
Technical - Artistic	f8	-.493	-.537	.636		Artistic	Technical
eigenvalues		3.619	2.848	1.786			
explanatory variance		40.208	31.647	19.840			
accumulated explanatory variance		40.208	71.855	91.695			

it is favored by subjects. Therefore, among the products with Zen imagery selected by the researcher, it is not the strong Zen implicature that attracts subjects but the vague Zen connotation, incurring strong aesthetic feelings and artistic senses, that makes the product lovable (Table 7).

Table 7. Product symbolism and preferences.

Q1 Strength of Zen sense												
Item	p4	p1	p2	p3	p8	p12	p9	p5	p10	p7	p6	p11
Mean	4.00	3.80	3.72	3.68	3.60	3.53	3.45	3.37	3.32	2.75	2.72	2.58

Q2 Religious Symbolism												
Item	p1	p12	p9	p3	p10	p4	p2	p5	p8	p7	p6	p11
Mean	4.58	4.58	4.38	4.22	4.10	3.97	3.95	3.60	3.55	3.07	2.08	2.02

Q3 Preference												
Item	p3	p6	p8	p2	p11	p5	p1	p4	p10	p7	p9	p12
Mean	3.68	3.55	3.52	3.48	3.47	3.40	3.35	3.23	2.98	2.82	2.57	2.32

Therefore, a conclusion can be drawn from the feedback of this questionnaire survey. The delivery of Zen through products can achieve a certain effect, but this does not mean that we have to give up the expression of beauty and artistry. On the contrary, through the creators' translation and the Zen or religious imagery, religious products may also be very artistic and aesthetic. Therefore, from the perspective of science and technology, the art of "decoding" and "recoding" brings about the theme of artistic creation as well as more possibilities and unexpected creativity.

References

1. Norman, D.A.: Emotional Design. Yuan-Liou Publishing, Taipei (2011). (Wang, H., Weng, Q., Zheng, Y., Zhang, Z. Trans.)
2. Suzuki, D.T.: Zen and Japanese Culture. Walkers Cultural Enterprises, Ltd., New Taipei (1992). (Lin, H. Trans.)
3. Suzuki, D.T., Liu, D.: Zen and Art. Heavenly Lotus Publishing, Taipei (1988)
4. Helen, W.: Zen in the fifties-interaction in art between east and west. 2nd edn. Artco Books, Taipei (2018). (Zeng, C., Guo, S., Chen, Y. Trans.)
5. Lin, R.T.: Cultural and creative product design: discussion on Kansei technology, user-friendly design and cultural and creative. Hum. Soc. Sci. Newslett. Q. 11(1), 32–42 (2009). [in Chinese, semantic translation]
6. Shi, S.-Y.: Zen and Enlightenment. Dharma Drum Publications Corp. Taipei (2001)
7. Shi, S.-Y.: Essentials of Buddhist Teaching. Dharma Drum Publications Corp., Taipei (2018)
8. Shi, S.-Y.: The Essentials of Practice and Attainment within the Gate of Chan. Dharma Drum Publications Corp, Taipei (2018)

Employing a User-Centered Elder and Youth Co-creation Approach for a Design of a Medication Bag: A Preliminary Study

Ding-Hau Huang[✉], Yu-Meng Xiao, and Ya-Yi Zheng

Institute of Creative Design and Management, National Taipei University of Business, Taoyuan, Taiwan
{hauhuang,alicezheng}@ntub.edu.tw

Abstract. Medication errors are a common problem in healthcare, particularly in elderly individuals, who often have multiple prescriptions due to chronic diseases. At present, the design of most commercial drug packaging do not take into account the deteriorating sensory, nervous, and musculoskeletal functions of the elderly, leading to errors in taking medications as a consequence of visual and/or behavioral impairments. Attempt have been made to optimize the designs of healthcare products and services, for example, medication bags using experience-based co-design (EBCD), by involving different stakeholders such as elderly individuals; however, the elderly have mostly played passive roles during those processes. Although the elderly were consulted, they rarely in fact participated in the product design process. This study therefore combined EBCD and design thinking to establish a user-centered elder and youth co-creation (EYCC) model, the process of which included five workshops: engage, empathize, define, ideate, and prototype and test. The participants of this study included 9 individuals between the ages of 61 and 74, who themselves or their family members were suffering from chronic diseases and were taking multiple types of medications, and 8 young designers. All participants cooperated together on the redesign of medication bags. The results of this study show that the workshops in the EYCC model improved the general self-efficacy of a number of the elderly participants and significantly improved the empathy of the young designers; however, improvements in drug compliance of the elderly participants were not found. The positive findings of this study can serve as a reference for future co-creation procedures involving the elderly and youth.

Keywords: User-centered design · Co-creation · Medication error · Medical packaging design

1 Introduction

According to the World Health Organization [34], a third of deaths worldwide are caused by drug misuse. Suffering from one or more chronic diseases becomes more prevalent as one ages; thus, multiple prescriptions are required, increasing the complexity of medication use [22] and the chance of medication errors. In the US, approximately 100,000

© Springer Nature Switzerland AG 2021
P.-L. P. Rau (Ed.): HCII 2021, LNCS 12771, pp. 311–329, 2021.
https://doi.org/10.1007/978-3-030-77074-7_25

medication errors occur each year [12]. At present, the designs of most commercial drug packaging do not take into account the deteriorating sensory, nervous, and musculoskeletal functions of the elderly, leading to errors in taking medications as a consequence of visual and/or behavioral impairments. There is ongoing research on how to optimize the designs of medication packaging for the elderly with the aim of reducing medication errors. For example, healthcare services have utilized experience-based co-design (EBCD), a joint design and test method involving in-depth interviews and observations that enables patients and relevant stakeholders to conduct group discussions and sharing. However, patients have mostly played passive roles during these processes, and though they are consulted, they in fact rarely participate in the product design process. Therefore, the resulting designs may deviate from the needs of actual users. Research has also shown that involving elderly individuals in the design process can provide clearer definitions of their usage needs that result in higher user satisfaction [10, 14]. Design thinking is a user-centered approach that has already been applied to designing various products and services [26]. It combines multiple fields and the co-design characteristics of different stakeholders, the procedures and operation tools of which can be freely adjusted depending on the objective. The objective of this study was to investigate the influence of the elder and youth co-creation (EYCC) model, which was developed by merging EBCD and the five steps of design thinking, on the drug compliance and general self-efficacy of elderly individuals as well as on the empathy of young designers.

2 Literature Review

2.1 Medication Use of the Elderly and Medication Packaging Designs

To improve the safety of medication use and establish correct medication concepts in Taiwan, the Ministry of Health and Welfare proposed the Five Core Competencies of Medication Use: (1) clearly expressing one's physical condition, (2) carefully reading medication labels, (3) knowing the method and time of medication use, (4) being in control of one's own body, and (5) building a good relationship with doctors and pharmacists [27]. For elderly patients who take multiple kinds of medications for chronic diseases, identifying medication correctly and avoiding adverse reactions are crucial [6]. In other words, the combination of taking multiple medications and poor drug compliance can easily lead to adverse responses as a result of interactions among the drugs [9]. Taking inappropriate medications can also easily cause harmful effects and even death. In order to assess and gauge drug compliance, a number of researchers have developed self-management questionnaires aimed at patients with chronic diseases, such as the Drug Attitude Inventory (DAI), developed by Hogan [17], and the self-reported Morisky Adherence Questionnaire (MAQ), created by Morisky [21]. Thompson, Kulkarni, and Sergejew [29] subsequently optimized the DAI and MAQ and developed the Medication Adherence Report Scale (MARS), which can be used for various diseases and medications. It has been applied to different clinical environments and served as a basis for clinical research involving the assessment of drug compliance.

As elderly individuals age, their sensory organs begin to exhibit varying degrees of deterioration, resulting in a gradual loss of sensory abilities. For example, the retinas lose their ability to discern images. With consideration to the physiological needs of the elderly in mind, medication packaging designs should use text, graphic symbols, and layouts that convey correct information about medication use, especially regarding potency and side effects [33]. Many organizations have begun developing graphic symbols that can enhance the understanding and the memory of users with regard to medication-related information, such as the United States Pharmacopeia Dispensing Information (USP) and the International Pharmaceutical Federation (FIP) graphic symbols. Warm colors such as red and orange, which are colors with longer wavelengths that can remedy the impact of the yellowing of the eye lens among the elderly, are also recommended in packaging designs. By catering to the changes in the vision of the elderly, their recognition of medication can be improved [33]. Furthermore, the text on medication packaging is the primary means of conveying medication information to the elderly, such as the name of medication, labels, instructions for use, and side effects. Due to the fact that it may be difficult for elderly patients to read very fine print, special consideration should be given to font size on packaging. The color of the text should also sharply contrast the background color so that medication information can be more easily read and understood.

2.2 Design Thinking, Empathic Design, and Co-design

Originating from the thinking logic of designers in product design, design thinking is a methodology centered on user needs. It emphasizes problem analysis beginning with the needs of people before creation. This approach is already widely used in commerce and technology and hadn't been applied to medical services until recent years. Some of the more well-known results include the Center for Innovation and the Embrace infant warmer. A number of different steps or procedures have been proposed by organizations for design thinking; the following five steps are the most common: (1) empathize, which involves observing and thinking from the user's perspective as well as understanding their pains and gains; (2) define, which involves understanding and excavating user needs based on the empathy in the first step to pinpoint the true cause of the problem, identify the most important needs, and combine relevant knowledge into insights; (3) ideate, which involves taking the defined problem out of the conventional framework to seek a more effective solution; multiple feasible design plans can be developed via teamwork before one is selected; (4) prototype, which involves finding the most feasible of the multiple solutions, creating a concrete prototype for the product or service, and then using the result as a medium of communication with the team or target consumers; and (5) test, which involves placing the prototype in a usage scenario for target consumers for actual testing [24].

The means of effectively achieving empathy is the key to user-centered designs. Empathy is an innate ability to think from the perspective of another, feel what they feel, and then help them. It promotes a greater understanding of the emotions of others and facilitates emotional communication [25]. Empathy can be divided into cognitive empathy and affective empathy (or emotional empathy). The easiest and most common way of gauging empathy is using self-report measures. Researchers have developed a variety of questionnaires regarding empathy. Spreng et al. [25] indicated that a consensus should be reached among various scales of empathy and then multiple empathy factor items in the existing scales should be merged into a set of highly correlated items, thereby forming a single empathy factor. They therefore developed the Toronto Empathy Questionnaire (TEQ), which has been widely used with good reliability and validity.

Co-design is an approach that requires the joint participation of relevant stakeholders in a design process [16]. In medicine, however, mostly only single viewpoints, such as those from patients, are included in design processes, and different types of stakeholders are rarely invited to participate together. The EBCD proposed by Bate and Robert is a medical service planning, assessment, and improvement method in which the co-design process includes patients, family caregivers, and medical personnel. This method has been used to improve the treatment of brain cancer patients and the nursing experience of medical personnel [4]. So far, the EBCD has been used in over 60 items in six countries and has been proven to effectively improve patient experiences [18]. This method includes eight phases in four steps (see Fig. 1): capturing experiences, understanding experiences, improvement, and follow-up [4] [23] [30]. To support EBCD research conducted by the National Health Service (NHS), free online toolkits were developed so that EBCD researchers could readily learn the steps of co-design [28]. However, there has been a lack of cooperation that would lead to co-design improvements, which is a challenge experienced by most EBCD items.

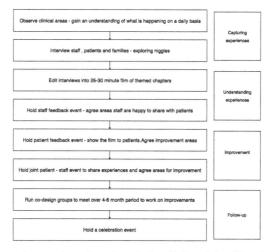

Fig. 1. Stages in experience-based co-design (Point of Care Foundation, 2017)

2.3 Elder and Youth Co-creation

The rapidly growing elderly population will become one of society's greatest burdens in the future. Whether this can be turned into a positive for society remains to be seen. Relevant research has shown that the participation of elderly individuals in the design process provides clearer definitions of their usage needs, resulting in greater user satisfaction [10, 14]. However, most elderly individuals who participated in research and development were passive. They were mainly consulted and merely validated instead of actively solving their own problems. In recent years, it has been advocated both domestically and overseas that the elderly should take a more active role in society and interact with youths to co-create value. For example, the Bron van Doen Social Design Labs developed design thinking processes and tools that have been used in the NIEUW/OUD project promoted in Eindhoven, the Netherlands, during the past three years. In this project, young local creative workers cooperate with the residents of the Vitalis Berckelhof elderly home to co-create product designs, literary works, and artistic photography as well as pass down culinary inheritances. In the past few years, this type of initiative has also gained traction in Taiwan, leading to the establishment of the first elderly-youth co-creation base in Asia. Via the participation of different generations, it has strengthened intergenerational communication, captured a multitude of vigor in design and innovation, and created a new generation of products and services for senior citizens.

A benefit of co-design is that stakeholders can increase the adoption rate, significance, and sustainability of services [32]. Having the elderly become co-creators is of value to not only consumers of technology but also to increase the use of technology [19]. Previous research has shown potential benefits of involving elderly individuals as co-designers of innovation concepts and of having prior contact with the content of interest to facilitate ideation [15].

2.4 General Self Efficacy

Common in psychosocial and health research, the concept of self-efficacy was first proposed by Bandura and is the most effective variable in predicting healthy behavior [3]. Self-efficacy refers to the belief in one's own ability to complete a designated task [2], reflecting the overall confidence that an individual displays in facing and dealing with difficulties [7]. Individuals with higher self-efficacy set higher goals for themselves, constantly try, are unyielding, and are readily willing to face new challenges even after previous failures [2]. Self-efficacy is a critical motivating factor that stimulates creative behavior [13], displaying an individual's awareness of whether they can effectively interact with the environment and confidently achieve a goal or expected behavior. The Stanford Patient Education Research Center was the first to create self-management items for patients with chronic diseases. Based on self-efficacy theory, they developed a series of intervention measures to improve the health behavior of patients by enhancing their sense of self-efficacy and in turn improving their quality of life [20].

3 Methodology

3.1 Participants

The participants of this study included 9 elderly women between the ages of 61 and 74, who themselves or their family member suffered from chronic diseases and were taking medications long-term, 8 design majors with a bachelor's or master's degree, including the researcher, and 1 teacher experienced in design thinking, who led the various workshops. The elderly participants first took a pretest using the MARS and were divided based on the results into a low-score group, a medium-score group, and a high-score group. Each group contained 3 elderly participants and 2–3 young designers. All of the participants had signed an informed consent form (NTNU-Form-05/12).

3.2 Data Collection Method

Pretest and posttest questionnaires were respectively administered before and after the study to the both elderly and young groups of participants and then a paired sample t-test was applied to the pretest and posttest results. For the elderly participants, we employed (1) the MARS developed by Thompson et al. [29] and (2) the General Self Efficacy Scale (GSES) developed by Zhang & Schwarzer [35]. The MARS contained 10 question items with "yes" or "no" responses. Except for Items 7 and 8 being forward-scored (yes = 1 point and no = 0 points), all other items were reverse scored. A higher total score indicated better drug compliance. The GSES assessed whether the self-efficacy of the participants differed after they took part in the activities of innovative design production. The items in the GSES were measured on a four-point Likert scale: 1 = Not at all true, 2 = Hardly true, 3 = Moderately true, and 4 = Exactly true. A higher total score from the ten question items indicated better general self-efficacy.

For the youth designers, we applied the Toronto Empathy Questionnaire (TEQ) designed by Spreng, McKinnon, Mar, and Levine [25], which contained 16 question items, 8 of which were negative items and reverse-scored, measured on a five-point Likert scale: 0 = Never, 1 = Rarely, 2 = Sometimes, 3 = Often, and 4 = Always. A higher total score indicated greater empathy.

3.3 Elder and Youth Co-creation Model

This study combined the four steps of EBCD, namely, capturing experiences, understanding experiences, improvement, and follow-up, with the five steps of design thinking to develop an elder and youth co-creation (EYCC) model. This model included five workshops: engage, empathize, define, ideate, and prototype and test (see Fig. 2). The application of the model helped gain an in-depth understanding of the difficulties that elderly individuals may encounter when taking medication, and then solutions were devised. In each workshop, the elderly were active participants and took part in completing the tasks.

Engage Workshop. The goal of this workshop was to introduce elder and youth co-creation cases and elucidate why thoughtful medication packaging designs were needed. Through design thinking experiences, which also built up the creative confidence of the elderly participants, young designers and the elderly participants familiarized themselves with each other.

Step 1–1 Workshop introduction: The design teacher introduced the participants and explained why thoughtful medication packaging designs were needed and important.

Step 1–2 Elderly and youth co-create case study: Using existing elder and youth co-creation cases as examples, young designers and elderly participants understood the importance of elder and youth co-creation, which also helped the elderly build up their confidence in their creativity.

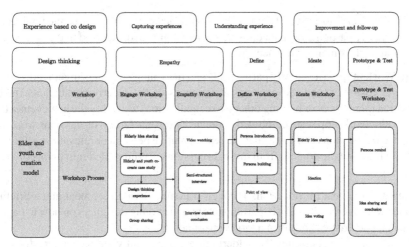

Fig. 2. Elder and youth co-creation model (EYCC model)

Step 1–3 Design thinking experience: During this step, the elderly were asked to design medication packaging for one another with the assistance from the young designers. Using the 5 steps of design thinking, the elderly participants interviewed and empathized with one another to identify problems, explore solutions, and brainstorm various possibilities. In the end, they created a conceptual prototype and shared their final concepts one at a time.

Step 1–4 Group sharing: The elderly participants shared the medication packaging design that they made for another elderly participant.

Empathy Workshop. The objective of the empathy workshop was to provide the young designers an understanding of the habits and problems of the elderly participants in different groups regarding medication.

Step 2–1 Video watching: A medication safety video from the Health Promotion Administration of the Ministry of Health and Welfare was shown to the elderly participants at the beginning of this workshop to give them an understanding of correct medication administration and help them reflect on whether they were making medication errors in their daily life.

Step 2–2 Semi-structured interview: Each young designer interviewed one to two elderly participants using the Five Core Competencies of Medication Use Scale developed by Chen [18] as the interview outline. This semi-structured interview included open questions and multiple-choice questions so that the elderly participants had more room to express their own views and opinions. Based on the responses of the elderly participants, their potential reasons as well as deeper thoughts and motives were explored. During this process, the young designers record the "yes" or "no" responses of the elderly participants on the questionnaire and noted any other content provided (e.g., reasons for using pill organizers).

Step 2–3 Interview content conclusion: The notes and contents from the elderly participants in a group were discussed and compiled by the young designers in said group and then categorized and named. The young designers then shared the final results with the elderly participants and confirmed whether any errors existed in the content and whether any adjustments were needed.

Define Workshop. The elderly participants and the young designers together discussed one target customer, and the personal experiences of the elderly participants (taking medication themselves or helping others take medication) as well as the achievements of the empathy workshop were combined to create a persona. Group discussions were then conducted on the pains and gains that the representative persona or relevant stakeholders may have had, and the content from these discussions was compiled into points of views (POVs).

Step 3–1 Persona Introduction: Through watching a video segment and a briefing, guidance was given to the elderly participants and the young designers on how to capture persona characteristics and form a target customer.

Step 3–2 Persona building: The elderly participants and young designers discussed what type of representative (e.g., an individual, group member, or family member with dementia) would serve as the target customer group and formed that persona. The written content included personality traits, basic attributes, medication habits, medication time, and source of medication.

Step 3–3 Point of view: Once the persona was created, the participants brainstormed the feelings and pains of the persona and what needs had yet to be satisfied based on the descriptions of the persona. These contents were compiled into POVs.

Step 3–4 Prototype (Homework): Before the ideate workshop, the elderly participants selected one POV and used pictures or words to draw or describe a concept.

Ideate Workshop. The objective of the ideate workshop was to have the young designers ideate multiple concepts based on the design concepts created by the elderly participants in Step 3-4, create sketches of them, and share them with their group, following by choosing a concept to be prototyped.

Step 4–1 Elderly Idea sharing: The elderly participants in each group shared the POV that they had chosen for ideation and the results of their ideation. At the same time, the young designers point it out of corresponding POV content.

Step 4–2 Ideation: Based on the results of Step 4–1, each young designer ideated several concepts and presented the concepts through drawings or writing. Finally, the design concepts that were developed were shared in each group.

Step 4–3 Idea voting: The elderly participants and the young designers together voted on one or more design concepts, then divided the work among them. The elderly participants and the young designers created different design concepts and completed them before the prototype and test workshop.

Prototype and Test Workshop. In the prototype and test workshop, the design concepts created by each group were shared with the other groups. Design feedback from different elderly participants were obtained as references for future revisions.

Step 5–1 Persona reminder: Before the elderly participants in each group shared their design concepts, the teacher explained the personas of each group and the pains that groups wished to resolve.

Step 5–2 Idea sharing and conclusion: Sharing which POV was used to create their prototype, the elderly participants in each group presented their ideas and concepts as well as explained the reasons for their designs. The young designers then each shared their conceptual prototypes. During the process, the elderly participants gave suggestions for revisions to the prototypes and concepts created by other elderly participants and young designers. The youth designers recorded and compiled this feedback.

4   Result and Discussion

4.1   Workshop Results

Engage Workshop. The purpose of the engage workshop was to boost the confidence of the elderly participants and help all participants be familiar with each other. The five steps of rapid experience design thinking were employed.

Step 1–1 Workshop amin introduction: The design teacher explained the purpose of the workshop and helped the elderly participants to understand the importance of medication packaging designs as well as the problems that existing packaging could cause them.

Step 1–2 Elderly and youth co-create case study: Through existing elder and youth co-creation cases, the elderly participants learned that other elderly individuals had already taken part in co-creation cases in Taiwan and that they themselves could participate in design as well.

Step 1–3 Design thinking experience: The elderly participants interviewed and empathized with one another, while the young designers helped them record main points, which included pill organizers being too small and being less convenient than sachets as well as the need for graphic symbols to help them read medication instructions. Conceptual drawings of the defined needs were made (see left image of Fig. 3).

Step 1–4 Group sharing: The elderly participants shared their packaging design ideas that they made for another elderly participant (see right image of Fig. 3). The final design concept was to add graphic marks to the pill sachets that one generally uses and to sort medication by date and time.

Empathy Workshop. The objective of the empathy workshop was to understand the medication habits and problems of the elderly participants among different groups. After the engage workshop, many of the elderly participants designed their own pill organizer case or bag at home.

Step 2–1 Video watching: After the participants watched the medication safety video from the Ministry of Health and Welfare, some of the elderly participants realized that they had medication habits that needed breaking. In this case, some of the elderly participants mentioned that they took their medication not at a fixed time but instead at a time that was convenient for them. Furthermore, they could not read information that was in English on medication packaging, and the bags given by smaller clinics which contained their medication may not have included all relevant information, such as side effects.

Fig. 3. Conceptual drawing (left) and result sharing (right)

Step 2–2 Semi-structured interview: The young designers interviewed the elderly participants and noted the main points (see Fig. 4). The results included 57 main points, such as taking sleeping pills to sleep better and purchasing medication with the aid of pharmacists.

Step 2–3 Interview content conclusion: The young designers discussed, compiled, and categorized their notes made with the elderly participants in their group, and the elderly participants confirmed whether the content was correct. All of the notes were divided into 16 categories, such as medication-taking conditions and doctor consultation (see Fig. 5).

Going abroad	The rest of the medicine	Consult a doctor	How the take medic
Receive medicine situation	The reason for taking medication	Avoid multiple medications	Pay attent to the precaution medicin
Maintain the habit of taking medications	Don't buy over-the-counter drugs	Fixable dosage	Pay attent to the bod blood pressure a blood sug
Fixable medicine taking time	The information on medicine bag	easy to carry with separate packing	The medic is placed i fixed posit

Fig. 4. Interview process   **Fig. 5.** Compilation of interview content

Define Workshop. In the define workshop, the elderly participants and the young designers in each group engaged in defining a character, designing a persona, and further confirming the design perspectives.

Step 3–1 Persona Introduction: After watching a video segment of an elderly individual taking medication incorrectly and an explanation of persona characteristics, the elderly participants pointed out the problems and characteristics of the persona in the video and gained an understanding on how the create a persona.

Step 3–2 Persona building: The young designers and elderly participants together discussed and created a persona. In one case, the representative persona was 80-year-old Mr. Dai, who had post-operative dementia, easily accepted the medication recommended by others, and did not pay attention to the instructions printed on medication packaging. He did not have any good medication habits, he did not take his medication at a fixed time, and he needed to be urged by his family members or caregiver to take his medication. In addition, his caregiver helped to sort his medication in pill organizers but was worried that the pill organizer cases or bags were not sealed well enough to prevent moisture from damaging the medication.

Step 3–3 Point of view: Using the persona, the young designers and the elderly participants wrote four POVs, including the need for correct and convenient pill organization, because clear organization and medication preservation were important to Mr. Dai, as well as the need for clear labels explaining medication instructions, because understanding the side effects and medication instructions were important to Mr.Dai.

Step 3–4 Prototype (Homework): The elderly participants selected one POV and used pictures or words to draw or describe their design concept. In one case, one elderly participant created a medication bag with graphical aids regarding medication time. The other elderly participant wrote down the needs that elderly individuals had when taking medicine. For example, memory loss and poor vision among elderly individuals meant that packaging designs should have larger font sizes. Family members should also accompany elderly patients to the doctor to give the doctor a better understanding of the patient's medical history and provide other family members with information about the side effects of medication and any translations into Chinese of information about the medications.

Ideate Workshop. Based on the concept in Step 3-4 of the define workshop, the participants brainstormed and developed the medical packaging design concept of their group.

Step 4–1 Elderly idea sharing: Both of the elderly participants designing medication bags brainstormed based on POV1 (correct and convenient pill organization) and POV2 (clear instruction labels for medication). The elderly participant who used text descriptions brainstormed based on POV4 (reminders and encouragement to take medication).

Step 4–2 Ideation: Based on the results of Step 4–1, the young designers ideated, drew, and shared the results with their group. The design concepts included having different colored medication bags for morning, noon, and night, which also contained small sachets of the pills needed for each day in order to prevent the wrong medication or the wrong doses from being taken. The medication bags were labeled with the instructions and graphical symbols for each type of medication so that elderly individuals would understand the symbols and would be willing to take their medication. As the text on existing medication bags are often too light in color and too small in font size, increasing the contrast, such as the use of vibrant colors and larger font sizes, is needed in visual presentation. The medication bags were designed with two sides; the front was designed by the elderly participants and had medication information in large font, whereas the back was designed for the patient's children, family members, or caregivers and had the same font size as existing medication packaging as well as space to record medication conditions. Thus, caregivers could record patient conditions and side effects in this space, which could then be communicated with the doctor during the next visit to the clinic (see Fig. 6).

Step 4–3 Idea voting: The elderly participants and the young designers together voted on the design concepts that they created, and the results were then integrated into a single medication bag. The elderly participants and young designers then completed a prototype based on their own understanding before the prototype and test workshop.

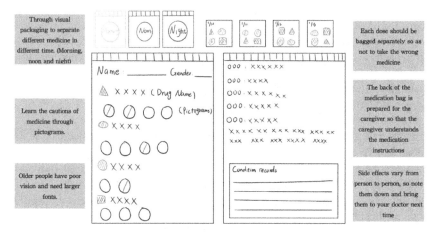

Fig. 6. Concept brainstorming results

Prototype and Test Workshop. After the ideate workshop, the elderly participants and young designers each designed their own prototypes. The prototype workshop mainly involved sharing their prototypes and giving one another suggestions.

Step 5–1 Persona reminder: The elderly participants recalled the pains of the persona that had been created by their groups and the things that their designs were based on.

Step 5–2 Idea sharing and conclusion: First, the elderly participants shared their designs and the POVs that the designs were based on, and then the young designers shared their prototype concepts. As the elderly participants shared, the young designers noted the keywords of the ideas of the elderly participants regarding the prototype that they had made as well as the suggestions made by the other elderly participants. Their suggestions included the following. The pill organization sachets designed by an elderly participant did not have a good seal, thereby preventing the long-term preservation of medication. The pill organization sachets displayed little information, which could lead to medication errors. Dates and the time of the next visit to the doctor could be added to the medication bag. The medication bags could be designed to be the same color as the medication to facilitate identification. The medication bags had a lot of text, and the graphical symbols were not obvious; thus, reminders and notes could be highlighted.

The design concepts of the young designers suggested that the pill organization sachets in the medication bags could also have different colors, the concept being that if pill organization sachets were visually appealing, it would also encourage elderly patients to take their medication (see Fig. 7).

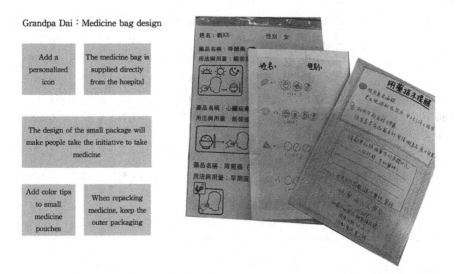

Fig. 7. Design concept and elderly feedback.

4.2 Questionnaire Results

SPSS 26.0 was employed to test the various variables. The influence of the design thinking-incorporated EYCC model on the drug compliance of the elderly participants after taking part in medication packaging design was verified using a paired sample t-test. The results revealed no significant differences between the pretest and posttest results regarding drug compliance ($p > 0.05$) (see Table 1). With regard to general self-efficacy, the results did not present any significant differences ($p > 0.1$) except for Pair 2 (If someone opposes me, I can find the means and ways to get what I want.) and Pair 8 (When I am confronted with a problem, I can usually find several solutions.) ($p < 0.1$). Further analysis of the general self-efficacy scores of the elderly participants (see Table 2) revealed a mean increase of 4.9 points in the overall score (see Table 3), indicating that the design thinking-incorporated EYCC model improved the general self-efficacy of some of the elderly participants after they took part in the design of medication packaging.

With regard to empathy, the test results displayed significant differences ($p < 0.05$) in Pair 5 (I enjoy making other people feel better.), Pair 7 (When a friend starts to talk about his/her problems, I try to steer the conversation towards something else.), Pair 8 (I can tell when others are sad even when they do not say anything.), and Pair 13 (I get a strong urge to help when I see someone who is upset.). There were no significant

Table 1. Paired Sample t-test result of drug adherence in the pretest and posttest.

Drug Adherence

Question	Pretest-Posttest	M	Std.	t	df	sig.
Pair 1	A1-A1H	0.111	0.333	1	8	0.347
Pair 2	A2-A2H	0.111	0.601	0.555	8	0.594
Pair 3	A3-A3H	−0.222	0.667	−1	8	0.347
Pair 4	A4-A4H	0.222	0.441	1.512	8	0.169
Pair 5	A5-A5H	−0.111	0.782	−0.426	8	0.681
Pair 6	A6-A6H	0	0.5	0	8	1
Pair 7	A7-A7H	−0.333	0.707	−1.414	8	0.195
Pair 8	A8-A8H	0	0.5	0	8	1
Pair 9	A9-A9H	−0.111	0.333	−1	8	0.347
Pair 10	A10-A10H	−0.222	0.441	−1.512	8	0.169

[*] indicates sig. <0.05

Table 2. Paired Sample t-test result of General-Self Efficacy in the pretest and posttest.

General Self-Efficacy

Question	Pretest-Posttest	M	Std.	t	df	sig.
Pair 1	B1-B1H	−0.333	1.118	−0.894	8	0.397
Pair 2	B2-B2H	−0.778	1.202	−1.941	8	0.088
Pair 3	B3-B3H	−0.667	1.323	−1.512	8	0.169
Pair 4	B4-B4H	−0.333	0.886	−1.155	8	0.282
Pair 5	B5-B5H	−0.444	1.014	−1.315	8	0.225
Pair 6	B6-B6H	−0.333	1	−1	8	0.347
Pair 7	B7-B7H	−0.333	0.707	−1.414	8	0.195
Pair 8	B8-B8H	−0.556	0.726	−2.294	8	0.051
Pair 9	B9-B9H	−0.556	1.13	−1.474	8	0.179
Pair 10	B10-B10H	−0.556	1.014	−1.644	8	0.139

[*] indicates sig. <0.1

differences in the other question items (p > 0.05) (see Table 4). Further analysis of the personal empathy results of the young designers (see Table 5) revealed a mean increase of 4.75 points in the overall score, thereby indicating that the design thinking-incorporated EYCC model improved the empathy of the young designers.

Table 3. Mean for General Self-Efficacy of elderly in the pretest and posttest.

General Self-Efficacy			
Elder	Pretest	Posttest	Std.
1	25	30	5
2	25	30	5
3	12	36	24
4	27	28	1
5	25	32	7
6	36	36	0
7	39	36	-3
8	35	36	1
9	30	34	4

Table 4. Paired Sample t-test result for Empathy in the pretest and posttest.

Question	Pretest-Posttest	M	Std.	t	df	sig.
Pair 1	C1-C1H	−0.5	0.756	−1.871	7	0.104
Pair 2	C2-C2H	−0.375	1.302	−0.814	7	0.442
Pair 3	C3-C3H	−0.25	1.165	−0.607	7	0.563
Pair 4	C4-C4H	−0.375	0.744	−1.426	7	0.197
Pair 5	C5-C5H	−0.75	0.707	−3	7	*0.02
Pair 6	C6-C6H	−0.375	0.744	−1.426	7	0.197
Pair 7	C7-C7H	−0.75	0.707	−3	7	*0.02
Pair 8	C8-C8H	−0.75	0.886	−2.393	7	*0.048
Pair 9	C9-C9H	−0.25	1.165	−0.607	7	0.563
Pair 10	C10-C10H	0.625	0.916	1.93	7	0.095
Pair 11	C11-C11H	0.25	1.035	0.683	7	0.516
Pair 12	C12-C12H	−0.125	0.835	−0.424	7	0.685
Pair 13	C13-C13H	−0.5	0.53452	−2.646	7	*0.033
Pair 14	C14-C14H	0.5	1.19523	1.183	7	0.275
Pair 15	C15-C15H	0	0.92582	0	7	1
Pair 16	C16-C16H	−0.375	1.06066	−1	7	0.351

* indicates sig. <0.05

Table 5. Mean for empathy level of youth designers in the pretest and posttest.

No.	Pretest	Posttest	Std.
1	50	49	−1
2	54	55	1
3	52	49	3
4	46	51	5
5	43	59	16
6	36	41	5
7	36	41	5
8	39	43	4

4.3 Discussion

The objective of this study was to determine the influence of the EYCC model on the drug compliance and general self-efficacy of elderly participants as well as on the empathy of young designers.

The study results revealed that regardless of the degree of drug compliance of the elderly participants before they took part in designing medication packaging, the EYCC model did not improve their drug compliance, due perhaps to the fact that the five workshops were completed over a period of 1.5 months. This could have been too short of a period to affect drug compliance, which involves personal beliefs and habits. Our results were similar to those derived by Unni, Shiyanbola, and Farris [31], who investigated the correlation between the drug compliance and beliefs of elderly individuals over a period of two years. Their results indicated that as age increases, drug compliance and beliefs no longer change with time. They did note however that the drug compliance of elderly individuals with low drug compliance in the beginning improved over time.

The EYCC model improved the general self-efficacy of some of the elderly participants after they took part in designing medication packaging. The questionnaire items "If someone opposes me, I can find the means and ways to get what I want." and "When I am confronted with a problem, I can usually find several solutions." presented significant differences, demonstrating that the elderly participants may have discovered that they could come up with different solutions to the medication problems of their peers during the workshops. Regarding the empathy of the young designers, the results showed that participation in these workshops enabled them to understand the joys and concerns of the elderly and recognize the fact that once children become independent and form families of their own, the elderly not only have their own focuses in life but must also take care of themselves. Abeyaratne et al. [1] pointed out that after their own workshops, the participants who were students reported that they could better understand the problems of the elderly and empathize with them.

Through persona creation, the elderly participants in each group drew on their own experiences or those of other elderly individuals in their life to make the persona authentic. The opinions expressed by the elderly participants helped with subsequent designs made by all stakeholders. Groups who shared their prototype designs with one another could prevent potential conflicts early on in the process of product development. From the perspective of each stakeholder, it can ensure that the intervention measures and their implementations are more acceptable and feasible. With elderly individuals participating in each stage of the design process, it was demonstrated that with effective guidance, the elderly also have design capabilities and that their creativity is not limited by age. In a study on the creativity of elderly individuals participating in the design of mobile healthcare applications, Davidson and Jensen [8] indicated that although the collaborating elderly individuals in their study did not have any experience, they not only took part in the design process but also developed a relatively creative product. The conclusions of that study were similar to those of this study. In the serious workshops, the elderly individuals mentioned their need for caring interactions with their family and children. Past studies similarly described that elderly individuals viewed relationships with their friends or family as the second most important things in their lives [11].

5 Conclusion

In the face of aging societies, there is a pressing need for designers to develop products from the perspective of the elderly. This study employed an EYCC model, which combines EBCD and the five steps of design thinking, to engage in designing medication packaging. The results of this study demonstrate that although the workshops did not significantly improve drug compliance among the elderly participants, general self-efficacy of some of the elderly participants and the empathy of the young designers improved. Our findings can serve as a reference for elder and youth co-creation procedures in the future.

References

1. Abeyaratne, C., Bell, J.S., Dean, L., White, P., Maher-Sturgess, S.: Engaging older people as university-based instructors: A model to improve the empathy and attitudes of pharmacists in training. Curr. Pharmacy Teach. Learn. 12(1), 58–64 (2020)
2. Bandura, A.: Self-efficacy: toward a unifying theory of behavioral change. Psychol. Rev. 84(2), 191 (1977)
3. Bandura, A., Freeman, W.H., Lightsey, R.: Self-efficacy: The exercise of control (1999)
4. Bate, P., Robert, G.: Toward more user-centric OD: Lessons from the field of experience-based design and a case study. J. Appl. Behav. Sci. 43(1), 41–66 (2007)
5. Bate, P., Robert, G.: Experience-based design: From redesigning the system around the patient to co-designing services with the patient. BMJ Qual. Saf. 15(5), 307–310 (2006)
6. Chang, W.J., Chen, L.B., Hsu, C.H., Lin, C.P., Yang, T.C.: A deep learning-based intelligent medicine recognition system for chronic patients. IEEE Access 7, 44441–44458 (2019)
7. Chen, G., Gully, S.M., Eden, D.: Validation of a new general self-efficacy scale. Organ. Res. Methods 4(1), 62–83 (2001)
8. Davidson, J.L., Jensen, C.: Participatory design with older adults: an analysis of creativity in the design of mobile healthcare applications. In: Proceedings of the 9th ACM Conference on Creativity & Cognition, pp. 114–123) (2013)
9. Davies, E.A., O'mahony, M.S.: Adverse drug reactions in special populations–the elderly. Br. J. Clin. Pharmacol. 80(4), 796–807 (2015)

10. Demirbilek, O., Demirkan, H.: Universal product design involving elderly users: a participatory design model. Appl. Ergon. **35**(4), 361–370 (2004)

11. Farquhar, M.: Elderly people's definitions of quality of life. Soc. Sci. Med.**41**(10), 1439–1446 (1995)

12. FDA.   https://www.fda.gov/drugs/information-consumers-and-patients-drugs/working-red uce-medication-errors 2019/8/23

13. Ford, C.M.: A theory of individual creative action in multiple social domains. Acad. Manag. Rev. **21**(4), 1112–1142 (1996)

14. Glass, T.A., De Leon, C.M., Marottoli, R.A., Berkman, L.F.: Population based study of social and productive activities as predictors of survival among elderly Americans. Bmj **319**(7208), 478–483 (1999)

15. Harrington, C.N., Wilcox, L., Connelly, K., Rogers, W., Sanford, J.: Designing health and fitness apps with older adults: examining the value of experience-based co-design. In: Proceedings of the 12th EAI International Conference on Pervasive Computing Technologies for Healthcare pp. 15–24 (2018)

16. Hickey, G., et al.: Guidance on co-producing a research project. Southampton: INVOLVE (2018)

17. Hogan, T.P., Awad, A.G., Eastwood, R.: A self-report scale predictive of drug compliance in schizophrenics: reliability and discriminative validity. Psychol. Med. **13**(1), 177–183 (1983)

18. Hsiang-Lan, C.: Perceptions on Safety Medication of Caregivers Working in Disability Welfare Institutions. National Defense Medical Center School of Public Health Master's thesis (2016)

19. Lee, J.M., Hirschfeld, E., Wedding, J.: A patient-designed do-it-yourself mobile technology system for diabetes: promise and challenges for a new era in medicine. Jama **315**(14), 1447–1448 (2016)

20. Lorig, K.R., et al.: Evidence suggesting that a chronic disease self-management program can improve health status while reducing hospitalization: a randomized trial. Medical Care, 5–14 (1999)

21. Morisky, D.E., Green, L.W., Levine, D.M.: Concurrent and predictive validity of a self-reported measure of medication adherence. Medical Care 67–74 (1986)

22. Mortazavi, S.S., et al.: Defining polypharmacy in the elderly: a systematic review protocol. BMJ open 6.3 (2016)

23. Piper, D., Iedema, R., Gray, J., Verma, R., Holmes, L., Manning, N.: Utilizing experience-based co-design to improve the experience of patients accessing emergency departments in New South Wales public hospitals: an evaluation study. Health Serv. Manag. Res.**25**(4), 162–172 (2012)

24. Plattner, H.: An Introduction to Design Thinking Process Guide. https://dschool-old. stanford.edu/sandbox/groups/designresources/wiki/36873/attachments/74b3d/Mode Guide-BOOTCAMP2010L.pdf (2010)

25. Spreng, R.N., McKinnon, M.C., Mar, R.A., Levine, B.: The Toronto Empathy Questionnaire: Scale development and initial validation of a factor-analytic solution to multiple empathy measures. J. Pers. Assess. **91**(1), 62–71 (2009)

26. Stickdorn, M., Hormess, M.E., Lawrence, A., Schneider, J.: This is Service Design Doing: Applying Service Design Thinking in the Real World. O'Reilly Media, Inc (2018)

27. Taiwan Food and Drug Administratio. http://mohw.gov.whatis.com.tw/. Accessed 21 Apr 2016

28. The King's Fund 2011. http://www.kingsfund.org.uk/projects/ebcd. Accessed 15 Oct 2013

29. Thompson, K., Kulkarni, J., Sergejew, A.A.: Reliability and validity of a new Medication Adherence Rating Scale (MARS) for the psychoses. Schizophrenia Res. **42**(3), 241–247 (2000)

30. Tsianakas, V., Robert, G., Maben, J., Richardson, A., Dale, C., Wiseman, T.: Implementing patient-centred cancer care: using experience-based co-design to improve patient experience in breast and lung cancer services. Supportive Care Cancer **20**(11), 2639–2647 (2012)
31. Unni, E.J., Shiyanbola, O.O., Farris, K.B.: Change in medication adherence and beliefs in medicines over time in older adults. Global J. Health Sci. **8**(5), 39 (2016)
32. van der Wal, M.H., Hjelmfors, L., Mårtensson, J., Friedrichsen, M., Strömberg, A., Jaarsma, T.: Variables related to communication about prognosis between nurses and patients at heart failure clinics in Sweden and the Netherlands. J. Cardiovasc. Nurs. **33**(2), E1–E6 (2018)
33. Wang, L.: Pharmaceutical packaging design for the elderly patients. Packaging Eng. (8), 62–67 (2018)
34. World Health Organization.: Medication errors (2016)
35. Zhang, J.X., Schwarzer, R.: Measuring optimistic self-beliefs: a Chinese adaptation of the general self-efficacy scale. Psychologia: Int. J. Psychol. Orient. (1995)

Co-creating Experience in Engaging Customers with Product Development: A Case Study of Hair Products

Mei-lin Huang[✉]

Graduate School of Creative Industry Design, National Taiwan University of Arts, New Taipei City, Taiwan

Abstract. Today, despite consumers having more product and service choices than ever before, they seem dissatisfied. Firms invest in greater varieties of products but are less able to differentiate themselves. Growth and value creation have become the dominant themes for managers (Prahalad and Ramaswany 2004). Breaking through the status quo, bringing creativity to new products that are more relevant to the market, both to online and offline networks, the users are no longer merely the recipients of value, but have become participants who are capable of creating value, breaking the traditional communication model. The experience of co-creating, sharing, is the beginning of creativity, and the real source of creativity is "relationship". Under this assumption, this research aims to investigate the product development of co-creation with customers. This study takes the development of hair growth products as an example. In order to achieve this goal, this research combines brand literature with research surveys and analyzes it by incorporating a customer co-creation experience, customer brand engagement, and customer satisfaction into the market research. The research is a comprehensive research model that integrates customer feedback into the co-creation experience, customer participation in brand product development, customer satisfaction, and the commercial value of the product.

Keywords: Value co-creation · Customer experience · Customer satisfaction

1 Introduction

Traditional product design and development was led by companies, however, companies still cannot satisfy customers and sustain profitable growth. Companies can no longer act autonomously, designing products, developing production processes, crafting marketing messages, and controlling sales channels with little or no interference from consumers. In the internet era, customers dissatisfied with available choices are armed with the means to provide direct feedback; consumers want to interact with firms and this relationship co-creates value (Prahalad and Ramaswany 2004). Usually, brands communicate through traditional marketing and advertising to build brand awareness and heavily discounted promotions to push sales. Through the impact of the internet, users are no longer the value recipients, users can create value through interacting with

© Springer Nature Switzerland AG 2021
P.-L. P. Rau (Ed.): HCII 2021, LNCS 12771, pp. 330–339, 2021.
https://doi.org/10.1007/978-3-030-77074-7_26

a company. Companies were using linear value chains to create value, which put value into a products or service, through online tools, consumers participate in the design of products before the products are launched, and companies can focus on target consumers to interact with said products that then draw in more customers through interaction with those target customers. Can these recognitions of a product be used as an effective marketing activity? This research is the result of one such creation experience for a brand and the customer-product development.

The current research is a comprehensive research model that integrates customer feedback into a co-creation experience, customer participation in brand product development, customer satisfaction, and the commercial value of the product. The focus of the research is to create value with consumers and reach target consumers through survey. The questionnaire design is based on product functional components and consumer needs. When consumers answer the questionnaire based on their own needs for this product, subsequent product designs are produced in accordance with the requirements of the consumers. In this time, the company can foster a deep emotional resonance with the consumer base through the experience of "co-creating marketing". The basis of this marketing strategy emphasizes the provision of consumer experiences in form of senses, emotions, actions, and related values as a tool for co-creation and execution. Through effective communication, consumers are left with a deep impression and the establishment of this memory aids in the achievement of marketing goals. The products used in this research case were hair shampoo and hair growth essence.

2 Literature Review

2.1 Market Value

According to the Biotechnology Industry Research Center of Taiwan Economic Research Institute (2016), the global market for external use hair health product is about 2.3 billion U.S. dollars. It will grow at a compound annual growth rate (CAGR) of 4.8% from 2015 to 2020. Estimated to reach a market of 2.9 billion US dollars in 2020, the market size of Taiwan's hair products is approximately NT$1.7 billion, with an annual growth rate of 3.3%, reaching 2 billion in 2020.

2.2 Advertising Methods

In the beauty industry, most brands use endorsement advertising for brand awareness. Traditional advertising is costly and it is still difficult for brands to build awareness. Where communication theory is concerned, the cognitive process of product advertising or spokesperson advertising includes three levels: technical (do you see?), semantic (do you understand?), and effect (are you moved by it?) (Lin et al. 2015, 2017). When consumers first see the product packaging, they perceive the appearance on "technical level", then understand the meaning and connotation of the "semantic level", finally they reach the "effect level" of emotional connection (Lin et al. 2015, 2017). These three levels correspond to the experience of product development co-creation. The questionnaire is designed to function as the coding process; the decoding process for consumers are are (1) whether the consumer sees it and triggers the sensory perception of the product's interest and appearance; (2) whether the consumer understands the meaning of the product and enters the cognitive thinking mode; (3) whether the consumer is moved and takes action in response to the psychological activity of emotional connection (Lin Rongtai and Li Xianmei 2015). For consumers, there are three key steps to understand the meaning of goods: attracting attention (recognition), correct cognition (understanding), and emotional connection (reflection), (2007; Lin et al. 2015, 2017).

Recognition, as situational perception, indicates whether product design can attract an audience; understanding, as the perception of artistic conception, indicates whether consumers can understand the meaning it conveys; reflection, as a psychological feeling, indicates whether consumers are moved by the effect of the product and then make purchases.

2.3 Market Survey Questionnaire Design

Prahalad and Ramaswany (2000) observed that with the development of the internet, people can get information easier than before, users can share experience about the products and services easily, and can also gather people with relevant interests to virtual communities in order to co-create products with a company. Through the impact of the internet, the target consumer age group (over 45 years old) easily adapted to the use of an online questionnaire, users naturally used social media as a communication tool. In the post-epidemic era, this research study shows how product co-creation with consumers can build relationships. Hypothesizing product characteristics, ingredients, product effects, consumer demands, and economic experiences will positively affect customer satisfaction, which in turn affects consumers' preference for products and affects purchase behaviors. Which variables do consumers value? This research shows how brands can learn through the co-creation process, constructing a relationship between consumers and products, conveying product information, establishing communication channels, and targeting potential customers call for action.

2.4 Value Co-creation Stands Out

Only companies who center creating value with their customers can obtain true business performance (Prahalad and Ramaswamy 2004), and co-creation is an interactive function focused on the creation of this kind of value. The behaviors of companies and customers can be categorized by fields (providers, associates, customers), and the interaction between them is direct or indirect, leading to different forms of value creation and co-creation. The conceptualization of value creation expands understandings of how value emerges and how to manage value creation; it also emphasizes the key role of direct interaction for value in co-creation opportunities. The theory of engagement (Pansari and Kumar 2017) suggests that the focus of marketing is customer engagement. The point of emphasizing co-creation initially started with customer engagement, the concept of co-creation appeared in multiple company-driven environments, such as virtual (Füller 2010), learning (Desai 2010), product development, and innovation (Ramaswamy 2011; Rowley et al. 2007). We can observe a change in the handling of co-creation, the role of the customer ranges from the structured phasing of the customer experience to co-design, participating in self-service and finally co-production of services (Prahalad 2004; Vargo and Lusch 2004). The conclusion emphasized the value of joint production, the company benefitted "by enabling customers to carry out their own value creation activities" (Normann and Ramirez 1993).

2.5 Perceptual Technology and Humanity

With the rise of the experience economy, the core value of products has evolved from the satisfaction of user needs and functions to the embodiment of consumers' pursuit of self-identity and lifestyle. Verganti (2012) believes that meaning comes from users and interactions between products, products can provide a platform for users to have interpretive spatial value. The created space is defined as the customer's physical or mental activities, practices and experiences, such as the realization of value through the following method: use and mental state (Heinonen et al. 2010; Grönroos and Ravald 2011). Customers are value creators, and the value they experience will improve or worsen. Value creation becomes a structured process in which companies and customers can define roles. When value is seen as the value creation of the customer in use, theoretically and managerially, through the accumulation process, the company as a service provider can promote the value creation of the customer through the following methods, and produce resources and processes that represent potential value or expected use value for the customer. However, the customer is the person who builds and experiences value through integration with the resources and processes in the customer's own social environment. In short, the customer is the value creator, and the company is the value promoter.

2.6 Consumer Culture Theory

Consumer culture theory proposed that customers create experience resources and market interaction through sharing and co-creation (Arnould and Thompson 2005). Value is always co-created, (Akaka et al. 2015; Grönroos and umar, etc.) consumer experience is social, gaining benefits through participating in co-creation activities. Social experience helps consumers build connections with like-minded people, enhancing their sense of belonging (Hussain et al. 2019; Nambisan and Baron 2009, 2010; Zhang et al. 2015). Participants seek more pragmatic benefits, which is essentially economic. Economic experience refers to the quality of products and services that consumers obtain by participating in creative activities or avoiding product-related risks (Etgar 2008; Verleye 2015). Füller (2010) shares that this kind of experience is for externally motivated clients who seek to be independent of the results of co-creation.

2.7 Definition of Involvement

Under certain circumstances, there will be special interest or attention paid to certain things (Antile 1984), an individual approach based on their own needs, values, and interests. When the product is related to the needs or values of the consumer, product involvement will occur (Cohen 1983). Because the product is different, the degree of consumer involvement is different. With regard to the subjective awareness and importance of the product (Bloch and Richins 1983), consumers will actively collect product information related to their own level of involvement (Miquel et al. 2002).

2.8 Originality/Value

Current research is a preliminary attempt to portray the results of customers co-creation experience in actual product development. The research is to study the relationship between customers who are interested in a specific product and the experience of co-creation and customer brand participation. Co-creation cannot be ignored, because under the right conditions, co-creation can help companies create value and reduce risks in areas such as strategy, innovation, and new product development. Consumers participating in the survey may also become the first customers to purchase new products, the company can develop special plans for customers participating in the survey to establish special relationships.

3 Research Theory

3.1 Research Scope and Objects

This research used qualitative study and context analysis, the survey was conducted in online, through Facebook advertising to reach target consumers. In the questionnaire, subjects' background was requested, including age, sex, and hair needs. An introduction and several pictures of hair essence and shampoo were presented before the main questions, identifying lifestyle, buying power, and interest in hair related products. Participants were age 18 to 60, loved outdoor activities, purchased related products such as hair growth shampoo, and lived in the Metropolitan North District; the survey lasted for one month and 263 responses were collected. Questions which are shown below:

1. Regarding hair essence product ingredients: scalp care/strengthening hair roots/active ingredient free/testosterone free
2. Regarding shampoo: silicone-free added ions/scalp regulation through essential oils
3. Five purchase options
4. Consumers purchase intentions

3.2 Data Analysis

An exploratory factor analysis was conducted with SPSS 20, related measurement tables were studied, the questionnaire are hair growth essence and shampoo-related products. The survey questionnaire uses product composition and product ingredients to direct product creation. The questionnaire was designed to capture these measurements. All of them were recorded on a five–point Likert scale ranging from strongly disagree "1" to strongly agree "5", (1) scalp revitalization (2) strengthening hair roots (3) active ingredient free (4) testosterone free (5) shampoo features silicone-free added ions (6) scalp regulation through essential oils.

In the questionnaires returned, the percentages of male and female demands on product design are as follows: women care about product ingredients (1) 66.2% (2) 38.2% (3) 54.4% (4) 12.5% (5) 12.5% (6) 53.8% Men care about ingredients (1) 33.8% (2) 60.3% (3) 43.0% (4) 87.5% (5) 87.5% (6) 38.5%, the figures show that men and women attach different levels of importance to product design. Men pay more attention than women to (2) strengthening hair roots and (5) shampoo without silicone. Women pay more attention to (1) scalp revitalization and (6) scalp regulation through essential oils. Product development focuses on items 1–6 listed in the table (Table 1).

Table 1. Product preferences sorted by gender

Product features			1	2	3	4	5	6	Total
Gender	Female	Number	49	26	43	2	2	7	129
		Female %	38.0%	20.2%	33.3%	1.6%	1.6%	5.4%	100.0%
		Preference %	66.2%	38.2%	54.4%	12.5%	12.5%	53.8%	48.5%
		Subtotal %	18.4%	9.8%	16.2%	0.8%	0.8%	2.6%	48.5%
	Male	Number	25	41	34	14	14	5	48.5%
		Males %	18.8%	30.8%	25.6%	10.5%	10.5%	3.8%	100.0%
		Preference %	33.8%	60.3%	43.0%	87.5%	87.5%	38.5%	50.0%
		Subtotal %	9.4%	15.4%	12.8%	5.3%	5.3%	1.9%	50.0%
	Other	Number	0	1	2	0	0	1	4
		Others %	0.0%	25.0%	50.0%	0.0%	0.0%	25.0%	100.0%
		Preference %	0.0%	1.5%	2.5%	0.0%	0.0%	7.7%	1.5%
		Subtotal%	0.0%	0.4%	0.8%	0.0%	0.0%	0.4%	1.5%
Total		Number	74	68	79	16	16	13	266
		Total %	27.8%	25.6%	29.7%	6.0%	6.0%	4.9%	100.0%
		Preference %	100.0%	100.0%	100.0%	100.0%	100.0%	100.0%	100.0%
		Subtotal %	27.8%	25.6%	29.7%	6.0%	6.0%	4.9%	100.0%

There are 4 age groups: (1) seventeen people age 18–35, (2) one hundred sixty people age 36–45, (3) sixty-seven people age 46–55, and (4) nineteen people over 56 years old. 36–45 year old's are the most willing to buy with 60.8%, followed by 46–55 years old with 25% willingness to buy. It shows that young adults are the ones who care about hair issue and also have the most purchasing power (Table 2).

Responses to question 3 regarding the demand for the purchase amount of the product (1) under 1,000 NT, (2) over 3,000 NT, (3) over 6,000 NT, (4) 12,000 NT, and (5) over

Table 2. Product preferences sorted by age group

			2	3	4	5	Total
Age	1	Number	0	7	7	3	17
		18–35 years %	0.0%	41.2%	41.2%	17.6%	100.0%
		Preference %	0.0%	7.1%	6.9%	6.0%	6.5%
		Sub-total %	0.0%	2.7%	2.7%	1.1%	6.5%
	2	Number	5	62	63	30	160
		36–45 years %	3.1%	38.8%	39.4%	18.8%	100.0%
		Preference %	38.5%	63.3%	61.8%	60.0%	60.8%
		Sub-total %	1.9%	23.6%	24.0%	11.4%	60.8%
	3	Number	6	23	24	14	67
		46–55 years %	9.0%	34.3%	35.8%	20.9%	100.0%
		Preference %	46.2%	23.5%	23.5%	28.0%	25.5%
		Sub-total %	2.3%	8.7%	9.1%	5.3%	25.5%
	4	Number	2	6	8	3	19
		> 56 years %	10.5%	31.6%	42.1%	15.8%	100.0%
		Preference %	15.4%	6.1%	7.8%	6.0%	7.2%
		Sub-total %	0.8%	2.3%	3.0%	1.1%	7.2%
Total			13	98	102	50	263

15,000 NT, found that with both men and women they preferred option (3), this can be considered for pricing strategy when commercially listing the product.

Responses to question 4 regarding willingness to buy the products were recorded on a five–point Likert scale ranging from strongly disagree "1" to strongly agree "5" (Table 5) the result is women's buying intentions are (3) 54.5% (4) 52.9% (5) 36%, respectively, and men's buying intentions are (3) 44.4% (4) 46.2% (5) 60%, both men and women have high willingness to buy, but the percentage of men expressing strong willingness is higher than that of women, the findings also provide some practical insights for business. This co-creation of value can make the company more aware of subsequent advertising strategies, such as targeting male consumers, and effective advertising (Table 3).

Therefore, the following conclusions can be made based on this questionnaire survey: (1) hair products with stem cells have not yet become a well-known product or technology

Table 3. Purchasing intentions sorted by gender

			2	3	4	5	
Gender	Female	Number	2	54	55	18	129
		% of female	1.6%	41.9%	42.6%	14.0%	100.0%
		PI %	15.4%	54.5%	52.9%	36.0%	48.5%
		Subtotal %	0.8%	20.3%	20.7%	6.8%	48.5%
	Male	Number	11	44	48	30	133
		% of male	8.3%	33.1%	36.1%	22.6%	100.0%
		PI %	84.6%	44.4%	46.2%	60.0%	50.0%
		Subtotal %	4.1%	16.5%	18.0%	11.3%	50.0%
	Other	Number	0	1	1	2	4
		% of other	0.0%	25.0%	25.0%	50.0%	100.0%
		PI %	0.0%	1.0%	1.0%	4.0%	1.5%
		Subtotal %	0.0%	0.4%	0.4%	0.8%	1.5%
Total		Number	13	99	104	50	266
		Gender %	4.9%	37.2%	39.1%	18.8%	100.0%

in Taiwan, (2) both men and women are willing to buy this product, (3) overall response to the concept of product is fairly positive. These results and conclusions serve as the reference material for the required analysis.

3.3 Empirical Analysis and Results

This study explores consumer adoption of value co-creation, based on statistical analysis and the subsequently obtained results. Consumers give feedback on the product ingredients and create value together so that the company can understand the characteristics of the products that consumers value and use this as the basis for product development. According to the analysis of consumer purchasing intentions in the questionnaire, there is a high proportion of consumers who are willing to purchase goods. Both men and women have the same enthusiasm with this product. Also, this case study shows that the co-creation experience between customers and brands is a key element of customer creativity. Firs, co-creation is a form of collaborative creation, initiated by the company, aimed at achieving innovation and obtaining co-creation value through communication with customers. Second, a rich synergy: co-creation combines management and marketing methods, using network tools to analyze trends and a combination of processes related to innovation, knowledge and team decision-making. Third, "relationships": co-creation emphasizes the importance of focusing on the quality of interactions between CRMs rather than advertising. Fourth, the learning process: through this study, if a company wants to gain a brand influence, the brand needs to blend knowledge and process to form an overall co-creation framework, not just to achieve co-creation.

4   Conclusions

Customer purchasing is affected by product effects, product ingredients, and prices. Current research has studied the relationship between male and female customers' emphasis on products, commodity pricing and customers' purchasing power through co-creation experience. Survey results show that product composition, price and product effects, as well as co-created experience, customer satisfaction, and purchase intention after customer participation have a positive impact. In terms of practical significance—this research can help the company's product development managers better understand the customer experience in co-creating products and formulating participation strategies for product launches. Because consumers participate in product formulations and consumers' willingness to buy products can be formulated according to the results of survey, customer experience can become product development strategies for management. Co-creation between firms and customers, as well as production and consumption, is about successfully tapping into the collective intelligence of consumers. Aided by information technology, which makes interaction spaces like online user communities possible, co-creation allows for a continuous process in which products are tuned to consumer needs.

5   Limitation of the Study

The design of this questionnaire is biased towards the relationship between product components and function, not including "emotional appeal" or the value of people who care for the "hi-tech" qualities of pursuing functions or the "hi-touch" taste of appealing to emotional needs. Future research may also investigate the effect of high-tech to high-tech feeling (from hi-tech to hi-touch), can also produce co-creation value. Injecting "emotion" into "design" and bringing "experience" into "life" is the focus of future product design with a sense of quality (Klegin and Caldwell 2012). Furthermore, inviting consumers into te further design product story, may moderate the value of co-creation on engagement behaviors.

References

Arnould, E.J., Thompson, C.J.: Consumer culture theory (CCT): twenty years of research. J. Consum. Res. **31**(4), 868–882 (2005)

Grönroos, C., Voima, P.: Critical service logic: making sense of value creation and co-creation. J. Acad. Mark. Sci. **41**, 133–150 (2013). https://doi.org/10.1007/s11747-012-0308-3

Chan, C.-S., Chang, T.C., Liu, Y.: Investigating Creative Experiences and Environmental Perception of Creative Tourism: The Case of PMQ (Police Married Quarters) in Hong Kong

Etgar, M.: A descriptive model of the consumer co-production process. J. Acad. Mark. Sci. **36**(1), 97–108 (2008)

Rowley, J., Kupiec-Teahan, B., Leeming, E.: Customer community and co-creation: a case study. Mark. Intell. Plan. ISSN 0263-4503

Hussain, K.: The role of co-creation experience in engaging customers with service brands (2019)

Hussian, K., Jing, F., Junaid, M.: The dynamic outcome of service quality: a longitudinal investigation. J. Serv. Theory Pract. **29**(4), 513–536 (2019)

Miquel, S., Caplliure, E.M., Aldas-Manzano, J.: The effect of personal involvement on the decision to buy store brands. J. Prod. Brand Manag. **11**(1), 6–18 (2002)

Lin, R., Li, H.-L.: Cross-cultural communication in design collaboration (2020)

Verleye, K.: The co-creation experience from the customer perspective: its measurement and determinants. J. Serv. Manag. **26**(2), 321–342 (2015)

The effect of personal involvement on the decision to buy store brands

Research on Upper Extremity Rehabilitation Product Use Needs and Development Suggestions

Lan-Ling Huang[✉], Chih-Wei Lin, Chi-Meng Liao, and Tao Yang

Design Innovation Research Center of Humanities and Social Sciences Research Base of Colleges and Universities in Fujian Province, Fujian University of Technology, No. 3 Xueyuan Road, University Town, Minhou, Fuzhou 350118, Fujian, China

Abstract. Stroke is the second leading cause of death and the third leading cause of disability. In order to restore activities of daily living, all patients need to receive occupational therapy. Rehabilitation products are an important equipment for rehabilitation treatment. The purpose of this research is to investigate the current situation, problems and needs of upper extremity rehabilitation products, and to propose design development suggestions for rehabilitation products design. The results of this study can be summarized as follows: (1) Types of upper extremity rehabilitation products used in clinic can be divided into four categories based on the technology: Traditional rehabilitation products, Digital rehabilitation products, Rehabilitation products with desktop virtual reality, Immersive virtual reality product. (2) The problems of clinical upper extremity rehabilitation products were proposed: expensive to maintain, single function and limited therapeutic actions, inconvenient to store, and cannot give the treatment progress suggestions to each patient. (3) All therapists think that virtual reality games intervention therapy can promote the interest and motivation for stroke patients. (4) The space applied of the rehabilitation treatment room is one of the factors that must be considered in rehabilitation products design. (5) Four development suggestions of rehabilitation product design were proposed: to provide a visualized expected treatment progress, to design home-based rehabilitation products, to assist patients and their family members in psychological development after stroke, Provide affordable rehabilitation products for families with different income levels. Results of this research will provide references for the rehabilitation product industry, medical industry, and improve social welfare.

Keywords: Upper extremity · Rehabilitation product design · Design suggestions

1 Introduction

The world's population is ageing: older persons are increasing in number and make up a growing share of the population in virtually every country, with implications for nearly all sectors of society, including labors and financial markets, the demand for goods and

© Springer Nature Switzerland AG 2021
P.-L. P. Rau (Ed.): HCII 2021, LNCS 12771, pp. 340–350, 2021.
https://doi.org/10.1007/978-3-030-77074-7_27

services such as housing, transportation and social protection, as well as family structures and inter-generational ties [1].

Worldwide, cerebrovascular accidents (stroke) are the second leading cause of death and the third leading cause of disability [2–4]. According to statistics, stroke no longer only occurs in elderly people, and approximately 10% to 15% of all strokes occur in adults aged 18 to 50 years [5]. After a stroke, most stroke patients have symptoms of upper extremity movement defects [6]. 85% of patients suffer from upper extremity dysfunction in the early stage of stroke, and 40% of patients still suffer from upper extremity dysfunction after the chronic stage [7]. A nationwide survey conducted in 22 provinces in China found that nearly 40.5% of elderly people aged 65 and above have difficulties in completing important activities of daily living (IADLs), of which 22.5% can only perform limited activities of daily living [8, 9]. In order to restore the patient's ability to live independently, all patients need to receive Occupational Therapy.

Rehabilitation products are one of the important tools in the occupational treatment process for stroke patients. At present, most of the traditional rehabilitation products used in clinical practice is static, and there is no immediate feedback information generated to the user. In the rehabilitation process, the occupational therapist usually requires the patient to repeatedly operate the traditional rehabilitation products. Many patients often feel bored due to repeating the same operation, and have a repulsive or negative mentality, which leads to poor rehabilitation effects. In order to improve this situation, make rehabilitation activities interesting and more dynamic. Many clinical therapists have applied commercially available virtual reality game systems (for examples: Nintendo Wii, Kinect for XBOX360™, XaviX, HTC Vive, Rapael Smart Glove®, etc.) in rehabilitation treatment, and found that they can improve patients' motivation for treatment [10–12]. Regarding the effectiveness of treatment, some studies have found that digital rehabilitation games interventions in upper extremity rehabilitation treatment have significant effects, and some studies indicate no significant difference [13–15].

This research applies the ergonomics design concepts (human, product and environment) to survey current situations of rehabilitation rooms and needs of rehabilitation products, and to confirm that the suggestions of occupational therapists on the design and development of rehabilitation products.

2 Method

This research includes three parts as follows: (1) to survey the types of upper extremity rehabilitation products used in hospitals, (2) to survey the current situations of rehabilitation room, and (3) to confirm the use needs and development of upper extremity rehabilitation products from occupational therapists. The survey details of each part was described as follows:

2.1 Part 1: The Types of Upper Extremity Rehabilitation Products Used in Clinic

Authors conducted field survey to observe the types of upper limb rehabilitation products used by stroke patients or occupational therapists, and summarized the types. A total four rehabilitation treatment centers in hospitals were surveyed (i.e. Department of physical medicine and rehabilitation in National Taiwan University Hospital, Department of rehabilitation Medicine in Chung Shan Medical University, Department of rehabilitation Medicine in Kaohsiung Medical University, Fujian University of Traditional Chinese Medicine Affiliated Rehabilitation Hospital).

First, the authors explained the purpose of this research to the rehabilitation therapist. The occupational therapist arranged observation time and locations. The authors went to the designated place to observe the use status of the subjects using different types of upper limb rehabilitation products. Please note that the first paragraph of a section or subsection is not indented. The first paragraphs that follows a table, figure, equation etc. does not have an indent, either.

2.2 Part 2: The Current Situations of Rehabilitation Room in Hospitals

In this part, the purpose of this research is to understand the relevance of rehabilitation treatment rooms and rehabilitation products. A preliminary field survey was conducted on the space applications of rehabilitation rooms in Taiwan and Fujian Province.

2.3 Part 3: The Design Needs and Development Suggestions of Upper Extremity Rehabilitation Products

Authors conducted expert interview to occupational therapists. Three questions were asked as follows: (1) the problems of the existing upper extremity rehabilitation products used in clinic, (2) the design needs of the upper extremity rehabilitation products, (3) development suggestions of the upper extremity rehabilitation products. Authors recorded the interview results for subsequent analysis.

3 Result and Discussion

According to the above surveys, the research proposes results and discusses as below.

3.1 The Types of Upper Extremity Rehabilitation Products Used in Hospitals

Four rehabilitation departments were investigated. Table 1 shows the types of upper extremity rehabilitation products that are common used, or used in some rehabilitation hospitals. Some upper extremity rehabilitation products are used in part hospitals, which are indicated by "*". This research classifies upper extremity rehabilitation products based on motor functions and technologies.

Classification Based on the Motor Functions of Rehabilitation Products. According to rehabilitation medicine, proximal is ranges of active joint motion for the shoulder and elbow joints. Distal is ranges of active joint motion for the forearm and wrist joints. These products can be classified as proximal motions, distal motions, and multiple motions.

D1–D12 are used to train gross motors for stroke patients who a need for upper extremity rehabilitation to convalescent levels of Brunnstrom stages III to V.

D13–D20 are used to train fine motors for stroke patients who a need for upper extremity rehabilitation to convalescent levels of Brunnstrom stages V to VI.

D25–D31 are used to train complex motors (such as Proximal/ Distal/ Cognitive functions) for stroke patients who a need for upper extremity rehabilitation to convalescent levels of Brunnstrom stages V to VI. The patient is usually asked to interact with the game task and make corresponding actions according to the game tasks.

Classification Based on Technologies Applied in Rehabilitation Products. According to the technology applied of rehabilitation products as the classification basis, it can be divided into four types as follows: (a) traditional rehabilitation products, (b) rehabilitation products with digital feedback, (c) desktop virtual reality applied in rehabilitation, and (d) immersive virtual reality applied in rehabilitation.

(a) Traditional rehabilitation products. These rehabilitation products (D1–D20 in Table 1) are active rehabilitation, static, and there is no immediate feedback information generated to the user. Part of them are focused on training the strength of upper extremity muscles.

(b) Rehabilitation products with digital feedbacks. Rehabilitation products (D22–D24) are designed with digital text feedbacks, such as number of operations, operating time, etc.). D24 is an improved design from traditional equipment (D13). D24 is an improved design of traditional equipment. It uses signal feedbacks (such as lighting and location, etc.) to guide interaction with users and to complete game tasks.

(c) Desktop virtual reality applied in rehabilitation. Based on the types of virtual reality, studies [16, 17] proposed that virtual reality falls into three major categories: text-based, desktop, sensory-immersive virtual reality and immersive virtual reality. Text-based networked VR involves real-time environments described textually on the Internet where people interact by typing commands and "speak" by typing messages on their computer keyboards. Desktop virtual reality is an extension of interactive multimedia involving three dimensional images and add to the experience of interactive multi-media without being considered immersive. Sensory-immersive VR, involves a mixture of hardware, software and concepts which allow the user to interact with and in a three dimensional computer generated "world" [16].

D25–D30 can be classified as desktop virtual reality. They project the game scene to the user through a computer, TV or suspended projection screen, and ask the patient to complete the game task by operating the controller. Compared with traditional rehabilitation products, desktop virtual reality provides a variety of game scenarios, tasks, and voice functions to interact with patients in real time, attracting patients' attention to game tasks.

D28–29 was originally designed for leisure and entertainment. It is used in rehabilitation therapy, indicating that the entertainment effect of the product can bring psychological benefits to patients.

(d) Immersive virtual reality applied in rehabilitation. Immersive virtual reality includes head mounted displays (HMD or 'eye-phones') which provide 3D vision of 200 degrees horizontally and 120 degrees vertically [18]. D17 provides a completely immersive situational experience. The therapist assists the patient to wear the head-mounted display, and the patient holds the controller and operates it to complete the task

Table 1. Types of active upper extremity rehabilitation products used in clinic

Motor functions	Technologies applied in active rehabilitation products	Upper extremity rehabilitation products		
A. Gross motor (Proximal motions)	a. Traditional rehabilitation products (no digital feedback)	D1	D2	D3
		D4	D5	D6
		D7	D8	*D10
		*D11	D12	
B. Fine motor (Distal Motor)		D13	D14	D15
		D16	D17	D18
		D19	*D20	
	b. Rehabilitation products with digital feedback	D21	D22	*D23
		*D24		

(*continued*)

Table 1. (*continued*)

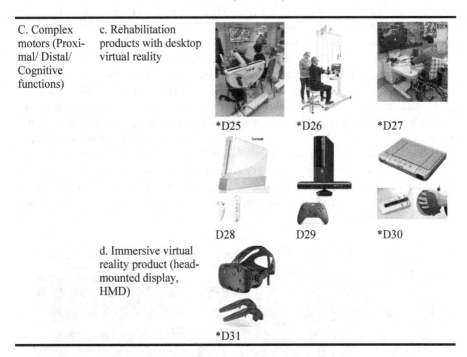

C. Complex motors (Proximal/ Distal/ Cognitive functions)	c. Rehabilitation products with desktop virtual reality	*D25	*D26	*D27
		D28	D29	*D30
	d. Immersive virtual reality product (head-mounted display, HMD)	*D31		

Proximal is ranges of active joint motion for the shoulder and elbow joints.
Distal is ranges of active joint motion for the forearm and wrist joints.
Some upper extremity rehabilitation products are used in part hospitals.

and achieve hand movement training. Immersive virtual reality can provide a scene that is highly similar to their home for patients to practice daily life functions.

From the survey results of the types of rehabilitation products mentioned above, it is known that with the development of technology and 5G technology, the functions and interactivity of rehabilitation products are changing. The design and development of rehabilitation products also incorporate the psychological needs of patients into design considerations.

3.2   The Current Situations of Rehabilitation Room in Hospitals

Three hospitals were surveyed in Taiwan and in Fujian. They are Fujian university of traditional Chinese medicine affiliated rehabilitation hospital (FUTCMSRH), Kaohsiung medical university (KMU), and Kaohsiung veterans general hospital (KVGH).

Figure 1 is the virtual reality training area of the rehabilitation treatment room in FUTCMSRH. The space is spacious and the space usage function is clear. Each treatment device is allocated to a specific location for storage, which is convenient for the therapist or patient to pick and place.

Figure 2 (left) is rehabilitation room in KMU. There are many types of treatment equipment, and limited space. Figure 2 (right) is rehabilitation room in KVGH. Many tables and chairs are set up in this space, and treatment equipment is stored on the right storage shelf to avoid clutter in the active area.

During the peak period of treatment, the therapist and patients will be full of the treatment room, and the interaction between users, treatment equipment and furniture may affect the quality of treatment. Appling the perspective of ergonomics design, it can be found that the types of patients/therapists activities in space and the storage of equipment are closely related to the application of the rehabilitation room space. Therefore, in order to improve the quality and services of clinical rehabilitation treatment, it is necessary to consider the needs of patients and therapists in the design of rehabilitation products and interior space design. For example, the storage location or height of the equipment should be considered for easy access or storage for disabled persons. Types of rehabilitation products with similar functions can be designed as integrated rehabilitation equipment.

Fig. 1. Rehabilitation room (left) in Fujian university of traditional Chinese medicine affiliated rehabilitation hospital, and the storage positions of rehabilitation products (right).

3.3   The Needs of Design and Development of Upper Extremity Rehabilitation Products

A total of ten occupational therapists were interviewed. Seven therapists are in Taiwan hospitals and three therapists in Fuzhou city, China. The survey results are summarized as follows: (1) the problems of the existing upper extremity rehabilitation products used in clinic, (2) the design needs of the upper extremity rehabilitation products, (3) development suggestions of the upper extremity rehabilitation products.

The Problems of the Existing Upper Extremity Rehabilitation Products. The therapists proposed five main problems: (1) Product imported from abroad is difficult and expensive to maintain. (2) Traditional rehabilitation products have single functions and limited therapeutic actions. (3) There are many types of small rehabilitation products, which is inconvenient to store. (4) Most of rehabilitation products cannot record the

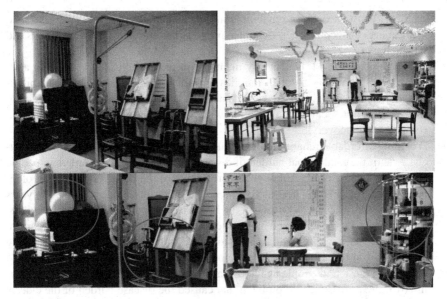

Fig. 2. Rehabilitation rooms in Kaohsiung medical university (left), and Kaohsiung veterans general hospital (right).

treatment progress for each patient. (5) Rehabilitation products are not easy to clean or maintain.

The Design Needs of the Upper Extremity Rehabilitation Products. The therapists proposed four main needs: (1) Improve interaction in the use of rehabilitation products. (2) Increase the connection between rehabilitation products and daily life situations. (3) Traditional rehabilitation products must be improved in design and interactivity. (4) Rehabilitation products combined with diverse scenes conform to local culture and meet the emotional needs of patients of different ages. Through the tasks in the scene, provide patients with collective functional training (such as motor function, cognitive function and language function training).

Development Suggestions of the Upper Extremity Rehabilitation Products. Three suggestions are proposed:

(1) Design a rental service system for rehabilitation products for rehabilitation. Rehabilitation treatment is a long-term treatment program. The therapist usually encourages patients to go back home and perform alternative treatment activities every day to increase the efficiency of recovery of motor functions. Recommendation: According to the patient's disability, classify the rented rehabilitation products and provide them for use. Track the patient's recovery process at home through the product.

(2) The progress of treatment is visualized for easy understanding by patients or family members. Use the advantages of big data to provide patients with predictable goals and promote motivation for treatment. The visualized treatment progress

can provide useful data for rehabilitation doctors and occupational therapists, and help patients understand their treatment progress and improve treatment effects and satisfaction.

(3) Assisting patients and their family members in psychological development after stroke. Stroke patients and their families are unfamiliar with stroke and rehabilitation knowledge and may feel worried and afraid. Therefore, it is recommended to conduct psychological communication with patients and their families, and build confidence in recovery.

(4) Providing affordable rehabilitation products for families with different income levels. 23% of the total global burden of disease is attributable to disorders in people aged 60 years and older. Although the proportion of the burden arising from older people ($\geq$60 years) is highest in high-income regions, disability-adjusted life years (DALYs) per head are 40% higher in low-income and middle-income regions, accounted for by the increased burden per head of population arising from cardiovascular diseases, and sensory, respiratory, and infectious disorders. The leading contributors to disease burden in older people are cardiovascular diseases (30·3% of the total burden in people aged 60 years and older) [19]. According to statistics in 2015, the primary health problem that caused the disease burden of the elderly in China was stroke (35.9 million DALYs, accounting for 27% of the total disease burden of elderly people aged 60 and over) [1]. Therefore, it is recommended to conduct psychological communication with patients and their families, and build confidence in recovery. It can be seen that the elderly have a heavy financial burden on diseases. Therefore, it is recommended to provide affordable rehabilitation products for the elderly in different economic situations.

4   Conclusion

The purpose of this research was to survey current situations of rehabilitation rooms and needs of rehabilitation products, and to confirm that the suggestions of occupational therapists on the design and development of rehabilitation products. The survey results can be summarized as follows:

(1) According to the technology applied of rehabilitation products as the classification basis, it can be divided into four types as follows: (a) traditional rehabilitation products, (b) rehabilitation products with digital feedback, (c) desktop virtual reality applied in rehabilitation, and (d) immersive virtual reality applied in rehabilitation. The development of technology and 5G technology, the functions and interactivity of rehabilitation products are changing. The design and development of rehabilitation products should incorporate the psychological needs of patients into design considerations.

(2) Regarding the needs of clinical rehabilitation products and treatment rooms. Appling the perspective of ergonomics design, it can be found that the types of patients/therapists activities in space and the storage of equipment are closely related to the application of the rehabilitation room space. It is necessary to consider the needs of patients and therapists in the design of rehabilitation products and interior

space design. (3) Regarding design development for rehabilitation product, four suggestions are proposed: (a) Provide a visualized expected treatment progress, based on the patient's recovery situation. (b) Design home-based rehabilitation products and track the progress of patients' treatment. (c) Assist patients and their family members in psychological development after stroke. (d) Provide affordable rehabilitation products for families with different income levels. Results of this research will provide design references for the rehabilitation product design industry and the medical industry.

Acknowledgment. This study was supported by the Fujian Provincial Social Science Planning Project with grant No. FJ2018B150.

References

1. United Nations: World Population Ageing. United Nations, New York (2017)
2. World Health Organization. https://www.who.int/news-room/fact-sheets/detail/the-top-10-causes-of-death. Accessed 21 Jan 2021
3. Hatem, S.M., et al.: Rehabilitation of motor function after stroke: a multiple systematic review focused on techniques to stimulate upper extremity recovery. Front. Hum. Neurosci. **10**, 1–22 (2016)
4. WHO's Global Health Estimates. https://www.who.int/data/gho/data/themes/mortality-and-global-health-estimates. Accessed 21 Jan 2021
5. George, M.G.: Risk factors for ischemic stroke in younger adults a focused update. Stroke **51**, 729–735 (2020). https://doi.org/10.1161/STROKEAHA.119.024156
6. Gowland, C., DeBruln, H., Basmajian, J.V., Piews, N., Burcea, I.: Agonist and antagonist activity during voluntary upper-limb movement in patients with stroke. Phys. Ther. **72**, 624–633 (1992)
7. McCrea, P.H., Eng, J.J., Hodgson, A.J.: Biomechanics of reaching: clinical implications for individuals with acquired brain injury. Disabil. Rehabil. **24**, 534–541 (2002)
8. World Health Organization. https://apps.who.int/iris/bitstream/handle/10665/194271/978924 5509318-chi.pdf?sequence=5. Accessed 21 Jan 2021
9. Hu, J.: Old-age disability in China: implications for long-term care policies in the coming decades. Pardee RAND Graduate School. https://pdfs.semanticscholar.org/d398/9a604c 600a7e13b5c1675388d4c33ab61f53.pdf?_ga=2.21341756.640292759.1612768484-136209 8902.1608982838. Accessed 21 Jan 2021
10. Baldominos, A., Saez, Y., Pozo, C.G.: An approach to physical rehabilitation using state-of-the-art virtual reality and motion tracking technologies. Proc. Comput. Sci. **64**, 10–16 (2015)
11. Kurzynski, M., et al.: Computer-aided training sensorimotor cortex functions in humans before the upper limb transplantation using virtual reality and sensory feedback. Comput. Biol. Med. **87**, 311–321 (2017)
12. Stewart, J.C., Yeh, S.C., Jung, Y., Yoon, H., Whitford, M., Chen, S.Y.: Intervention to enhance skilled arm and hand movements after stroke: a feasibility study using a new virtual reality system. J. Neuroeng. Rehabil. **4**(21), 6 (2007)
13. Joo, L.Y., et al.: A feasibility study using interactive commercial off-the-shelf computer gaming in upper limb rehabilitation in patients after stroke. J. Rehabil. Med. **42**(5), 437–441 (2010)

14. Piron, L., Tonin, P., Piccione, F., Iaia, V., Trivello, E., Dam, M.: Virtual environment training therapy for arm motor rehabilitation. Presence **14**, 732–740 (2005)
15. Chen, M.H., et al.: A controlled pilot trial of two commercial video games for rehabilitation of arm function after stroke. Clin. Rehabil. **29**(7), 674–682 (2015)
16. Loeffler, C.E., Anderson, T.: The Virtual Reality Casebook. Van Nostrand Reinhold, New York (1994)
17. Moore, P.: Learning and teaching in virtual worlds: implications of virtual reality for education. Australas. J. Educ. Technol. **11**, 91–102 (1995)
18. Winn, W.: A Conceptual Basis for Educational Applications of Virtual Reality. (HITL Report No. R-93-9). University of Washington, Human Interface Technology Laboratory, Seattle, WA
19. Prince, M.J., et al.: The burden of disease in older people and implications for health policy and practice. Lancet **385**(9967), 549–562 (2015)

The Impact of Cross-Cultural Trade Conflicts on the Product Design Strategies of Chinese Home Appliance Enterprises

Honglei Lu[1,2(✉)] and Yen Hsu[1,2]

[1] The Graduate Institute of Design Science, Tatung University, No. 40, Sec. 3, Zhongshan N. Rd., Taipei 104 10461, Taiwan
[2] Department of Design, College of Art, The Graduate Institute of Design Science Tatung University, No. 111, Jiulong Road, Hefei 230601, Anhui, China

Abstract. The signing of a memorandum by the President of the United States in March 2018, the U.S. coordinates the use of the clause mechanism to protect the U.S. trade interests, making the home appliance industry, which is already quite competitive in the market, more directly face alternative competition from imports from other countries. Therefore, the trade conflict has a significant impact on the development of China's home appliance industry. This study examines the product design strategies of Chinese home appliance companies. The study also analyzes and summarizes the opportunities and threats that may arise from the changing environment under the trade dispute, and proposes strategies for Chinese home appliance companies to respond. SPSS analysis was conducted on 24 samples. Factor analysis was conducted on the samples based on strategic factors to investigate the impact of the trade conflict on the product design strategies of Chinese home appliance companies and the types of product design strategies they adopted to respond to the trade conflict. The main findings of the study are: 1) the impact of the US-China trade war on Chinese home appliance companies includes three aspects: production cost, reduction of exports to the US, and cooperation between the US and Chinese industries; 2) six product design strategies were summarized based on the analysis of interviewed companies' data and manufacturers' interview information; 3) it is worth noting that the product design strategies adopted by the case home appliance companies are related to the characteristics and specifications of the case companies. Therefore, these research results will provide a valuable reference for the sample companies in the home appliance industry to respond to trade conflicts and future competitive design strategies, as well as a reference for product design strategies for companies in other fields to respond to cross-cultural management.

Keywords: Trade conflict · Home appliance companies · Product design strategy

1 Introduction

Since China's accession to the World Trade Organization (WTO), China has achieved rapid economic growth by opening up its economy and focusing on trade in services,

© Springer Nature Switzerland AG 2021
P.-L. P. Rau (Ed.): HCII 2021, LNCS 12771, pp. 351–370, 2021.
https://doi.org/10.1007/978-3-030-77074-7_28

and is now the world's two largest economies and trading nations along with the United States [37] At the same time, the international influence of the Chinese appliance industry is growing, with more and more Chinese appliance products being respected (trusted) by American consumers. As China is the largest trading partner of the U.S., the U.S. has a growing trade deficit with China, but the large trade deficit has hurt U.S. interests and led to escalating trade conflicts between China and the U.S. In March 2018, the U.S. President signed a memorandum of findings, and the U.S. coordinated the use of the clause mechanism to protect U.S. trade interests based on 'section 301 In March 2018, the U.S. President signed a memorandum of findings, and the U.S. coordinated the use of the provisions to protect U.S. trade interests by announcing tariffs on Chinese exports to the U.S. and restrictions on Chinese investments and mergers and acquisitions in the U.S. under section 301 [25]. According to the Chinese Ministry of Commerce's administrative announcement, the United States has imposed tariffs on $300 billion worth of Chinese goods to the United States. The tariff increase mainly targets some products of "Made in China 2025" [12], including air conditioners, refrigerators, washing machines and other products. According to the "2019 Annual Report on China's Home Appliance Industry" jointly released by the China Household Electric Appliance Research Institute and the Household Electric Appliance Industry Information Center, the data shows that the export scale of China's home appliance industry in 2019 was 303.4 billion, with air conditioning showing a downward trend of 4.82% year-on-year due to the aggravation of the trade conflict with the United States. From the export data, the U.S.-China trade dispute makes the original market competition is quite fierce home appliance industry, the need to directly face the low price of imported products from other countries, affecting the competitiveness of the U.S. market. Therefore, the U.S.-China trade conflict has had a significant impact on the development of China's household appliance industry.

Over the years, Chinese home appliance brands have continued to innovate technology and improve design, and have gradually gained a greater influence in the global market voice. The Chinese home appliance industry has the world's richest and most complex industrial chain and is now the world's largest home appliance production base [23] Therefore, the U.S. considers China as a strategic competitor and the relationship between China and the U.S. will have a huge impact not only on their own economies but also on the global economy [33, 36] The tariff increase and the ensuing market competition from Korean, Japanese, and European home appliance companies have further increased the level of competition from Chinese home appliance companies in the U.S. market. Chinese home appliance companies need to revamp their strategic direction from a model-oriented approach to quality-oriented design innovation to find differentiated markets.

In response to the vicious economic development brought about by the trade conflict between China and the United States, Wang (2020) analyzes the major industries in China affected by the trade dispute between China and the United States and argues that the government should try to obtain more window periods in the process, and should mitigate the impact by using defensive clauses and balancing taxes, promoting the balance of trade in goods through economic regulation, firmly not giving up the policy of supporting industrial technology, insisting on reciprocity in mutual investment rules, and following WTO-related principles, and actively adjusting the industrial structure, rent and tax incentives for R&D [11]. We should not give up the policy of supporting industry and technology, insist on reciprocity of mutual investment rules and follow the relevant WTO principles, and actively adjust the industrial structure, rent and tax incentives for research and development. At the same time, we will make use of the "Belt and Road" policy to realize the transformation of enterprises, break the trade barriers of the United States, and promote the balance of imports and exports between the two countries. In addition, Lu Yong xiang (2019) believes that enterprises need to rapidly adjust their industrial structure, promote the transformation and upgrade of various industries, accelerate the transformation of China's creation, enhance innovation and design capabilities in key areas, accelerate the enhancement of international competitiveness, prioritize the research and development of smart products, smart manufacturing, and smart management, etc.; make full use of industrial loans, update equipment, strengthen internal management systems, and adopt excellent management professionals to develop products with international appeal. The government should directly participate in coordinating the innovation and development of the home appliance manufacturing industry, assisting enterprises in constructing an innovative design talent cultivation system, upgrading industrial technology, developing new products, raising awareness of intellectual property rights and national standards, conducting international cooperation, developing high value-added and low-cost products, or assisting products in obtaining international certification, in order to enhance the products of enterprises. International competitiveness [8].

Therefore, the managers and industrial designers of these companies' R&D departments must constantly respond to the challenges of the changing market environment [30]. Sung and You (2009) argue that design plays a key role in enhancing corporate competitiveness and product design strategies [10]. Cooper (1999) also pointed out that new product development strategy is the basis for all new product development in a company [20]. Cheng-Lein Teng (2009) emphasizes that design and general corporate strategy are interdependent and should be considered as one of the key points of corporate strategy [2]. Based on the above overview, it is also the motivation of this study to evaluate the aspects of product design strategies to respond to the impact of trade conflicts. Many scholars believe that starting from product development and design can enhance the competitiveness of enterprises; not only tangible products need design, but also intangible services need the concept of design for overall planning [13, 14, 38]. Therefore, the design of R&D strategy is part of the strategy to respond to the trade dispute between China and the United States, and is as important as the production and marketing strategy [14]. In a fast-changing environment, companies must not only launch new products quickly, but also choose the right technology and products to meet the changing needs of customers and threats from competitors [6]. Therefore, this study intends to analyze the impact of the U.S.-China trade conflict on the Chinese home appliance industry, i.e., to analyze the opportunities and threats of the changing U.S.-China trade environment for Chinese home appliance companies in the U.S. market, and to propose product design strategies in response. In order to explore the most suitable design strategies for Chinese home appliance companies under the cross-cultural trade conflict, and to make the product design strategies fit the strengths of home appliance companies while minimizing the competitive threats. The results of the study will provide a reference for the design strategies of Chinese home appliance manufacturers in response to trade conflicts, and enhance the value and competitiveness of products and enterprises in response to different environmental changes under cross-cultural trade conflicts.

2   Literature Review

The home appliance industry refers to the industry engaged in "manufacturing of household appliances" and "manufacturing of audio-visual electronic products". Due to the variety of related products, this study adopts Min-Ta Lin's [5] viewpoint and classifies the home appliance industry as "enterprises that provide electronic and electrochemical products and equipment for household or personal use". According to the membership list of the Home Appliance Association of the People's Republic of China (2017–2020), there are currently 4,971 domestic manufacturers engaged in the production and sale of home appliances, most of which do not produce a single product, and the products sold by each manufacturer are diversified. According to the industrial survey conducted by the Bureau of Statistics of the Ministry of Economic Affairs, the domestic electric appliance industry includes large enterprises and small and medium-sized enterprises, among which the production of large home appliances such as air conditioners, refrigerators and washing machines is dominated by large enterprises, while the rest of small home appliances have many small enterprises and more brands. There are many factors that affect a country's foreign trade policy, such as tariff, culture and relative strength distribution, etc. This study focuses on the impact of trade conflicts on Chinese home appliance companies in the sections of tariff, culture and design strategy.

2.1   Tariff Factors

Using a computable general equilibrium (CGE) model, World Bank analysts made an economic forecast: a 25% tariff increase on all Chinese imports into the US would reduce world exports by 3% and global products by 1.7%. The analysis shows that 'an escalation of U.S.-China tariffs could reduce global exports by up to 3 percent ($674 billion) and global revenues by up to 1.7 percent ($1.4 trillion), with losses across all regions'. "The largest revenue decreases are in China and the United States, up to 3.5% ($426 billion) and 1.6% ($331 billion), respectively. The most affected sectors included agriculture, chemicals, and transportation in the United States; and point equipment, machinery, and other manufacturing in China" [24].

Lau (2019) argues that if China's account with the US will be cut in half due to high import tariffs, then the direct loss in China's GDP will be at least 0.43%. If we take into account the profile effects, the trade conflict will result in a 1.12% reduction in China's GDP [31]. PWC (2018) argues that in addition to the direct impact of trade conflicts, trade frictions between China and the U.S. may also affect China's direct investment in the U.S. Therefore, it is increasingly important to explore the tariff issue in trade conflicts.

2.2 Cultural Factors

Culture is the language of self-actualization [16]; Alexander Wendt's theory of international political culture suggests that culture (including norms, institutions, collective identities, and shared knowledge) determines national identity, interests, and behavior. Alexander Wendt's theory of international political culture argues that culture (including norms, institutions, collective identities, and shared knowledge) determines the identity, interests, and behavior of nations. Huntington's (2000) "Clash of civilizations" or cultural realism argues that "civilization is a cultural entity...... The shift of power among civilizations is leading and continues to lead to the resurgence of non-Western societies and the increasing assertion of their own cultures". Pomeranz and Topik (2017) argue that the conflict between the United States and China is largely rooted in cultural differences between the two cultural entities or societies, and that the outcome of the conflict reflects changes in relative power or power between the two powers [28]. Geographical factors together create different "regions" [35]; and different regions have constraints on regional/national interactions, the formation of collective identities, and the diffusion and internalization of institutionalization processes or norms at the global level [9]. Therefore, exploring the context of different cultural factors is an important consideration in studying the impact of the U.S.-China trade conflict.

2.3 Product Design Strategy

Products are the goods that an enterprise sells to customers and are the core of the enterprise's survival and development, bringing growth and market competitiveness to the enterprise. Products are mainly produced due to the market demand, to attract the attention, desire, purchase or use of consumers, and to meet the needs of consumers of tangible goods and intangible services [3]. Therefore, enterprises enter the market competition with products, and without excellent products, there is no condition for the survival of enterprises [21]. Oakley (1990) argues that product design strategy and business strategy are interdependent [34]. In addition, Mozota and Clipson (1990) by Porter's point of view clearly pointed out that design has three basic strategies: (1) design to reduce costs, in the use of cost advantages as the main competitiveness of the company; (2) design to build an image, in the use of a good image as the company's competitive niche; (3) design for the user, in order to enhance the competitiveness of the enterprise by becoming a professional in a single market [15]. Castellion and Markham (2013) argue that product design strategy is the way in which a company conducts product design [17]. Lee (1999) observed that design strategy can be seen as a design theory and practice that is specific to a firm's strategy to reflect future marketing and technological advances [4]. Enterprises need to determine the design goals and directions due to market, technology and other environmental trends, to assess the design factors and resources to make a comprehensive application, designed as a corporate design staff or design matters can follow the guidelines, but also to help the product, communication, environment, identification design guidelines. Therefore, product design strategy refers to a complete set of action plans and resource allocation in response to market changes in product design, through the product design strategy to help companies gain a competitive advantage [13]. For example, Kelley (1992) proposed "The Strategic Palette" based on successful cases, which summarizes 12 strategic factors (function, brand, sales, marketing, system, service, strategic alliance, law, standard, development, material and manufacturing) that affect product success [29]. Depending on the product and development environment, different combinations of design strategies can be assembled for different factors. Song and You (2009) further developed and integrated 10 design strategy attributes based on the above design strategy factors (namely, improving the company's product image, developing unique product features, designing a good human-machine interface, reducing product production costs, conforming to specifications and standards, making it easy to manufacture and maintain, enhancing promotional effectiveness, considering environmentally friendly design, enhancing social and cultural performance, and improving product quality The product design strategy is composed of the following: (1) design the product to meet the design objectives through the implementation of the design strategy. Based on the above strategic concepts [10], Teng (2001) summarizes 13 design strategies (product serialization, product diversification, product line expansion, product line verticalization, market objectives, new technology innovation, cost reduction, sharing, added value, shape value, shape image, product color, and green design). These strategies are applied in a variety of ways to enhance the competitiveness of the company's products, either as a whole or as a single product [1]. Lu (1993) divides products into three levels: functional, operational, and aesthetic, and uses the strategies that should be used in the product life cycle and the design concept of product concept. Design strategy is a way to achieve innovative design goals [7]. By utilizing the R&D capabilities of a company, the design team can implement design strategy by analyzing the impact of customer needs and competitors on achieving corporate performance goals [26].

According to Sung and You (1999), design strategy is a practical response to corporate innovation activities. marketing information and adaptability, enhancing product development capabilities, enhancing corporate brand image, enhancing technical cooperation, improving R&D departments, developing new target markets, increasing investment in design and R&D, developing unique product shapes, enhancing corporate product design image, improving product diversity, designing good human-machine interfaces, developing unique product functions, emphasizing socio-cultural performance, and emphasizing environmental design) [27]. The product design strategy is defined as a product design strategy based on the following principles Product design strategy is defined as a strategy based on a set of related strategic attributes that aims to achieve the company's innovation goals and help the company gain a competitive advantage in the market environment by providing a unique product design strategy.

In summary, product design strategy determines a firm's ability to survive in the face of future challenges, and new product development is closely related to product design strategy activities [18, 22, 32]. Conducting effective product design strategies is the key to survival and maintaining competitive advantage [19]. According to online industry data, home appliance companies in the U.S. may face threats such as weakened product competitiveness and declining revenues. The effectiveness of appliance companies' product design strategies in response to trade disputes will also affect the future transformation and upgrading of the global appliance industry. Therefore, the results of this study will also provide valuable reference for Chinese home appliance companies to cope with the market changes in the cross-cultural context.

3   Methodology

The study of product design strategies of home appliance companies in response to trade conflicts was conducted in two stages: questionnaire survey and in-depth interviews.

In the first stage, a questionnaire survey was used to investigate the product design strategies of home appliance companies: including literature review, questionnaire design, selection of survey respondents, and inviting experts to make suggestions on the questionnaire questions to analyze the specific situation of product design strategies of home appliance companies in response to trade conflicts.

In the second phase, the first phase of the questionnaire was sent to the respondents through WeChat and mail. 24 copies were returned, and due to the impact of the new epidemic, 8 pilot companies were selected for in-depth interviews in combination with WeChat voice remote interviews. The content of the interviews included information on the business units of the interviewed companies, the positive and negative impacts of the trade conflict on the interviewed companies, their strengths and weaknesses compared to their competitors, and their response measures, especially in terms of product design practices. The interviews were then organized into written records and comparative analyses were conducted to analyze the data related to the implementation of product design strategies of home appliance companies.

The purpose of the literature review is to examine product development design strategies in the home appliance manufacturing industry. This study uses a questionnaire survey designed by Yen Hsu (2012) to integrate product design strategy factors based on the perspective of Kelly and Song and You (1999).

The research survey was conducted from November 2020 to January 2021. Based on the enterprise database of China Household Electric Appliances Industry Association, 80 R&D departments of household electric appliance enterprises were selected as the sample. The survey was adjusted based on the results of the survey, because of the epidemic we chose to send the questionnaire through Questionnaire Star, and finally 24 questionnaires were returned, (33.3% of the total survey form) based on the 24 enterprises questionnaires that were returned for data recording. The first part of the questionnaire asked respondents about product design innovation strategies, and the relative importance of these strategies was evaluated using a seven-point Likert scale. These strategies were evaluated on a seven-point Likert scale, where 7 to 1 means strongly agree, agree, somewhat agree, average, somewhat disagree, disagree, and strongly disagree, respectively. The evaluation factors include: strengthening technical cooperation, improving R&D department, improving product quality, emphasizing environmental design, easy manufacturing and maintenance, strengthening product development capability, developing new target markets, improving product diversity, reducing product production cost, improving corporate brand image, emphasizing social and cultural performance, developing unique product shapes, developing unique product functions, designing good interactive experience, improving product design and development process, and improving corporate brand image. Improve product design and development process, enhance corporate product design image, strengthen marketing consultation and adaptability, and increase investment in design and R&D.

Factor analysis was conducted on 24 questionnaires and Bartlett's spherical test showed that the variables were not unrelated to each other ($X2 = 227.164$, $df = 153$, p < 0.001), while the KMO was .801, indicating that this data is suitable for factor analysis. Using the eigenvalue greater than one rule, four factors should be taken, and according to the scree plot, four factors should be taken. The four factors were extracted by the principal axis method, and the orthogonal axis was performed by the maximum variation method (Table 1).

Table 1. Explain the total amount of variation

Rotation squared and load capacity			
Principal components	Total	Variant %	Cumulative %
1	3.998	22.211	22.211
2	3.145	17.475	39.686
3	2.966	16.480	56.165
4	2.480	13.776	69.941

From this, the first factor contains 8 evaluation strategies, and its content is mostly related to the emphasis on technology to enhance corporate image, so it is named as technology advantage strategy. The second factor contains 6 evaluation strategies mostly related to product quality experience, so it is named as customer-oriented strategy. The third factor contains 4 items related to product development process and market target, so it is named as imitation strategy. The fourth factor contains 4 items related to production cost and production efficiency, so it is named as cost advantage strategy. These four factors can explain 69.941% of the variation, among which 22.21% of the variation can be explained by the technological advantage type, 17.47% of the variation can be explained by the customer-oriented type, 16.48% of the variation can be explained by the imitation type, and 13.77% of the variation can be explained by the cost advantage type. Then, the factors were analyzed in clusters and it was found that the firms could be divided into four clusters. For the product design strategy, it was found that there were 9 firms in the technology advantage strategy group, 3 firms in the customer orientation strategy group, 7 firms in the cost advantage strategy group, and 5 firms in the imitation strategy group, as shown in Table 2. Two companies from each of the four strategy groups were then selected for further interviews on product design strategies in response to trade conflicts. The interviews included data on product type and product business type.

The interviewees were mainly managers of R&D and design departments and directors of industrial design, and the interviewees were: Midea (A), Haier (B), Xiaomi (C), Little Bear Electric (D), Auma (E), AUX (F), Xin bao (G), and Granz (H), which are referred to by letters for convenience. Table 3 lists the information of the companies, the interviewer's duties, and the main products of the interviewed companies.

Semi-structured interviews were used for this study. The questions included basic departmental data, positive and negative effects of trade conflicts on product design strategies, relative strengths and responses of the interviewed companies, and product design measures.

Table 2. Cluster analysis of observed date.

Classification of the groups	Product development strategic groups				
	Technical advantages	Customer orientation	Cost advantages	Imitation	Total
Firms	9	3	7	5	24
Percentage	37.5%	12.5%	29.1%	20.8%	100%

Table 3. Company names, units, titles of the interview subjects, and major products of the companies interviewed.

Company name	Unit interviewed	Title	Major products
A	Lifestyle Design Department	Deputy Director	Refrigerators, washing machines, microwave ovens, fans, dishwashers, induction cookers, rice cookers, water dispensers, water heaters, water purification equipment, humidifiers, etc.
B	Lifestyle Design Department	Manager	Air conditioning, electric fans, heaters, etc.
C	Industrial Design Department	Product Manager	Tablet PCs, TV sets, air cleaners, washing machines, air conditioners, etc.
D	Industrial Design Department	Senior Designer	Humidifiers, electric steamers, egg cookers, electric hot pots, etc.
E	Lifestyle Design Department	Designers	Refrigerators
F	R&D department	Designers	Air conditioners, commercial air conditioners
G	Product Design Department	Designers	Coffee maker, whisk, toaster, etc.
H	Lifestyle Design Department	Designers	Microwave ovens, electric steamers, steam ovens, air conditioners, refrigerators, washing machines, dishwashers

The notes recorded during the interviews were converted into text and tables, and then the basic data of the companies, including product design and product development

Table 4. Type of business form of the surveyed companies

	Technical advantages		Customer orientation		Cost advantages		Imitation	
Companies	A	B	C	D	E	F	G	H
Year founded	1968	1984	2010	2006	2002	1986	1995	1978
Size mold (large, medium, small)	Large	Large	Large	small	small	small	Medium	Medium
OBM/ODM/OEM	OBM	OBM	OBM	OBM	ODM	ODM	OEM	OEM
2020 capital	6700	3000	7000	150	60	85	400	240
2020 employees	101826	87447	15000	4000	9000	30000	15000	35000

strategies, were analyzed and derived. Table 4 shows the business types of the four strategic directions of the interviewed companies.

As can be seen from Tables 3 and 4, A and H are the two oldest companies in the home appliance industry; B, G and F are companies established after the 1980s, of which B is the largest; C, D and E are companies established after 2000, of which C is a technology innovation company, and smart hardware based on the Internet and combined with software experience is in a rapid growth stage.

Technical Advantages. Enterprise A is an old home appliance enterprise in China, a large international enterprise integrating consumer electrical appliances, HVAC, robotics and automation systems, intelligent supply chain, and chip industry. Starting from home appliances, it has gradually transformed into equipment manufacturing and global R&D and production layout, with more than ten brands such as Midea and Welling. Overseas markets are mainly in the Americas, the Middle East, Africa and the Asia Pacific region; its main products are household air conditioners, commercial air conditioners, refrigerators, washing machines, rice cookers, water dispensers, induction stoves and other household appliances. We have the largest and most complete air conditioning industry chain in China, and the largest and most complete small home appliance group and kitchen appliance group in China. Meitei Kitchen Appliances Manufacturing Co., Ltd. is mainly responsible for the research and development of kitchen appliances.

Founded in 1984, B Corporation is the world's number one brand of large home appliances and a leading global provider of good living solutions. From "the world's No.1 brand of home appliances" to "the world's only eco-brand", the Group's Casadei, together with Samsung and LG, is the world's No. 1 white goods brand, with the Qingdao Hai gao Innovation and Design Center being responsible for most of the R&D and manufacturing work.

Customer Orientation. Founded in 2010, C Enterprise is a global mobile Internet enterprise focusing on the research and development of smart hardware and electronic products, as well as an innovative technology enterprise focusing on high-end smartphones, Internet TV and smart home ecological chain construction. The company has gradually increased its voice in the global smart product market and has formed a greater influence, moving from the independent development of a single product to building an entire ecosystem and using it as a platform to gradually develop its organizational structure, production, operations, services and brands. This is a very different path from other home appliance companies. The Xiaomi Group has great flexibility in building an open organization, designing user experience, and expanding resources, and creating many new spaces. Xiaomi currently has an independent design department except for the TV, other core products and ecological chain are an industrial design platform, including industrial design department, cell phone design department, CMF (Color, Material and Finishing) department. Its main products are smartphones, washing machines, air conditioners, TVs, purifiers and other smart household appliances.

D enterprise has a complete product development, production and sales system, is the creator of Moe home appliances, is also the practitioner of Moe life. The long-tailed market has forged the growth of Bear Electrical, which is a leading company in the field of creative small appliances in China by actively expanding into domestic household appliances, mother and child appliances and personal health appliances. Its main products include humidifier, yogurt maker, electric lunch box, egg cooker and so on.

Cost Advantages. E is a company specializing in the research and development, manufacturing and sales of refrigerators, with the world's leading refrigerator core production technology and equipment. The company's products are mainly in the refrigerator category. For 11 consecutive years, the company has been the top exporter of refrigerators in China, and is currently the largest ODM manufacturer of refrigerators in mainland China.

Founded in 1986, the company has ten manufacturing bases in the fields of home appliances, electrical equipment and medical care. F Enterprise is committed to becoming an Internet-based, intelligent home appliance industry group, adhering to the corporate mission of "creating intelligent life, achieving excellence in the workplace", and is committed to building a green and intelligent industrial advantage. AUX Corporation is actively participating in the "One Belt, One Road" market, and its product categories are mainly air conditioning.

Imitation. G Enterprise has always focused on the global small home appliance market and is a company with several specialized products and more than a dozen accessory companies in electrical, electronic, die-casting, injection molding, hardware, mold, spraying, printing, etc. Xinbao mostly focuses on OEM/ODM business, and its customers are mostly low-end customers with strong price sensitivity. In OEM business, it is not much different from general enterprises, and the main competitive means is cost factor; Xinbao has established technical research and development strategic alliance with international small home appliance groups, and its products include electric kettle, electric coffee maker, egg beater, blender and more than 2000 models and series. The products include electric kettles, electric coffee makers, egg beaters, blenders and more than 2000 models.

H Enterprises is a leading global provider of integrated healthy home appliances and smart home solutions, and is one of the leading companies in China's home appliance industry with extensive international influence. Since the establishment of the company, it has been focusing on manufacturing innovation and development, and by 2019 Grants products and services are supplied from Guangdong, China to over 200 countries and regions worldwide.

Combined with Table 4, enterprises with a market capitalization of more than $200 billion are classified as large-scale companies, enterprises with a market capitalization balance of $20 billion to $100 billion are classified as medium-sized enterprises, and enterprises with a market capitalization of less than $20 billion are classified as small-sized enterprises. Enterprises A, B and C with market capitalization greater than $200 billion are classified as large-scale enterprises; enterprises G and H with market capitalization between $20 billion and $100 billion are classified as medium-sized enterprises; enterprises D, E and F with market capitalization less than $20 billion are classified as small-sized enterprises.

In terms of corporate form, Table 4 shows that A, B, C and D are classified as OBM type because the proportion of their main business exceeds 70%, among which B's Casady ice and wash products are in the first rank of global white household appliances with Samsung and LG in Korea. All other enterprises except E and F are white home appliance product line superior to black home appliance product line, and the proportion of ODM business exceeds other business income and owns independent brand. G and H are mainly small home appliance start-ups that have transformed into home appliance enterprises, and have a higher proportion of R&D for OEM type products with their rich product lines.

As shown in Table 3, A, B, and C have more types of products, including refrigerators, air conditioners, washing machines, and other household appliances; E and F have a single product category, with E's main product being refrigerators and F's product being air conditioners; D, G, and H have more types of products, with small household appliances being the main product category.

4   Analysis and Discussion

From the interviews, the impact of the US-China trade war on Chinese home appliance companies includes: production costs, reduction of exports to the US, and cooperation within the US-China industry. The information on the negative impact on the interviewed enterprises is summarized in Table 5.

Table 5. Impact of trade conflict on case enterprises

Case Companies\Impact Projects	A	B	C	D	E	F	G	H
Product Production Cost	−	−	o	o	−	−	−	−
Export to the United States	−	−	o	o	−	o	−	−
Corporate Industry Cooperation	o	o	o	o	−	o	−	−

Note: − Negative Impact; o No effect

Production Costs. Due to the increase in import tariffs, enterprises A and B have been actively acquiring local enterprises in the U.S. in the past few years to expand their market presence in the U.S. Because of the increase in tariffs, the cost of importing products to the U.S. has increased, which is not conducive to the price competitiveness of the products of enterprises A and B in the U.S. and Samsung, LG, Panasonic, etc. in Japan and Korea, and their competitive advantages have been affected. Work. The increase in tariffs has increased the cost of orders from E, F, G, and H to U.S. customers, and the rapid rise of the Southeast Asian OEM business chain has resulted in many orders for U.S. exports to Southeast Asian OEMs, which has had a significant negative impact on the business of these two companies.

Reducing exports to the United States. The negative impact is due to the fact that companies A and B are large and well-known international home appliance companies with high product line categories and global share, as well as a certain market share in the U.S. home appliance sector. C and D enterprises mainly focus on the domestic demand in China and Southeast Asia market, so there is no impact. Enterprises E and F have a single product category, while Enterprise C, as the leading refrigerator exporter, has certain customers in the U.S. refrigerator market and therefore has a negative impact. F has no impact because of its small market share in the U.S. The global manufacturing industry is undergoing large-scale changes, and trade conflicts have led to higher tariff costs. The changes in the tax system have made it more difficult for G and H companies to use low-cost OEMs. Therefore, it has a negative impact.

U.S.-China Corporate Industry Cooperation. As a result of trade conflicts, the U.S. government's ban on flexibility and quantity control on Chinese imports to the U.S. is gradually increasing, favoring the export of Japanese, Korean, and European and U.S. home appliance companies to the U.S., adding a lot of uncertainty to the cooperation of Chinese home appliance companies in the U.S. industry. In the long run, there will be a negative impact on the cooperation between enterprises A, B, and E exporting some products to the U.S. and local enterprises in the U.S. However, due to China's continuous introduction of policies such as "One Belt, One Road" and "Internal Circular Economy", enterprises C, D, and F can take this opportunity to accelerate their development. Enterprises G and H are mainly engaged in OEM structure, therefore, the cooperation in order cooperation within the industry is hindered due to trade conflicts, thus causing negative impact.

The majority of respondents believe that the escalation of trade conflicts has brought about significant changes in the control of product development costs, increased attention to design departments by market executives, and product standardization. To further understand the product strategy approaches of the respondents, this study used the 18 product design strategies summarized by Yen Hsu (2017), and asked the respondents about the possible design strategies to confirm whether they thought they had such strategies in response to trade conflicts, and summarized them in Table 6.

Further statistics on the adoption of product design strategies in response to trade conflicts were collected from the respondents. The six design strategies used by all respondents were: improving product quality; improving product production costs; making it easier to manufacture and maintain; enhancing marketing information and adaptability; enhancing the company's product design image; and exploring new target market segments. In terms of enterprise size and cluster type, there seems to be some correlation

Table 6. Comparison of product design strategies adopted by the surveyed companies

	Technical advantages		Customer orientation		Cost advantages		Imitation	
	A	B	C	D	E	F	G	H
Reducing the production cost of products	V	V	V	V	V	V	V	V
Making it easy for manufacturing and maintenance	V	V	V	V	V	V	V	V
Including additional value to products	V	V	V	V	V	V	V	
Uplifting product quality	V	V	V	V	V	V	V	V
Improving product design and development procedure	V	V	V	V	V			
Enriching marketing information and responding ability	V	V	V	V	V	V	V	V
Reinforcing technical cooperation	V	V	V	V		V		
Increasing product diversity	V	V	V	V	V		V	V
Exploring new target market sections	V	V	V	V	V	V	V	V
Design better human machine interface	V	V	V	V	V			
Developing unique product functions	V	V	V	V	V	V	V	V
Increasing R&D budgets	V	V				V	V	

(continued)

Table 6. (*continued*)

	Technical advantages		Customer orientation		Cost advantages		Imitation	
	A	B	C	D	E	F	G	H
Reinforcing R&D division in counterparts	V	V	V					
Developing unique product forms	V	V	V	V			V	V
Uplifting design images of the company	V	V	V	V	V	V	V	V
Uplifting brand images of the company	V	V	V	V	V	V		
Emphasizing social and cultural performance	V	V	V	V				
Taking environmental design into consideration	V	V	V					

Note: "V" indicates that the company adopted the specific design strategy

between the design strategies and the competitive strategy attributes of the enterprises; the analysis is as follows:

Technical Advantages. Table 6 shows that Technical Advantages has a very diverse approach to the market, with respondents from companies A and B using all 18 product design strategies. Combining Table 4, it is assumed that these two companies are both industry giants and have a similar corporate structure. From the perspective of business form, both companies are OBM type; and the product development categories are the same, both include large appliances and small appliances. From the interview, we know that there are slight differences in the application of strategies. With a complete service system, high brand recognition and accelerating internationalization, Company A has strengthened its strategy of developing new markets for small home appliances and after-sales service. In the China market, Company A continues to maintain its leading position in certain strong products, and to enhance its competitiveness by providing mutual production support with its counterparts. In terms of strategy implementation, Company A has leveraged its long-established brand image and reputation, relied on the advantages of the industrial chain in the Pearl River Delta, and continued to adjust its sales and service system, actively developed new products with forward-looking concepts, improved production processes, strengthened product price advantages, improved production automation efficiency, and actively explored new markets as a response strategy. At the same time, we cooperate with foreign design institutions to master design technology and design trends, develop new functions and increase the added value of products, and further enhance the image of our products.

Company B launched the U.S. market as early as 1999, with a complete international service system, high brand awareness, rapid internationalization, early foreign market deployment and operation, and rapid strategic response. With the increase in trade tariffs, the company is actively acquiring GEA appliances in the U.S., which is conducive to the construction of overseas manufacturing bases, so the strategy tends to respond flexibly. The main strategy of Company B is to improve the automation and production efficiency of U.S. manufacturing plants, enhance service channels and quality, adjust sales markets, strengthen the ability to develop new markets and after-sales services, and at the same time reduce the production cost of products, make products easy to manufacture and repair, improve product quality, improve the design and development process, and enhance the image of the company's products. With the deepening of the [Belt and Road] policy, companies like AB are also trying to seize this opportunity and seek to expand their market presence in other regions.

Customer Orientation. Table 6 shows that the Customer Orientation type of enterprises is to meet customer needs and increase customer value as the starting point of business operation. These types of enterprises pay special attention to the study of customers' consumption ability, consumption preference and consumption behavior, and pay attention to the innovation of existing product development and marketing methods to dynamically adapt to customer needs. As the home appliance products of enterprises C and D are mainly for the domestic market in China, they are not affected by the trade conflict too much. The small home appliance market is a typical long-tail market, and the long-tail market has forged the growth of Enterprise D. Enterprise D is focusing on online channels, and its market is mainly in the strategic deployment of China's domestic product market. Its strategy is to quickly deploy lifestyle appliances and mother and child appliances with its Moe product strategy. Respondents from these two companies also generally believe that strategies to reduce production costs, make it easier to manufacture and repair, improve product quality, strengthen market information collection and response capabilities, and enhance the image of the company's product design are essential. At the same time, we also believe that companies need to adhere to the "user experience as the core to produce products that customers prefer". C enterprise size volume and technical advantages type similar to the practice. The strategy is diversified; it tends to be flexible and responsive.

Cost Advantages. Table 6 shows that this type of companies have adopted strategies to improve product quality; reduce product production costs; facilitate manufacturing and maintenance; enhance marketing information and adaptability; enhance corporate product design image; explore new target markets; enhance corporate brand image; increase product value added; and develop unique product shapes. However, the company's system is different, and there is a big difference in the design strategy other than the above, compared with Table 4, it is presumed that the reason for the difference is that Enterprise E mainly focuses on exporting refrigerators as the main core business, and the product development category is relatively single. During the interview, we learned that refrigerator export is the brand feature of Enterprise E, which has been the champion of refrigerator export for 11 consecutive years, focusing on the foreign sales market, while Enterprise F has some products of OBM type, mainly air conditioners, with a variety of small home appliances, focusing more on the domestic market in China.

Imitation. Table 6 shows that this type of companies adopt strategies such as improving product quality; reducing product production costs; making it easier to manufacture and maintain; enhancing marketing information and adaptability; improving the image of corporate product design; exploring new target market segments; increasing the diversity of product lines; and developing unique product features. G and H enterprises' current product strategy is mainly based on light technology to go sales volume strategy, and their R&D departments are mainly dominated by market imitation product planning. Less investment in design and R & D, the development of unique product modeling: product homogenization is becoming more and more serious, product iteration is fast, companies from the perspective of saving R & D costs to consider. In terms of product diversity, we understand that the increase in tariffs has led to an increase in product costs, and that a reasonable improvement in product diversity can control production costs; G and H are OEM-type companies, with business orders mainly from ODM and OEM, and relatively little business in designing human-machine interfaces. G and H have not adopted the strategy of unique product functions, emphasizing social and cultural performance, and emphasizing environmental protection design, we speculate that the reason is that the difference in company system is small and the business form has certain limitations, and they are more inclined to OEM, and emphasizing environmental protection design will increase the cost of OEM; these companies have complete product production line support, and there is pressure on the operation of enterprise R&D capital, so they mainly focus on OEM business. OEM business is the main business.

5   Conclusions

After comparing the case companies A, B, C, D, E, F, G, and H, it was found that the product design strategies adopted by Chinese home appliance companies could be classified according to their characteristics: Technical Advantages, Customer Orientation, Cost Advantages, and Imitation. Among these four types, Technical Advantages, Cost Advantages, and Imitation are all affected to different degrees. The findings are summarized as follows:

1. The impact of trade disputes on Chinese home appliance enterprises: product production costs, reduction of exports to the U.S., and cooperation between the Chinese and U.S. industries in three areas, and the negative impact on the interviewed enterprises in different degrees. The impact on the business of exporters to the U.S. is relatively large, while the short-term impact on companies that are actively acquiring companies to accelerate the establishment of manufacturing bases in the U.S.

is not significant, and the export of products to companies with market strategies in Southeast Asia, Europe and Africa is not much affected. Case related companies need to adjust their strategies as soon as possible.

2. By collating the data of the interviewees, a total of 18 product design strategies were found: strengthening technical cooperation, improving R&D departments, improving product quality, emphasizing eco-friendly design, making it easy to manufacture and maintain, strengthening product development capabilities, developing new target markets, improving product diversity, reducing product production costs, improving corporate brand image, emphasizing social and cultural performance, developing unique product shapes, developing unique product functions, designing good interactive experiences, improving product design and development processes, enhancing corporate product design image, strengthening marketing consultation and adaptability, and increasing investment in design and R&D. Respondents generally believe that: improving product quality; improving product production costs; making it easier to manufacture and repair; enhancing marketing information and adaptability; enhancing the image of product design; and improving product diversity are essential strategies for companies to adopt in their product design strategies in response to trade conflicts.

The product design strategy of the case company is related to the characteristics and specifications of the case company. Enterprises A and B are very diverse in their response measures, including 18 product design strategies, which are presumed to be due to similar scale of development and strong emergency response capability, but also slightly different in strategy implementation. E and F are cost-advantaged companies, and their design strategies are slightly different from each other due to the single product development category and different product directions. G and H enterprises belong to the imitation type, because the business is mainly OEM type, this trade dispute has a very big impact on them, such enterprises need to accelerate their own industrial transformation and upgrading through this process.

Cross-cultural trade conflict is also a challenge and an opportunity. Companies need to adopt different product design strategies, to serve any heterogeneous cultural conflicts under cross-cultural conditions, and to seize the opportunity to accelerate industrial upgrading. Only with the integration of cross-cultural management can business operations be smooth and competitiveness be enhanced.

References

1. Teng, C.-L.: Design Strategy: Product Design Management Tools and Competitive Tools. Asia Pacific Book Publishers (2001)
2. Teng, C.-L.: Design strategy discourse. J. Des. **5**(2) (2009)
3. Hu, Y.: The road to success for new product development. Qual. Control Mon. **36**(11), 63–66 (2000)
4. Lee, S.-F.: The meaning, value and influencing factors of corporate image strategy. Taichung Bus. Coll. J. (1999)
5. Lam, M.: Retail Evolution of China's Home Appliance Market. National Taiwan University Graduate Institute of Business Administration (1994)

6. Lin, M.-J.J., Lee, B.: Study on the impact of new product development strategy and market information processing mechanism on the performance of new products. J. Sci. Technol. Manag. (2000)

7. Luh, D.-B.: Establishment of a theoretical model of product view in stages. Ind. Des. **22**(4), 213–223 (1993)

8. Lu, Y.: Innovative design competitiveness study. J. Mech. Des. **36**(1) (2019)

9. Shang, H.: Peace and the evolution of the modern international system. Int. Polit. Stud. **2** (2019)

10. Sung, T., You, M.: An Empirical study on design strategy and design performance of information firms in Taiwan. J. Des. **4**(1), 47–59 (1999)

11. Wang, J.: The evolution of the path of trade friction between China and the United States and strategies for coping with it. Collection **16** (2020)

12. Wu, X.: Made in China 2025. Teach. Chin. Univ. **008**, 9–11 (2015)

13. Hsu, Y., Chang, W.: A study of the product design related strategy of Taiwanese home appliance industries dealing with entering WTO. J. Des. **9**(1), 1–12 (2004)

14. Baxter, M.: Product Design: A Practical Guide to Systematic Methods of New Product Development. Chapman & Hall, London (1995)

15. Borja de Mozota, B., Clipson, C.: Design as a strategic management tool. In: Design Management: A Handbook of Issues and Methods, pp. 73–84 (1990)

16. Bottici, C., Challand, B.: The Myth of the Clash of Civilizations. Routledge, London (2013)

17. Castellion, G., Markham, S.K.: Perspective: new product failure rates: influence of argumentum ad populum and self-interest. J. Prod. Innov. Manag. **30**(5), 976–979 (2013)

18. Cooper, R., Press, M.: The Design Agenda: A Guide to Successful Design Management. Wiley, Hoboken (1995)

19. Cooper, R.G.: Industrial firms' new product strategies. J. Bus. Res. **13**(2), 107–121 (1985)

20. Cooper, R.G.: Product Leadership: Creating and Launching Superior New Products. Da Capo Press, Cambridge (1999)

21. Creusen, M.E.: Research opportunities related to consumer response to product design. J. Prod. Innov. Manag. **28**(3), 405–408 (2011)

22. Driva, H.: The role of performance measurement during product design & development in a manufacturing environment. University of Nottingham (1997)

23. Dutta, S., Geiger, T., Lanvin, B.: The global information technology report 2015. Paper Presented at the World Economic Forum (2015)

24. Freund, C., Ferrantino, M., Maliszewska, M., Ruta, M.: Impacts on global trade and income of current trade disputes. MTI Pract. Not. **2** (2018)

25. White House: Presidential Memorandum on the Actions by the United States Related to the Section 301 Investigation. White House (2018)

26. Hsu, Y.: Comparative study of product design strategy and related design issues. J. Eng. Des. **17**(4), 357–370 (2006)

27. Hsu, Y.: Linking design, marketing, and innovation: managing the connection for competitive advantage. Int. J. Bus. Res. Manag. **3**(6), 333–346 (2012)

28. Huntington, S.P.: The clash of civilizations? In: Crothers, L., Lockhart, C. (eds.) Culture and Politics, pp. 99–118. Palgrave Macmillan US, New York (2000). https://doi.org/10.1007/978-1-349-62965-7_6

29. Kelley, L.: The strategy palette. Commun. Art **99**(7), 134–139 (1992)

30. Kim, N., Im, S., Slater, S.F.: Impact of knowledge type and strategic orientation on new product creativity and advantage in high-technology firms. J. Prod. Innov. Manag. **30**(1), 136–153 (2013)

31. Lau, L.J.: The sky is not falling! Econ. Polit. Stud. **7**(2), 122–147 (2019)

32. Marxt, C., Hacklin, F.: Design, product development, innovation: all the same in the end? A short discussion on terminology. J. Eng. Des. **16**(4), 413–421 (2005)

33. Novikov, S.V., Iniesta, D.S.V.: China-United States trade war, its impact on the global economy and the onset of the global economic downturn. Amazonia Investiga **8**(22), 3–5 (2019)

34. Oakley, M.: Design Management: A Handbook of Issues and Methods. Blackwell Publishing, Hoboken (1990)

35. Pomeranz, K., Topik, S.: The World that Trade Created: Society, Culture, and the World Economy, 1400 to the Present. Routledge, London (2017)

36. Qiu, L.D., Zhan, C., Wei, X.: An analysis of the China–US trade war through the lens of the trade literature. Econ. Polit. Stud. **7**(2), 148–168 (2019)

37. Sukar, A., Ahmed, S.: Rise of trade protectionism: the case of US-Sino trade war. Transnatl. Corp. Rev. **11**(4), 279–289 (2019)

38. Thomas, R.J.: New Product Development: Managing and Forecasting for Strategic Success. University of Texas Press, Austin (1993)

Cultural Differences and Cross-Cultural Communication

Deep Learning Model for Humor Recognition of Different Cultures

Rosalina Chen and Pei-Luen Patrick Rau[✉]

Department of Industrial Engineering, Tsinghua University, Beijing, China
rpl@mail.tsinghua.edu.cn

Abstract. In recent years, significant improvements have been made in the field of sentiment analysis, particularly in computational humor. In this research work, an AI-based cross-cultural humor recognition model has been developed and implemented. The model consists of a Convolutional Neural Network that can assess whether a given sentence is humorous or not and whether the sentence has a western or Chinese type of humor. The model has been trained and tested over the created dataset composed of 463314 English sentences and 111614 Chinese sentences. The initial model setting reached an accuracy of 64,48%. The analysis of the obtained results showed the importance of three main contributors to the model accuracy, namely, the dataset variety, dimension and model's hyperparameters. Finally, these contributors were optimized by various tests, resulting in the final model obtaining an accuracy of 96,73%. The flexibility of the model allows applications in several areas such as private social media for cross-cultural communication, cross-cultural marketing and even fake news detection.

Keywords: Natural Language Processing · Sentiment analysis · Humor recognition · Deep learning · Neural networks

1 Introduction

As human-computer interaction systems are gaining importance in our lives, there is a need for intelligent systems to be able to dynamically interact with people. If computers are ever going to communicate naturally with humans, they must be able to use humor. However, computational humor is far from state of the art. Over the last decade, online platforms enabled people to generate and share a vast amount of data. According to IBM, every day about 2.5 quintillion bytes of data are generated, 80% of which is unstructured data with textual data as the majority [1]. It is here that we see an opportunity, designing a Deep Learning (DL) model, able to understand extensive, nonlinear, or even incomplete data, that helps companies assessing different types of humor from user-generated textual data. Aa a matter of fact, humor plays an important role in social interaction, and if correctly used, it can become a source of competitive advantage for any company's cross-cultural communication strategy.

P.-L. P. Rau (Ed.): HCII 2021, LNCS 12771, pp. 373–389, 2021.
https://doi.org/10.1007/978-3-030-77074-7_29

2   Literature Review

Humor recognition is the ability of the machine to assess if a given content has humor or not. Studies on humor in human-computer interaction are still at early stages. Research can be found mainly in pure Natural Language Processing (NLP) field. This is partly because humorous content uses complicated and ambiguous language features. In particular, it requires a deep semantic interpretation of irony and sarcasm. However, interfaces will definitely need computational humor and other cognitive abilities as machines are becoming an integral part of our lives.

On the one hand, the majority of humor recognition early studies are related to binary classification using a set of given linguistic rules [2, 3]. These works have been focusing on an approach called lexicon-based. Lexicon based approach splits the sentence into small sentences or words then counts the number that each small sentence or words show up, creating a model of a bag of words. Then the subjectivity of each word is assessed from an existing lexicon database, which is a database of existing tagged words prerecorded by researchers. Finally, the overall subjectivity or humor of the analyzed text is found.

On the other hand, recent approaches use Artificial Intelligence (AI), more precisely Deep Learning (DL) a subcategory of Machine Learning (ML) which is part of AI. This approach takes the corpus or sentences that are previously labelled positive or negative by researchers, then trains their built classifier on that corpus and finally when given a new sentence the trained classifier will be able to classify the new sentence as either positive or negative, hence humorous or not.

The lexicon-based approach is easier to implement, but the AI approach is more accurate. DL models, in particular, Neural Networks (NN) can understand the subtleties of humor better because they do not analyze text at the surface value, but they are able to create an abstract representation of what they have learned. For instance, paper [3] utilizes deep learning to recognize humor in Yelp's reviews. In particular, they built both the traditional approach lexicon-based model, using a bag of words and two deep learning models using Recurrent Neural Network (RNN) and Convolutional Neural Network (CNN). Among the three models, CNN outperformed the other models in accuracy. Further validation of Convolutional Neural Network in humor detection can be seen in the paper [5] where the famous series The Big Bang Theory has been successfully analyzed using CNN.

All the mentioned studies are related to English corpus, which has undergone major developments in recent years. Furthermore, the increasing importance of cognitive capabilities in human-computer interaction is starting to spread among all countries. As proof of this, in 2019, the Iberian Languages Evaluation Forum (IberLEF) proposed a challenge in the humor recognition of Spanish tweets [6]. To solve this task, the author adapted a model found in [7], which uses a combination of CNN and RNN layers to detect Italian tweets. In the same year, similar studies were conducted in Japan and China related to sentiment analysis of Chinese slangs [8] and emoticons [9] on Weibo's tweets.

3 Methodology

The methodology used to conceptualize, develop and implement the model consists in three main phases: Problem identification, Data acquisition, Modelling and Results analysis.

Problem identification includes a literature review where an extensive analysis of the current literature has been done. The research was proposed to identify problems, opportunities and criteria in Humor, AI and Marketing. Consequently, concepts, state of the art and feasibility of computational humor in these areas have been evaluated. Overall, the literature shows that three trends are converging:

- Marketing is deploying significantly more humor than in the past.
- Companies are already adopting AI applications and have high expectations on Sentiment Analysis.
- Computational humor is gaining importance.

Looking closely at humor recognition models, there are mainly two approaches, the lexicon-based approach and deep learning-based approach. Advantages and Disadvantages of both methods have been discussed, and a critical comparison of methods found out that CNN outperforms other methods. Attempts of generalization have not been accomplished yet.

Data acquisition is carried out after a strong understanding of the literature. The problem is a multiple classification issue with unsupervised learning, where learning from human behavior is required. Data requirements have been assessed to address the problem effectively. Therefore, the dataset needed to be of a certain size and type. In particular, there was a continuous loop between data collection, preprocessing and requirements checking to guarantee a suitable dataset to train the model.

The model generation has been done using a functional approach to program in Python language in Jupiter Notebook environment, while results analysis revealed insights to improve the model. The functional approach is modular compared to using a procedural way of writing programs. The principle is to keep the created functions separated from the data, so the debugging process is limited to zones. Moreover, functions can be stored into variables, that can be used as a parameter or return as a type of data. In this way, the debugging process is reduced hence giving more space to testing and evolution of the created model. Since the aim is to develop and universal humor recognizer, this is just the first implementation, and therefore next step indicates the possibility of further improvements iteration that may change both datasets and/or part of the model or even the direct application and testing in marketing tools.

4 Dataset Creation

In order to build and properly test the model, two datasets are required, one with English corpus and one with Chinese corpus. The datasets are then associated with labels and preprocessed to be fed into the created convolutional neural network model.

The English dataset is composed of mainly one-liners, news, proverbs, headline and wiki. The length of these sentences is ranged between 1 and 200 characters.

In total, there are 463314 sentences:

- 231657 are positive: the jokes are downloaded from an open dataset on Kaggle [10]
- 231657 are negative: the news is downloaded from 2007 to 2010 from Conference ACL 2016 [11]

Precautions have been taken to avoid problems that could happen in the testing of the model. In particular, when data are collected from very different pools and then labelled positive or negative, the model could result in high-performance discrepancy because of the differences among the training pools. In order to avoid this kind of domain discrepancies problem, given that positive data were already labelled in the open-source Kaggle project, negative sentences were chosen by selecting sentences that contained the majority of words that appear in positive samples. In other words, the criteria used to select data is that news items have content composed of words that have been used in the positive data. Moreover, the maximum length of 200 was set accordingly as it was assessed to be the highest length allowed by computational power.

For the Chinese dataset, since the majority of the dataset with the necessary requirements are privately owned, a Chinese dataset has been created by extracting them from social media. One liner jokes have been chosen because they are humorous but at the same time do not have too complex language structure. The created dataset is composed of jokes and news. In total, there are 111614 sentences:

- 57654 are positive: the jokes are extracted from a Chinese website called ZOL [12]
- 53960 are negative: the news is extracted from Tsinghua Natural Language Processing Lab [13]

Compared to the English dataset, the creation of this dataset required to search for the right website from where to extract jokes. For the negative labelled samples, on Github Tsinghua open data was available. While for jokes websites, the decision was between Weibo and ZOL. The latter was chosen mainly because of the density of jokes and the structure of the website. Also, in this case, since the origin of the data is from very different pools. In order to avoid a potential misleading high performance of the model due to this domain gap, the selection criteria are similar to the collection of English data. News composed of the words that have been used in jokes has been selected randomly with chosen probability 1%.

4.1 Pre-processing

After retrieving and creating the two datasets, each dataset was split into two parts: a training set and testing set. The training set is further split into a validation set. The training set is the portion where our model learns from, where our model leans how to interprets informal language, spaces, spelling mistakes, emoticons, sarcasm and other ambiguous characteristic. Then the validation set is the part of the training process that prevents the model from overfitting data. The training helps the model to fit the weights while validation helps to tune the parameters accordingly. Finally, the testing set is used by the model to test itself by comparing its predicted labels to actual labels, therefore calculating the accuracy of the model and assessing model's performance when given new raw data.

Pre-processing is required since Neural Networks cannot take strings directly as inputs. We have to vectorize our inputs. Therefore, converting sentences into numerical representations or vectors is necessary. The steps applied are:

- **Word tokenization:** is the process that splits the words into letters and spaces which are called tokens. For example:

I have a pen, I have an apple
['I', 'have', 'a', 'pen', ',', 'I', 'have', 'an', 'apple', '.']

- **Lower case:** transforming all the words into lower case. This step is controversial. On the one hand, transforming all the upper-case words into the lower case could result in losing the meaning of the word. On the other hand, the problem in leaving the upper case is that the model does not recognize the same word if it is written in upper/lower case. From a computational perspective, the advantage of keeping this process is higher than the disadvantages because it affects the accuracy of the model. Continuing the preview example:

['I', 'have', 'a', 'pen', ',', 'I', 'have', 'an', 'apple', '.']
['i', 'have', 'a', 'pen', 'i', 'have', 'an', 'apple']

- **Remove stopwords:** removing all the words which don't have meaning in a sentence, for instance 'the'. Continuing the previous example:

['i', 'have', 'a', 'pen', 'i', 'have', 'an', 'apple']
['have', 'pen', 'have', 'apple']

- **Creation of index dictionary:** The dictionary is composed by key = word and value = index number. This is required to transform the list of tokenized words into integer, so that at the input of the embedding layer, the model can understand in which line he is going to map. Continuing the previous example:

['have', 'pen', 'have', 'apple']
[1, 2, 1, 3]

- **Padding:** It converts each list into a matrix and pad it. This is necessary to ensure consistency in our input's dimensionality. It will pad each sequence with a zero until it reaches the max possible sequence length. Continuing the previous example:

[1, 2, 1, 3]
If padding is set at 5
[[0, 1, 2, 1, 3], [4, 5, 55, 3, 6] ...]

5 Convolutional Neural Network Model

Figure 1 shows the overall structure of the created Convolutional Neural Network. The first layer is the input layer, where data are loaded from the preprocessed pickle and json files and fed into the Convolutional Neural Network. The set maximum length is 200. Then, there is the embedding layer, which is the part that is giving every word significance, the dimension is 200×200, the first dimension is set due to the input layer compatibility, while the second dimension is chosen arbitrarily, and it is the vector's meaning dimension. Tests have also been conducted with 50 and 100; however, 200 revealed to have the best result, as shown in the discussion section. Conv1D indicates convolutional layers with one dimension. Convolutional layers are paired with Max-pooling layers, and this is a usual practice due to dimensionality issues. Finally, Fully Connected layers concatenate the previous layers. Its dimension is 1000×1 to guarantee compatibility with the pooled layers. Dropout is not visible, speeds up the process and helps in preventing overfitting. Generally, this is how data flows in a Neural Network, where each layer is transformed in several layers deep of computation. More in detail, the network is generated by adding layers one on the top of the other. In total the CNN is composed of 15 layers:

Layer (type)	Output Shape	Param #	Connected to
input_1 (InputLayer)	[(None, 200)]	0	
embedding (Embedding)	(None, 200, 200)	80000200	input_1[0][0]
conv1d (Conv1D)	(None, 199, 200)	80200	embedding[0][0]
conv1d_1 (Conv1D)	(None, 198, 200)	120200	embedding[0][0]
conv1d_2 (Conv1D)	(None, 197, 200)	160200	embedding[0][0]
conv1d_3 (Conv1D)	(None, 196, 200)	200200	embedding[0][0]
conv1d_4 (Conv1D)	(None, 195, 200)	240200	embedding[0][0]
global_max_pooling1d (GlobalMax	(None, 200)	0	conv1d[0][0]
global_max_pooling1d_1 (GlobalM	(None, 200)	0	conv1d_1[0][0]
global_max_pooling1d_2 (GlobalM	(None, 200)	0	conv1d_2[0][0]
global_max_pooling1d_3 (GlobalM	(None, 200)	0	conv1d_3[0][0]
global_max_pooling1d_4 (GlobalM	(None, 200)	0	conv1d_4[0][0]
concatenate (Concatenate)	(None, 1000)	0	global_max_pooling1d[0][0] global_max_pooling1d_1[0][0] global_max_pooling1d_2[0][0] global_max_pooling1d_3[0][0] global_max_pooling1d_4[0][0]
dropout (Dropout)	(None, 1000)	0	concatenate[0][0]
dense (Dense)	(None, 128)	128128	dropout[0][0]
dropout_1 (Dropout)	(None, 128)	0	dense[0][0]
dense_1 (Dense)	(None, 2)	258	dropout_1[0][0]

```
Total params: 80,929,586
Trainable params: 929,386
Non-trainable params: 80,000,200
```

```
Total params: 1,765,795,586
Trainable params: 929,386
Non-trainable params: 1,764,866,200
```

Fig. 1. Overall structure of the CNN

- **Input layer:** The first layer of the model is where preprocessed data are loaded, at this stage both training data and labels, as well as test data and labels, are already nicely sorted into integers, which are the only values our model can understand. After the preprocessing, the chances are that data are of different lengths; therefore, a standard sentence length needs to be chosen, the length has been set to 200. In particular, if a sentence is longer than the set value, the model cut it to the chosen length, and if it is shorter than the chosen value, it pads it to the chosen length.

- **Embedding layer:** Following the input layer, there is the embedding layer. This layer is one of the most important in the model since it has the ability to encode words' meaning, intrinsic relations and cultural features. In order to make the Neural Network learn, word embeddings have been used. A word embedding is a word converted into a multi-dimensional space vector. The idea is that words with similar sentiment and/or meaning will have similar vector direction into space. In order to convert a word into a vector, several axes have been considered. The number of axes considered is equal to the dimensions we want the words to have. The more dimensions we consider, the more meanings we are encoding into the embedding. A word embedding is able to understand tons of relationships between words based on raw elements. Hence the right tool to encode semantic and cultural features into the model. In fact, word embeddings use the idea of context because the same word used in a different context could have a different meaning. As a result, the closeness of vectors resembles the closeness of words that those vectors represent. In addition, vectors carry on operations, and these operations also have meanings. The method used consists of the decomposition of a co-occurrence matrix. A co-occurrence matrix is a matrix where both rows and columns correspond to different words, and the entry at any given point in the matrix just counts how many times those two words happen in the same context. These matrixes are built by researchers and online contributors. The Glove has been used, which stands for global vector for word representation. There are several advantages in using Glove, such as the transfer learning benefits; therefore, the complexity is reduced due to the fixed size of the embeddings and finally contributors constantly improve Glove to include more features.

- **5 Convolutional and Max-pooling:** A Neural Network learns the features layers by layers in an incremental manner. It extracts lower lever features such as letter, then little higher-level features such as words, then higher lever features such as sentences and the relationship among them. As highlighted in the literature review, Convolutional Neural Networks have outperformed other techniques in Sentiment Analysis. After the embedding layer, there is the alternation of convolutional and Max-pooling layers. When added to a model, Max-pooling reduces the dimensionality by reducing the output from the previous convolutional layer. Therefore, Max-pooling layers are used to reduce the dimensionality of the data while maintaining the main features of the sentence. In particular, features are extracted by filters and then inserted into the final sentence representation. The main decision to model in this part is the choice of the filters because a convolution means the transformation of the layer's input through the chosen filter. Therefore, several filters are needed to slice and map the input sentence. Five types of filters (2, 3, 4, 5, 6) have been chosen. Their dimensions are 2×200, 3×200, 4×200, 5×200 and d 6×200. Each filter scrolls over the whole embedding layer, from the first row until it covers the last row. This process creates each row of the convolutional layer. Thus, for the 2 × 200 filter, it takes 199 scrolls, and its size does not change, which always remains 200. For 3 × 200 filters, 198 scrolls are needed and the same for the remaining filters. In order to extract the intrinsic and semantic features and consequently building a features map, an activation function is required. A Rectified Linear Unit (ReLU) has been used. The function is very simple and easy to compute; it identifies an identity for all positive values while zero for all negative values. Even though most of the previous works used Sigmoid activation function, ReLU has been chosen because it solves some drawbacks of the Sigmoid function, which will be deeper discussions in the discussion section.

- **Concatenation, dropout and dense:** Concatenation is necessary after convolution and max-pooling layers. This improves dramatically the network's ability to mine and capture the intrinsic characteristic of a sentence. Through deeper convolutions, the model is able to incrementally cover sentences, and finally, a global summary of the sentence characteristics is created. The first fully connected layer is what in Fig. 1 is called Concatenate. In general, using a fully connected layer is a computationally cheap way of learning nonlinear combinations. This layer flattens all the previous connected layers into a 1-dimensional vector, and then it is fed into a Dense Layer with 128 nodes, as there were 128 dimensions. They then output to a 2-node layer, the last layer, which is densely connected, and this last layer is characterized by an activation function, the SoftMax function. SoftMax function takes the vectors of values and transforms it into a vector of output probabilities. In particular, it calculates e (natural number) to the power of every element in the vector, divided by the sum of those values. This is carried out for each value of each vector, resulting in a probability output vector. Therefore, the Softmax function pulls apart the results into a value between 0 and 1, which sums up to 1, which represents a list outputs' probabilities. It is worth mentioning that the obtained probabilities are not the degree of humor, but how likely the sentence given in input is a joke. In order to have good training, the model needs an optimizer and a loss function. The optimizer improves the learning rate, which specifies how fast we want our network to train. In particular, Adam optimizer has been used, which is a common choice in deep learning applications. It stabilizes the learning rate and updates it while the process processes. At the same time, the loss function is used to understand if there is overfitting of data by randomly turning on and off different pathways in our network. The binary cross-entropy function is a common practice to calculate loss, it holds off on testing against test data, and it helps to find the differences between the predicted output and expected output. Dropouts have been set as a precaution to the overfitting of data. It randomly activates and deactivates paths in the network that forces the network to generalize instead of memorizing values.

5.1 Model's Outputs

The current User Interface (UI) is very simple, as shown in Fig. 2. The percentages displayed in output are the likelihood of whether the sentence is a joke or not. If the value is lower than 50%, it means that the given sentence is not a joke; otherwise, it is likely to be a joke. Generally, the sentence "I have a pen, I have an apple" does not seem humorous. However, the model detects it as humorous because the embeddings of the model consider the context. In fact, that sentence is humorous because of viral content in 2016, and therefore the model has classified it as humorous. This example shows that the model considers not only direct jokes but also more subtle ones. Moreover, regarding Chinese jokes, the model detects efficiently jokes composed from the switching of different tones characters. In fact, that joke cannot be understood by people who are not familiar with the Chinese language.

Fig. 2. User interface and outputs examples

6 Discussion

The above proposed and implemented model for humor recognition and datasets are built by firstly tokenizing the input text data with arbitrarily chosen sentence's length, and the dimension of the embedding (200×200). The embeddings have been built using Glove. Then, the embeddings were fed to the convolutional and max-pooling layers. Finally, the outcome of the convolution and max-pooling were then flattened into a one-dimensional vector. Once the model has been trained, it has been tested against the test set. The final dataset split is 90% training and 10% testing. In order to understand the performance of the model and whether the Neural Network is functioning properly, the following indicators have been considered:

- **Accuracy:** which is defined as the ratio between the correct predicted labels over the total labels. In other words, how correct the model is when it is fed with new data. The higher this metric is, the better.
- **Loss:** which depending on the loss function chosen, it has a different definition. Overall, it indicates the error in the model per each training and testing iteration. Therefore, the lower this metric is, the better.

It is fundamental to keep track of these two Key Performance Indicators (KPI) in order to avoid **overfitting**. Overfitting happens when the model memorizes values from the dataset rather than learning from the dataset. To assess whether the model is overfitting or not, the comparison between training accuracy and testing accuracy have been made. The model is overfitted when the training accuracy surpasses greatly the testing accuracy and/or there is a clearly different trend between training and testing curves.

Various experiments have been conducted on the available datasets. The aim was to reach higher accuracy with an acceptable loss. According to the analyzed literature, an accuracy of over 90% can be considered good. However, even though it is one of the main KPI to compare with the previous works, for a complete benchmark assessment, the model should be fed in input with the same dataset used in the literature.

Once the model has been trained, evaluation has been done comparing against the test labels. The first naïve approach resulted in an accuracy of 64,5%, which can be considered a poor performance. After various experimentation, the model was able to reach an accuracy of 96,73%. In the next sections, the choices that led to an increase in performance are discussed.

Table 1. Test accuracy and loss with naïve model setting

Epoch	1	2	3	4	5	6	7	8	9	10
Test accuracy	0,5088	0,4952	0,5939	0,5968	0,6143	0,64976	0,6747	0,6608	0,6432	**0,6448**
Test loss	0,0306	0,0089	0,0079	0,00133	0,00129	0,00144	0,00115	0,0101	0,0106	**0,0140**

6.1 Results Analysis

Table 1 reports the values of the testing Accuracy and Loss for the first naïve model setting. Figure 3 and 4 show both testing and training Accuracy and Loss of the model. Focus is given on testing values because it represents the actual behaviour of the model when it is given new data. The first naïve setting obtained a final accuracy of **64,5%** and a loss of **1,4%**. As anticipated earlier, an accuracy around or over 90% is considered good; therefore, the found accuracy is not acceptable. Overall, the trend of both training and testing curves for the two metrics are regular and in line with the expectation and literature. It can be observed that with the increase of the testing accuracy, there is a decrease in testing loss and vice versa. Similarly, this trend is present also in training Accuracy and Loss. However, observing the curves closely, in Fig. 3, there is an increasing discrepancy between training and testing values, the gap is about 21,70%. This led to the understanding that the model is slightly overfitted despite the incredibly low loss value, which indicates the contrary. This result is counterintuitive since precautions have been taken to avoid overfitting. The dropout technique has been used, to randomly shut down paths in the neural network and allow the model to find new connections and generalize, therefore not memorizing the datasets. Nevertheless, this was not enough, therefore changes have been made to increase the performance of the model.

Table 2 reports the values of the testing Accuracy and Loss for the optimized model setting. Figure 5 and 6 show both testing and training Accuracy and Loss of the model. Compared to the previous model history shown, here we can observe a dramatic increase in performance. Several parameters have been changed to reach this substantial increase in performance. In a Neural Network, although the majority of the parameters can be learned automatically by the model from the data. There are few parameters that need to be set in advance, these higher-level parameters are called Hyperparameters.

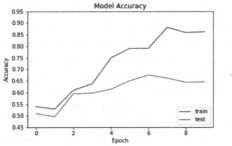

Fig. 3. Training and testing Accuracy

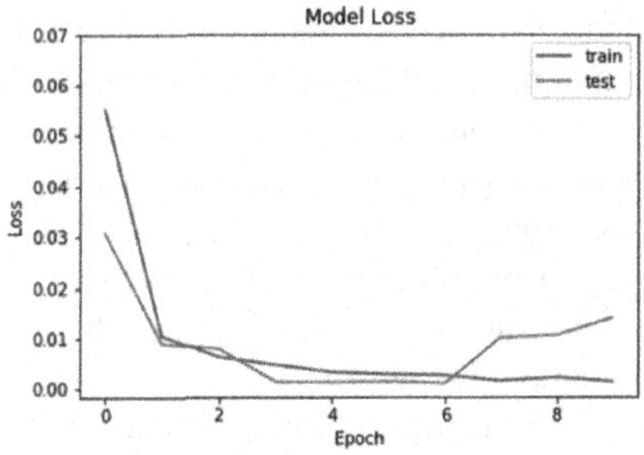

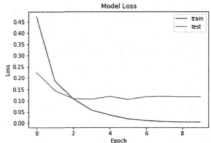

Fig. 4. Training and testing Loss

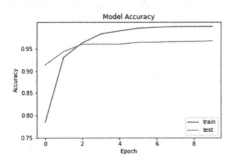

Fig. 5. Training and testing Accuracy optimized

Fig. 6. Training and testing Loss optimized

Table 2. Test Accuracy and Loss with optimized model

Epoch	1	2	3	4	5	6	7	8	9	10
Test Accuracy	0,9137	0,9434	0,9598	0,9600	0,9597	0,9641	0,9644	0,9656	0,9660	**0,9673**
Test Loss	0,2245	0.1442	0.1078	0.1055	0.1180	0.1038	0.1168	0.1178	0.1161	**0.1163**

As revealed by the literature, there is not ruled to determine which combination of parameters are the best in order to obtain higher accuracy. The decision is the outcome of the trial and error approach. However, extensive research has been done on the optimization of hyperparameters. This was beneficial to understand the relationship between all the parameters and thus for computational time since the training and testing of each epoch last about 2 h, therefore with 10 epochs, the computer needs to process for 20 h straight. Figure 5 displays the accuracy of the optimized model, where the final test accuracy is **96,73%**. While Fig. 6 shows the final loss to be **11,63%**. The smooth increasing accuracy without fluctuation indicates a good learning process. However, it can be seen that the model reaches the highest accuracy and lower loss after 8 h of training and testing, around epoch 4. After that epoch, the model's accuracy improves slightly, but the test loss remains stable. In the following figures, a comparison between the first naïve setting and the optimized model has been carried out.

On the one hand, Fig. 7 displays the test Accuracy of the two settings, naïve and optimized respectively. It can be observed that parameters' tuning has been very effective. In fact, there is an increase of +32,23% in the final testing Accuracy. On the other hand, Fig. 8 displays the test Loss of the two settings. The final testing Loss of the optimized model is about +10,23% higher than the final testing Loss of the naïve model. As pointed out earlier, the lower is the loss, the better. The naïve model loss is incredibly low; however, the optimized model loss is still in an acceptable range. Moreover, another difference between the two cases is the learning process. The naïve setting has a gradual learning process but with fluctuations while the optimized setting reaches a plateau at a higher speed and without fluctuations. Due to the reasons described, we can conclude that the optimized model is not overfitting.

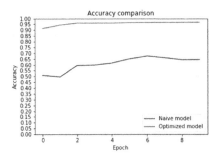

Fig. 7. Comparison between testing Accuracy

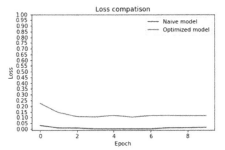

Fig. 8. Comparison between testing Loss

6.2   Performance Improvement Process

In this paragraph, the choices that led to performance improvement are explained. The relation between all the Hyperparameters and the model performance is not intuitive. In fact, Hyperparameters need to be selected very carefully they depend on each other in unknown ways, therefore optimizing one factor at the time and combining the result is not beneficial. After building the model with a naïve standard-setting, wanted parameters have been changed and tested with a set of possible reasonable values. After setting the new hyperparameters, the model accuracy was recorded for each interaction. Finally, the score was compared to the previous best score and updated if it has higher accuracy. This is how the model reached 96,73% accuracy. The modified parameters between the naïve setting and the optimized setting are:

• **Number of layers and filters:** The number of layers is one of the main features of the architecture of the network. In each layer, the number and size of filters are set. The latter is fundamental to grasps patterns and cultural semantic features. The more layers the model has, the more sophisticated it is and the more it is able to model more extensive linguistic features. The initial model was characterized by fewer layers compared to the optimized model, in particular, 11 versus 15. The increased number of layers increases the overall complexity, which could lead to even more overfitting compared to the naïve setting. However, this did not happen thanks to the following parameters tuning.

- **Activation function:** Firstly, experimentation with Sigmoid activation function has been carried since a binary output was needed, which worked well with few layers of the naïve model setting. However, with the increase in the number of layers in the network, this activation function was not suitable anymore because it has the problem of vanishing gradients. The issue was solved using the above described ReLU activation function for convolutional and fully connected layers. While for the output layer Softmax function have been used in both naïve and optimized setting. Which functions well since the output depends on the entire set of elements in the input.

- **Learning rate:** The decision for the learning rate was initially taken gradually and manually for the naïve model setting. Usually, the learning rate ranges from 0–1, and it is defined as the rate the model to learns. The problem with a large learning rate is that the training process is faster; however, the solution found could not be the optimal one. While the problem with choosing a small learning rate is that it takes a considerably higher number of epochs to train and sometimes it gets stuck. Instead, for the optimized model, an optimizer to tune the learning rate of the model has been used. The utilized optimizer is Adam optimizer, which in literature have been proven to have the best performance on several NLP's problems.

- **Embedding size:** Different embedding sizes have been used. In particular, the naïve model started with dimensions 50, then upgraded to dimension 100. Finally, in the optimized model, the dimension was set to 200. The idea is to encode more meaning possible in order to grasp the subtle features of humor.

- **Number of epochs and batch sizes:** The batch size is how we group the data. The dataset has been split into training and testing, with a portion for validation. However, in each section, we can group data into batches. For instance, training data can be made of one big batch or usually, the data are divided into small batches. At the same time, the epochs indicate the time, the number of iteration that the model will work on the whole dataset. The number of epochs depends on computational power. For the naïve model, the model was trained and tested over 10 epochs with batches of 200 samples. While for the optimized model, the training and testing still last 10 epochs but with batches of 512. The chosen number of epochs is small compared to ones read in literature, they are usually with a high number of epochs, ranging from 100 to 1000 or more. In contrast, batch size ranges from 10 to 1000. Therefore, the batch size for the optimized model is average size. For further developments epochs of 150, 300 could experiment, however, this needs powerful computational power or smaller datasets.

6.3 Limitations

Despite the good results obtained in terms of accuracy, the built model does have limitations. Limitation lays on both the dataset and the modeling part. The biggest limitation is the built **dataset**. The model needs to learn from vast and variety of data. However, the datasets built are characterized by a low variety of jokes. In particular, they are composed mainly of one-liners or short jokes. For this purpose, a much wider variety and is required. Therefore, the model handles better short sentences compared to longer ones, despite setting the input to higher lengths. Moreover, as anticipated in the literature and in the model development, assumptions and generalizations have been made in order to be able to develop the model. Humor is defined by oxford dictionaries as: *"The quality of being amusing or comic, especially as expressed in literature or speech"*, hence on one hand jokes are sufficient to define the concept of humor, but on the other hand, humor is necessary to define a joke. As a result, in this work humor have been detected using jokes despite the different meaning the two concepts have, and thus a limitation of the work. However, this approach is in line with the previous literature analyzed in this field. Indeed, jokes are metrics to measure humor.

Furthermore, there is still the problem of the differences among the domain of the data. Including more data points from the different website have been considered, however finding websites with sentences that utilize the words already included in the datasets to limit the new domain discrepancy is hard. A solution to this issue could be access to the datasets that researchers have used in the literature. This would be beneficial to assess the model properly. Moreover, for future development, the model is aimed to assess also the degree of humor therefore, the datasets should be labelled accordingly. However, this opens other limitations because defining the degree of humor is an extremely subjective task. From a feasibility point of view, it requires more research on humor classification in the field of psychology.

Secondly, about the modelling part, extensive research has been conducted for the implementation of the model. The general guidelines for architecture and parameters tuning suggestions have been found. However, there are no specific method or rules to determine a neural network; the trial and error approach is limiting. In fact, this approach sometimes leads to no or very low convergence of the training phase. Despite the functional approach taken, this could negatively affect the scalability of the model.

Lastly, in regard to the word meaning representation. Although the improvements of the last decade in encoding more and more words meanings into understandable computer values, this is still far from advanced. The humor not only depends on the language or the cultural elements but also on a variety of **subjective and qualitative factors** that strongly change the perception of the degree of humor among individuals. To overcome this issue, more and more researchers and students are improving Word2vec and more recently Glove and Concept net.

6.4 Implications and Potential Applications

The value of the created model lays in its modeling, modularity, methodology and high accuracy and therefore suitable for an existing marketing application and transversal applications. Indeed, the ultimate model aim is to assess human humor in an automatic way, without using traditional human featured tools. This can be beneficial to a variety of applications. If the right dataset can be found or created, the created model can be trained to also detect anger or other types of human subjective characteristics and therefore extended to the broader sentiment analysis in the extraction of the opinions that users express on the web regarding different products and services. This can lead to, on the one hand, a measure of brand perception and on the other hand, the creation of a psychographic profile web every user. Two main fields have been identified for the potential applications: private (in communications applications) and public (such as web analytics in reviews, social media, customer services).

In detail, the first category refers to the communication applications such as WhatsApp, WeChat and Emails. An integrated deep learning classification application in these platforms is useful in cross-cultural communication. For instance, when during a cross-cultural conversation there are sentences difficult to interpret, the integration of the model in the above-mentioned application can help the user understand the intent of the interlocutor, which is particularly useful in business conversations. In addition, for mails, the model if integrated with the system can classify whether a mail is a spam or not, and as reiterated previously, the main difficulty will be finding/creating and cleaning the right dataset on with the model can train.

The second category, which is also the most significant, refers to using the model as integrated with web analytics on social media platforms, to scan and mine over reviews, tweets and any text content we want to extract meaningful data for a company. The model is sufficiently trained, is able to recognize a variety of subjective human factors such as humor, anger, happiness and even love. The ultimate goal for marketing applications is the creation of a customers' holistic profile, namely psychographic profile, undercover a user's interest but also the reasons behind why people buy certain products than other, therefore allowing the introduction the right product to the right people.

In addition, the most important traversal application of the built model is probably fake news detection on social media, users on the web trust and share articles shared by their friends and in an incredible amount of time, there is not fact-checking behind an article. Facebook [14] is renovated to tackle this issue using sentiment analysis. Their methodology includes the classification of words that are often used in fake news titles. It has been classified that the categories most used in fake news click baits are words related to the sentiment of anger, fun and love. As a result, Facebook is currently building open-source pre-trained model on their dataset allowing transfer learning and therefore each person who wants to develop fake news detection models can use their dataset and build on top of what Facebook has built. With the availability of those pre-trained models and datasets, our model could be a fair and simple approach as part of fake news detection. As a support of this, in the Open Data Science Conference 2018, J. Venkat [15] actually proposed a model for fake news detection using Convolutional Neural Networks including pre-trained embeddings, he chooses World2vec, and max-pooling layers, which is where the model piece together the important features extracted. His results obtained an excellent accuracy of 99,8%, however, he stressed the limitations of his model, which similar to ours, lays on the datasets, in fake news detection field is very challenging to collect available datasets, in fact even humans find it difficult to labels what is fake from what is not accurately.

7   Conclusion

Scientists have long believed that human subjectivity features are characteristics that no computer will ever be able to understand and reproduce. This belief has been undermined, and recent studies have shown the possibility of including humor detection in a variety of applications. Therefore, with the right implementation choices, human subjectivity is no more an issue for computers. In fact, promising results have been reached.

The purpose of this research work is to conceptualize, develop and implement a deep learning model that is able to recognize humor of different cultures, that can be utilized for marketing purposes as an automatic detection method from user-generated content. As a result, a convolutional neural network has been created on Jupiter Notebook using a functional approach to programming based on Keras with TensorFlow backend.

The model consists of a Convolutional Neural Network composed of 15 layers. The first layer is the input layer, where the set maximum length for a given sentence is 200, this is followed by an embedding layer, which dimension is 200×200. Then, there are five Convolutional layers paired with five Max-pooling layers, on which five filters (2, 3, 4, 5, 6) are deployed with dimensions 2×200, 3×200, 4×200, 5×200 and 6×200. Finally, Fully Connected layers concatenate the previous layers, its dimension is s 1000 $\times$ 1 to guarantee compatibility with the pooled layers, and lastly, there is an output layer with SoftMax function. This function pulls apart the results given from preview layers into a value between 0 and 1, that sums up to 1, which represents the likelihood of a given sentence to be humorous or not. The main Hyperparameters choices for the model are ReLU as activation function and binary cross-entropy as the loss function. Adam optimizer has been used to stabilize and update the learning rate, while dropout has been set to prevent overfitting of data by randomly turning on and off different pathways in the model. Therefore, the model receives in input any sentences, and as output, it gives four information, whether the sentence is humorous or not, whether the sentence is western or Chinese type of humor. The use of deep learning model is fundamental to grasp the subjectivity that characterizes humor. During the development of the model, the main difficulties understood how to model such a human subjectivity, and during the implementation phase, the debugging process.

The model has been trained and tested over the created dataset composed of 463314 English sentences and 111614 Chinese sentences. The initial model setting reached an accuracy of **64,48%**, which is a low performance compared to literature. The analysis of the obtained results shows the importance of three main elements that impact on the model accuracy, namely, the dataset variety, dimension and hyperparameter choices. Hyperparameters considered are the number of layers, filter size, embeddings size, activation function, loss function and optimizer. From the initial naïve model setting, the model has been modified and optimized, reaching a final accuracy of **96,73%**. This performance is remarkable, however, there are limitations to the model. The major limitation is the created dataset, despite the acceptable size, the dataset is characterized by low variety. Finding the proper dataset is still a challenge in sentiment analysis field due to its subjective peculiarity, and building one has computational power limits. Moreover, despite the improvements in encoding words' meaning, the methodology is not mature yet. Furthermore, another limitation discussed is that humor and jokes might not have the same meaning. On the one hand, analyzing the definition of humor, jokes are sufficient to define the concept of humor, on the other hand, humor is necessary to define a joke. As a result, in this work humor have been detected using jokes despite the different meaning the two concepts have. However, this approach is in line with the analyzed literature.

Indeed, humor depends on subjective factors, cultural elements but also on a variety of individual differences in perception and jokes are just one type of metric to measure it. The developed model represents a good base for further development. Moreover, the value of the model lays not only on the high accuracy reached but also its flexibility to transversal applications. In particular, it can be trained on another type of dataset, adjusted accordingly to detect anger, fear, and other types of very subjective and human characteristics. More extensively, it could also be developed to detect fake news, as discussed previously. Overall, all the mentioned applications have great potential, and as the literature has revealed, computational sentiment analysis is the field that researchers and businesses are enthusiasts about.

References

1. DOMO. https://www.domo.com/solution/data-never-sleeps-6
2. Litman, D., Purandare, A.: Humor: prosody analysis and automatic recognition for F*R*I*E*N*D*S*. In: Proceedings of the 2006 Conference on Empirical Methods in Natural Language Processing, Sydney (2006)
3. Kiddon, C., Brun, Y.: That's what she said: double entendre identification. In: Proceedings of the 49th Annual Meeting of the Association for Computational Linguistics (ACL), Portland, pp. 89–94 (2011)
4. Oliveira, L.D., Rodrigo, A.: Humor Detection in Yelp Reviews (2015)
5. Bertero, D., Fung, P.: A long short-term memory framework for predicting humor in dialogues. In: North American Chapter of the Association for Computational Linguistics: Human Language Technologies, NAACL HLT 2016 - Proceedings of the Conference, pp. 130–135 (2016). https://doi.org/10.18653/v1/n16-1016
6. Giudice, V.: Humor detection in Spanish tweets with character-level convolutional RNN. In: ASPIE96 at HAHA (IberLEF 2019) - CEUR Workshop Proceedings, pp. 165–171 (2019)
7. Cignarella, A.T., Frenda, S., Basile, V., Bosco, C., Patti, V., Rosso, P.: Overview of the EVALita 2018 task on irony detection in Italian tweets (IRONITA). In: CEUR Workshop Proceedings (2018). https://doi.org/10.4000/books.aaccademia
8. Li, D., Rzepka, R., Ptaszynski, M., Araki, K.: Convolutional Neural Network for Chinese Sentiment Analysis Considering Chinese Slang Lexicon and Emoticons, pp. 2–5 (2019)
9. Li, D., Rzepka, R., Ptaszynski, M., Araki, K.: Emoticon-aware recurrent neural network model for Chinese sentiment analysis. In: 2018 9th International Conference on Awareness Science and Technology (ICAST 2018), pp. 161–166 (2018). https://doi.org/10.1109/ICAwST.2018.8517232
10. Kakkle Open Source Project. https://www.kaggle.com/abhinavmoudgil95/short-jokes
11. Europarl: First Conference on Machine Translation (WMT16). https://www.statmt.org/wmt16/translation-task.html
12. XiaoHua ZOL. https://xiaohua.zol.com.cn/
13. THU NLP Lab. https://github.com/thunlp/THUCTC
14. Zuckerberg, M.: Facebook to start detecting and labeling fake news (2017). https://whatsnewinpublishing.com/zuckerberg-pivots-facebook-to-start-detecting-and-labeling-fake-news/
15. Venkat, J.: Detection & Classification of Fake News Using Convolutional Neural Nets (2018)

Expressing Agreement in Swedish and Chinese: A Case Study of Communicative Feedback in First-Time Encounters

Anna Jia Gander$^{(\boxtimes)}$, Nataliya Berbyuk Lindström , and Pierre Gander

University of Gothenburg, 405 30 Göteborg, Sweden
{anna.jia.gander,nataliya.berbyuk.lindstrom,
pierre.gander}@ait.gu.se

Abstract. This paper is a case study that explores how communicative feedback for expressing agreement is used by Swedish and Chinese speakers in first encounters. Eight video-recorded conversations, four in Swedish and four in Chinese, between eight university students were analyzed. The findings show that both Swedish and Chinese speakers express agreement more multimodally than unimodally. The Swedish participants most frequently use unimodal vocal-verbal feedback *ja* and *nej* (equivalent to *yes* and *no* in English respectively), unimodal gestural feedback, primarily head nod(s), multimodal feedback *ja* in combination with head nod(s) and up-nod(s). The Chinese participants most commonly use unimodal vocal-verbal feedback *dui* and *shi* (equivalent to *yes* in English), followed by the unimodal gestural feedback head nod(s), and multimodal feedback *dui* and *shi* in combination with head nod(s) to express agreement. Also, the findings indicate that the expression of agreement varies between both cultures and genders. Swedes express agreement more than Chinese. Females express agreement more than males. The results can be used for developing technology of autonomous speech and gesture recognition of agreement communication.

Keywords: Cross-cultural communication · Agreement · Communicative feedback · Swedish · Chinese · Cultural/gender differences · First-time encounters

1 Introduction

In social communication activities, understanding the counterpart's expressions of agreement through social signals (such as vocal-verbal unimodal feedback *hm, yeah, no,* gestural unimodal feedback a head nod/shake, or multimodal feedback *yes* in combination with a head nod) is a fundamental aspect for achieving mutual understanding in interaction. Expressing agreement indicates "the situation in which people have the same opinion, or in which they approve or accept something" (Cambridge University Press 2020). People can share convergent or divergent opinions, proposals, goals, reciprocal relations, attitudes and feelings through social signals of agreement and disagreement (Khaki et al. 2016). Recognizing and interpreting such social signals is an essential

© Springer Nature Switzerland AG 2021
P.-L. P. Rau (Ed.): HCII 2021, LNCS 12771, pp. 390–407, 2021.
https://doi.org/10.1007/978-3-030-77074-7_30

capacity for social life, and for developing Intelligent Systems, from dialogue systems to Embodied Agents capable of sensing, interpreting and delivering social signals (Poggi et al. 2011). Thus, research on multimodal human interaction in naturalistic settings from various communication contexts is essential to enhance understanding of social signals and its applications in digital technology development.

Earlier research has shown that people express agreement in different ways, and culture is one of the factors which influences it. In high-context cultures such as China and Malaysia, face saving leads to avoidance of expressing disagreement and saying 'yes' doesn't necessarily mean agreement, which is the opposite to low context cultures such as Germany and the U.S. (Hall 1959; Kevin 2004). The research of Swedes' communication has shown that they give much feedback in interactions, signaling seeking consensus and avoiding conflict (Pedersen 2010). Gender is another factor influencing agreement expression. The research of Malaysian Chinese agreement strategies shows that females tend to express more agreement in interactions than the males (Azlina Abdul 2017).

Few studies in cross-cultural communication research focus on analysis of real-time interactions. Currently, there is little research with a focus on multimodal communication and cross-cultural comparison between Sweden and China, although they have increasing contacts and collaborations in both academia and industry, which motivates why more research is needed in this area.

2 Background

Social signals are perceivable stimuli that, either directly or indirectly, convey information concerning social actions, social interaction, attitudes, social emotions and social relations (Vinciarelli et al. 2012). Expressing agreement indicates "the situation in which people have the same opinion, or in which they approve or accept something" (Cambridge University Press 2020). Agreement and disagreement are expressed by specific social signals, through which people in the interaction express, for example, whether they share the same opinions, whether they accept each other's proposals, whether they have convergent or divergent goals, attitudes and feelings, how their reciprocal relations are, and how to predict the development and outcome of the discussion or negotiation.When people argue in a discussion or negotiation, it is important for people to understand whether, what, and how they agree or disagree.

In order to perceive and understand the social signals of agreement and disagreement, one must be able to catch and interpret the relevant feedback expressions, which are often expressed in different communicative modalities – words, gestures, intonation, facial expressions, gaze, head movements, and posture (Gander 2018; Poggi et al. 2011). One may look at the other while smiling and nodding to express agreement, while shaking head and saying "no" to signal disagreement. Gestural unimodal expressions of agreement (such as head nod, finger thumb up) have been researched in different datasets such as TV live political debate (Poggi et al. 2011), dyadic dialogues about the likes and dislikes on common topics such as sports, movies, and music (Khaki et al. 2016), and human-computer interaction through gestures via touchless interfaces (Madapana et al. 2020). Expressions of agreement have also been explored for developing an implementation and training model of agreement negotiation dialogue, for example, with a dataset of dialogues where people have contradictory communicative goals and try to reach agreement, in particular focusing on the sequence of communicative acts and the vocal-verbal unimodal means (Koit 2016). Mehu and van der Maaten (2014) stressed the importance of using multimodal and dynamic features to investigate the cooperative role of pitch, vocal intensity and speed of speech to verbal utterance of agreement and disagreement. Although researching both visual and auditory social signals of agreement and disagreement has obtained growing attention, studies on multimodal communication of agreement and disagreement still focus more on the role of nonverbal behavior in social perception (e.g., Mehu and van der Maaten 2014) rather than the combination of multimodalities in expressing agreement and disagreement. This paper will investigate both gestural unimodal, vocal-verbal unimodal, and their combined multimodal communicative behaviors to contribute to big data corpus of naturalistic interactions. Such empirical research findings can be used for developing and evaluating statistical models of agreement and disagreement in interactions for instance, and also for designing technology of autonomous speech and gesture recognition of agreement communication.

Besides modality issues, there have been also cultural and gender concerns of agreement communication. As known, understanding the counterpart's expressions of agreement is a fundamental aspect for achieving mutual understanding in cross-cultural communication activities. Across cultures, people express agreement in different ways. The research about the Malaysian Chinese agreement strategies shows that interlocutors tend to complete and repeat part of the previous speaker's utterance and give positive feedback, which is explained by striving for supporting speaker's face and harmony rooted in Chinese cultural values (Azlina Abdul 2017). The research also emphasizes that female speakers tend to express more agreement than male speakers. In high-context cultures, for example, China and Malaysia, face saving leads to avoiding expressing open disagreement with interlocutors and saying 'yes' doesn't necessarily mean agreement; whereas, a more direct communication like "no, I don't think so" or "sorry, I don't agree" is common in low context cultures, such as Germany and the U.S. (Hall 1959; Kevin 2004). The Swedes are known for giving much feedback in interactions, signaling consensus seeking and conflict avoidance (Pedersen 2010). Few studies in cross-cultural communication research focus on analysis of real-time interactions. Also, there is little research focus on multimodal communication and cross-cultural comparison between Swedish and Chinese communication patterns, though contacts between Chinese and Swedes are increasing. In recent years, Chinese direct investments in Sweden have tripled and are expected to increase further (Weissmann and Rappe 2017). Sweden and China have many international collaborations in both academia and industry. However, few studies of real-time interactions have assessed similarities and differences between Swedish and Chinese cultures in general, with little focus on expressing agreement in particular. We are going to address this in the present paper.

3   Research Aim and Questions

This paper aims at exploring and comparing how Chinese and Swedes express agreement in interactions. More specifically, it focuses on the multimodal communicative feedback for expressing agreement, including vocal, gestural and combination of vocal and verbal feedback expressions.

Two specific research questions will be addressed:

1. What feedback expressions are used for expressing agreement in the Swedish and the Chinese conversations?
2. What similarities and differences can be observed between the Swedish and the Chinese expressions of agreement?

4 Methodology

4.1 Data Collection and Participants

Our data consist of four Swedish-Swedish and four Chinese-Chinese first-time encounters' dyadic conversations between eight university students in Sweden. The conversations were in languages of Swedish and Mandarin Chinese respectively and were video- and audio-recorded in a university setting. Each conversation lasts approximately ten minutes. The participants were 23–30 years old, four males and four females. They were native speakers of Swedish and Mandarin Chinese. The participants did not know each other at the time of the recordings. They were invited to get acquainted and interact with one another, with no limitation on the topics. Anonymity was emphasized in the research project. The participants were asked for permission to use the pictures from the recordings for publications. All participants gave written consent for their participation in the research and the use of their data.

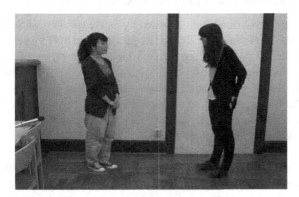

Fig. 1. Recording of the encounter.

Two participants (see, e.g., Fig. 1) were paired in a classroom and asked to interact with each other for about 10 min after the research assistant switched video-cameras on and left the room. The interactions were audio- and video-recorded by three video cameras (left-, center-, and right-position) with interlocutors standing face-to-face to capture the participant's communicative movement as much as possible. The cameras were placed off the participants' field of vision to minimize possible effects of video recording on participants' behavior.

Each participant was involved in two mono-cultural dialogues, in which the gender of the other interlocutor varied. The first seven minutes of each dialogue were studied, ˜ because the shortest dialogue was around seven minutes long. Within the time given, the use of agreement expressions was compared between the Chinese and the Swedish speakers. An overview of the data is presented in Table 1.

Table 1. Overview of the recordings (C = Chinese, S = Swedish, f = female, and m = male).

Recording	Participants	Analyzed length	Language
Dial.1	Cf1–Cf2	7:00 min. each	Chinese
Dial.2	Cf1–Cm2	(56 min in total)	
Dial.3	Cm1–Cf2		
Dial.4	Cm1–Cm2		
Dial.5	Sf1–Sf2		Swedish
Dial.6	Sf1–Sm2		
Dial.7	Sm1–Sm2		
Dial.8	Sm1–Sf2		

4.2 Transcription, Coding, and Inter-coder Rating

The recordings were transcribed and annotated by native speakers of Swedish and Chinese, according to the Gothenburg Transcription Standard (GTS) version 6.4 (Nivre et al. 2004) and the MUMIN Coding Scheme for the Annotation of Feedback, Turn management and Sequencing Phenomena (Allwood et al. 2007). This allows inclusion of such features of spoken language as pauses and overlaps as well as comments on communicative body movements and other events. Feedback expressions through various modalities, their communicative act units, and their communicative functions were annotated. The following conventions from GTS and MUMIN were used in the paper (see Table 2).

Given the differences and connections between the perception modalities (e.g., visual modality, auditory modality, and haptic modality) and production modalities (e.g., gestural input is perceived through visual modality, vocal-verbal input through auditory modality, tactile input through haptic modality) (cf. Gander 2018), this paper focuses on the perception-sensory perspective. Thus, the researched agreement expressions are classified into three categories:

1. Vocal-verbal unimodal: verbal expressions only
2. Gestural unimodal: bodily communication, including head nods, facial displays, etc.
3. Vocal-verbal + gestural multimodal: combinations of (1) and (2)

The annotation was carried out by one annotator, and inter-coder reliability checking was performed independently by another annotator. The inter-coder agreement

Table 2. Transcription and MUMIN coding conventions.

Symbol	Explanation
$Cf1	Speaker (Chinese, female)
[ ]	overlap brackets; numbers used to indicate the overlapped parts
{j}a	j is not pronounced
/, //, ///	a short, intermediate and a long pause respectively
:	lengthening
< >, @ < >	comments about non-verbal behavior, comment on standard orthography, other actions, clarifications
FB	feedback
VFB	vocal-verbal feedback
GFB	gestural feedback
CPUE/A	contact, perception, understanding, emotion/attitude
CPU	contact, perception, understanding

on the annotation of feedback and its communicative functions is moderate, with a free-marginal $\kappa = 0.58$.

By using the coding scheme (Allwood et al. 2007), the basic communicative function of feedback was coded as CPU (i.e., contact, perception and understanding). A code of CPU means 'I want to and am able to continue the conversation and I perceive and understand what you have communicated.' In addition, we also coded other communicative functions of feedback in our data, such as effective and epistemic stances which include emotion, attitude, agreement, acceptance, and evaluation etc. (Allwood et al. 2000; Paggio et al. 2017; Wessel-Tolvig and Paggio 2013; Gander 2018). There, agreement differs from acceptance in that agreement indicates acceptance, whereas acceptance does not indicate agreement. One can accept but may not necessarily agree. Acceptance is the necessary but not sufficient condition for agreement. In this paper, we focus specifically on agreement. CPU + agreement refers to that 'I want to and am able to continue the conversation and I perceive and understand what you have communicated' which is meant by CPU (only) plus 'I agree with you on what you have said'.'

During the coding work, we identified CPU + confirmation too, and distinguished it from CPU + agreement, with a consideration of whether it is confirming a truth or fact or agreeing on an opinion. We do not want to simplify a truth or a fact as an opinion, and we do not want to simply take 'confirmation' as one type of 'agreement'. Accordingly, we define the criterion of identifying CPU + agreement and CPU + confirmation as below.

The code of CPU + agreement occurs when the interlocutor is agreeing on some subjective opinions that the other interlocutor has expressed in the previous utterance which is not necessarily the truth or fact; while, CPU + confirmation is coded when the interlocutor is confirming what the other interlocutor has said is true. As has been stated and reasoned above, in this research we do not consider confirmation as one type of agreement. Instead, we take agreement to be based on shared views or opinions (not necessarily truth or fact), whereas confirmation to be associated with facts or truths.

4.3 Data Analysis

Occurrences of agreement feedback were identified and categorized according to modality (see Sect. 4.2, vocal-verbal unimodal, gestural unimodal, and vocal-verbal + gestural multimodal). Further, interactional sequences were qualitatively analyzed in order to examine the communicative functions of agreement feedback in the interactions. Additionally, quantitative analysis of frequencies and percentages of feedback occurrences were performed, comparing cultures (Swedish/Chinese) and genders. Because of the limited size of the samples, analysis of differences was made directly on the observed frequencies in the samples rather than being statistically tested, and generalizations were not attempted.

5 Results

The first research question aims to identify the agreement expressions in Swedish and Chinese. The second research question aims to explore the similarities and differences between the Swedish and the Chinese agreement expressions. Results are presented using raw frequencies and percentages of agreement feedback. Excerpts are extracted from the transcriptions.

5.1 General Overview of Agreement Feedback

This section presents a general overview of the types of agreement feedback in the data. Table 3 shows that the Swedish participants in the sample use more of both unimodal and multimodal agreement feedback than the Chinese participants (110 compared to 54 occurrences). Compared to the Chinese, the Swedes also use more vocal-verbal expressions of agreement (52 versus 17) and more multimodal feedback (41 compared to 27).

Regarding gender, both the Chinese and the Swedish female speakers use more agreement expressions than male (81% versus 19% in the Chinese sample and 67% versus 33% in the Swedish sample), including all the three types of modalities of agreement expressions. In the Chinese sample, the share of gestural unimodal and multimodal feedback expressions is much higher in the females than in the males (90% and 89% compared to 10% and 11% respectively). In addition, Table 3 also shows that the Swedish females use more agreement expressions than the Chinese females (74 compared to 44), so as the Swedish males compared to the Chinese males (36 versus 10).

Table 3. Overview of agreement feedback used by the Swedish and Chinese participants (FB = Feedback, M = Male, F = Female).

FB	Chinese			Swedish		
	M	F	Total	M	F	Total
Vocal-verbal unimodal	6 35%	11 65%	17 100%	15 29%	37 71%	52 100%
Gestural unimodal	1 10%	9 90%	10 100%	6 35%	11 65%	17 100%
Vocal-verbal + gestural multimodal	3 11%	24 89%	27 100%	15 37%	26 63%	41 100%
Total:	10 19%	44 81%	54 100%	36 33%	74 67%	110 100%

5.2 Swedish and Chinese Unimodal Vocal-Verbal Feedback Expressions for Agreement

Regarding unimodal agreement feedback expressions, we will look at the Swedish and the Chinese data respectively.

In the Swedish data, the identified Swedish unimodal vocal-verbal feedback expressions can be grouped into three categories: 1) *yes* and its variations, for example, *ja* (*yes*), *{j}a* (*yeah*), repetitions of *{j}a* (*yeah*); *{j}a: precis* (*yes exactly*); *jo* (*sure*); 2) *m* and its variations, for example, *m, m:* (lengthening); 3) *no* and its variations (e.g., when interlocutor agrees with preceding negative statement), such as *no, no no* (see Table 4). The Swedish participants show a tendency of expressing agreement using primarily *{j}a* and its variations, with an occurrence of 69% of all the unimodal vocal-verbal feedback expressions. The Swedish participants also use unimodal words *no* and *m* with their variations to express agreement, with frequencies of 17% and 14% respectively (see Table 4).

The data presented in Table 4 also shows that the Swedish females use more vocal-verbal feedback expressions of *yes* and *m* and the variations than the Swedish males, with occurrences of 50% versus 19% and 12% versus 2%, respectively. In the meanwhile, the Swedish females and males show similar preferences in using *no* and its variations, with frequencies of 9% and 8%.

To present an example from our data, in Excerpt 1 below, Swedish female 2 expresses agreement while listening to Swedish female 1 about Swedish female 1's experiences with home care. Swedish female 2 agrees with Swedish female 1 on that some people one meet at the workplace people can be nice while others can be avoided, by using unimodal vocal-verbal feedback expression *{j}a*.

Table 4. Unimodal vocal-verbal feedback (VFB) agreement expressions used by the Swedish participants

Swedish Unimodal VFB	English translation	M	F	Total
Yes and its variations				
{j}a; {j}a:; ja; hja (ingressive); {j}a ja; {j}a {j}a; {j}a ja ja ja ja ja ja ja; {j}a eller hu{r}; {j}a: precis; precis; {j}a precis precis; {j}a precis jo verkli{j}en; {j}a jo precis; ja ja men precis; {j}a precis nä; hja exakt jo; kjo; {j}a jo	yeah; yeah:; yes; ha; yeah yes; yeah yeah; yes yes yes ... yes yes; eah that's right; yeah: exactly; exactly; yeah exactly exactly; yeah exactly sure that's true; yeah sure exactly; yes but exactly; yeah exactly no; yeah exactly sure; yeah sure	10 19%	26 50%	36 69%
No and its variations				
nä nä; näe; nä men visst; nä nä de{j} e klart; näe precis; nä	no no; no; no exactly; no no of course; no exactly	4 8%	5 9%	9 17%
M and its variations				
m; m:; m nä; m jo {j}a	yes; yeah:; yeah no; yeah sure yeah	1 2%	6 12%	7 14%
Total:		**15 29%**	**37 71%**	**52 100%**

Excerpt 1. (SwDial.5) With some you work great, with others you don't.

1 $Sf1   de beror ju så mycke på va man få fö utbyte me personen liksom så att
it depends so much on what kind of connection you get with the person like so that

2   / e // det e ju lite e e ja jobba hemtjänsten innan / å där e de ju också lite
/ eh // it's a little eh eh I worked home care before also / and there it's also a little

3   de att man träffar [väldigt mycke människor] / å vissa funkar man jättebra me
like you meet [1 a lot]1 of people / and with some you work great with

4   / å de e jättetrevlit // å andra e lite så hä bara / måste ja gå dit ida
/ and it's very nice // and others are a little like / do I have to go there today

5 $Sf2   < [{j}a]>
< [1 yeah]1 >
@ < vocal-verbal feedback; CPU + agreement >

Then, turning to the Chinese data, Table 5 presents the unimodal vocal-verbal feedback expressions used by the Chinese participants when expressing agreement.

Table 5. Unimodal vocal-verbal feedback (VFB) agreement expressions made by the Chinese participants.

Chinese Unimodal VFB	English translation	M	F	Total
Yes and its variations				
dui, dui na shi	*yes it is/ right*	4 23%	10 59%	14 82%
M and its variations				
en, a	*yeah; ah*	1 6%	1 6%	2 12%
No and its variations				
bu shi	*no, (it is) not*	1 6%	0 0%	1 6%
Total:		**6 35%**	**11 65%**	**17 100%**

Dui and its variation *dui na shi* (equivalent of *yes* and *right* in English) are the most common (82%) agreement expressions in the Chinese data. The Chinese female speakers give more agreement feedback than the Chinese males (65% compared to 35%).

Excerpt 2 below illustrates the Chinese female speaker 1 agreeing with the Chinese female 2 on that the work in China is faster tempo and more task oriented compared to that in Sweden:

Excerpt 2. (ChDial.1) Work tempo and task orientated.

1   $Cf2   bu hui xiang zhe tou er gan jue guo de xiang bi jiao chong shi
            it was not like here (I) feel life is more occupied and (task) oriented

2   $Cf1   < **dui** > you ya li ma
            < **right** > *pressure and drive isn't it*
            @ < **vocal-verbal feedback; CPU + agreement** >

Comparing the data of Table 4 and Table 5, one can observe that in both Swedish and Chinese conversations the vocal-verbal feedback *yes* and its variations are the most common unimodal agreement expressions. Also, in both Swedish and Chinese data, the female speakers give more feedback than the males. Besides, there is more variety in feedback expressions used in the Swedish sample compared to the Chinese data.

5.3 Swedish and Chinese Unimodal Gestural Feedback Expressions for Agreement

The Chinese and the Swedish unimodal gestural feedback expressions for agreement will be analyzed in this section. An overview of the data is presented in Table 6.

Table 6. Swedish and Chinese unimodal gestural feedback (GFB) agreement expressions

Swedish Unimodal GFB	M	F	Total	Chinese	M	F	Total
a nod; nods	2	8	10	a nod; nods	0	8	8
shake	1	0	1	smile	1	1	2
Total	**3**	**8**	**11**	**Total**	**1**	**9**	**10**

As can be seen from Table 6, both Swedish and Chinese speakers use unimodal gestural expressions relatively seldom, compared to how they use unimodal vocal-verbal expressions (presented in the previous section). Nod(s) are the most common unimodal feedback expression to communicate agreement in both Swedish and Chinese samples.

Excerpt 3 below illustrates the Swedish male speaker 2 agreeing with the Swedish male speaker 1 on that travelling is better to do after having completed education. A possible reason for just nodding (of Sm2) in this excerpt can be that the speaker (Sm2) does not want to interrupt an apparently lively monologue of the other speaker (Sm1).

Excerpt 3. (Sw Dial.8) Travelling funds.

1    $Sm1   nä de tycke inte ja bättre å göra nä du e klar kanske då
no I don't think so it's better to do when you're finished maybe then

2   å sen när du har gjort lumpen också får du lite pengar
and then when you have (...) done boot camp too you get a little money

3   där ifrån också så de har du mer i reskassa
from there too so it's eh do you have more travelling funds

4    $Sm2   @ < **head movement: nods** >
@ < **gestural feedback; CPU + agreement** >

Similarly, in the Chinese data, by nodding the female speaker 2 shows agreement on that if there is an opportunity to work then one should work rather than studying (see Excerpt 4).

Excerpt 4. (Ch Dial.3) Work or study?

1    $Cm1   < you ji hui jiu gong zuo >
< *(if) there is an opportunity (I'll) work* >
@ < gaze down to the left >

2    $Cf2   < **head nods** >
@ < **gestural feedback; CPU + agreement** >

5.4 Swedish and Chinese Multimodal Feedback Expressions for Agreement

In this section, the Swedish and the Chinese multimodal feedback expressions for agreement will be presented respectively. We will start with the Swedish data, followed by the Chinese data.

The Swedish multimodal feedback expressions are grouped in the same categories as the vocal-verbal unimodal expressions (see Sect. 5.2), namely, (1) *yes* and its variations, (2) *no* and its variations, and (3) *m* and its variations. Bodily movements used with each vocal-verbal expression comprise the corresponding multimodal unit. Table 7 presents an overview of the Swedish multimodal expressions identified in the data.

Table 7. Multimodal feedback (VFB + GFB) agreement expressions used by the Swedish speakers.

VFB	English translation	GFB	M	F	Total
***Yes* and its variations + bodily movement**					
{j}a; ja; ja ja ja ja; ja det ä det; ja det är ja; jo	*yeah; yes; yes yes yes yes; yes yes sure; it eh yes; sure*	nods	5	15	20
ja; {j}a	*yes; yeah*	up-nod	0	2	2
{j}a	*yeah*	tilts	4	0	4
{j}a precis ja; {j}a {j}a precis absolut	*yeah exactly yes; yeah yeah exactly absolutely*	nods	1	2	3
Subtotal:			**10** **24%**	**19** **46%**	**29** **70%**
***No* and its variations + bodily movement**					
nä men man {j}a	*no but you yeah*	nods	1	0	1
näe näe näe exakt	*no no no exactly*	tilt	1	0	1
nä nä nä; m nä	*no no no; yeah no*	shakes	0	2	2
nä	*no*	nod + tilt	0	1	1
{j}a nä	*yes no*	tilt + eyebrow rise	1	0	1
nä de tycke inte ja	*no I don't think so*	eyebrow frown	2	0	2
Subtotal:			**5** **12%**	**3** **8%**	**8** **20%**
***M* and its variations + bodily movement**					
m; m ja; m jo; m yeah	*yeah; yeah yes; yeah sure; yeah yeah*	nods	1	3	4
Subtotal:			**1** **2%**	**3** **8%**	**4** **10%**
Total:			**16** **38%**	**25** **62%**	**41** **100%**

The most common multimodal feedback expressions for agreement employed by the Swedish participants are *yes* and its variations + nods as well as *no* and its variations + bodily movements including shakes, eyebrow movements and tilts (with frequencies of 70% and 20%).

An example is presented in Excerpt 5. The Swedish male speaker 1 expressed agreement with the Swedish male speaker 2 on randomly selected group work, by using the multimodal expression of *yeah* combined with head tilt.

Excerpt 5. (Sw Dial.8) Not cool.

| 1 | $Sm2 | man kan bli tilldelad å sånt men de bli ju väl typ så hä typ / |
| | | *you could be assigned and such but then it's something like /* |

| 2 | | a plocka upp skit från [1< vägarna å sånt >]1 |
| | | *yeah pick up crap from [2< the roads and stuff like that >]2* |

3	$Sm1	[1 < {j}a > / de e inte så kul]1
		[2 < **yeah** > / *that's not very cool*] 2
		@ < **head: tilt** >
		@ < **vocal-verbal and gestural feedback; CPU + agreement** >

Then, in the Chinese data, the Chinese multimodal feedback expressions for agreement can be grouped in two categories: 1) *yes* and its variations + bodily movement and 2) *no* and its variations + bodily movement (see Table 8).

The majority of multimodal agreement expressions in the Chinese sample are contributed by the Chinese female speakers. As a variation of *yes* (see Table 8), the vocal-verbal component *en* (equivalent to *yeah* in English) combined with the gestural component nod(s) comprised the most common multimodal feedback expression for agreement, which is only used by the females. Chuckle, nod and smile gestural feedback expressions combined with the vocal-verbal feedback *dui*, *shi* or *en* (equivalent to *yeah* in English) comprise the most common multimodal agreement expressions.

Table 8. Multimodal feedback (VFB + GFB) agreement expressions used by the Chinese speakers

VFB	English translation	GFB	M	F	Total
Yes and its variations + bodily movement					
dui; dui dui dui dui	right; right... right	nods	0	2	2
dui	right	chuckle + nod	0	1	1
dui	right	smile + up-nod	0	1	1
dui dui; dui	right right; right	smile + nods	0	1	1
a dui	ah right	smile	1	0	1
shi	yes	nods	0	1	1
en shi	yes it is	nods	0	1	1
ying gai shi	should be	nods	0	1	1
en	yes	nod(s)	0	11	11
en	yes	chuckle + nods	0	1	1
Subtotal:			**1** **5%**	**20** **90%**	**21** **95%**
No and its variations + bodily movement					
bu shi	no/not	shake	0	1	1
Subtotal:			**0** **0%**	**1** **5%**	**1** **5%**
Total:			**1** **5%**	**21** **95%**	**22** **100%**

Excerpt 6 presents an example of the Chinese female speaker 2 agreeing with the Chinese male speaker 1 on the open mindedness of their families regarding what they are allowed to do or not to do. The speaker Cf2 first provided a verbal feedback *en* (equivalent to *yeah* in English), followed by head nods and a lengthy pause which reinforced her vocal-verbal expression of agreement.

Excerpt 6. (Ch Dial. 3) Openness of families.

1 $Cm1   dan shi wo shi mei you wo men jia dou hen kai fang
            but I do not have my families are open-minded

2 $Cf2   < en /// > wo ma ye shi peng you ///
            < yes /// > my mother has a friend (like this) too ///
            @ < head nods >
            @ vocal-verbal and gestural feedback; CPU + agreement

6 Discussion

Regarding modalities, the data has shown that both the Swedish and the Chinese speakers use vocal-verbal unimodal, gestural unimodal and multimodal expressions for expressing agreement. Unimodal gestural feedback expressions (e.g., head, shake, and smile) are less frequently used, compared to unimodal vocal-verbal and multimodal feedback expressions. This possibly indicates that vocal-verbal means are prioritized over the gestural means when people express agreement, and that simply nodding, for example, is not enough to ensure or reinforce agreeing in both cultures. Nevertheless, in both Swedish and Chinese samples, nodding is still the most common unimodal gestural expression of agreement, which is in line with previous research (Andonova and Taylor 2012). The findings also show that both Swedish and Chinese participants use head nod(s) in combination with *yes* of their own respective languages as the most common multimodal feedback expressions for showing agreement, which is consistent with earlier research (Lu and Allwood 2011). Also, the Swedish participants use more agreement feedback expressions both unimodally and multimodally than the Chinese participants. This is consistent with the prior studies which have noted the importance of consensus in Sweden (Havaleschka 2002) and the tendency that giving much feedback in interactions indicates listening and agreeing (Allwood et al. 2000; Tronnier and Allwood 2004) and avoiding conflict (Daun 2005; Pedersen 2010).

Besides the features of modalities for expressing agreement, both Swedish and Chinese female participants in this research use more agreement feedback than the males. This finding supports the research of Deng et al. (2016) on that females tend to have higher emotional expressivity and provide more feedback than males. In addition, although it has been found that both the Swedish and the Chinese female speakers express agreement more often than the males, this gender difference was bigger in the Chinese sample compared to the Swedish sample. This finding is in line with that of Azlina Abdul's (2017) research, in which the Chinese females are found giving considerably more agreement feedback than the Chinese males. Further, as Sweden is considered to be a society in which gender roles overlap to a higher extent than in China, it might be reflected in more similarities between the Swedish male and female participants in expressing agreement (Hofstede 1994).

The data has also shown that the Chinese participants provide considerably fewer agreement feedback expressions than Swedes. It can potentially be explained by that in Sweden interactions are often informal in particular among students at a university context, due to a relatively low power distance (Hofstede 1994). On the contrary, in China, a differentiation is clearly made between polite/impolite and insider/outsider communication, the former often occurring between strangers while the latter between close acquaintances (Fang and Faure 2011). As Chinese culture is a high context culture, many Chinese people are cautious in communicating with 'outsiders' by, for example, limiting their contact to brief and functional communicative exchanges (Hall 1959), which might be reflected in the Chinese speakers in our research being less willing to express themselves compared to the Swedes.

Because these findings are based on a particular communication context of first encounters and between a limited number of participants, the present research can be seen as a case study for the future work with more participants and varied contexts to investigate agreement expressions, patterns and strategies in Swedish and Chinese or Western and Eastern cultures further.

7 Conclusion

This paper explored and compared how Chinese and Swedes express agreement in first-time encounter's conversations by using social signals of communicative feedback. In this paper, eight audio- and video-recorded interactions between four Swedish and four Chinese participants were analyzed. The identified agreement feedback expressions were classified into three types according to which modalities are involved, namely, the vocal-verbal unimodal, the gestural unimodal, and the vocal-verbal + gestural multimodal expressions. Two research questions were investigated, regarding respectively what feedback expressions are used for expressing agreement in the Swedish and the Chinese conversations and what similarities and differences can be observed between the Swedish and the Chinese expressions of agreement.

About the feedback expressions used for expressing agreement, we have found that participants from both samples use both unimodal and multimodal feedback expressions. The Swedish participants most frequently use unimodal vocal-verbal feedback *ja* and *nej* (equivalent to *yes* and *no* in English respectively), unimodal gestural feedback, primarily head nod(s), multimodal feedback *ja* in combination with head nod(s) and up-nod(s). In the Chinese data, the most commonly used feedback is unimodal vocal-verbal expression *dui* and *shi* (equivalent to *yes* in English), followed by the unimodal gestural expression head nod(s), and multimodal expression *dui* and *shi* in combination with head nod(s) to express agreement.

Regarding the similarities between the Swedish and the Chinese samples, this case study suggests that both Swedish and Chinese speakers express agreement more multimodally than unimodally, most often with *yes* in combination with nod(s). The infrequent use of unimodal agreement expressions possibly indicates that simply nodding, for example, is not enough to reinforce agreeing. Also, in both samples the females showed more agreement than males. On the differences between the Swedish and the Chinese data, the Swedish participants expressed more agreement than the Chinese, which can be explained by the value the Swedes put on reaching consensus and avoiding conflict. Also, the gender difference between males and females in expressing agreement has been found bigger in the Chinese sample than the Swedish one. This might be because that Sweden is considered to be a society in which gender roles overlap to a higher extent, which could be reflected in more similarities between male and female swedes in social behaviors.

This case study suggests that the expression of agreement varies between cultures and genders. Both Swedish and Chinese speakers express agreement more multimodally than unimodally. Females express agreement more than males. Swedes express agreement more than Chinese. In spite of the limitations of single communication context and number of participants, this paper contributes to an increased understanding of Swedish and Chinese cultural similarities and differences in expressing agreement. The empirical findings can contribute to cross-cultural multimodal communication research and intercultural practices such as business negotiation and educational collaboration. It can also contribute to big data corpus and automatic analysis of agreement communication, which may help designing technology of autonomous speech and gesture recognition, and developing machine socio-cultural training.

Acknowledgements. The authors would like to thank the participants for their time and engagement.

References

Allwood, J., Cerrato, L., Jokinen, K., Navarretta, C., Paggio, P.: The MUMIN coding scheme for the annotation of feedback, turn management and sequencing phenomena. Lang. Resour. Eval. **41**, 273–287 (2007)

Allwood, J., Traum, D., Jokinen, K.: Cooperation, dialogue and ethics. Int. J. Hum.-Comput. Stud. **53**, 871–914 (2000)

Andonova, E., Taylor, H.: Nodding in dis/agreement: a tale of two cultures. Cogn. Process. **13**, 79–82 (2012)

Azlina Abdul, A.: Agreement strategies among Malaysian Chinese speakers of English. 3L, Language, Linguistics, Literature, 23 (2017)

Cambridge University Press: Cambridge Dictionary Online (2020)

Daun, Å.: En stuga på sjätte våningen: Svensk mentalitet i en mångkulturell värld. B. Östlings bokförlag Symposion, Eslöv, Sweden (2005)

Deng, Y., Chang, L., Yang, M., Huo, M., Zhou, R.: Gender differences in emotional response: Inconsistency between experience and expressivity. PLOS ONE **11**, e0158666 (2016)

Fang, T., Faure, G.O.: Chinese communication characteristics: a Yin Yang perspective. Int. J. Intercult. Relat. **35**, 320–333 (2011)

Gander, A.J.: Understanding in real-time communication: micro-feedback and meaning repair in face-to-face and video-mediated intercultural interactions, Doctoral dissertation, University of Gothenburg (2018)

Hall, E.T.: The Silent Language. Doubleday, Garden City (1959)

Havaleschka, F.: Differences between Danish and Swedish management. Leadersh. Organiz. Dev. J. **23**, 323 (2002)

Hofstede, G.: Cultures and organizations: software of the mind. Organiz. Stud. **15**(3), 457–460 (1994)

Kevin, A.: Culture as context, culture as communication: considerations for humanitarian negotiators. Harv. Negot. Law Rev. **9**, 391–409 (2004)

Khaki, H., Bozkurt, E., Erzin, E.: Agreement and disagreement classification of dyadic interactions using vocal and gestural cues. In: IEEE International Conference on Acoustics, Speech and Signal Processing (ICASSP), Shanghai, pp. 2762–2766 (2016)

Koit, M.: Developing a model of agreement negotiation dialogue. In: Proceedings of the International Joint Conference on Knowledge Discovery, Knowledge Engineering and Knowledge Management (IC3K 2016), pp. 157–162. SCITEPRESS – Science and Technology Publications, Setubal, Portugal (2016)

Lu, J., Allwood, J.: Unimodal and multimodal feedback in Chinese and Swedish monocultural and intercultural interactions (a case study). NEALT Proc. Ser. **15**, 40–47 (2011)

Madapana, N., Gonzalez, G., Zhang, L., Rodgers, R., Wachs, J.: Agreement study using gesture description analysis. IEEE Trans. Hum.-Mach. Syst. **50**(5), 434–443 (2020)

Mehu, M., van der Maaten, L.: Multimodal integration of dynamic audio-visual cues in the communication of agreement and disagreement. J. Nonverbal Behav. **38**(4), 569–597 (2014)

Nivre, J., et al.: Göteborg transcription standard. V. 6.4. Gothenburg: Department of Linguistics, Göteborg University (2004)

Paggio, P., Navarretta, C., Jongejan, B.: Automatic identification of head movements in video-recorded conversations: can words help? In: Proceedings of the Sixth Workshop on Vision and Language, pp. 40–42. Association for Computational Linguistics, Valencia, Spain (2017)

Pedersen, J.: The different Swedish tack: an ethnopragmatic investigation of Swedish thanking and related concepts. J. Pragmat. **42**, 1258–1265 (2010)

Poggi, I., D'Errico, F., Vincze, L.: Agreement and its multimodal communication in debates: a qualitative analysis. Cogn. Comput. **3**, 466–479 (2011)

Tronnier, M., Allwood, J.: Fundamental frequency in feedback words in Swedish. In: The 18th International Congress on Acoustics (ICA 2004), pp. 2239–2242. International Congress on Acoustics, Kyoto, Japan (2004)

Vinciarelli, A., Pantic, M., Heylen, D., Pelachaud, C., Poggi, I., D'Errico, F.: Bridging the gap between social animal and unsocial machine: a survey of social signal processing. IEEE Trans. Affect. Comput. **3**(1), 69–87 (2012)

Weissmann, M., Rappe, E.: Sweden's approach to China's belt and road initiative: still a glass half-empty. The Swedish Institute, Stockholm, Sweden (2017)

Wessel-Tolvig, B.N., Paggio, P.: Attitudinal emotions and head movements in Danish first acquaintance conversations. In: 4th Nordic Symposium on Multimodal Communication, pp. 91–97. Northern European Association for Language and Technology, Gothenburg, Sweden (2013)

The Strategic Advantages of Artificial Intelligence System for Product Design Teams with Diverse Cross-Domain Knowledge

Yen Hsu[1]([✉]) [iD] and Yu-Houng Chaing[2]

[1] The Graduate Institute of Design Science, Tatung University, Taipei, Taiwan
erickshi@ms1.hinet.net
[2] Department of Industrial Design, Tatung University, Taipei, Taiwan

Abstract. New product development is often promoted and managed by enterprises in the form of projects. A new product development project involves a knowledge-intensive process and a series of complex team-working procedures. Therefore, enterprises can establish new product development process or model through practical experience of projects, which can not only serve as the basis for continuous learning and progress of R&D organizations, but also serve as the benchmark for the management of new product development activities. In this study, Construction Ontology-based NPD Process Recommendation Smart System (ONPS) consistent knowledge base architecture. ONPS assist the company, department quickly build and easy to maintain the body of knowledge; at the same time build a graphical user interface for presenting Find knowledge in knowledge, enhance the efficiency of reuse of knowledge. And with three desktop computers as a case study; the original will-depth interviews and expert designers to take advantage of this study ONPS to build ontologies validation framework; and requested the original expert designers use SUS ease of use in the assessment of verification graphical user interface. The results showed that ONPS is feasible, the corporate sector can help quickly build a structure consistent body of knowledge, reasoning ability and possess the knowledge, easy to maintain, but also have a high degree of scalability.

Keywords: Product design team · Artificial Intelligence · Cross-domain diversity knowledge

1 Introduction

1.1 New Product Development Activities

New product development is often promoted and managed by enterprises in the form of projects. A new product development project involves a knowledge-intensive process and a series of complex team-working procedures. Therefore, enterprises can establish new product development process or model through practical experience of projects, which can not only serve as the basis for continuous learning and progress of R&D

© Springer Nature Switzerland AG 2021
P.-L. P. Rau (Ed.): HCII 2021, LNCS 12771, pp. 408–419, 2021.
https://doi.org/10.1007/978-3-030-77074-7_31

organizations, but also serve as the benchmark for the management of new product development activities (Cheng and Chen 2013). Enterprises usually establish or refer to some standard processes to ensure the smooth implementation of projects, including many standard procedures and forms published by the International Organization for Standardization (ISO), such as ISO 9000 Family Quality Management standard, ISO 14001 Environmental Management System and so on. Enterprises can also ensure the record of the R&D process and intellectual property rights. However, these standard operating procedures and forms are very complicated, which often add a lot of workload to inexperienced team members, and because new product development projects are often unique, the same set of standard procedures or practices cannot be fully applied to all projects (Bell 2007; Mathieu et al. 2017). Moreover, if the project completely follows the same set of development standards and does not adjust according to its characteristics, it may cause the increase of labor cost or a waste of time, or even lead to the failure of the project (Park et al. 2006; Kaur and Sengupta 2013).

To avoid the above phenomenon, enterprises or R&D organizations usually adopt flexible tailoring according to the characteristics of new product development projects to flexibly adjust the process, definition and practice of new product development standards, so as to make it a new process that can be referred to or adopted by project members. For example, "Project Process Tailoring" (PPT) is a kind of tailoring (Ginsberg and Quinn 1995). Demirors et al. (2000); Williams and Cockburn (2003) also put forward arguments like "Process Tailoring Patterns" (PTP), but a review of these arguments or practices showed that the process tailoring is insufficient in most cases, because in project practice, consideration should also be given to "rules of expertise and experience in the domain" (Park et al. 2006), "tailor's personal competency experience" (Clarke and O'Connor 2012; Martínez-Ruiz et al. 2012), or the "system tools or approaches" used (Park et al. 2006) to ensure the quality of the tailored process. For this reason, when a new product development project is "series outputs of the same product platform", "extension of existing products", "improvement of existing products" and other common new product development activities (Ulrich and Eppinger 2015), the standard process should be simplified and tailored appropriately. However, if the process is not properly tailored due to lack of knowledge, ability or experience and the standards must be observed, additional costs may be caused by unnecessary work. Conversely, if the tailoring is too strong to reach the specific project process standards, it may lead to the failure to systematically preserve project knowledge and R&D intellectual property (Park and Bae 2013; Samuel et al. 2015).

Based on the above research background and motivation, the main objectives of this study include building product design and development knowledge, applying ontological skills in new product development projects and process tailoring, developing knowledge query system, and guiding project team members to carry out project activities smoothly. Hence, the purposes of this study include:

(1) to collect, sort out and confirm project knowledge of new product development and design activities with the R&D project personnel or project system form as the entry window.
(2) to apply ontological techniques to new product development project experience and knowledge to build an intelligent information system for knowledge accumulation

in new product design and development process, so as to guide the project team members to carry out the project activities smoothly.

(3) to evaluate the system usability of the ontological knowledge system through the project team and related members of the case enterprise.

2 Literature Review

2.1 Ontology and Web Ontology Language (OWL)

In recent years, knowledge management has gained increasingly more attention and has become an important way to manage knowledge resources within various organizations. Through knowledge acquisition, sharing, application and development, organizations enhance their ability to control the intangible resource of knowledge, so as to improve their operational efficiency and competitiveness. Among various topics on knowledge management, ontology-related research has been published continuously. Ontology is "the study of which nouns represent real beings and which nouns represent only a concept" (Hsiao et al. 2016). By clear definition of the relevant concepts, correlations, rules and limitations of domain knowledge and explicit description of the content of a knowledge domain, knowledge in this domain can be reused continuously. Besides, because ontology is established in line with fixed rules and clearly defined terms, knowledge can also be shared with others. To improve and effectively establish domain ontology, researchers in ontological engineering have conducted a lot of research on ontology development process, ontology life cycle and ontology development methodology, and have also provided many ontology development instruments and language for establishing ontology (Lai and Liou 2007; Hsiao et al. 2016).

Ontology is applied to capture knowledge in a specific domain and give a general interpretation to the knowledge in the domain, so as to provide users and computers with a common vocabulary and framework definition, so that users can have a common understanding and cognition of the concept of the knowledge in the domain (McGuinness and Van Harmelen 2004). In recent years, scholars in AI and information technology-related domains have also begun to apply ontology to knowledge representation, that is, by using the basic elements of ontology--concepts and the relations between concepts, as the knowledge model to describe the real world. Targeting this trend, World Wide Web Consortium (W3C) also began to define many ontology-related languages, such as RDF, DAML+OIL and OWL (Compton et al. 2012; Díaz et al. 2016).

Ontology is mainly used to integrate and process a vast amount of knowledge, whose function is to capture knowledge in a specific domain and give a general interpretation to the knowledge in the domain, so as to provide users and computers with a common vocabulary and framework definition, so that users can have a common understanding of the knowledge in the domain (Díaz et al. 2016). Therefore, for knowledge-intensive, experience-oriented and highly reusable R&D projects, ontology is worth exploring and applying. In an R&D organization, project-related knowledge is required to be modeled in order to effectively integrate past project knowledge and make it the experience of the organization (Simperl et al. 2012).

However, instruments are required in product development process to help capture and represent project knowledge so that project implementation and required knowledge acquisition can be presented. Ontology is used in many domains mainly because it can express computer languages (e.g., XML) that explicitly describe semantics and relationships, that is to say, ontology provides project members and computers with a common vocabulary and framework definition, so that project members can have a common understanding of the knowledge of software process. Furthermore, ontology can systematically elucidate software process domain concept, and in this study, conceptual structure is molded to establish the ontology model of the software process, and further analysis and reasoning are completed to feedback the project members so that they can share project knowledge (Simperl et al. 2012).

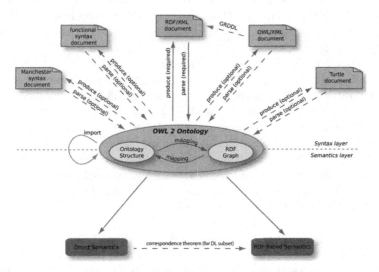

Fig. 1. The structure of OWL 2 (W3C, 2012)

Figure 1 gives an overview of the OWL language, showing its main building blocks and how they relate to each other. Web Ontology Language (OWL) is an example. OWL is a semantic markup language developed by W3C working group. OWL ontology language is designed for applications that process information rather than just present it to humans (Mcguinness and Van Harmelen 2004). OWL makes its content easier to be interpreted by machine than XML, RDF, and RDF Schema (RDF-S) by providing more terms with formal semantics. OWL language is described through a series of document files, each of which has a different purpose (McGuinness and Van Harmelen 2004):

(1) XML provides a surface syntax for structured document files, but it is unable to impose any semantic constraints on the meaning of these documents.
(2) RDF is a data model about objects (or resources) and their relationships, and provides simple semantics for the data model which can be expressed in XML syntax.
(3) RDF Schema is a vocabulary that describes properties and classes of RDF resources, and provides semantics about the hierarchical structure of these properties and classes.

(4) OWL adds more terms for describing properties and classes, such as disjoint-ness between classes, cardinality (e.g. exactly one), equivalence, richer types of properties and property characteristics (such as symmetry).

2.2 Model of Project Knowledge and Knowledge Sharing

Project management background knowledge is divided into "generalized knowledge" and "situational knowledge" (Xu and Ramesh 2008; Lorenz 2014). Generalized knowledge means explicit knowledge that can be clearly expressed, such as organizational regulations or operating procedures and instructions contained in working guidelines (Lukyanenko et al. 2017). Situational knowledge refers to the ways in which knowledge is generated and the situation where it is applied, such as project factors like products, processes and people to various situations, tailoring operation and tailoring strategy results of response measures (Hollauer and Lindemann 2017).

This study probes into project knowledge. During the project development, project personnel mostly use the organization's existing background knowledge and resources, and will generate new knowledge after analysis and use, which can serve as the foundation of the organization's future project knowledge. Users' permission to use the knowledge is based on their role in the enterprise, and each role has a different depth or level of knowledge requirements. Thus, the task assigned to the role may also change the user's permission. Because task, role and conceptual knowledge also influence each other, the knowledge to which they have access must be shared to achieve the goals of the enterprise. There are three modes of knowledge sharing:

(1) Role knowledge sharing: Roles represent the collection of duties in the enterprise as roles are assigned to different tasks. The enterprise achieves its goals via the cooperation or interactions among the roles. Hence, knowledge required by other roles can be shared through different interactions among the roles. For example, the role of the R&D engineer requires knowledge of bicycle R&D as well as knowledge of the product in the mass production stage to design good products that are easy to manufacture.

(2) Task knowledge sharing: When roles perform the tasks assigned, each task has its own knowledge supporting the completion of the task. Due to its interaction with other tasks, knowledge that supports that task can be shared at execution time. For example, priority of tasks. In the design of a bicycle, the frame must be designed first, followed by the transmission system. Hence, when a designer is designing a transmission system, he may need to refer to the relevant knowledge or principles of bicycle frame design.

(3) Conceptual knowledge sharing: The knowledge concept layer constructed by the ontology establishes the relations between knowledge via relations. The cognition of knowledge concepts varies among different people, so there are a lot of cognitive differences that arise in enterprises or virtual enterprises. Such problems can be coped with by establishing relations between conceptual knowledge. For example, in terms of synonyms or subsets, different knowledge concepts correspond to different entity knowledge to share conceptual knowledge.

3 Research Method

3.1 Research Architecture

The research architecture is divided into three stages, namely case knowledge collection and analysis, ONPS intelligent knowledge system establishment, system design and usability evaluation. New product design and development cases were selected, and members of new product development project organizations of enterprises in ICT-related industries announced by the Office of Accounting and Statistics of Executive Yuan were selected as the research objects. The interviewees include project managers and project members in related departments of new product design and development, which are companies that develop, design and manufacture physical end-consumer products. Besides, enterprises with better R&D scale and implementation of product quality assurance system were preferred, as these enterprises are better at utilizing R&D resources under the existing organizational management structure, which can positively help the implementation of product design and development activities, and can obtain more representative research results.

Data on new product design and development project related knowledge was collected, and a semi-structured question table was established via the relevant literature review, and was used as the subject of case study data collection or interview. The research objects were expected to be the team members of the new product design and development project of the case enterprise in the database. In this study, one enterprise product design and development project organizations and 3 product development and design cases were contacted, and subsequently the R&D process, specifications, forms, drawings, documents, interview transcripts, etc., were collected as widely as possible.

The product development case knowledge was analyzed and classified according to "requirement analysis", "functional decomposition", "behaviour analysis", "behaviour evolution", "corresponding structure" and "design change". Interview questions are as follows:

What is the market positioning and demand of the "case product"?
What are the user groups of the "case product"?
What are the user requirements of the "case product"?
What are the relevant design specifications for the "case product"?
What < function > will the "case product" use to satisfy < requirements > ?
Can the < function > be decomposed into different subfunctions?
What is the < behaviour > corresponding to < function > ?
What are the evolution processes of < behaviour > and < structure > ?
What is the < structure > corresponding to < behaviour/function > ?
What design changes have been made between < product x > and < product y > ?

4 Analysis and Discussion

4.1 ONPS (Ontology-Based NPD Process Recommendation Smart System)

In this study, Protégé, open-source software of ontology knowledge base developed by SMI (Stanford Medical Informatics) Center of Stanford University with Java, is used to establish ONPS (Ontology-based NPD Process Recommendation Smart System) Knowledge System. With excellent design and numerous plug-in programs, Protégé has become the most widely used software for establishing and maintaining ontology (Noy et al. 2001). As a result, Protégé has received strong support from academic, government and enterprise users who use Protégé to build knowledge-based solutions in such domains as biomedicine, e-commerce and organization modeling. Protégé's plug-in architecture can be used to build simple and complex ontology-based applications.

Developers can integrate the output of Protégé with rule systems or other problem solutions to build various intelligent systems. Protégé has the following features:

(1) it gains active support from a strong community of users and developers who raise questions, write documentation and contribute plug-ins;
(2) it supports the latest OWL 2 Web Ontology Language and RDF specification from W3C;
(3) it can be extended and provides a plug-and-play environment, making it a flexible foundation for rapid prototyping and application development. The Protégé ontology modeling software is shown in Fig. 2.

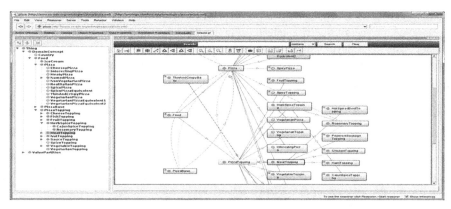

Fig. 2. Protégé ontology modeling software (Stanford University 2016)

4.2 HermiT

HermiT is an ontology reasoning engine and a module extended from Protégé. It is a SWRL-based ontology reasoner written with Web Ontology Language (OWL). Given an OWL file, HermiT can determine if the ontology is consistent, identify and reason out the inclusion relationships between more classes. W3C is fully compatible with OWL 2 direct semantics as standardized reasoning. HermiT is based on a new "HypertaBleau" calculus that supports a wide range of performance standards and novel optimizations to improve reasoning real-world ontology, providing more efficient reasoning than the first publicly available OWL reasoning of any previously known algorithm. It might take minutes or hours to complete classification in the past, while HermiT takes seconds to do it now. In addition to the standard OWL 2 reasoning tasks for implication checking, HermiT supports a variety of specialized reasoning services, such as class and property classification, as well as OWL 2 standards for various external features, like DL security rules, SPARQL queries, and explanatory charts (Glimm 2014).

4.3 ONPS System Architecture Building

Define data categories: Protégé supports user-defined data types, hence declarative knowledge, procedural knowledge, situational knowledge and strategic knowledge are added to the page of data types in this study. Besides, an annotation can be added to each data type to indicate the description of the data type.

Define related properties: After the basic type hierarchy architecture is formed and data types are defined, we will be able to define the relation property and data property of the links between the various types. Data property will be used to define each piece of knowledge, and what, how, why, context, when and where are used in each piece of knowledge. In declarative knowledge, information such as size, material and color of the product is used.

Adopt or not: It means whether instances of actual behaviour evolution are being adopted or not when the data is used to distinguish between key tasks. There will be multiple instances of actual behaviour evolution, but only one will actually be adopted. This instance will correspond to behaviour, so we need to distinguish which instance of behaviour evolution is finally adopted. Hence, we must use "adopt" data to judge. To make a judgment, we only need to know whether it is adopted, thus the data range will be a Boolean value. This will be used for reasoning by SWRL semantic rules. At last, the data range of each data property is defined and restricted, and the data types are classified according to knowledge types:

(1) Requirement: Requirement analysis in key tasks. All analysis instances of requirements are instances concerning function, so the representation of SWRL is as follows:

Rule: Requirement_Analysis(?req) - > Requirement(?req).

(2) Function: Functional decomposition in key tasks. All instances of sub-functions resolved are instances concerning function. Therefore, the representation of SWRL is:

Rule: Functional_Decomposition(?func) - > Function(?func).

(3) Structure: Structural decomposition in the key task. All substructure instances decomposed are instances concerning structure. Therefore, the representation of SWRL is:

Rule: Structural_Decomposition(?struc) - > Structure(?struc).

(4) Behaviour: Analysis of actual behaviour evolution and expected behaviour in key tasks. All the evolved instances with "adopt (true)" are instances concerning behaviour. Therefore, the representation of SWRL is:

Rule: Actual_Behaviour_Evalution(?behav),   Adopt(?behav,   true)   -   >
Behaviour(?behav).

Rule: Expected_Behaviour_Analysis(?behav),   Adopt(?behav,   true)   -   >
Behaviour(?behav).

Taking A1 project of Case Company A as an example, this study sets domain as the input project. Therefore, an A1 class (PCCAS) is established under domain, and A1 is classified according to the part class, component and component individual according to the structure provided by the framework. After all the class and subclass structures are established, the instance is then built for each component under A1. From the Front-Cover component under Cover component class, we can also see the details of how the FrontCover class is used, that is, how it relates to other instances or classes. The other component instances will be built in the same form. Furthermore, reasoner (reasoning engine) is found in the tool column of Protégé in this study. It can be seen that its options are very simple: start reasoner to start reasoning; synchronize reasoner to synchronize the reasoned results to the current ontology knowledge base; stop reasoner and explain inconsistent ontology, and manual logic explanation is required when a problem occurs. Also, you can see in the red arrow that HermiT reasoning engine is selected, as shown in Fig. 3.

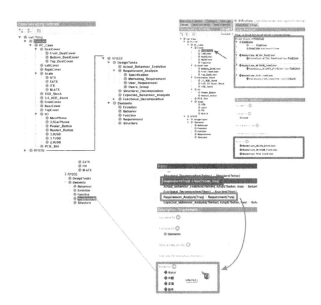

Fig. 3. ONPS knowledge system design and establishment (Source: this study)

Finally, when ONPS knowledge system is established, an.owl ontology file is generated, which will be read later by the user interface and server access. Through the above process, the ontology knowledge base of A1 case product has been successfully established via ONPS framework, and the reasoning engine can be used for reasoning normally. In the process of reasoning, the reasoning engine will also check the logic of the entire ontology for errors. If no error is reported, it means that all the axioms of the ontology knowledge base can operate normally, which also validates the feasibility of ONPS framework.

4.4 System Usage and Interface Usability Evaluation

In this study, 5 product design experts were invited to analyze the user interface of ONPS knowledge system based on SUS. The results showed that the average score of ONPS in SUS was 50.2, which did not meet the standard of 68 points in general interface usability in the literature. In line with the score range of SUS scale, it was close to "acceptable" (OK) grade. However, the experts interviewed generally believe that ONPS can provide substantial help to designers, especially new designers, by enabling them to quickly understand the previous project knowledge in product development and design department, so as to help them quickly fit into the team or into the situation. ONPS also preserves the connotative knowledge of experienced designers. Hence, it facilitates senior designers' searching for the details of past projects. Compared with the known management system of most companies' design projects, it has a better effect for designers (Fig. 4).

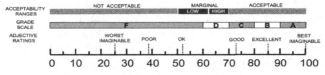

Fig. 4. SUS score intervals (Bangor et al. 2009)

5 Conclusion

This study is conducted via case study, system design and application evaluation, and data is collected for knowledge related to new product design and development projects. A semi-structured question table was established through relevant literature review, which was used as the subject of case study data collection or interview. The research objects were expected to be the team members of the new product design and development project of the case enterprise in the database. A total of 3 product development and design cases were contacted in this study. Product development case knowledge was classified and analyzed according to "requirement analysis", "functional decomposition", "behaviour analysis", "behaviour evolution", "corresponding structure", "design change", etc. Subsequently, ONPS system was established by using OWL language to describe and build the knowledge ontology architecture, and on this basis, SWRL's HermiT module was applied to describe the relationship between instances and apply inference rules.

Further, 5 product design experts were invited to analyze the user interface of ONPS knowledge system based on SUS. The results showed that the average score of ONPS in SUS was close to "acceptable" (OK) grade. However, the experts interviewed generally believe that ONPS is very helpful to designers, especially new designers, because it enables them to quickly understand the previous project knowledge in product development and design department. Moreover, senior project members for new product design and development projects should be able to transform with ease the project situation into situational knowledge, so as to facilitate the transfer and reuse of knowledge. However, less experienced project members need the situational knowledge of previous projects and member transformation retained by the organization to avoid improper project tailoring as much as possible. Relatively, when the organization fails to retain the situational cases completely, i.e. the information related to past projects, and the knowledge acquired by the relevant personnel is incomplete, they can only follow the suggestions of generalized knowledge, and cannot properly refer to the tailoring rules or apply the tailoring suggestions to the problems.

References

Bangor, A., Kortum, P., Miller, J.: Determining what individual SUS scores mean: adding an adjective rating scale. J. Usability Stud. **4**(3), 114–123 (2009)

Bell, S.T.: Deep-level composition variables as predictors of team performance: a meta-analysis. J. Appl. Psychol. **92**(3), 595 (2007)

Cheng, C.C., Chen, J.-S.: Breakthrough innovation: the roles of dynamic innovation capabilities and open innovation activities. J. Bus. Ind. Market. **28**(5), 444–454 (2013)

Clarke, P., O'Connor, R.V.: The situational factors that affect the software development process: towards a comprehensive reference framework. Inf. Softw. Technol. **54**(5), 433–447 (2012)

Compton, M., et al.: The SSN ontology of the W3C semantic sensor network incubator group. Web Semant.: Sci. Serv. Agents World Wide Web **17**, 25–32 (2012)

Demirors, O., Demirors, E., Tarhan, A., Yildiz, A.: Tailoring ISO/IEC 12207 for instructional software development. Paper presented at the Proceedings of the 26th Euromicro Conference. EUROMICRO 2000. Informatics: Inventing the Future (2000)

Díaz, M., Martín, C., Rubio, B.: State-of-the-art, challenges, and open issues in the integration of Internet of Things and cloud computing. J. Netw. Comput. Appl. **67**, 99–117 (2016)

Ginsberg, M., Quinn, L.: Process Tailoring and the Software Capability Maturity Model: Technical Report CMU/SEI-94-TR-024. Software Engineering Institute. Carnegie Mellon University, Pittsburgh, Pennsylvania (1995)

Glimm, B., Horrocks, I., Motik, B., Stoilos, G., Wang, Z.: HermiT: an OWL 2 reasoner. J. Autom. Reason. **53**(3), 245–269 (2014)

Hollauer, C., Lindemann, U.: Design process tailoring: a review and perspective on the literature. Paper Presented at the International Conference on Research into Design (2017)

Hsiao, C., Ruffino, M., Malak, R., Tumer, I.Y., Doolen, T.: Discovering taxonomic structure in design archives with application to risk-mitigating actions in a large engineering organisation. J. Eng. Des. **27**(1–3), 146–169 (2016)

Kaur, R., Sengupta, J.: Software process models and analysis on failure of software development projects. arXiv preprint arXiv:1306.1068 (2013)

Lai, C.-Y., Liou, W.-C.: A service-oriented architecture for constructing ontology-based learning objects repository. Paper Presented at the Multimedia Workshops, 2007. ISMW 2007. Ninth IEEE International Symposium on (2007)

Lorenz, R.D.: The flushing of ligeia: composition variations across Titan's seas in a simple hydrological model. Geophys. Res. Lett. **41**(16), 5764–5770 (2014)

Lukyanenko, R., Parsons, J., Wiersma, Y., Wachinger, G., Huber, B., Meldt, R.: Representing crowd knowledge: guidelines for conceptual modeling of user-generated content. J. Assoc. Inf. Syst. **18**(4), 2 (2017)

Martínez-Ruiz, T., Münch, J., García, F., Piattini, M.: Requirements and constructors for tailoring software processes: a systematic literature review. Softw. Qual. J. **20**(1), 229–260 (2012)

Mathieu, J.E., Hollenbeck, J.R., van Knippenberg, D., Ilgen, D.R.: A century of work teams in the journal of applied psychology. J. Appl. Psychol. **102**(3), 452 (2017)

McGuinness, D.L., Van Harmelen, F.: OWL web ontology language overview. W3C Recommendation **10**(10), 2004 (2004)

Noy, N.F., Sintek, M., Decker, S., Crubézy, M., Fergerson, R.W., Musen, M.A.: Creating semantic web contents with protege-2000. IEEE Intell. Syst. **16**(2), 60–71 (2001)

Park, S.-H., Bae, D.-H.: Tailoring a large-sized software process using process slicing and case-based reasoning technique. IET Softw. **7**(1), 47–55 (2013)

Park, S., Na, H., Park, S., Sugumaran, V.: A semi-automated filtering technique for software process tailoring using neural network. Exp. Syst. Appl. **30**(2), 179–189 (2006)

Samuel, B. M., Watkins, L., Ehle, A., Khatri, V.: Customizing the representation capabilities of process models: understanding the effects of perceived modeling impediments. IEEE Trans. Softw. Eng. **41**(1), 19–39 (2015)

Simperl, E., Bürger, T., Hangl, S., Wörgl, S., Popov, I.: ONTOCOM: a reliable cost estimation method for ontology development projects. Web Semant.: Sci. Serv. Agents World Wide Web **16**, 1–16 (2012)

Ulrich, K., Eppinger, S.: Product Design and Development. McGraw-Hill Education, New York (2015)

Williams, L., Cockburn, A.: Agile software development: it's about feedback and change. IEEE Comput. **36**(6), 39–43 (2003)

Xu, P., Ramesh, B.: Using process tailoring to manage software development challenges. IT Prof. **10**(4), 39–45 (2008)

Memorability of Japanese Mnemonic Passwords

Kosuke Komiya[(⊠)] and Tatsuo Nakajima

Department of Computer Science and Engineering, Waseda University, Tokyo, Japan
{kosukekomiya,tatsuo}@dcl.cs.waseda.ac.jp

Abstract. Password authentication is the most commonly used mechanism for user authentication. However, its vulnerability to different attacks such as dictionary attacks or brute force attack is well known. The users often use password authentication in insecure ways, such as using weak passwords or reusing passwords, which leads to password crackings. Though these problems are apparent, the trade-offs between password strength and password memorability prevent users from using strong passwords. To realize high password strength and memorability, the use of mnemonic passwords is suggested. However, due to its characteristic that the users must use English sentences, this password-generation strategy is not widely used in countries such as Japan, which do not use English as their native language. Therefore, we introduce Japanese mnemonic passwords, which are passwords using password-generation techniques optimized for Japanese users. We conducted a user study to explore the memorability of Japanese mnemonic passwords. We discuss the types of errors made by the participants and how Japanese mnemonic passwords' usability can be enhanced. We also discuss how this strategy can be used in other non-English-speaking countries.

Keywords: Mnemonic password · Japanese

1 Introduction

Password authentication is used as a primary authentication system for various services. However, password authentication systems are constantly being exposed to various security risks, such as password cracking or leaks. One promising approach to realize a secure password authentication system is using a password generator and manager. There are several studies in the literature reporting that the use of a password manager is not very common [1, 2]. Instead, people generally create passwords themselves, using techniques such as combining words and numbers or replacing the alphabets of a word with some other letters or numbers, which makes their password vulnerable to dictionary attacks.

Considering password managers are not commonly used, an alternative password-generation technique that can ensure both password strength and memorability is needed. Using mnemonic passwords is an ideal approach to realize both strength and memorability. Mnemonic passwords are generated by concatenating the initials of each word in a sentence. For example, the user chooses a memorable random sentence, "I live in Tokyo with a dog and two cats", and generates a mnemonic password, "IliTwada2c". A study has shown that mnemonic passwords have the same memorability as that of the

© Springer Nature Switzerland AG 2021
P.-L. P. Rau (Ed.): HCII 2021, LNCS 12771, pp. 420–429, 2021.
https://doi.org/10.1007/978-3-030-77074-7_32

passwords naively selected by the participants and the same strength as that of randomly generated passwords [3]. Thus, mnemonic passwords have a high degree of security owing to their randomness and memorability.

We carried out a preliminary research on university students to check whether mnemonic passwords are commonly used in Japan. The results of the study reveal that mnemonic passwords are not widely used among Japanese university students.

This study proposes "Japanese mnemonic passwords", which are mnemonic passwords optimized for Japanese users. Using our password-generation technique, by remembering a sentence in the Japanese language (6–8 characters long) and two numbers, the user can generate a 14–18-alphabet long password. In Sect. 2 of this paper, we introduce related works. In Sect. 3, we describe the details of the preliminary study we conducted to check whether mnemonic passwords are widely used among Japanese students. Section 4 introduces our approach, explaining how Japanese mnemonic passwords are generated. Sections 5 and 6 explain the user study we conducted, and finally, in Sect. 7, we discuss the challenges faced in the use of Japanese mnemonic passwords, which we discovered during our study, and briefly discuss future works.

2 Related Work

Johannes et al. [4] analyzed the strength of the mnemonic passwords and showed that mnemonic passwords generated from sentences obtained through web crawling have the required similarity with mnemonic passwords created by humans. Further, on comparing 18 password-generation rules, they showed that different password-generation rules have different degrees of password strength and that the strength of mnemonic passwords grows linearly with the increase in password length.

Yan et al. [3] showed that mnemonic passwords' memorability is the same as that of the passwords that were selected naively. However, Aiping et al. [5] pointed out that mnemonic passwords' recall rate is low and analyzed what kind of mistakes are made during the recall process. Another study showed that providing extra instructions for generating mnemonic passwords leads to low usability, due to increased time in password generation and recall [6]. These findings suggest that high-complexity password-generation instructions may lead the mnemonic password strategy to be unacceptable to users.

In a study carried out in Indira Gandhi National Open University, on 202 teachers, staff, and university students, 38% of the participants answered they have used mnemonic passwords before [7]. In contrast, in our preliminary study, conducted on 84 Japanese university students, only 3.6% of the students said they have used mnemonic passwords before. This difference, we assume, can be attributed to that greater familiarity with the English language in India compared with Japan.

3 Preliminary Study

We carried out a preliminary study to examine the prevalence of mnemonic passwords among Japanese students. 84 participants were enrolled in the study from the Department of Computer Science and Engineering, Waseda University. We asked the participants the following three questions through an online form.

Q1. Do you know mnemonic passwords?

Q2. Mnemonic passwords are passwords that are created by concatenating the initials of each word from a sentence. This type of password realizes high password strength and memorability. Do you know such types of passwords?

Q3. Have you ever used mnemonic passwords before?

We asked Q1 and Q2 to see if the term *mnemonic password* was recognized by the Japanese students. Table 1 shows the number of Yeses for each question.

Table 1. The number of yeses for Question 1, 2, and 3.

Question	Number of yeses (84 participants)
Q1	3
Q2	19
Q3	3

It is clear from Table 1 that the term *mnemonic password* is not widely known to Japanese students, whereas 19 students (22.6%) knew the mnemonic password-generation strategy. However, only three students (3.6%) answered that they have used mnemonic passwords before. Therefore, there exists a gap between the actual usage rate and the recognition rate of mnemonic passwords.

4   Our Approach

We assumed that the lack of familiarity with the English language caused the low usage rate of mnemonic passwords in our study. Therefore, we introduce Japanese mnemonic passwords, which entirely use the characteristic of Japanese characters, and thus are optimized for Japanese users. By improving the usages of Japanese mnemonic passwords, we aim to encourage secure password generation among Japanese users. Figure 1 shows the three types of Japanese characters: hiragana, katakana, and kanji.

Fig. 1. Three types of Japanese characters.

Hiragana and katakana are syllabaries, and each letter represents a syllable. For example, the letter "か" in the figure represents the "ka" sound. On the other hand, kanji are ideograms, which represent a meaning. For example, the letter "可" in the figure represents "possible." In the Japanese writing system, we combine these three types of characters to construct a sentence.

Most of the kanji can be read in more than two ways, and each reading is categorized into *on-yomi* and *kun-yomi* (*yomi* means the way of reading). For example, the kanji "雨", which means *rain,* can be read in three ways, "あめ(ame)," "あま(ama)", and "(u)." The readings *ame* and *ama* are categorized as *on-yomi,* and *u* is categorized as *kun-yomi.* Using this characteristic, you can generate a mnemonic password that has a longer length than its original sentence.

Japanese mnemonic passwords are generated in the following order.

1. Think up a Japanese sentence with 6–8 characters. More than half of the characters used in the sentence must be kanji.
2. Chose the order from "*on-yomi* to *kun-yomi*" or "*kun-yomi* to *on-yomi.*"
3. Pick the initial alphabet of each character in the sentence. If the character is kanji, pick each reading from *on-yomi* and *kun-yomi* and order it as described in instruction 2. If the kanji has more than one *on-yomi* or *kun-yomi*, choose either one.
4. Pick two numbers and place one at the beginning and the other at the end.

Figure 2 shows an example.

Order: on-yomi to kun-yomi, numbers: 1 and 3
悲 → on-yomi: hi, kun-yomi: kanashii
し → shi
ん → n
だ → da
貴 → on-yomi: ki, kun-yomi: totoi
婦 → on-yomi: hu, kun-yomi: onna
人 → on-yomi: jin, kun-yomi: hito
Password is 1hksndkthojh3

Fig. 2. Example of the Japanese mnemonic password-generation process.

In Fig. 2, the Japanese sentence "悲しんだ貴婦人" is converted to a Japanese mnemonic password. The letters "悲," "貴," "婦," "人" are kanjis, and letters "し," "ん," "だ" are hiraganas. The initials of the two readings from each kanji and the initial of each hiragana are lined up, and two numbers are added at the beginning and end to form a Japanese mnemonic password "1hksndkthojh3". As you see from the password-generation process, the length of the Japanese mnemonic password is expected to be

$$2(\text{the two numbers}) + 2 * \text{numbers of kanjis} + \text{number of hiraganas(or katakanas)}$$

We included the condition "More than half of the characters used in the sentence must be kanjis" in the instructions so that the length of the generated password would be above a certain level. If the user follows the instructions, the length of the Japanese password would be 12–18 characters. In this example, a password of 13 alphabets was generated from a Japanese sentence comprising seven letters.

5   User Study

We conducted a user study with 13 participants (age $24-27$, 9 males, and 4 females, Japanese native speakers). Each participant was asked to think up a sentence and generate a Japanese mnemonic password on the first day. On the 4th, 7th, and 14th day, the participants were asked to input the sentence thought of and the password reproduced from the sentence through a smartphone app. Each participant moved on to the next day only if he/she could input the correct sentence and the password in three attempts. After the 14th day, we asked participants a few questions through a questionnaire and recorded their responses through an interview.

5.1   Results

The participants in our study generated passwords with an average length of 13.69 characters from a Japanese sentence with an average length of seven characters. Twelve out of 13 participants were able to input the correct sentence and the password until the 14th day, whereas only one participant failed to correctly recall the password in three attempts on the 4th day and thus could not move on to the next day.

Table 2. The number of errors on each day.

Day	Number of errors
4th	8
7th	1
14th	1

Table 2 shows the number of errors made on each day. The most number of errors were made on the 4th day, and only one participant made an error on the 7th and 14th day.

The types of errors made could be categorized into the following:

- *Conversion error*: When a participant makes an error in the conversion process. For example, forgetting to convert a letter or converting using a wrong *on-yomi* to *kun-yomi* order.

- *Wrong word*: When a participant recalls a wrong word in a sentence.
- *Wrong reading*: When a participant reads a kanji wrong.
- *Wrong number*: When a participant puts a wrong number on the start or end of the mnemonic password.

Table 3 shows the number of each error type made throughout the study.

Table 3. Number of each error type.

Error type	Number of errors
Conversion error	7
Wrong word	1
Wrong reading	1
Wrong number	1

Most of the errors were conversion errors. Reading a letter correctly or deciding which reading is *on-yomi* and which reading is *kun-yomi* is sometimes difficult even for the Japanese native speakers because there are more than 2,000 kanjis in the Japanese language, and some of these readings are not common in everyday usage. Furthermore, there was only one instance of the wrong word error throughout the study, suggesting that it was not difficult for the participants to recall the 6–8-character long Japanese sentence.

5.2 Questionnaire and Interview

The following questions were asked in the questionnaire. We also asked for additional comments for each question. However, two of the 13 participants did not answer the interview.

Q1. Do you usually use password managers? (Yes/No/Do not know password managers)
Q2. Do you think your password usage is secure? (1: Strongly disagree – 5: Strongly agree)
Q3. Were the sentence and the generation strategy easy to memorize? (1: Strongly disagree – 5: Strongly agree)
Q4. Would you think it would be more difficult to memorize if the sentence were a little longer (9–12 characters)? (1: Would be more difficult, 2: Would have no difference, 3: Would be easier)
Q5. Would you use this generation strategy in the future? (1: Strongly disagree – 5: Strongly agree)

As can be seen in Fig. 2, only 3 participants (23.1%) use password managers. However, this number does not include the number of participants who answered that they store their passwords in their browser. As per Fig. 3, about half of the participants (46.1%)

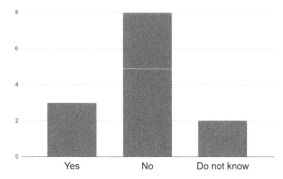

Fig. 3. Do you usually use password managers?

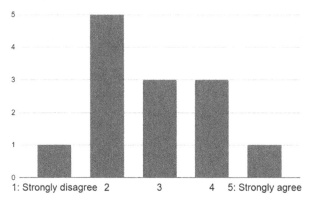

Fig. 4. Do you think your password usage is secure?

answered that they do not think their password is secure. Some participants said that they use password managers or store passwords in the browser, but they are not sure whether storing passwords in such applications is safe. Some participants mentioned the risk of having the master password or the computer cracked (Fig. 4).

As is clear from Fig. 5, 9 participants answered that the Japanese mnemonic password was easy to memorize, whereas 4 answered that it was not. Considering that 12 of the 13 participants could recall the sentence and the password, we assumed that it was easy to memorize the Japanese mnemonic password. However, some participants commented that they are not sure whether they could memorize the sentence if they had to memorize multiple passwords. Therefore, one participant commented that Japanese mnemonic passwords could be used as the master password for a password manager.

We asked Q4 to see if we have the chance of generating longer passwords by using longer sentences. However, more than half of the participants answered that it would be more difficult to memorize a longer sentence (Fig. 6).

Figure 7 shows that more than half of the participants answered that they would not use this password-generation strategy. The two main reasons behind this were: 1) The conversion process was troublesome. 2) There is no need to change the current password usage.

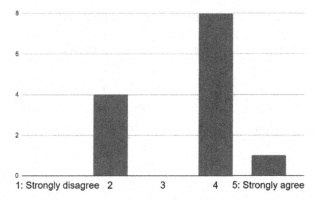

Fig. 5. Were the sentence and the generation strategy easy to memorize?

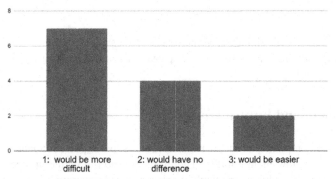

Fig. 6. Would you think it would be more difficult to memorize if the sentence were a little longer?

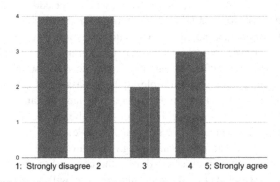

Fig. 7. Would you use this generation strategy in the future?

6 Deviation of the Used Alphabets

Table 4 shows how frequently each alphabet appeared in the passwords of the 13 participants. Alphabets k, m, n, and s appeared more than 10% of times, whereas alphabets e, f, g, o, p, and z rarely appeared. Additionally, alphabets l, q, v, x will not appear in Japanese

Table 4. Frequency of each alphabet used in the passwords

Alphabet	Frequency	Alphabet	Frequency	Alphabet	Frequency
a	4.61%	J	1.97%	s	15.79%
b	1.97%	K	15.13%	t	3.95%
c	1.32%	L	0.00%	u	2.63%
d	4.61%	M	11.18%	v	0.00%
e	0.66%	N	13.16%	w	2.63%
f	0.66%	O	0.66%	x	0.00%
g	0.66%	P	0.66%	y	1.97%
h	7.89%	Q	0.00%	z	0.66%
i	5.26%	R	1.97%		

passwords because these alphabets are never used to represent Japanese syllables. When the deviation of the alphabets used in the passwords is too large, it would be easy to crack the passwords.

7 Discussion and Future Work

Due to the high complexity of the conversion process, we expected Japanese mnemonic passwords to have low memorability. Therefore, we did not include processes such as converting to a special character, which is common when generating English mnemonic passwords. However, the result shows that using a single Japanese mnemonic password would have high memorability. Therefore, we may consider additional conversions to make the password even stronger. However, judging from the responses to questionnaire and the participant comments, Japanese mnemonic passwords do not have high usability due to the complex conversion process. Moreover, it is unclear whether the Japanese mnemonic passwords are memorable when the users need to use multiple passwords. Therefore, an additional study with multiple Japanese passwords is needed.

Using *on-yomi* and *kun-yomi* to gain two password lengths can be used to generate a long password from a short sentence. However, from the number of conversion errors made and the interview comments, we see that Japanese native speakers are not used to distinguishing *on-yomi* and *kun-yomi*, though they are aware of the readings. This resulted in the participants feeling stress during the conversion process. Additionally, directly using the initial alphabets of *on-yomi* and *kun-yomi* will cause a large deviation of the alphabets used in the passwords, which will make the Japanese mnemonic passwords weak.

Japanese mnemonic passwords are passwords optimized for Japanese speakers. However, the generation strategy can be optimized for different language speakers. For example, Chinese pinyin uses 22 initials when representing the pronunciation of a Chinese character. Moreover, a Chinese character includes tones as attributes. By combining these characteristic, for example, converting the character "馬" (pronounced "ma" with the fourth tone) to "m4", it is possible to create a generation strategy optimized for Chinese speakers.

In this study, we introduced Japanese mnemonic passwords that are optimized for Japanese users. The study showed that Japanese mnemonic passwords have high memorability for a single password. However, we could not realize high usability due to the complexity of the conversion process.

References

1. Hoonakker, P., Bornoe, N., Carayon, P.: Password authentication from a human factors perspective: results of a survey among end-user. In: Proceedings of the Human Factors and Ergonomics Society 53rd Annual Meeting (2009)
2. Alkaldi, N., Renaud, K.: Why do people adopt, or reject, smartphone password managers? In: EuroUSEC 2016: The 1st European Workshop on Usable Security (2016)
3. Yan, J., Blackwell, A., Anderson, R., Grant, A.: Password memorability and security: empirical results. IEEE Secur. Priv. **2**(5), 25–31 (2004)
4. Kiesel J., Stein B., Lucks S.: A large-scale analysis of the mnemonic password advice. In: NDSS (2017)
5. Xiong, A., Ge, H., Proctor, W.R.: An empirical study of mnemonic password recall errors, Who are you?! In: Adventures in Authentication Workshop (2020)
6. Yang, W., Li, N., Chowdhury, O., Xiong, A., Protoctor, W.R.: An empirical study of mnemonic sentence-based password generation strategies. In: Proceedings of the 2016 ACM SIGSAC Conference on Computer and Communications, pp. 1216–1229 (2016)
7. Kumar, N.: Password in practice: an usability survey. Glob. Res. Comput. Sci. **2**(5), 107–112 (2011)

The Research of Willingness to Use Japanese Apps and TAM

Tzuhsuan Kuo[✉]

Fuzhou University of International Studies and Trade, Fuzhou, P.R. China

Abstract. Due to the advent of the Internet and the global village era, the rapid advancement of information technology and networks in recent years, as well as the rapid development of the global Internet and the continuous improvement and development of artificial intelligence equipment, we continue to accept "Internet +" and "artificial intelligence" in our lives. + " influence. Especially when we are learning foreign languages, we cannot do without the APP function of mobile phones and the APP of smart devices such as tablets for learning. This kind of learning method has become an unstoppable trend and trend. Especially since last year, because of the relationship between the new coronavirus, the entire learning environment and the teaching environment have had a great impact and changes. What has not happened so far has met us in this epidemic, which has changed the entire learning style and variations of teaching forms. Therefore, online learning and online teaching have become a very important learning and teaching method today.

The main purpose of this article is to explore the current development of Japanese APPs? And the motivation of college students choosing Japanese APPs? Based on the results of the questionnaire survey, The spssau questionnaire analysis is used. Seven scales are used to detect the correlation analysis between learners' perception of ease of use and perceived usefulness of system usage and content design before use.

In addition, it also analyzes the interactivity and willingness to continue using and uninstalling the Japanese APP system after use. The result is that the reliability between the seven variables is high and the system design and content design are perceptively easy to use and useful. It positively affects college students' willingness to continue using Japanese apps. On the contrary, the design and content of the system negatively affect the willingness to use uninstall.

This article analyzes the current research trends of the development of Japanese learning apps, and uses the current research literature to discuss its phenomenon and learners' experience. At the same time, through the results of collating relevant literature, we hope to provide suggestions for future Japanese APP designers, as well as provide learners' needs for Japanese APP applications, and provide suggestions and related issues for perfect improvement and development.

Keywords: M-Learning · Foreign language APP · Learning network APP

© Springer Nature Switzerland AG 2021
P.-L. P. Rau (Ed.): HCII 2021, LNCS 12771, pp. 430–441, 2021.
https://doi.org/10.1007/978-3-030-77074-7_33

1   Introduction

After the "mobile learning" trend began in 2013, the large-scale emergence of mobile education has caused the application of foreign language learning websites and APPs to spring up. Among them, foreign language learning applications account for a large proportion of APPs. The main reason is that learners of foreign language resources have a strong demand and most learners believe that using APP to learn foreign languages is a convenient and free learning application that can improve learning efficiency in a short time. Especially since last year, because of the relationship between the new coronavirus, the entire learning environment and the teaching environment have had a great impact and changes. What has not happened so far has met us in this epidemic, which has changed the entire learning style and teaching. Variations of form. It is also a new era trend that replaces traditional learning methods and teaching methods. In this regard, both students and teachers must properly use the "mobile learning" method in the preparation of action learning and action teaching, take learning out of the classroom out of the classroom, and focus on the application of mobile APP all the time to achieve better learning results.

2   References Related to Foreign Language Action Learning

Learners are based on the characteristics of technological advancement and adapting to changes in the learning environment as the characteristics of the "Technology Acceptance Model" (TAM) (Davis 1989). The above scholars have proposed perceived usefulness and perceived ease of use. Perceived ease of use affects the attitude of using new technologies. These two factors are the most important factors explaining or predicting users' acceptance and use of technology (Davis 1989; Davis et al. 1989; Huang and Huang 2013).

Technology acceptance refers to the recognition, approval, acceptance and continuous using of a certain information technology or system (Zhang and Yan 2012), and perceived usefulness refers to the degree to which users believe that new technology can improve personal knowledge or work efficiency, perceived ease of use refers to the degree to which users think new technologies are easy to use. This model assumes that the actual use of the system is determined by the willingness to use the behavior. The willingness to use is determined by the user's attitude toward using the system and perceived usefulness. The user's attitude toward using the system is determined by the perceived usefulness and perceived ease of use determines together, perceived ease of use affects perceived usefulness, and perceived usefulness mediates the influence of perceived ease of use on intention to use (Yan 2011).

Based on Zhang Wanzhen's (2015) view of adopting the acceptance model of technology, including five dimensions of perceived ease-of-use, perceived usefulness, perceptual learning effectiveness, system characteristics, and continuous using intention, this paper explores how college students use M-learning when learning English vocabulary and adopting acceptance factors of using M-learning. From the previous studies, we have found that the correlation between the effectiveness of perceptual learning and the intention of continuous using is very high. This result shows that whether students are willing to use the new style of class mainly depends on whether they feel that learning is effective, and whether the "perceived learning effectiveness" is The key to whether students are willing to continue to use new technology mobile learning resources.

M-Learning has various characteristics such as convenience, immediacy, suitability, mobility, ubiquity, and portability (Zhang and Yan 2012). In addition to the convenience that is not affected by time and space, learners can use different sensory experiences such as sight, hearing and touch to assist in learning.

Chang et al. (2003) believes that there are three essential elements of M-learning, namely the mobile learning device, the communication infrastructure, and the learning activity model. The main equipment of "online learning" is a desktop computer, so the main equipment of "mobile learning" is PDA (Personal Digital Assistant), smart phone and telephone.

Shepherd (2001) proposed: M-Learning was not only digitized, it also had mobile characteristics, so mobile learning took a step forward and could learn anytime and anywhere without restricted by the desktop computer environment, a portable mobile learning device and the wireless network environment provides opportunities for information at your fingertips. In other words, M-learning allows learners to enter the information network to obtain information when they need it most or at the most appropriate time. It is a flexible learning method and learning environment. Gao (2001) summarized M-learning into six characteristics: 1. The urgency of learning needs 2. The initiative to acquire knowledge 3. The mobility of the learning field 4. The interactivity of the learning process 5. Teaching Contextualization of activities 6. Readiness of teaching content.

3   The Status Quo of Foreign Language Action Learning

Looking back at the last school year of 2019 due to the impact of the epidemic, all regions called for "stopping classes without stopping school". Therefore, online education has become an essential tool for everyone and an important channel for learning. Foreign language learning is nothing more than using APP resources to participate in classroom teaching and learning. The following Table 1: The number of active users of online Japanese learning apps in December 2019 is TOP5 Additionally, in terms of the usage rate of Japanese learning apps, on 2020-08-11 Internet data learned that most people think of Japanese, and most people think of Sakura, Anime, travel and many other words, Japan can be said to be a country very close to me. There are a large number of tourists visiting every year, so it is of course important to know a little basic Japanese, so try to find the ranking of Japanese learning as shown in the table below: Among them, Ryoryo Japanese is recommended because it included listening, speaking, reading, and writing exercises, as well as training for license exams, news NHK's daily listening practice, and a great mobile learning platform. In addition, Baidu Translate is also one of the frequently used APPs. Simple translation sentence patterns and word interpretation and pronunciation can provide learners with a great platform for Japanese translation.

From the above, it can be learned that whether it is Japanese or foreign language learning apps, there are various forms, rich content, detailed classification, and can meet the various needs of users, whether it is vocabulary, grammar, listening, speaking, or

Table 1. Japanese learning APP TOP5 ranking

Ranking	APP name	Feature
1	Japanese dubbing	Japanese Fun Dubbing is a fun dubbing software. App is specially created for Japanese lovers, providing users with fun Japanese dubbing methods, allowing users to easily practice and learn spoken Japanese through the software
2	Ryoryo Japanese	Ryoryo Japanese app is a very well-known Japanese language learning platform in China. This platform can help friends who want to learn Japanese learn Japanese content online. This platform provides various stages of learning from basic Japanese to proficiency in Japanese
3	Today's Japan APP	Today's Japan APP is a learning software tailored for Japanese learning enthusiasts. Today's Japanese software makes Japanese learning easy. Japanese kana syllabary, Japanese grammar, and Japanese test analysis help you learn Japanese easily
4	Concise Japanese	Concise Japanese is a practical language learning software that allows users to learn Japanese from scratch. It includes some videos and lectures, and can teach you writing skills
5	IQiyi Video	IQiyi Video is a client software of iQiyi that focuses on video playback. IQiyi Video included all the movies, TV series, variety shows, animation, music, documentaries and other ultra-clear (720P), 1080P, 4K video content of iQiyi

Update on August 11, 2020
Reference: https://shouyou.kuai8.com/rank/riyxx/

courses, studying abroad, songs, movies You can easily learn foreign languages with just a mobile phone. Using APP to learn is interesting, so Japanese learning apps can inspire users' enthusiasm for learning. Moreover, they also have a variety of functions that can not only supervise user learning, but also attract users' attention of more freedom and ease. Compared with traditional classrooms, users do not have to be afraid of learning Japanese because of making frequent mistakes in front of everyone. Users who do not want to disclose their identities can also choose to communicate anonymously. Online learning provides a variety of teaching modes, and users can choose courses that suit them according to their needs.

In addition, there are more English learning apps in foreign language teaching, with more forms, and richer content than small language learning apps. This phenomenon shows that the developers of foreign language learning apps need to pay more attention to the development and design of small language learning apps. In the form of the foreign language dictionary tool APP, the content of the dictionary has high homogeneity and the similarity of function settings is also high.

4   Model Theory Construction and Hypothesis Establishment

The model theory in this paper is mainly constructed on the basis of two models. (1) In 1995, Goodhue constructed a sub-model from a huge model of TPC, which is the mission technology adaptation model. The theory includes the relationship between technical characteristics, task characteristics, use behavior and use performance. The model pointed out that the task characteristics and technical characteristics will affect the task matching degree, task technical matching and use behavior of the technology used by users, while significantly affecting the use performance. That is to say, if software is used by users and the functions provided by it matches the tasks they want to complete, the higher the performance of the software will be. In addition, Goodhue also conducted research on another model in the same year (1995). The research showed that tasks, technology and personal characteristics all significantly affect the degree of task technology matching. Therefore, combining these two task technology adaptation theories with foreign language apps, the "technology" in the model refers to "Japanese apps" in this article. "Technical features" refer to the functions provided by "Japanese software" to help college students learn Japanese. "Task" mainly refers to "learning". The "task characteristics" mentioned here focus on promoting college students to use Japanese software for learning more. Users refer to college students who use software to help complete tasks. Their user characteristics include gender, grade, and professional experience. These characteristics affect whether users use the software and whether they continue to use it. "Technical matching" refers to the degree of adaptation of the software to help users complete a set of tasks. Usage refers to the behavior of using software to complete tasks, and usage performance refers to the degree to which users achieve their goals after using the software. When software is used, the user can feel the performance of the software, by comparing the performance after use and the expected size before use, it will improve or reduce the user's next use expectations, and will wake up whether the experiencer will continue in the future use.

(2) According to the "Technology Acceptance Model" TRA theory, it can be divided into the following three characteristics: 1. "Behavior intention" is the factor that can control individual behavior most. The individual's feelings about the behavior goal and the feelings of the people who are important to him determine the behavior intention, and both act on the individual to make the behavior corresponding to it. 2. TRA theory is only used on the premise of voluntary behavior. As a general model to explain human behavior, it is widely used and has good predictive effects. 3. The TRA theory is not aimed at a certain type of specific behavior, so there is no corresponding scale developed for a certain type of specific behavior.

The "Technology Acceptance Model" proposed by Davis in this research is a streamlined technology acceptance model and adjusted after deleting the original usage attitude. It mainly emphasizes that external variables lead to users' willingness to use by influencing users' perceived usefulness and perceived ease of use. Perceived usefulness here refers to the subjective feeling that college students think Japanese software is easy. Its external variables include personal characteristics, system design characteristics, organizational structure, task characteristics, and other forms of perception. This paper selects system quality, content quality and interactivity as external variables based on the characteristics of foreign language apps.

According to the task technology adaptation theory, it can be known that the use of Japanese software and its performance depend to a large extent on the matching degree of the user's needs and the functions provided by the software. When the functions of its Japanese software can meet the needs of supporting users' tasks, the software is the main motivation before it is accepted and used. On the contrary, if you are not satisfied with the software after use, the software will be uninstalled or not used continuously. Therefore, system quality, content quality, and interactivity are selected as external variables of the integration task technology adaptation model and technology acceptance model, and the following hypotheses are proposed:

First of all, there are the following six assumptions in terms of "task technology adaptation".

- Hypothesis 1: The system quality of foreign language apps positively affects the perceived ease of use of foreign language apps by college students.
- Hypothesis 2: The system quality of foreign language apps positively affects the perceived usefulness of foreign language apps by college students.
- Hypothesis 3: The content quality of foreign language apps positively affects the perceived ease of use of foreign language apps by college students.
- Hypothesis 4: The content quality of foreign language apps positively affects the perceived usefulness of foreign language apps by college students.
- Hypothesis 5: The interactivity of foreign language apps positively affects the perceived ease of use of foreign language apps by college students.
- Hypothesis 6: The interactivity of foreign language apps positively affects the perceived usefulness of foreign language apps by college students.

Moreover, it is generally believed that if college students can quickly learn how to use a foreign language APP, users may not be able to continue to use this software. However, if a college student thinks that the APP is very laborious to operate, it is likely to cause him to uninstall the software, so the following hypothesis is proposed:

- Hypothesis 7: Japanese students' perceived ease of use of Japanese apps has a positive impact on college students' continued use of the software.
- Hypothesis 8: Japanese students' perceived ease of use of Japanese apps negatively affects college students' intention to uninstall the software.
- Hypothesis 9: Japanese students' perceived ease of use of foreign language apps positively affects their perceived usefulness of the foreign language software.

5 Empirical Analysis of Foreign Language Mobile Learning Applications

5.1 The Current Situation of Japanese Undergraduates Using Foreign Language Apps

In this study mainly refers to the related scale designs that have been used in previous literatures such as Peng (2018), Zhang (2015), Guo (2016), and then summarizes the characteristics of the system quality, content quality, and perceived ease of use. Perceived usefulness, interactivity, continuous use and willingness to uninstall these 7 variables. There are at least 5 questions for each variable. Among them, in order to investigate the motivation and experience of learning in more detail, emphasizing the relevance of perceived ease of use and perceived usefulness to other aspects, relevant analysis was carried out, and at the same time, the classification data was added to the scene. Below are the variables of using Japanese apps, the number of years of using Japanese apps, and the installation of several foreign language apps. The main purpose is to understand the motivations of college students before choosing Japanese apps. At the same time, you can learn your experience and opinions after use. And analyze the continuity and satisfaction degree of the use of Japanese apps by college students. This can be explored and verified from the uninstall software. The above analysis is a breakthrough point different from the previous literature.

In this study, we used the questionnaire method of the questionnaire star to ask the students of the Japanese department to do the questionnaire during the Japanese major class in a few minutes before class. The number of samples distributed was about 90 people, 86 valid questionnaires were all valid questionnaires, and the recovery rate was 95%. In this questionnaire, the distribution of gender, years of Japanese learning, years of grade, and Japanese APP learning structure.

From the gender of the sample, there are relatively more "females" in the sample, with a ratio of 79.07%, and relatively few males. It can be seen that most of the Japanese majors are girls. From the perspective of the number of years of Japanese learning, because the sample is for the lower grades, more than 6 of the sample is the number of years of learning "1 year", and the proportion of learning for 2 years is 30.23%. From the distribution of grades, most of the samples are "sophomore", the proportion is 53.49%. And the proportion of large copies is 46.51%. In the sample, the percentage of "Yes" who has used Japanese apps to learn Japanese is 86.05%, which shows that most students still use it in their studies. However, there are still a small number of students who have never used Japanese apps.

Analysis of the use of Japanese APP by Japanese students. 81.4% of the students who used the APP and have used the Japanese APP in the past six months accounted for 81.4%. It is known that most people have the habit of using APP to study. There are several Japanese APPs installed on the mobile phones, 2 of them accounted for 45.35% (39 people) and 1 of them accounted for 32.56% (28 people). There are 3 for 15 people. A small number of 4 people installed four. The longest use of the Japanese APP is 24 people for less than one month, 21 people for 1–2 months, 12 people for 3–5 months, and 10 people for 6–11 months. There are 19 people in 1–2 years. It can be seen that Japanese students use the APP more in a short time. In addition, are you satisfied with the current Japanese apps? Generally and relatively satisfied, more than 80%. It can be seen that the current Japanese Learning APPS can meet the needs of students.

Table 2 There are relatively many "vocabulary" in the sample, with a proportion of 43.02%. Another 31.40% of the samples are translations. There are relatively few listening and speaking and comprehensive classes. It can be seen that most of the Japanese apps are used in vocabulary, and then in translation.

From the multiple-choice questions is to known that Japanese apps are used when needed (73.26%) and after class (38.37%). In addition, as a self-study application (29.07%), during class (22.09%), Before class (20.93%). This shows that most people use it after class and when needed. Therefore, Japanese apps can play the role of a teacher to help learning after class.

Table 2. Application classification of using Japanese Learning apps

Item	Frequency	Percentage
Vocabulary	37	43.02%
Translations	27	31.40%
listening and speaking	5	5.81%
Comprehensive	5	5.81%
All of the above	12	13.95%
Total	86	100.0%

5.2 Motivation Analysis of Japanese Undergraduates Using Japanese Apps

This research refers to the scale design that has been used in the literature in previous years. The measurement items of the system quality, content quality, interactivity, perceived usefulness, perceived ease of use, willingness to continue, and willingness to uninstall 7 variables, the measurement of each variable There are at least five questions. In order to be more in line with the characteristics of Japanese software that can improve students' Japanese listening, speaking, reading and writing skills, some listening, reading, and writing measurement items are added. The variables required to be measured in the model are as follows:

In order to understand the motivation of Japanese undergraduates for using Japanese APPs, first determine the reliability and stability of this questionnaire. Therefore, the SPSSAU software analysis of the questionnaire star is used to obtain the Cronbach α coefficient to test the reliability of the entire questionnaire content and the reliability of the subscales of each dimension. The result process is as follows:

The reliability coefficient of the questionnaire in Table 3 is 0.990, which is greater than 0.9, which indicates that the reliability of the data in this questionnaire is very high. Regarding the "CITC value", the CITC values of the analysis items are all greater than 0.4, indicating that the analysis items have a good correlation and the reliability level is good. In summary, the reliability coefficient value of the research data is higher than 0.9, which indicates that the reliability of the data is of high quality.

In order to understand the relationship between variables and continuous use intentions, the commonly used Pearson is selected below. Simple correlation coefficient method to analyze the correlation between variables. The following are the hypotheses proposed when the model was constructed:

Table 3. The reliability coefficient of the questionnaire

Variable	Questions	Cronbach α	Cronbach α	Inspection standard
System quality	11	0.969		
Content quality	10	0.986		
Interactivity	3	0.949		
Perceived usefulness	6	0.979	0.990	≥0.9
Perceived ease of use	6	0.969		
On the willingness continue	6	0.981		
On the willingness to uninstall	6	0.959		

First of all, there are the following six assumptions in terms of "task technology adaptation". The test object is hypothesis 1 to hypothesis 6 in chapter three.

Table 4. Correlation between motivation and variables for using Japanese apps

Hypothetical item	Correlation	Hypothetical	
System quality and perceived ease of use	Pearson	0.575**	Valid
System quality and perceived usefulness	Pearson	0.538**	Valid
Content quality and perceived ease of use	Pearson	0.702**	Valid
Content quality and perceived usefulness	Pearson	0.694**	Valid
Interactivity and perceived ease of use	Pearson	0.775**	Valid
Interactivity and perceived usefulness	Pearson	0.746**	Valid

Table 4 Hypothesis 1, the correlation between system quality and perceived ease of use is 0.575** through Pearson's test, indicating a significant correlation. It can be explained that the beautiful design, start-up speed, convenience and fluency of the system use quality, etc., positively affect the operation and learning of the Japanese APP, which is easy to complete or easy to complete. Hypothesis 2 The correlation between system quality and perceived usefulness is 0.538** through Pearson's test, indicating a significant correlation. It can be explained that the start-up speed, convenience and fluency of the system have a positive impact on using Japanese APPs to learn, etc., and they can benefit a lot from the listening, reading and writing of the system APP. Hypothesis 3 Whether the content quality of foreign language apps has a positive impact on the perceived ease of use of foreign language apps by college students. The Pearson test showed 0.702** to indicate a significant correlation. Therefore, knowing the quality of APP content will make learners feel the ease of use. Hypothesis 4 The content quality of foreign language apps positively affects the perceived usefulness of foreign language

apps by college students. The Pearson test showed 0.694** to indicate a significant correlation. Hypothesis 5 and 6 of the correlation between the interactivity and perceived usefulness and perceived ease of use of foreign language apps showed 0.775** and 0.746** to indicate significant correlation.

It is generally believed that if college students can learn how to use a certain Japanese APP quickly, users may not be able to continue using this software. However, if a college student thinks that the APP is very laborious to operate, it is likely to cause him to uninstall the software, so the following hypothesis is proposed:

The test object is hypothesis 7 to hypothesis 9 in chapter three.

Table 5. Analysis of the use of linear regression analysis to detect the perceived ease of use of Japanese APPs

Hypothetical item	Regression coefficient	Hypothetical
The perceived ease of use has a positive impact on continued use of the software	0.576, t = 2.463, P = 0.016	Valid
The perception ease of use negatively affects intention to uninstall the software	1.38, t = 4.656, P = 0.000	Valid
The perceived ease of use has a positive impact on perceived usefulness of software	(0.895, t = 2.932, P = 0.000	Valid

Table 5 is an analysis of the use of linear regression analysis to detect the perceived ease of use of Japanese APPs, the continuous use of software, the uninstall intention of the software, and the perceived usefulness of foreign language software. Generally speaking, Hypothesis 7 is easy to use Japanese apps, and the regression coefficient (0.576, t = 2.463, P = 0.016) means that proficiency in Japanese apps is easy to have a positive effect on the continued use of the software. Assuming 8 is easy to use Japanese APP, the regression coefficient (1.38, t = 4.656, P = 0.000) means that the more proficient in using Japanese APP software, the less easy to uninstall it is a negative effect. Hypothesis 9 is easy to use Japanese APP. The regression coefficient (0.895, t = 2.932, P = 0.000) means that the easier the Japanese APP is to use, the more useful it is to use the system, which is a positive influence.

6 Research Findings and Conclusions

From the above questionnaire, we know the current situation of Japanese-language college students' use of Japanese-language apps. We know that almost all college students who study Japanese have had experience with Japanese-style apps. And the parts of speech and translation are the most frequently used, which shows that the parts of speech and translation are the most urgent needs of students in learning Japanese. Secondly, the longest continuous use of Japanese apps is one month. It can be seen that Japanese students use apps more in a short time. It can be seen that the use of Japanese apps is to meet short-term needs such as school exams. And the ability test is also widely used. In addition, more than 80% of the satisfaction requirements of the current Japanese APP are known and relatively satisfied. It can be seen that the current Japanese APP can meet the needs of students. In addition, there are more cases where Japanese apps are used when needed (73.26%) and after class (38.37%). In addition, they are used as self-study apps (29.07%), during class (22.09%), and before class (20.93%).

The hypothesis proposes that the correlation between system quality and perceived ease of use presents 0.575**, indicating a significant correlation. It can be explained that the beautiful design, start-up speed, convenience and fluency of the system use quality, etc., positively affect the operation and learning of the Japanese APP, which is easy to complete or easy to complete. The correlation between system quality and perceived usefulness is 0.538**, indicating a significant correlation. It can be explained that the start-up speed, convenience and fluency of the system have a positive impact on using Japanese APPs to learn, etc., and they can benefit a lot from the listening, reading and writing of the system APP. Whether the content quality of foreign language apps has a positive effect, the perceived ease of use and detection of foreign language apps by college students showed a significant correlation of 0.702**. Therefore, knowing the quality of APP content will make learners feel the ease of use. The content quality of foreign language apps has a positive impact on the perceived usefulness of foreign language apps of college students, showing 0.694**, indicating a significant correlation.

In addition, in the analysis of whether the various variables affect each other, linear regression analysis is used to detect the Japanese students' perceived ease of use of Japanese apps, the continuous use of the software and the uninstall intention of the software, and the perceived usefulness of foreign language software. It is easy to use Japanese apps. The regression coefficient (0.576, t = 2.463, P = 0.016) means that proficient use of Japanese apps is easy to have a positive effect on the continued use of the software. It is easy to use Japanese APP.

Based on the findings and recommendations of this research, we know that a condition that can satisfy college students' learning Japanese APP is system design and content design. Motivation factors that perceive ease of use are important factors that affect college students' use. Therefore, in system design and content design, attention should be paid to the ease of use of the system, the convenience of functions and the importance of easy operation. The content design must present the usefulness of college students' learning needs. At the same time, the interactivity and correspondence of the Japanese APP platform in the process of use are also necessary design goals. Having the above conditions is a Japanese learning APP website that is easy to use and useful to meet the needs.

References

Wu, C., Zeng, W.: "Mobile learning system combining mobile phone and cloud", a special report from the Information Management Institute of Huwei University of Science and Technology, Taiwan (2008)

Tian, L., Du, X.: Research on mobile multimedia networked foreign language teaching based on APP technology. Curric. Educ. Res. (7), 99–100 (2015)

Lin, H., Xiao, X., et al.: Research and investigation on the use of Japanese learning APP by college learners. Foreign Lang. Teach. Res. 33 (2016)

Ding, S., Wang, X., Liu, J.: A review of foreign language learning APP research. Northeast Asia Foreign Language Forum (2020)

Ni, Y.: Exploration of the flipped classroom mode of higher vocational english based on mobile app. J. Harbin Vocat. Tech. Coll. (5), 64–66 (2016)

Wei, Y.: Investigation and research of educational apps in informal learning of college students. In: Li, Q. (ed.) Guangxi Normal University, Guilin, p. 5 (2016).

Ma, Y., Zhao, L., Li, N., Wang, S.: New mobile learning resources--an exploration of the development model of education APP. I China Electron. Educ. (4), 64–70 (2016)

Wang, Y., Qiu, J., Guo, C., Zhang, X.: Research on current language strategies for mobile learning Japanese for Japanese majors. Mod. Commun. (5), 183–184 (2016)

Guo, Z.: Research on the motivation of learning English action learning materials. Shih Hsin University's Master's Thesis in Graphic Communication and Digital Publishing, Taipei City (2016)

Weng, J.: Research on higher vocational practical english teaching model based on mobile phone-assisted teaching. Vocat. Educ. Forum (5), 66–70 (2017)

Pan, H.: Application classification and application advantages of foreign language education APP in foreign language learning. Inf. Rec. Mater. 18(1) (2017)

Guo, F.: Mobile teaching APP in college classroom teaching application research. Sci. Educ. Wenji (2) (2018). Issue 426.

Chang, F., Gao, F., Wang, M.: Investigation and research on usage status and functional requirements of Japanese language learning phonetic apps by college students. Exam. Wkly (2018)

Wang, H., et al.: The construction of a seamless learning model for basic foreign languages in the era of artificial intelligence. Basic Foreign Lang. Educ. (7), vol. 20, Issue 4 (2018)

Clark, R.C., Mayer, R.E.: E-Learning and the Science of Instruction. Pfeiffer, San Francisco (2003)

Chen, C.M., Chung, C.J.: Personalized mobile English vocabulary learning system based on item response theory and learning memory cycle. Comput. Educ. 51(2), 624–645 (2008)

Karrer, T.: eLearning Technology: What is eLearning 2.0 (2006)

Ahonen, M., Pehkonen, M., Syvanen, A., Turunen, H.: Mobile learning and evaluation. Interim report. Digital Learning 2 project. Working papers. University of Tampere: Hypermedia Laboratory (2004)

Abdous, M., Camarena, M.M., Facer, B.R.: MALL technology: Use of academic podcasting in the foreign language classroom. ReCALL J. 21, 76–95 (2009)

Arnone, P.M.: Motivational Design: The Secret to Produce Effective Children's Media. Oxford, Toronto, US (2005)

Safran, C., Helic, D., Gütl, C.: E-learning practices and web 2.0. In: Proceedings of ICL2007, CD Format, Villach, Austria, 26–28 September (2007)

Foreign language learning websites:

https://dy.163.com/article/F5K7FQJC05198R91.html?referFrom=baidu
https://finance.sina.com.cn/wm/2019-12-14/doc-iihnzhfz5801410.shtml
https://zhinan.qianzhan.com/detail/181221-b5508c3a.html

Cultural Differences Demonstrated by TV Series: A Cross-Cultural Analysis of Multimodal Features

Xiaojun Lai, Nan Qie, and Pei-Luen Patrick Rau[✉]

Department of Industrial Engineering, Tsinghua University, Beijing, China
rpl@mail.tsinghua.edu.cn

Abstract. TV series is one of the most popular entertainment media globally and is a representation of popular culture. It reflects a specific group's daily life culture and certain characteristics of society such as social norms. This paper improved and verified a new cross-cultural analysis method by analyzing facial expressions, original text features and audio features extracted from TV series datasets. We adopted the TV series from America, Japan, and Korea and extracted the textual features from the original text database rather than the translated one. We added the emotional frequency of text and part-of-speech frequency in the text modality. The emotional frequency of facial expressions and text were combined to explore the relation between nonverbal and verbal expressions. In addition, the feasibility of using audio features to further extend the new cross-cultural analysis method were explored. Overall, 1656 features extracted from 90 TV dramas were analyzed, including 42 facial features, 32 text features and 1582 audio features. The statistical results of the feature comparisons revealed the similarities and differences between the three countries and agreed with many existing theories, which resulted in traditional cross-cultural studies. Machine learning models of random forest and support vector machine were used for feature selection and classification to enhance the understanding of important features and conduct country classification.

Keywords: Cross-cultural analysis · Multimodal features · TV series

1 Introduction

Culture is complex and includes knowledge, beliefs, art, ethics, law, customs, abilities, and habits of members of a social group [1]. In a culture, individuals use similar means of interpreting things around them [2]. Cross-cultural research systematically compares the differences between different cultures to explain the occurrence, distribution, and reasons of cultural variation, thus solving complex cultural problems on a global scale [3]. Cross-cultural research has applications in many fields. For example, in management, cross-cultural training of employees can positively impact their self-esteem and confidence, acceptance of other cultures, interpersonal skills, adaptability, and performance [4]. There are four methods commonly used in traditional cross-cultural research:

© Springer Nature Switzerland AG 2021
P.-L. P. Rau (Ed.): HCII 2021, LNCS 12771, pp. 442–462, 2021.
https://doi.org/10.1007/978-3-030-77074-7_34

questionnaire surveys [5], interviews [6], observations [7] and experiments [8]. However, these methods are limited because of different cultural behavior definitions [9, 10] and different measurement techniques, equivalence deviations in concepts and empirical methods, difficulty in obtaining representative samples [11], and translation problems [12–14].

In addition to real-life scenarios, cross-cultural research can also be conducted based on media, especially popular culture products. Williams [15] suggested that culture can be used to refer to "the works and practices of intellectual and especially artistic activity." According to this meaning - culture as signifying practices - Storey [16] considered soap operas, novels, pop music and comics as examples of popular culture. Recently, some studies have started research on popular culture products from a cross-cultural perspective. For instance, Weber [17] analyzed the major decisions described in American and Chinese twentieth-century novels to study the impact of cultural backgrounds of decision-makers on their implicit choices in making decisions. In addition, Hatzithomas [18] analyzed the content of advertising samples in the largest circulation magazines in the UK and Greece, exploring the effect of avoiding uncertainty and individualism/collectivism on the use of various types of humor in print advertising in different cultures. However, the methods used in these studies require considerable labor costs or have the same limitations as traditional research methods.

Therefore, we aim to explore a new cross-cultural research method to alleviate this situation. Considering TV series as an example of popular culture, facial expressions, textual features, and audio features in media data can be automatically extracted with some technical tools. The results can be linked with cross-cultural theories to resolve the shortcomings of traditional cross-cultural research methods. Based on this idea, Xu [19] explored the significance of facial expressions and textual features and obtained positive results. For example, in the textual modality, American dramas used more personal pronouns than Japanese and Korean dramas, which agreed with Hofstede's cultural dimension theory: Americans (an individualistic culture) pursue individual values and interests, while Japanese and Korean (collectivist cultures) emphasize the self in the collective [5]. In addition, the result was consistent with the findings of Kashima et al. [20] that collectivist cultures had more pronoun-drop languages. Therefore, the cross-cultural research method based on multimodal features of TV series is worthy of continued exploration.

However, the existing work only used visual and textual modalities. The textual modalities adopted a translated text database (Chinese subtitles) and extracted the textual features from only two dimensions: emotional polarity and word frequency, which have some limitations. First, the quality of translation is often affected and the context might even be lost because of the difficulty of translating the text [12], the translator's familiarity with the text [13], and the translation process of different languages [14]. For example, in Japanese, the sentence pattern of "I think" (と思います) is often used to express non-absolute opinions. However, when translated into Chinese, they often become absolute opinions. Therefore, the results of the text analysis based on Chinese subtitles may not be completely reliable. Second, in addition to emotional polarity and word frequency, more dimensions can be analyzed in the text modality, such as the emotional frequency of text, syntactic structure, and part-of-speech frequency [21]. Third, emotions can also be perceived from audio features except for facial expressions and text [22]. Trompenaars [23] illustrated that emotional culture has exaggerated emotional expression, more interruption in spoken language expressions, and more fluctuation in tone, whereas neutral culture has introverted and implicit emotional expression, fewer interruptions in the communication process, and a lower and flatter pitch for speakers with higher social status. Therefore, audio features can be added to the multimodal model for cross-cultural analysis.

This study expanded and improved the cross-cultural research method based on multimodal features and compared it with traditional cross-cultural research results to further confirm the validity of the method and provide direction for future cross-cultural research. First, the TV series dataset was adopted from three culturally representative nations: America, Japan, and Korea, the original text database (i.e., English, Japanese, and Korean subtitles) were collected, and the emotional frequency of text and part-of-speech frequency in addition to emotional polarity and word frequency for statistical analysis among countries were extracted. Then, the emotional frequency of facial expressions and text were combined to explore the relation between nonverbal and verbal expressions in different countries. Moreover, audio features were extracted to analyze their differences between different countries. By comparing the multimodal features above, the theories and results of many existing traditional cross-cultural studies were verified and the validity of the improved method was confirmed. Finally, machine learning models - random forest and support vector machine (SVM) – were used for feature selection and classification, strengthening the understanding of important features and implementing classification based on country categories.

2   Literature Review

2.1   Limitations of Traditional Cross-Cultural Research Methods

Traditional cross-cultural research methods include questionnaires, interviews, observations and experiments. The questionnaire could be a paper questionnaire or an online questionnaire, and the latter is widely used. This method is simple to implement and has a low cost, flexible question form, high response rate, and fast collection time, particularly with the support of the Internet [24]. However, the questionnaire method, particularly the online questionnaire, has a large sample bias problem owing to the different attitudes of the individuals responding to the survey emails or errors in email addresses [25]. The reliability and validity of questionnaires are affected because of the translation problems of different languages, and ambiguous items need to be screened and removed. For instance, Spielberger [26] found that Singaporean Chinese understood "I am secretly quite critical of others" as criticizing others privately, and therefore regarded it as venting anger, while Western participants interpreted it as resentment but without comments, considering it to suppress anger. In addition, different cultures have a different scoring propensity, that is, there is a default reaction bias in cross-cultural comparison (the tendency to express consent rather than objection to the problem), which requires further evaluation and correction during the research process. For example, Smith et al. [27] found that the difference between the highest and lowest scores between countries accounted for 20% of the scale when analyzing managers' dependence on the level of work guidance resources in different countries.

In contrast, interview, observation and experimental methods have a higher reliability. The interview method provides a more detailed and in-depth understanding of the research content. However, it requires higher cost and skill of the interviewer, and the data are not uniform, making the analysis process complicated. The observation method is relatively simple and straightforward, and more realistic data can be obtained simultaneously. However, the time of occurrence of events and observation objects are under constraint, and it is also difficult to understand people's thoughts in depth. Despite the controllability, the experimental design and operation of the experimental method are complicated. Tajfel believed that the experimental process was the individual's reaction to the artificially designed situation and ignored the real social situation [28]. Moscovici believed that social knowledge existed in the interpersonal environment [29]; thus, the study of social representation was based on interviews and observations. In addition, the samples are difficult to find, and most of them are students, which vary from country to country and cannot reflect all social classes, and therefore lack representation [11].

2.2 Multimodal Features in Different Cultures

Different cultural groups have different facial expression habits, verbal expressions, and acoustic characteristics, and the following sections discuss these differences in more detail.

Facial Expression. Basic emotions in facial expressions include anger, happiness, sadness, disgust, surprise, and fear. Matsumoto later introduced the seventh basic emotion: contempt [30]. Ekman found that facial expressions are universal and provide sufficient clues to detect emotions [31]. Even when using an imposed-etic approach, all subjects from different countries (such as Brazil, Japan, and America) can recognize the facial expressions of basic emotions. In addition, the recognition accuracy rate is still high even when using facial expression materials from different countries [32].

Different cultures differ greatly in the expression of facial emotions. Koopmann et al. [33] found that Germans endorsed angry and sadness expressions more, while Americans endorsed contempt and disgust expressions more. In addition, the display of emotions of facial expressions is related to self-construal. The context is vital for those with interdependent self-construal, and thus it is necessary to suppress or express different emotions in different contexts. Based on the self-construal measurement of white American women and their records of watching short films of body stumps, Matsumoto et al. [34] found that people with independent self-construal do not avoid expressing negative emotions, while those with interdependent self-construal show fewer negative emotions.

Verbal Expression. Kashima et al. [20] found that people in individualist countries used the personal pronoun *I* more often, whereas collectivist cultures had more pronoun-drop languages. Besides, Agnew et al. [35] believed that the promise of a relationship might lead people to use more plural pronouns, which in turn enhanced their sense of commitment. In addition, Fitzsimmons et al. [36] found that the relationship expressed by *we* was closer than *she* and *I* by asking the participants to read the relationship described by pronouns. Different parts of speech are expressed at different frequencies in different languages. Semin et al. [21] found that in related and dependent-based societies such as India and Turkey, specific verbs were expressed more, whereas in individualist cultures such as the Netherlands, abstract nouns and adjectives were expressed more.

Acoustic Characteristics. Emotions can be recognized by speech, and they have cross-cultural consistency. Bostanov et al. [22] found that people communicate with each other by perceiving the acoustic characteristics of speech and understanding the emotions behind them. In addition, voice emotions can be recognized in a cross-cultural context [37, 38]. Different acoustic characteristics, such as pitch, duration, and intensity, can reflect different dimensions of emotions [39]. Trompenaars [23] believed that the acoustic characteristics of different cultures are different: emotional culture has exaggerated emotional expression, more interruption in spoken language expressions, and more fluctuation in tone, whereas neutral culture has introverted and implicit emotional expression, a more complete sentence of every speaker, and a lower and flatter pitch when the social status of the speaker is higher. Asia belongs to a neutral culture, while America is at an intermediate level.

2.3 Multimodal Analysis Based on Media Data

In the modal analysis, recently, researchers paid more attention to deep learning and multimodal fusion methods, which have been applied in various fields, such as emotional computing, event detection, personnel tracking, and video classification [40]. Among them, the most relevant to this study was sentiment analysis in emotional computing. The following sections introduce the study methods of a single modality, and the related research on multimodal fusion based on media data.

Methods of a Single Modality. First, for facial expressions, the advent of deep learning allows the automatic extraction of features. Xu et al. used a CNN to pre-train large-scale data for object recognition, which in turn was used for mood prediction [41]. Poria et al. [42] developed a convolutional recurrent neural network (RNN), in which CNN and RNN were stacked together for training, to extract visual features. Second, in terms of textual emotion, its recognition mainly uses the word envelope BoW of the big emotion dictionary for word modeling or statistical methods for annotating polar or emotional labels on a large dataset [43]. Third, pitch, intensity, speech rate, and speech quality in acoustic parameters play an important role in emotion recognition and analysis [42]. The OpenSMILE toolkit [44] can extract all the above key audio features. Deep learning has received increasing attention in audio classification research, for example, CNN is used to extract features from audio before classifiers are used for emotion classification [45].

Studies of Multimodal Fusion Based on Media Data. Multimodal fusion collects data from various modalities and combines them for analysis [40], improvs the overall results or the accuracy of the decision, and provides the remaining information compared to the method with a single modality. Different online video sharing platforms provide available datasets for multimodal sentiment analysis, including MOUD datasets (Spanish recorded videos) [46, 47], YouTube datasets (product review video collections on YouTube) [48, 49], and ICT-MMMO dataset (online video collection of English movie reviews on YouTube and ExpoTV) [50]. Morency et al. [48] developed a YouTube dataset to extract text polarity, visual "smiles," and "sight" times, and audio features, such as pauses and tones, and fused these features for emotion classification, which obtained an accuracy of 55.33%. Based on the dataset of Morency, Poria et al. [49] used features extracted by FSDK 1.7, image features extracted by GAVAM, such as facial feature points (FCP), audio features extracted by OpenEAR [51], and textual features extracted by SenticNet [52], and obtained 16.00% higher accuracy using the feature level fusion method compared to Morency's research.

Textual modality plays the most critical role in multimodal sentiment analysis, and its performance is significantly improved when combined with audiovisual features [50]. Based on the same dataset of MOUD, independent studies on multimodal sentiment analysis by Pérez et al. [46] and Poria et al. [47] found that an efficient and intelligent textual analysis engine can outperform visual and audio single-modal classifiers. In the study by Pérez et al. [46], the accuracies of the textual, visual, and audio modalities were 70.94, 67.31, and 64.85%, respectively, whereas in the study of Poria et al. [47], the accuracies of these three modes were 79.77, 76.38, and 74.22%, respectively.

In addition to video platforms, researchers have started using television dramas as research objects. Chen et al. [53] created the original Emotionlines dataset by capturing and grouping the dialogs in each episode of American drama "Friends". Words of multiple speakers were contained in each dialog, and their final emotional labels were annotated through a majority voting scheme, including emotional tags of pleasure, sadness, fear, anger, surprise, disgust, and neutral and non-neutral polar labels. Based on this dataset, Poria et al. [54] performed emotional recognition by adding audio modality and reserving context and obtained a 61.6% accuracy. Therefore, multimodal analysis based on TV dramas is feasible. However, prior studies on multimodal features of media data have not interpreted cultural differences. This study explores the differences between different cultures through the multimodal analysis of facial expressions, textual features, and audio features.

3 Methodology

3.1 TV Series Samples

TV series samples were collected based on the TV series list in Xu's study [19]. Xu [19] found that different types of dramas had no impact on the results of corresponding countries but enriched and expanded the scenarios of TV series. Therefore, this paper no longer analyzed the influence of drama types. Owing to copyright restrictions, only 30 Japanese and 30 Korean dramas' subtitles were collected. 30 American dramas from the TV series list were selected to ensure the consistency of the samples.

3.2 Extracting Textual Features from Subtitles

Natural Language Processing Tools. NLTK was chosen as the processing tool of English subtitles, which builds on Python programs and contains more than 50 corpora and vocabulary resources for classification, tokenization, stemming, parsing, and semantic reasoning [55]. Mecab developed by Kudo [56] was selected for the processing tool of Japanese subtitles. It does not depend on specific language, dictionary, and corpus but uses the CRF model to estimate parameters and performs well. Okt (Open Korean Text), written by Scah Hohyon Ryu in Scala, was adopted for the processing tool of Korean subtitles. It supports four functions: normalization, tokenization, stem extraction, and phrase extraction. With these tools, we completed tokenization of the subtitle data and got prepared for subsequent feature extraction.

SenticNet: Sentiment Dictionary. SenticNet API [52], which supports various languages, was used to extract emotional polarity and emotional tag features. For emotional polarity, both value (positive or negative) and intensity (a floating number range from -1.0 to $+1.0$) are available for a given concept. For moodtags, the sentics associated with the concept can be categorized with two related moodtags from the moodtags set: admiration, anger, disgust, fear, interest, joy, sadness, and surprise.

First, based on concept polarity, we roughly considered that each line's emotional polarity is the sum of the polarity of all the words in the line, therefore its value interval can overflow $[-1, 1]$. Thus, the polarity of each drama subtitle equals the mean of the polarity of all its lines. As an extension of polarity, all lines of each drama subtitle with polarity higher than 1.0 were returned as extremely positive emotional text data, whereas all lines with polarity below -1.0 as extremely negative emotional text data. Hence, the frequencies of extremely positive and negative emotional expressions for each drama were concluded as supplementary features of polarity. Second, the emotional frequency of the subtitles of each drama was expressed as the frequency of the words in the drama carrying the corresponding moodtags. Thus, we obtained the features of emotional polarity and emotional expression frequency for each drama subtitle. Third, the part-of-speech frequency features were extracted from the emotional subtitles (a combination of the two new emotional text datasets returned above). Using natural language processing tools, the frequencies of nouns, adjectives, verbs, and interjections of each drama were obtained.

Word Frequency Table of Native Language. To establish a cross-cultural word frequency table, we divided the 1st person pronoun proportion into the 1st person pronoun singular proportion and the 1st person pronoun plural proportion based on the various categories of Chinese word frequency table in Xu's study [19]. Then, two native speakers of English, Japanese, and Korean, respectively, were invited to search for corresponding words of each category in their cultures. The word frequency table of corresponding cultures was formed for statistical analysis (shown in Table 1). After tokenization using natural language processing tools, the word frequency table was used to perform word frequency statistics for each country.

Table 1. Word frequency table of English, Japanese, and Korean

Type	English	Japanese	Korean
personal pronouns	me, they, he, she, ...	彼, 彼女, あいつ, こいつ, ...	나, 너, 저, 우리, ...
1st ppron singular	I, me, my, mine	僕, ぼく, 私, わたし, ...	나, 내, 저, 자기, ...
1st ppron plural	we, us, our, ours	私達, 私たち, 我々, 僕たち, ...	우리, 저희, 우리들, 우리둘, ...
2nd ppron	you, your, yours	あなた, あんた, おまえ, 君ら, ...	너, 네, 자네, 너희, ...
kinship term	father, mother, dad, ...	父, お父さん, パパ, ぱぱ, ...	할머니, 할아버지, 외할머니, ...
approval	good, great, fantastic, fine, ...	賛成, OK, 了解, いい, ...	맞아, 그래, 좋아, 동의, ...
disapproval	terrible, bad, awful, ...	反対, いや, 嫌, だめ, ...	반대, 반대하다, 싫어, 안돼, ...
modesty	pardon, forgive, ...	うかがう, 窺う, 申し上げる, 参る, ...	소생, 소인, 저, 저희, ...
respect	sincerely, sir, madam, miss, ...	お父様, お母様, ご覧, ご存知, ...	부탁, 축하하다, 축하합니다, ...
plan	organize, arrange, prepare, ...	計画, プラン, 作戦, スケジュール, ...	하려하다, 하려고 한다, ...
rule	regulation, directive, order, ...	規則, ルール, 規範, 遵守, ...	규칙, 원칙, 규율, 법칙, ...
sureness	shall, will, want, exactly, ...	絶対, 必ず, 必須, 決して, ...	확정, 확실, 확실하다, 확정되다, ...
unsureness	maybe, perhaps, almost, ...	おそらく, たぶん, きっと, かも, ...	확실하지 않다, 불확실, 불확정, 글쎄, ...
musculinity	strong, courageous, ...	成功, 富, 勝利, 職位, ...	성공, 승리, 이기다, 이겼다, ...
femininity	kind, gentle, sensitive, ...	優しさ, 親しみ, 愛してる, 気遣う, ...	친절, 밀접, 다정, 친근, ...
self_express	I think, wanna, gonna, ...	思う, 感じる, 信じる, 考え, ...	착각하다, 생각하다, 착각했다, 생각했다
sex	coitus, mating, sexy, ...	セックス, エッチ, えっち, ヤる, ...	성, 섹스, 같이 자다, 잠자리하다, ...

3.3 Extracting Audio Features

The 'emobase2010' reference set in the OpenSMILE toolkit was used to extract audio features [44, 51]. The set represents a current state-of-the-art feature set for affect and paralinguistic recognition and is recommended as a reference for new emotion recognition. The results contained 1582 features. 21 statistical functions, such as mean, standard deviation, skewness, kurtosis, quartile, are applied on 34 LLD and 34 corresponding deltas coefficients and yields 1,428 features. In addition, for four pitch-based LLD and their four deltas coefficient contours, 19 statistical functions are applied and 152 features are obtained. Lastly, the number of pitch onsets (pseudo-syllables) and the total duration of the input (2 features) are appended. With this set, we generated a bash file on the Mac terminal to batch process a large number of audio files and obtained feature tables for each audio file. After sorting out the files according to the drama and country, we finally got the audio feature data set of 90 TV series.

3.4 Feature Selection and Classification

First, t-SNE was used to visualize the features in low-dimension space and judge the separability of data. All high-dimensional features were projected into 2D or 3D space for observation. The data can be divided in high-dimensional space if it was separable in low-dimensional space. Otherwise, it was inseparable in high-dimensional space or could not be projected into low-dimensional space. Second, feature selection was used to screen out the most statistically significant features from the original feature set to enhance the understanding of features. Algorithms that can obtain the feature coefficients or the feature importance attributes can be used as the base learner of the embedded selection, and at the same time, they can train the features to classify the cultures. Hence, we used random forest and SVM to achieve the goal. Since the feature quantity was considerably larger than the sample size of TV series, especially the audio features, over-fitting was easy to occur in the classification process. The random sampling process of random forests can reduce overfitting, while for SVM, penalty terms of L1 regularization can be added into the loss function to prevent overfitting and improve generalization. Besides, L1 rather than L2 can obtain sparse solutions that have less contribution to the model and can be used for feature selection and classification.

4   Results

4.1   Textual Features

Emotional Features. As shown in Table 2, the average polarity of American dramas was significantly higher than that of Japanese and Korean dramas. In addition, the frequency of extreme positive emotions (positive_rate) of American dramas was significantly higher than that of the other two countries, while the frequency of extreme negative emotions (negative_rate) of Korean dramas was significantly the highest among the three countries. Furthermore, with respect to specific emotional frequencies, American dramas expressed significantly the highest frequency of emotions of joy, surprise, admiration, and interest, and Korean dramas expressed significantly the highest frequency of emotions of fear, disgust, sadness, and anger, whereas the frequencies of all emotion types in Japanese dramas were significantly the lowest. All these results indicated that in terms of verbal expression, American dramas tended to express positive emotions, Korean dramas tended to express negative emotions, and Japanese dramas tended to show fewer emotions.

Word Frequency Features. As shown in Table 3, first, the frequency of personal pronouns (ppron) in American subtitles was significantly the highest among the three countries. However, the proportions of each ppron type (first-person pronouns singular: 1st_ppron, first-person pronouns plural:1st_pprons, and second-person pronouns: 2nd_ppron(s)) in all ppron in Korean dramas were the highest. The 1st_ppron of American dramas was significantly higher than that of Japanese dramas, while the 1st_pprons of Japanese dramas was significantly higher than that of American dramas. Second, American subtitles had significantly higher frequencies of expressing approval, disapproval, and opinions (self_express) than Japanese and Korean dramas. Third, the frequencies of expressing modesty and respect in Korean dramas were significantly the highest, whereas the frequency of expressing modesty in Japanese dramas was significantly higher than that in American dramas. Fourth, the frequencies of expressing plan, sureness, unsureness, and sex in American dramas were significantly the highest. Last, the frequency of expressing masculinity in Korean dramas was the highest, whereas American dramas expressed femininity terms most frequently.

Table 2. Statistical results of emotional polarity and specific emotional frequencies

Features	Mean			Stdev			Anova (p)	Wilcoxon rank sum test (p)		
	America	Japan	Korea	America	Japan	Korea		America-Japan	America-Korea	Japan-Korea
polarity	0.9123	0.0483	−0.182	0.1093	0.0253	0.0429	< 0.001	< 0.001	< 0.001	< 0.001
positive_rate	0.3896	0.0477	0.0501	0.0392	0.0112	0.0114	< 0.001	< 0.001	< 0.001	0.8187
negative_rate	0.0252	0.0233	0.1240	0.0035	0.0062	0.0125	< 0.001	0.167	< 0.001	< 0.001
admiration	0.1958	0.0443	0.1072	0.0072	0.0052	0.0097	< 0.001	< 0.001	< 0.001	< 0.001
anger	0.0256	0.0185	0.0654	0.0025	0.0035	0.0059	< 0.001	< 0.001	< 0.001	< 0.001
disgust	0.0538	0.0145	0.1309	0.0045	0.0026	0.0099	< 0.001	< 0.001	< 0.001	< 0.001
fear	0.0192	0.0066	0.0615	0.0024	0.0019	0.0060	< 0.001	< 0.001	< 0.001	< 0.001
interest	0.1404	0.0181	0.0752	0.0061	0.0042	0.0093	< 0.001	< 0.001	< 0.001	< 0.001
joy	0.1395	0.0245	0.0952	0.0059	0.0036	0.0093	< 0.001	< 0.001	< 0.001	< 0.001
sadness	0.0450	0.0221	0.0955	0.0029	0.0036	0.0086	< 0.001	< 0.001	< 0.001	< 0.001
surprise	0.0597	0.0162	0.0371	0.0030	0.0037	0.0034	< 0.001	< 0.001	< 0.001	< 0.001

Part-of-Speech Frequency Features. As shown in Table 4, the frequency of nouns in the emotional subtitles of American dramas was significantly lower than that of Japanese and Korean dramas, whereas the adjectives (adj) and verbs appeared significantly most in American dramas. The frequency of verbs in Japanese dramas was the lowest, whereas the frequency of adjectives in Korean dramas was the lowest. There was no significant difference in the frequency of interjections among the three countries.

Table 3. Statistical results of word frequencies

Features	Mean			Stdev			Anova (p)	Wilcoxon rank sum test (p)		
	America	Japan	Korea	America	Japan	Korea		America-Japan	America-Korea	Japan-Korea
ppron	0.1678	0.0262	0.0440	0.0105	0.0043	0.0049	< 0.001	< 0.001	< 0.001	< 0.001
1st_ppron	0.3736	0.3484	0.4210	0.0294	0.0435	0.0524	< 0.001	0.0242	< 0.001	< 0.001
1st_pprons	0.0764	0.0802	0.1039	0.0213	0.0417	0.0328	0.004	0.0004	0.0575	0.0038
2nd_ppron(s)	0.2944	0.2988	0.3463	0.0182	0.0673	0.0473	< 0.001	0.5692	0.0006	0.0042
kinship term	0.0039	0.0045	0.0127	0.0020	0.0026	0.0071	< 0.001	0.4553	< 0.001	< 0.001
approval	0.0173	0.0061	0.0101	0.0044	0.0017	0.0028	< 0.001	< 0.001	< 0.001	< 0.001
disapproval	0.0158	0.0035	0.0007	0.0018	0.0012	0.0003	< 0.001	< 0.001	< 0.001	< 0.001
modesty	0.0002	0.0026	0.0046	0.0001	0.0011	0.0014	< 0.001	< 0.001	< 0.001	< 0.001
respect	0.0056	0.0057	0.0094	0.0011	0.0019	0.0034	< 0.001	0.6627	< 0.001	< 0.001
plan	0.0026	0.0007	0.0006	0.0008	0.0003	0.0002	< 0.001	< 0.001	< 0.001	0.5895
rule	0.0005	0.0001	0.0004	0.0003	0.0001	0.0002	< 0.001	< 0.001	0.6520	< 0.001
sureness	0.0069	0.0008	0.0013	0.0012	0.0003	0.0003	< 0.001	< 0.001	< 0.001	< 0.001
unsureness	0.0089	0.0011	0.0014	0.0012	0.0004	0.0003	< 0.001	< 0.001	< 0.001	0.0008
masculinity	0.0012	0.0012	0.0027	0.0003	0.0006	0.0010	< 0.001	0.9470	< 0.001	< 0.001
femininity	0.0032	0.0011	0.0022	0.0009	0.0008	0.0006	< 0.001	< 0.001	< 0.001	< 0.001
self_express	0.0122	0.0020	0.0023	0.0016	0.0006	0.0006	< 0.001	< 0.001	< 0.001	0.0748
sex	0.0009	0.0000	0.0002	0.0008	0.0001	0.0003	< 0.001	< 0.001	< 0.001	< 0.001

Visualization of Text Features. Combining the four textual features above, for each drama, a 32-dimensional vector was obtained. Figure 1 shows an obvious clustering

Table 4. Statistical results of part-of-speech frequencies of emotional subtitles

Features	Mean			Stdev			Anova (p)	Wilcoxon rank sum test (p)		
	America	Japan	Korea	America	Japan	Korea		America-Japan	America-Korea	Japan-Korea
noun	0.1589	0.2808	0.4358	0.0136	0.0627	0.0150	< 0.001	< 0.001	< 0.001	< 0.001
adj	0.0465	0.0211	0.0087	0.0053	0.0047	0.0025	< 0.001	< 0.001	< 0.001	< 0.001
verb	0.1873	0.1092	0.1295	0.0072	0.0138	0.0117	< 0.001	< 0.001	< 0.001	< 0.001
interjection	0.0073	0.0088	0.0075	0.0031	0.0038	0.0052	0.348	0.2009	0.6414	0.1260

effect, because the distance between dramas from the same country was small, whereas the distance between dramas from different countries was large in the 2D space, suggesting that the dramas of different countries had apparent differences in the textual features. Figure 1 shows that the text features were separable, which can be used to classify the countries.

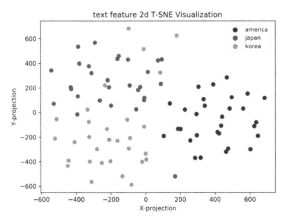

Fig. 1. T-SNE visualization of textual features

4.2   Relation Between Facial and Verbal Expression in Emotion

The results of emotional frequency of facial expressions in Xu's study [19] were adopted and combined with the text emotional frequency features. We denoted $w_{emotion_i}$ – the rate of nonverbal vs. verbal expression of $emotion_i$ by dividing nonverbal expression of $emotion_i$ by verbal expression of $emotion_i$. Table 5 shows that the $w_{emotion_i}$ of Japanese dramas were all significantly higher than those of American and Korean dramas, indicating that Japanese dramas focused more on nonverbal (facial) expressions than American and Korean dramas. Moreover, Table 6 shows that the emotional frequencies of facial expressions in Japanese dramas were significantly higher than those of text when expressing emotions of joy, fear, disgust, sadness, anger, and surprise, whereas American dramas were generally the opposite, except for the expression of joy. In addition, in the Korean TV series, the emotional frequencies of text were significantly higher than those of facial expressions for some emotions (e.g., joy and surprise), and vice versa when expressing other emotions.

Table 5. Statistical results of nonverbal vs. verbal expression of *emotion*ᵢ

Face/Text	Mean			Stdev			Anova (p)	Wilcoxon rank sum test (p)		
	America	Japan	Korea	America	Japan	Korea		America-Japan	America-Korea	Japan-Korea
joy	0.6299	3.2700	0.7489	0.1812	1.2878	0.2606	< 0.001	< 0.001	0.0823	< 0.001
fear	1.8354	5.2029	0.9436	0.9551	2.4920	0.5763	< 0.001	< 0.001	< 0.001	< 0.001
disgust	4.0012	13.6213	1.8973	0.5795	3.9676	0.3545	< 0.001	< 0.001	< 0.001	< 0.001
sadness	1.0620	3.4550	1.0278	0.2823	1.2222	0.3141	< 0.001	< 0.001	0.959	< 0.001
anger	1.4253	2.5302	0.7121	0.4043	1.0660	0.2600	< 0.001	< 0.001	< 0.001	< 0.001
surprise	3.7522	11.7642	7.5511	0.4597	3.9597	1.6342	< 0.001	0.2009	< 0.001	< 0.001

Table 6. Wilcoxon signed-rank test results (Null hypothesis: the median equals to one)

Face/Text	Joy	Fear	Disgust	Sadness	Anger	Surprise
America	< 0.001	< 0.001	< 0.001	0.349	< 0.001	< 0.001
Japan	< 0.001	< 0.001	< 0.001	< 0.001	< 0.001	< 0.001
Korea	< 0.001	0.192	< 0.001	0.586	< 0.001	< 0.001

4.3 Audio Features

A large number of audio features based on statistical functions were extracted by OpenS-MILE; however, no related literature in the cross-cultural research has illustrated their impacts. Therefore, we only visualized the features in low-dimensional space to judge whether the obtained audio features can reflect the differences between different countries. The final feature F0final__Turn_duration was removed because it has a significant correlation with the national category. Then, a feature vector of length 1581 was obtained for each drama. Figure 2 presents the distribution of audio features in 2D space where clustering effects can also be found: the distance between American and Japanese and Korean dramas was considerably large, while Japanese and Korean dramas were partially overlapped. Figure 2 also shows that the audio features are generally separable and can be used for multimodal fusion and classification.

4.4 Multimodal Feature Selection and Classification

There are obvious differences in the features of the three modalities - face expression [19], text, and audio - in different countries, and we sought to fuse these clues to select the important features and conduct classification. Visual modality included 42 facial

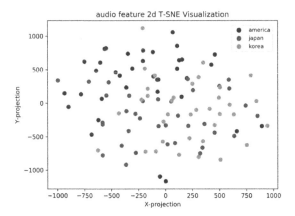

Fig. 2. T-SNE visualization of audio features

expression features [19]. Textual modality included 32 textual features (3 features of emotional polarity, 8 features of emotional frequency of text, 17 features of word frequency related to cultural theories, and 4 features of part-of-speech frequency). Audio modality included 1581 features after removing the F0final__Turn_duration feature.

Random Forest. First, 0.02 was set as the threshold to reduce the possibility of overfitting, and 19 features were screened out. As shown in Fig. 3, text features were the most important in the model, including polarity, frequency of extreme emotions (positive_rate and negative_rate), emotional frequency of text (text_rate of admiration, anger, disgust, interest, joy, and sadness), word frequency (approval, disapproval, rule, and sex), and part-of-speech frequency (noun, adj, and verb). The selection results indicated that America, Japan, and Korea had significant differences in these features, which was consistent with the results of the feature comparisons. The important features of audio modality were pcm_fftMag_mfcc_sma_de[0]_amean and IsqFreq_sma_de[1]_iqr2–3, and the important feature in the facial expression was the drooping frequency of the jaw (facerate_jawDrop). Next, the number of trees (n_estimators) was set to 100, and a classification model with 100% accuracy was obtained.

Fig. 3. Feature selection results of random forest (according to feature_importances_)

SVM. First, the L1 regularization penalty term was used in the SVM model to verify the feature selection and determine the important features that play roles in classification. A total of 19 features with non-zero coefficients were screened out by setting C = 0.5 (five features for American dramas, seven for Japanese, and seven for Korean). If a certain feature in the culture had a positive coefficient, the feature was higher in the corresponding culture, and vice versa. As shown in Fig. 4, the results of the feature coefficients are consistent with the results of the feature comparisons above. For instance, the coefficient of ppron was positive in America, indicating that American dramas used personal pronouns more frequently in verbal expressions. Moreover, the textual features played the most important role in the classification of all three countries: ppron and noun for American dramas, text_rate of disgust and joy for Japanese dramas, and polarity, negative_rate, disapprove, text_rate of disgust, and adj for Korean dramas. The important features of audio modality included features related to IspFreq, F0final, F0finEnv and pcm_loudness, and visual modality had no important features. Second, a 10-fold cross-validation was selected for the SVM model, and a 100% average accuracy of classification was obtained. In addition, LeaveOneOut was selected for cross-validation, which also had a 100% accuracy.

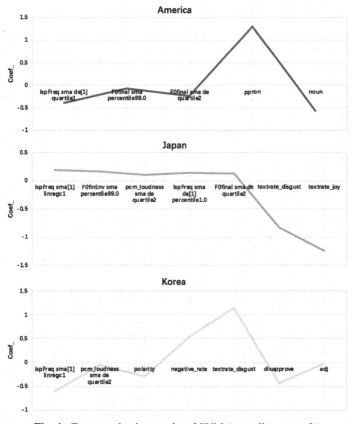

Fig. 4. Feature selection results of SVM (according to coef_)

5 Discussion

5.1 Verbal Expression

Emotion Expression. First, American dramas were more inclined to express more positive emotions, and Japanese dramas tended to show fewer emotions. Some consistent results can be found in traditional cross-cultural studies. Kitayama et al. [57] found that for Americans, the reported frequency of positive emotional states was higher than that of negative emotions, which was not the case for Japanese people in their daily lives. In addition, as a subjective and intrinsic attribute, emotion is also related to the self-construal in different cultures [58]. As a culture with independent self-construal, America is more inclined to highlight and express individual abilities, and thus emphasizes and increases the expression of positive emotions, while reducing and avoiding negative emotions. However, Japan is a culture of interdependent self-construal where the self's external attributes in society and the collective are more important. Therefore, it is more inclined to hide the intrinsic attributes of the self in Japan [59], and the expression of emotions is suppressed. Second, Korean dramas tended to express more negative emotions rather than positive emotions. Researchers have yet to agree on whether the self-construal of Koreans is independent [60] or interdependent [61]. Therefore, Korea can be regarded as a hybrid culture with independent and interdependent self-construal.

Word Frequency Related to Cultural Theories. The word frequency table used in this study was established based on Hofstede's cultural theory [62] and Hall's high-context and low-context theory [63]. First, the frequency of personal pronouns reflects the individualism vs. collectivism dimension. As an individualistic culture, Americans pursue individual independence, and self-idea can be fully expressed. In contrast, Korea and Japan belong to a collectivist culture (Korea is more collectivist than Japan), which emphasizes the relationship and value of individuals in the collective, and often needs to suppress individual ideas. Therefore, this theory suggests that American subtitles have the highest frequency of ppron. In addition, collectivism can also be reflected by plural pronouns [35, 36], which is consistent with the results that Korean dramas had the highest frequencies of 1st_pprons, followed by Japanese dramas, and finally American dramas.

Second, the frequencies of modesty and respect expression reflect the power distance index. America is inclined to a low-power distance culture that emphasizes equality for all, while Japan and Korea belong to a high-power distance culture where the hierarchy of power is clear and people with less power are accustomed to obeying and respecting those with higher power. Thus, words expressing self-humiliation or respect were used more frequently in verbal expressions in Japanese and Korean dramas.

Third, the frequencies of expressing approval, disapproval, and self-idea reflect the culture of high context or low context. Hall [63] proposed that, in the high-context cultures, most of the information is pre-stored in the social background or internalized in a person, and the expression is more indirect and vague; for low-context cultures, the message conveys most of the information and is more direct in expression. America is a low-context culture, while Japan and Korea are typical high-context cultures, which is consistent with the results that American dramas were more inclined to express approval, disapproval, and self-opinions than Japanese and Korean dramas.

Finally, the frequency of words related to sex reflects the degree of openness to sexual culture in the country. American drama subtitles have a significantly higher frequency of sexual words than Japanese and Korean subtitles, indicating that the sexual culture in America is more open than in Japan and Korea [64–66].

However, some controversies exist between these results and existing research results and theories. First, the frequencies of words related to masculinity and femininity reflect masculinity vs. femininity. The masculinity culture is more prominent in Japan, and the femininity culture is more prominent in Korea, while America is neutral. Second, the word frequencies of expressing plan and rule reflect the long-term orientation vs. short-term orientation, while the word frequencies of expressing sureness and unsureness reflect the uncertainty avoidance dimension. America has the lowest scores in these two dimensions, indicating that the corresponding word frequencies in American dramas should also be the lowest. However, opposite results were obtained for the three aforementioned dimensions. A possible reason for this is that these three cultural dimensions are more related to daily life behaviors and thus have a lower degree of matching with the verbal language. Thus, the results of verbal expression do not fully represent the cultural group's daily behavior, especially when the behavior is high-context.

Part-of-Speech Frequency. Semin et al. [21] found that in a culture dominated by relationship and dependence, emotion is a mark of relationship and is embodied more in specific verbs (e.g., A punches B), whereas in a culture where emotions serve self-identity, emotions are more of a self-marker and are expressed more as abstract nouns and adjectives (e.g., aggressive) by comparing the emotional terms, key events in life, and corresponding emotions. Therefore, America is a culture with independent self-construal in which emotions are more marked by individuals; thus, the frequency of abstract language in American subtitles (adjectives and nouns) should be higher. Conversely, Japanese culture tends to be interdependent self-construal, and the frequency of specific verbs should be higher. However, we found that only the frequency of adjectives in American emotional subtitles was significantly higher than that of Japanese emotional subtitles, while the results of the frequencies of nouns and verbs were different from those in previous research. The TV subtitles in this study did not focus on key events and emotions, and thus the nouns (e.g., school) and verbs (e.g., open) were not necessarily related to emotions.

5.2 Relation Between Facial and Verbal Expression in Emotion

According to Hall's theory [63], for low-context cultures, information in communication is mostly conveyed in verbal messages, whereas high-context cultures are more indirect and information is expressed more often through background and non-verbal methods. In this research, America is a low-context culture that requires more contextual information in verbal expressions in daily life, while Japan and Korea are high-context cultures where people understand the things happening at the potential level in advance, including their life, laws, and government; it is scarce for Japanese to correct other people's words or explain [34]. Verbal messages in American dramas contain more emotional information than facial expressions, reflecting the characteristics of a low-context culture. In contrast, Japanese dramas attended more to nonverbal expressions in all emotional types, indicating that most emotional information is conveyed in facial expression rather than verbal language in the high-context culture. However, in Korean subtitles, some emotions were more expressed through words, while some emotions were more expressed through facial expressions. In addition to facial expressions, high-context information can be stored in a social background, body language, etc. Therefore, we cannot conclude that Korean culture is not a high-context culture, however we can assume that its high-context characteristics do not mainly rely on facial expression.

5.3   Acoustic Expression

The results in Fig. 2 indicate that American dramas were quite different from Japan and South Korea in terms of audio, while Japanese and Korean dramas had more similar audio characteristics. This situation agreed with the theory of Trompenaars [23]: Japan and South Korea belong to the neutral culture, while America is at an intermediate level.

5.4   Multimodal Feature Selection and Classification

The results of feature selection and feature classification provided a better understanding of the different cultures. For example, in terms of textual features, a higher frequency of personal pronoun expressions corresponded to American culture, a lower frequency of verbal expressions of emotions of joy and disgust corresponded to Japanese culture, while a higher frequency of verbal expressions of negative emotions such as disgust corresponded to Korean culture. In addition, in terms of acoustic characteristics, combining the results of the two machine models, we found that the line spectrum frequency (lspFreq), mel-frequency cepstral coefficients (mfcc), pitch contour (F0finEnv), fundamental frequency (F0final) and loudness (pcm_loudness) played an important role in the classification of culture. The first three features are often used in the emotion recognition with audio features, and the latter two are mainly related to the basic properties of the audio.

6   Conclusion and Limitations

With the rapid development of the Internet, big data, and machine learning, an increasing amount of video data is uploaded and shared. Based on the multimodal features of the TV series dataset, we have improved and verified a new cross-cultural research method: by extracting the facial expression, original text and audio features in multimedia data of different cultures and conducting comparative analysis to understand the similarities and differences between different cultures. In addition, based on the random forest and SVM model, we further completed feature selection and classification and determined the features that played an important role in each culture to strengthen the understanding of features and identify the cultural category to which the data belongs. Four major conclusions were observed: 1) The textual features extracted from the original text database can reflect the cultural differences of different countries and verify cultural theories, including polarity, specific emotional frequency, word frequency and part-of-speech frequency. For example, American dramas frequently expressed positive emotions, whereas Japanese dramas tended not to express emotions in verbal messages. 2) The comparison between the facial and textual expressions of emotions reflected the differences in nonverbal (facial) and verbal expressions in different countries. For example, American dramas tended to express emotions verbally, while Japanese dramas mainly used non-verbal expressions and conveyed less emotional information in verbal messages. 3) Audio features can reflect cultural differences and can be applied to cross-cultural studies. 4) By combining the three modal features of facial expression, text, and audio, random forest and SVM were used to select important features and conduct classification, and an average accuracy of 100% was obtained. The results of this study can be extended to more areas related to cross-cultural research, providing a reference for the future direction of cross-cultural researchers. Moreover, this study can be applied in public life, such as human resources management of multinational companies and integration in the new cultural environment.

There are a few limitations of this study. First, owing to the small sample size and the large number of features, over-fitting was challenging to avoid. Hence, in future research, the sample size of the dataset must be increased to make the results more convincing and improve the reliability of the multimodal feature classifier. Second, the audio files were mixed with background music and other sound effects, such as motion, making the audio files not entirely suitable for the process of different culture communication, thus decreasing the reliability of the analysis results. In addition, the extracted features of audio are mainly used for emotion recognition in most cases, and they are too bottom to describe the characteristics of audio files. Therefore, unsupervised learning for emotion recognition can be used to obtain audio emotional features of different cultures for analysis.

References

1. Tylor, E.B.: Primitive culture: researches into the development of mythology, philosophy, religion, art, and custom. J. Murray **1**, 1 (1871)
2. Rohner, R.P.: Toward a conception of culture for cross-cultural psycholog. J. Cross-Cult. Psychol. **15**(2), 111–138 (1984)
3. Ilesanmi, O.O.: What is cross-cultural research? Int. J. Psychol. Stud. **1**(2), 82 (2009)
4. Black, J.S., Mendenhall, M.: Cross-cultural training effectiveness: a review and a theoretical framework for future research. Acad. Manag. Rev. **15**(1), 113–136 (1990)
5. Hofstede, G.: Culture's Consequences: International Differences in Work-Related Values. Sage, Beverly Hills (1980)
6. Chang, L.: Socialization and social adjustment of single children in China. Int. J. Psychol. **39**, 390 (2004)
7. Li, H.Z.: Culture and gaze direction in conversation. RASK **20**, 3–26 (2004)
8. Ji, L.J., Nisbett, R.E., Su, Y.J.: Culture, change and prediction. Psychol. Sci. **12**, 450–456 (2001)
9. Sternberg, R.J.: Culture and intelligence. Am. Psychol. **59**, 325–338 (2004)
10. Mundy-Castle, A.C.: Social and technological intelligence in Western and non-Western cultures. Universitas Univ. Ghana Legos **4**, 42–45 (1974)
11. Smith, P.B., Bond, M.H., Kağitçibasi, C.: Understanding Social Psychology Across Cultures. SAGE Publications Ltd. (2006)

12. Treisman, A.: The effects of redundancy and familiarity on translating and repeating back a foreign and a native language. Br. J. Psychol. **56**, 369–379 (1965)
13. Nida, E.: Toward a science of translation. E, J. Brill, Leiden, Netherlands (1964)
14. Miller, G.A., Beebe-Center, J.G.: Some psychological methods for evaluating the quality of translations. Mech. Transl. **3**, 73–80 (1956)
15. Williams, R.: Keywords. Fontana, London (1983)
16. Storey, J.: Cultural Theory and Popular Culture: An Introduction, 8th edn., p. 2. Routledge (2018)
17. Weber, E., Ames, D., Blais, A.-R.: "How do i choose thee? Let me count the ways": a textual analysis of similarities and differences in modes of decision-making in China and the United States. Manag. Organ. Rev. **1**(01), 87–118 (2005)
18. Hatzithomas, L., Zotos, Y., Boutsouki, C.: Humor and cultural values in print advertising: a cross-cultural study. Int. Mark. Rev. **28**(1), 57–80 (2011)
19. Xu, X.: A new cross-cultural research method based on multimodal features. [I], pp. 6–12. Tsinghua University, Beijing (2018)
20. Kashima, Y., Kashima, E.: Culture and language: the case of cultural dimensions and personal pronoun use. J. Cross-Cult. Psychol. **29**, 461–468 (1998)
21. Semin, G.R., Gorts, C.A., Nandram, S., Semin-Goossens, A.: Cultural perspectives on the linguistic representation of emotion and emotion events. Cogn. Emot. **16**, 11–28 (2002)
22. Bostanov, V., Kotchoubey, B.: Recognition of affective prosody: continuous wavelet measures of event-related brain potentials to emotional exclamations. Psychophysiology **41**(2), 259–268 (2004)
23. Trompenaars, F., Hampden-Turner, C.: Riding the waves of culture: understanding diversity in global business. Nicholas Brealey International (2011)
24. Smee, A., Brennan, M., Hoek, J., Macpherson, T.: A test of procedures for collecting survey data using electronic mail. In: Refereed WIP Paper, Australian and NewZealand Marketing Academy (ANZMAC) Conference Proceedings, 30 November– 2 December, pp. 2447–2452. University of Otago, Dunedin, New Zealand (1998)
25. Comley, P.: The Use of the Internet as a Data Collection Method, Media Futures Report. Henley Centre, London (1996)
26. Tanzer, N., Sim, C.Q.E., Spielberger, C.D.: Experience and expression of anger in a Chinese society: the case of Singapore. In: Spielberger, C.D., Sarason, I.G., et al. (eds.) Stress and Emotion: Anxiety, Anger and Curiosity, vol. 16, pp. 51–65. Taylor & Francis, Washington, DC (1996)
27. Smith, P.B., Peterson, M.F., Schwartz, S.H., et al.: Cultural values, sources of guidance and their relevance to managerial behavior: A 47-nation study. J. Cross-Cult. Psychol. **33**, 188–208 (2002)
28. Israel, J., Tajfel, H.: Context of Social Psychology: A Critical Assessment. Academic Press, London (1972)
29. Moscovici, S.: Society and theory in social psychology. In: Israel, J., Tajfel, H. (eds.) The Context of Social Psychology: A Critical Assessment, pp. 17–68. Academic Press, London (1972)
30. Matsumoto, D.: More evidence for the universality of a contempt expression. Motiv. Emot. **16**, 363–368 (1992)
31. Ekman, P.: Universals and cultural differences in facial expressions of emotion. In: Cole, J. (ed.) Nebraska Symposium on Motivation, vol. 19, pp. 207–282. University of Nebraska Press, Lincoln (1972)
32. Ekman, P., Friesen, W.V., O'Sullivan, M., et al.: Universals and cultural differences in the judgment of facial expressions of emotion. J. Pers. Soc. Psychol. **S3**, 712-71 (1987)
33. Koopmann-Holm, B., Matsumoto, D.: Values and display rules for specific Emotions. J. Cross-Cult. Psychol. **42**(3), 355–371 (2011)

34. Matsumoto, D., Kupperbusch, C.: Idiocentric and allocentric differences in emotional expression, experience and the coherence between expression and experience. Asian J. Soc. Psychol. **4**, 113–131 (2001)

35. Agnew, C., Van Lange, P., Rusbult, C., Langston, C., Insko, C.A.: Cognitive interdependence: Commitment and the mental representation of close relationships. J. Pers. Soc. Psychol. **74**, 939–954 (1998)

36. Fitzsimons, G.M., Kay, A.C.: Language and interpersonal cognition: causal effects of variations in pronoun usage on perceptions of closeness. Pers. Soc. Psychol. Bull. **30**(5), 547–557 (2004)

37. Scherer, K.R., Banse, R., Wallbott, H.G.: Emotion inferences from vocal expression correlate across languages and cultures. J. Cross-Cult. Psychol. **32**(1), 76–92 (2001)

38. Thompson, W.F., Balkwill, L.L.: Decoding speech prosody in five languages. Semiotica **158**, 407–424 (2006)

39. Besson, M., Magne, C., Schön, D.: Emotional prosody: sex differences in sensitivity to speech melody. Trends Cogn. Sci. **6**(10), 405–407 (2002)

40. Poriaa, S., Cambriac, E., Bajpaib, R., Hussaina, A.: A review of affective computing: from unimodal analysis to multimodal fusion. Inf. Fusion **37**, 98–125 (2017)

41. Xu, C., Cetintas, S., Lee, K., Li, L.: Visual Sentiment Prediction with Deep Convolutional Neural Networks (2014)

42. Poria, S., Chaturvedi, I., Cambria, E., Hussain, A.: Convolutional MKL based multimodal emotion recognition and sentiment analysis. In: Proceedings of ICDM, Barcelona (2016)

43. Mishne, G., et al.: Experiments with mood classification in blog posts. In: Proceedings of ACM SIGIR 2005 Workshop on Stylistic Analysis of Text for Information Access 19, pp. 321–327. Citeseer (2005)

44. Eyben, F., Wollmer, M., Schuller, B.: Openear—introducing the Munich open-source emotion and affect recognition toolkit. In: 2009 3rd International Conference on Affective Computing and Intelligent Interaction and Workshops, pp. 1–6. IEEE (2009)

45. Anand, N., Verma, P.: Convoluted feelings convolutional and recurrent nets for detecting emotion from audio data. Technical report, Stanford University (2015)

46. Pérez-Rosas, V., Mihalcea, R., Morency, L.: Utterance-level multimodal sentiment analysis. In: ACL, no. 1, pp. 973–982 (2013)

47. Poria, S., Cambria, E., Gelbukh, A.: Deep convolutional neural network textual features and multiple kernel learning for utterance-level multimodal sentiment analysis. In: Proceedings of EMNLP, pp. 2539–2544 (2015)

48. Morency, L., Mihalcea, R., Doshi, P.: Towards multimodal sentiment analysis: harvesting opinions from the web. In: Proceedings of the 13th International Conference on Multimodal Interfaces, pp. 169–176. ACM (2011)

49. Poria, S., Cambria, E., Howard, N., Huang, G.B., Hussain, A.: Fusing audio, visual and textual clues for sentiment analysis from multimodal content. Neurocomputing **174**, 50–59 (2016)

50. Weninger, F., Knaup, T., Schuller, B., Sun, C., Wollmer, M., Sagae, K.: Youtube movie reviews: sentiment analysis in an audio-visual context. Intell. Syst. IEEE **28**(3), 46–53 (2013)

51. Eyben, F, Wollmer, M., Schuller, B.: OpenSMILE-The Munich versatile and fast open-source audio feature extractor. In: Proceedings of ACM Multimedia (MM), Florence, Italy, pp. 1459–1462 (2010)

52. Cambria, E., Poria, S., Hazarika, D., Kwok, K.: SenticNet 5: discovering conceptual primitives for sentiment analysis by means of context embeddings. In: AAAI, pp. 1795–1802 (2018)

53. Chen, S., Hsu, C., Kuo, C., Ting-Hao, T., Huang, H., Ku, L.: Emotionlines: an emotion corpus of multi-party conversations (2018). arXiv preprint arXiv:1802.08379

54. Poria, S., Hazarika, D., Majumder, N., Naik, G., Cambria, E., Mihalcea, R.: MELD: a multimodal multi-party dataset for emotion recognition in conversation. In: ACL (2019)

55. Bird, S., Loper, E., Klein, E.: Natural Language Processing with Python. O'Reilly Media Inc. (2009)
56. Kudo, T.: Mecab: yet another part-of-speech and morphological analyzer (2006). https://mecab.sourceforge.net
57. Kitayama, S., Markus, H.R., Kurokawa, M.: Culture, emotion, and well-being: good feelings in japan and the United States. Cogn. Emot. **14**(1), 93–124 (2000)
58. Markus, H.R., Kitayama, S.: Culture and the self: implications for cognition, emotion, and motivation. Psychol. Rev. **98**(2), 224 (1991)
59. Fiske, A.P., Kitayama, S., Markus, H.R., Nisbett, R.E.: The cultural matrix of social psychology. In: Gilbert, D.T., Fiske, S.T., Lindzey, G. (eds.) The Handbook of Social Psychology, vol. 2, 4th edn., pp. 915–981. McGraw Hill, New York (1998)
60. Park, H.S., Levine, T.R.: The theory of reasoned action and selfconstrual: evidence from three cultures. Commun. Monogr. **66**(3), 199–218 (1999)
61. Singelis, T.M., Sharkey, W.F.: Culture, self-construal, and embarrassability. J. Cross-Cult. Psychol. **26**(6), 622–644 (1995)
62. Hofstede, G.: Culture's Consequences: COMPARING values, Behaviours, Institutions and Organizations Across Nations, 2nd edn. Sage, Thousand Oaks (2001)
63. Hall, E.T.: Beyond Culture. Anchor Books/Doubleday, Garden City (1976)
64. Abramson, P.R., Pinkerton, S.D.: Sexual Nature/Sexual Culture. [S.l.]: University of Chicago Press (1995)
65. Soh, C.S.: The Comfort Women: Sexual Violence and Postcolonial Memory in Korea and Japan. [S.l.]: University of Chicago Press (2008)
66. Robertson, J.: Takarazuka: Sexual Politics and Popular Culture in Modern Japan. [S.l.]: University of California Press (1998)

Trust and Trustworthiness in Northeast Asia

Xin Lei, Minjeong Ko, and Pei-Luen Patrick Rau[⊠]

Department of Industrial Engineering, Tsinghua University, Beijing, China
rpl@mail.tsinghua.edu.cn

Abstract. This study examines the levels of trust and trustworthiness in northeast Asia, including the northeast, southeast, and central regions of China as well as South Korea. Trust and trustworthiness were measured using a simulated two-tier supply chain experiment. The retailer obtains a demand forecast and reports a value to the supplier; then the supplier determines the production based on the retailer's report. The retailer has an incentive to inflate the report to guarantee sufficient production, whereas the supplier is motivated to deflate the retailer's report to avoid overproduction. Two participants took part in the experiment together every time: one played as a retailer, and one played as a supplier. In total, 160 participants joined the experiment: 40 from Liaoning (northeast China), 40 from Zhejiang (southeast China), 40 from Hubei (central China), and 40 from South Korea. The results showed that trust in Hubei was relatively high among the investigated regions and country, whereas no difference was found in the level of trustworthiness. Additionally, the trust level was steady through the entire cooperation stages. In contrast, the trustworthiness level was lower in the early stage than in the late stage in Liaoning and Hubei but not in Zhejiang and South Korea. Besides, trust and trustworthiness in Hubei were higher than in Liaoning and South Korea in the middle cooperation stage, and trust in Hubei was higher than trust in Zhejiang in the late cooperation stage.

Keywords: Trust · Trustworthiness · China · South Korea · Culture

1 Introduction

The effect of culture on trust and trustworthiness is critical in management science. International business has been increasing in recent years, reflecting increasing communication and relationship building between different cultures. It is crucial to understand how trust develops and how cultural background influences trust and trustworthiness. For instance, South Korean consumers favor their domestic brands, whereas Chinese and Japanese consumers are willing to purchase foreign brands. The reason may be that South Koreans have high trust in domestic brands but are suspicious of foreign brands. Hence, it is necessary to understand the cultural influence on trust when entering Korean markets. For another example, eBay fails in the Chinese market, but Taobao succeeds. eBay requires consumers to pay first and then wait for the product to be delivered. In contrast, Taobao comes up with a third-party payment application called Alipay. Consumers pay the money to Alipay; and only after they receive the products, Alipay transfers money to

© Springer Nature Switzerland AG 2021
P.-L. P. Rau (Ed.): HCII 2021, LNCS 12771, pp. 463–474, 2021.
https://doi.org/10.1007/978-3-030-77074-7_35

the seller. To some extent, eBay's failure in China is due to not understanding the trust of Chinese consumers.

Along with the benefits of international business, different cultures create challenging situations for international companies. Additionally, such problems exist in domestic markets where culture significantly varies between different regions within a country. For example, large-scale psychological differences have been discovered between northern Chinese and southern Chinese, suggesting regional differences in the independent-interdependent and analytic-holistic thinking within China [1]. Consequently, we suspect that trust and trustworthiness may also vary between different regions within China. To validate this suspicion, we select three representative regions in China: Liaoning, Hubei, and Zhejiang. The three regions involve northern Chinese, southern Chinese, and those of the Yangtze River basin which is the dividing line between northern and southern China. According to Talhelm et al.'s rice-wheat theory [1], Liaoning, which is located in Northern China, is expected to be more independent; Zhejiang, which is located in southeastern China and lies on the coast, is expected to be more interdependent; Hubei, which lies in the Yangtze river basin, is expected to be in the middle land.

China and South Korea share the traditional culture of Confucianism, which is a significant difference from western cultures. As a result, China and South Korea are usually expected to have a similar culture. While there are similarities between Chinese and Korean cultures, there are significant differences between the two cultures, including language, lifestyle, population, and even how to greet when people first meet. Hence, we assume that differences also exist in the trust and trustworthiness between South Korea and China. Therefore, in addition to the three representative regions in China, this study also investigates trust and trustworthiness in South Korea and aims to explore cultural differences in the level of trust and trustworthiness.

In this study, we employed a simulated supply chain experiment from Özer et al. [2] to measure trust and trustworthiness. In a two-tier supply chain, the retailer informs the supplier of the forecast demand, and then the supplier determines the production based on the retailer's report and sells all the products to the retailer. This experiment has been used in previous studies to investigate the differences in trust and trustworthiness between Chinese and Americans [3, 4] as well as the regional differences within China [5]. This study is aimed at benefiting people who are involved in international business with China and South Korea. The investigated regions occupy a prominent position in the global economy. Understanding trust and trustworthiness in these regions is expected to benefit entering Chinese and South Korean markets.

2   Related Studies and Research Questions

Trust and trustworthiness are fundamental components for dynamic interactions and able to facilitate transactions [6]. Over several decades, trust and trustworthiness have been defined in a variety of disciplines such as economics, sociology, management, psychology, and etc. [7]. Trust is an expectation that the other party will act in consideration of partner's interest, and trustworthiness is the behavior that assists the expectation [8, 9]. Flores and Solomon [10] said: "In an ideal case, because of trustworthiness, you trust someone, and someone's trustworthiness motivate you to trust him/her." Trust and trustworthiness are usually interchanged as if they had the same meaning [11]. The term trustor describes the party that makes the decision, while the term trustee describes the party wishing to be trusted [9]. Trustworthiness of a trustee has a significant impact on the trustor's trust in the trustee [12].

Previous studies have shown that Asian culture places more importance on interdependence, whereas western culture values independence more [13, 14]. According to Kelly and Thibaut [15], self-construal motivates to a transformation of the self and affects the level of cooperation; additionally, individuals with interdependent self-construal have high levels of cooperation because they do not only consider their own interests but also care about others' interests. Hence, we suspect that individuals with interdependent self-construal may behave more trustworthy and more trusting to promote cooperation with others. However, some other studies discover that American trustors have a propensity for trusting others more than Japanese [16]. Trust propensity is a dispositional willingness to rely on others [17]. It is influenced by the trust inherent in a society shaped by its culture [18]. Additionally, trust propensity is likely to be the most relevant trust antecedent in contexts involving unfamiliar actors [19]. Additionally, American trustors tend to be more trusting than Japanese [16, 20] and have higher trust index than Chinese [21]. Americans have more independent self-construal, whereas Chinese and Japanese have more interdependent self-construal. Thus, it can hardly reach a consensus on the relationship between trust and independent-interdependent self-construal.

China and South Korea have been representatives of interdependent individuals in many studies [21–23]. A more recent study, however, reminds us to rethink this idea and finds that China is the middle land between independent and interdependent self-construal [24]. According to the historical records and researches in various disciplines, differences in terms of dialect, customs and psychology exist between northern Chinese and southern Chinese [25–27]. Talhelm et al. [1] have discovered large-scale psychological differences within China: a history of farming wheat makes northern Chinese more independent, whereas farming rice makes southern Chinese more interdependent [1]. These agricultural legacies continue to affect people in the modern world [28]. Accordingly, this study selected three representative regions in China, including Liaoning located in northern China, Zhejiang located in southern China, and Hubei located in the Yangtze River basin which is the dividing line between northern and southern China. In addition, we also collected data of participants from South Korea for comparisons. To be more specific, we used region instead of culture to describe this independent variable. In summary, we propose the first research question in this study.

Research question 1: What are the differences in the level of trust/trustworthiness among these regions: Liaoning, Hubei, Zhejiang, and South Korea?

Reputation plays an essential role in building trust. Good reputation facilitates developing trust because the information conveyed by reputation helps to reduce social uncertainty among the recipients of reputation [29]. A higher reputation of the trustee is likely to stimulate a higher level of trust of the trustor [30]. For example, favorable perception of a firm's reputation leads to its enhanced credibility, which is a determinant of trust [31]. Anderson and Weitz [32] found that trust of a channel member is linked with the operator's reputation. When the operator is reputed to be fair in relationships, their dedication increases [33]. In other words, a poor reputation of the trustee will negatively affect others' trust in him/her. Hence, people are willing to invest in their reputation for higher trust in themselves. Cochard et al. [34] found that, in a seven-period investment game, both the amount sent and the percentage returned increased up to Period 5 and dropped sharply thereafter. This can be explained that people attempt to build their reputation by trusting and being trustworthy towards others for a long-term cooperation. Additionally, there is a high cost for opportunistic actions in the early stage, but the penalty for untrustworthiness is low in the late stage because the relationship is going to end soon [35]. Therefore, we assume that trust and trustworthiness are high in the early stage and then decline in the late stage. Here we propose the following research questions.

Research question 2: What are the differences in the level of trust/trustworthiness across the cooperation stages?
Research question 3: What is the interaction effect of region and cooperation stage on the level of trust/trustworthiness?

3   Method

3.1   Participants

One hundred and sixty participants took part in the experiment: 40 from Liaoning, 40 from Hubei, 40 from Zhejiang, and 40 from South Korea. All participants with an average age of 22 have lived in the target region for at least 10 years. Informed consent was obtained from all participants included in the study. The study was approved by the authors' institutional review board. A pair of participants from the same region, who did not know each other before the experiment, participated the experiment together. Participants did not know the information about their counterparts in the experiment.

3.2   Task

Before starting the experiment, two participants were randomly assigned a role: one participant played as a retailer, and one participant played as a supplier. The retailer purchased products from the supplier and then sold them all to customers. For the supplier, his production cost was 50 per unit, and his selling price was 100 per unit. For the retailer, her purchase cost was 100 per unit, and her selling price was 140 per unit. The supplier lost equally when he produced one more and one less, but the retailer would never lose money. The task was implemented using z-tree and z-leaf software [36].

Each task consisted of five stages, and each stage consisted of five rounds of transactions. Hence, each task had a total of 25 rounds of transactions, and the procedure of a round was as follows. First, the retailer got to know the forecast demand ξ, but the supplier only knew that the demand was a random variable that uniformly distributed from 100 to 400. Second, the retailer informed a demand forecast to the supplier R, which may differ from the actual value ξ. Third, the supplier determined his production P based on the retailer's reported demand. Forth, the actual consumer demand $D = (\xi + \varepsilon)$ came out after both parties made their decisions. Market uncertainty was considered in actual consumer demand as ε. Both the retailer and the supplier knew that ε was uniformly distributed from -75 to 75. Then, the retailer purchased $\min(D, Q)$ units of products from the supplier and sold all of them to customers. Thus, if the supplier produced more than the actual customer demand $(Q > D)$, he had to bear the loss of overproduction himself. Finally, the data of this round were presented to both the retailer and the supplier, including the retailer's report R, the supplier's production Q, the actual customer demand D, the actual sales $\min(D, Q)$, and their own profits. However, the supplier could not see the true demand forecast ξ from beginning to end. When a new round of transaction started, the data of historical transactions were presented to participants.

3.3 Procedure

The experiment was conducted in a controlled laboratory environment that was not affected by the surrounding environment. Two participants participated in the experiment simultaneously. They signed the informed consent form and provided personal information upon they arrived. Then the participants were informed of the experimental instructions and were required to solve several calculate problems associated with the experiment task to ensure that they understood it well. Next, after both participants correctly solved the calculation problem, they were instructed to perform the simulated supply chain task independently, in which the computer played as a retailer and the participant played as a supplier. The computer presented the true demand forecast to the participant, and then the participant decided his/her production. Each participant played 10 rounds with the computer. Subsequently, one participant played as the retailer, and the other played as the supplier to complete the experiment together. In a round, the retailer reported the forecast demand to the supplier, and the supplier determined the production. There were 25 rounds in the experiment. Finally, the participants were instructed to fill in the post-experiment questionnaire. The whole experiment lasted approximately 45 min.

3.4 Measures

The trustworthiness of the retailer was captured as $(R - \xi)$. The retailer had an incentive to inflate the demand forecast in her report to guarantee adequate product supply. A fully trustworthy retailer should provide a true forecast demand without inflation to the supplier. A larger value meant higher forecast inflation, reflecting a lower level of trustworthiness. The trust of the supplier was captured as $\overline{(Q^* - \xi)} - (Q - R)$. The trust of the supplier was related to the production adjustment from the retailer's report, but the supplier's production decision was influenced by two major factors: the trust of the supplier in the retailer and the supplier's risk attitude (i.e., risk-prefer, risk-neutral, and risk-averse). The supplier's risk attitude was captured by the production adjustment in the task with a computer counterpart $\overline{(Q^* - \xi)}$, with Q^* being the production decision. Hence, we used the adjusted production adjustment $\overline{(Q^* - \xi)} - (Q - R)$ to measure the supplier's trust. A larger value referred to higher adjustment in production, reflecting a lower level of trust.

4 Results

4.1 Research Question 1: Trust and Trustworthiness in Chinese Regions and South Korea

Research question 1 concerns the difference in the level of trust/trustworthiness among these regions: Liaoning, Hubei, Zhejiang, and South Korea. Table 1 lists the means and standard deviations of trust and trustworthiness in each region/country. The data did not pass the tests of normality and homogeneity of variance. The region was a between-subject variable; thus, the Kruskal-Wallis test was used to determine whether there were significant differences in the level of trust/trustworthiness among the studied regions. Additionally, the pairwise comparisons were conducted using the Wilcoxon rank-sum tests.

First, the results showed no difference in the trust level among these regions and country, $\chi_3^2 = 5.93, p = .115$. Yet, we found that trust in Hubei was significantly higher than trust in Zhejiang ($p = .032$) and marginally significantly higher than trust in Liaoning ($p = .086$). That's to say, trust in Hubei was relatively high among the studied regions. Second, the results revealed no difference in the trustworthiness level among these regions and country, $\chi_3^2 = 3.07, p = .381$; and the pairwise comparisons showed no difference (Figs. 1, 2).

Table 1. Means and standard deviations of trust and trustworthiness

Region	Trust		Trustworthiness	
	Mean	SD	Mean	SD
Liaoning	24.40	24.54	17.60	18.16
Hubei	11.15	15.41	7.55	19.73
Zhejiang	27.50	30.67	19.65	25.31
South Korea	18.46	19.70	18.84	22.00
Kruskal-Wallis test	$\chi_3^2 = 5.93, p = .115$		$\chi_3^2 = 3.07, p = .381$	

4.2 Research Question 2: Multi-stage Trust and Trustworthiness in Chinese Regions and South Korea

Regarding trust in Liaoning, no significant difference was observed among the five stages, $\chi_4^2 = 2.34, p = .673$. Yet, we found that trust in Stage 3 was lower than trust in Stage 1 ($p = .042$) in Liaoning. The results showed no difference in the trust level across the five cooperation stages in Hubei ($\chi_4^2 = 2.52, p = .642$), Zhejiang ($\chi_4^2 = 4.68, p = .322$), and South Korea ($\chi_4^2 = .68, p = .953$). Overall, the level of trust was steady through the cooperation stages in each region/country.

Regarding trustworthiness in Liaoning, the results revealed a significant difference among the five stages ($\chi_4^2 = 10.48, p = .033$). The post-hoc comparisons showed the

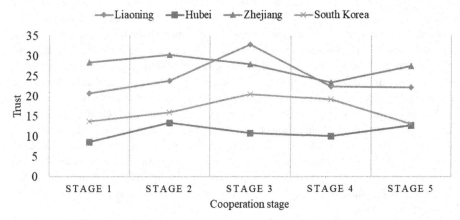

Fig. 1. Trust across five cooperation stages in each region/country

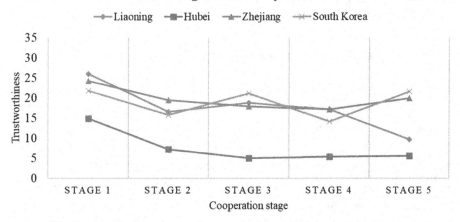

Fig. 2. Trustworthiness across five cooperation stages in each region/country

following results: trustworthiness in Stage 1 was lower than trustworthiness in Stage 4 ($p = .087$) and trustworthiness in Stage 5 ($p = .009$). Regarding trustworthiness in Hubei, the results revealed a marginally significant difference among the five stages ($\chi_4^2 = 7.85$, $p = .097$). The post-hoc comparisons showed the following results: trustworthiness in Stage 1 was lower than trustworthiness in Stage 2 ($p = .070$), Stage 3 ($p = .036$), Stage 4 ($p = .038$), and Stage 5 ($p = .097$). Additionally, no significant difference was discovered in trustworthiness level among the five stages in Zhejiang ($\chi_4^2 = 1.58$, $p = .813$) and South Korea ($\chi_4^2 = 7.00$, $p = .134$).

4.3   Research Question 3: Interaction Between Region and Cooperation Stage

Research question 3 concerns the interaction effect of region and cooperation stage on the level of trust/trustworthiness. Table 2 lists the means and standard deviations of the trust level in every stage in each region/country, and Table 3 lists the means and standard deviations of the trustworthiness level. The data did not pass the tests of normality and homogeneity of variance. The Kruskal-Wallis test was used to determine whether there were regional differences in the level of trust/trustworthiness in each stage. Then, the pairwise comparisons were conducted using the Wilcoxon rank-sum tests.

Regarding the trust level, the results showed no regional difference in each stage. Yet, the pairwise comparisons showed the following differences. In Stage 1, trust in Hubei was marginally significantly higher than trust in Zhejiang ($p = .076$). In Stage 3, trust in Hubei was significantly higher than trust in Liaoning ($p = .033$) and trust in South Korea ($p = .050$); trust in Hubei was marginally significantly higher than trust ($p = .052$) in Zhejiang. In Stage 4, trust in Hubei was significantly higher than trust in Zhejiang ($p = .033$).

Regarding the trustworthiness level, the results revealed no reginal difference in each stage. Yet, the pairwise comparisons showed the following differences. In stage 3, trustworthiness in Hubei was significantly higher than trustworthiness in South Korea ($p = .015$) and marginally significantly higher than trust in Liaoning ($p = .055$). In stage 4, trustworthiness in Hubei was marginally significantly higher than trust in Liaoning ($p = .079$) and trust in Zhejiang ($p = .083$).

Table 2. Means and standard deviations of multi-stage trust

Region	Stage 1		Stage 2		Stage 3		Stage 4		Stage 5	
	Mean	SD	Mean	SD	Mean	SD	Mean	SD	Mean	SD
Liaoning	20.67	35.68	23.87	37.51	32.82	48.84	22.44	35.51	22.19	45.43
Hubei	8.63	29.42	13.38	30.10	10.87	26.95	10.11	25.80	12.74	29.03
Zhejiang	28.37	50.35	30.28	45.89	27.90	38.21	23.42	34.90	27.53	34.20
South Korea	13.68	23.53	15.91	41.52	20.47	33.80	19.22	39.71	13.03	38.37
Kruskal-Wallis test	$\chi_3^2 = 4.64, p = .200$		$\chi_3^2 = 2.37, p = .499$		$\chi_3^2 = 6.85, p = .077$		$\chi_3^2 = 5.24, p = .155$		$\chi_3^2 = 2.30, p = .513$	

Table 3. Means and standard deviations of multi-stage trustworthiness

Region	Stage 1		Stage 2		Stage 3		Stage 4		Stage 5	
	Mean	SD	Mean	SD	Mean	SD	Mean	SD	Mean	SD
Liaoning	25.95	41.00	16.54	28.94	18.75	31.92	17.10	29.33	9.65	21.75
Hubei	14.81	31.38	7.21	25.02	4.94	24.12	5.29	26.13	5.49	26.53
Zhejiang	24.14	35.66	19.44	32.52	17.83	29.37	17.04	29.05	19.81	32.17
South Korea	21.71	28.04	15.79	33.81	21.09	29.49	14.05	33.40	21.55	33.33
Kruskal-Wallis test	$\chi_3^2 = 2.78, p$ $= .426$		$\chi_3^2 = 2.17, p$ $= .538$		$\chi_3^2 = 5.75, p$ $= .124$		$\chi_3^2 = 4.24, p$ $= .237$		$\chi_3^2 = 4.60, p$ $= .203$	

5 Discussion

Research question 1 concerns the difference in the levels of trust and trustworthiness among the studied Chinese regions and South Korea. According to our results, Hubei was relatively high on trust among these regions, and Hubei was also high on trustworthiness on average. Additionally, from the post-experiment interviews, we discovered that participants from Hubei tended to be more cooperative and to value the counterpart's interests, which was consistent with the experimental results. In contrast, Zhejiang was relatively low on trust and trustworthiness on average, suggesting that participants from Zhejiang were more cautious and exhibited more self-protective behaviors. A possible explanation is that the developed private enterprise in Zhejiang influences the trust building process. It is estimated that family-owned enterprises account for 90% of private business in Zhejiang [37]. People prefer to do business transactions with close family members and insiders in their social networks, and they develop business networks through guanxi bonding. Guanxi literally refers to interpersonal relationship or social connections, but the trust and trustworthiness studied in this paper was targeted at strangers instead of acquaintances.

Research question 2 concerns the differences in the levels of trust and trustworthiness across cooperation stages. According to the literature findings, we assumed that the trust and trustworthiness would be high in the early stage and then decline in the late stage. However, the results did not support our assumption and even indicated an opposite direction. Trustworthiness was relatively low in the early stage but increased in the late stage in Liaoning and Hubei. A possible explanation is that participants were gradually developing trust through the cooperation, and the retailers cared about the suppliers' interests so that they did not take opportunistic behaviors. Additionally, the trust level was steady through the entire cooperation in these regions and country. Trust and trustworthiness were interdependent in this experimental setting. The suppliers' trust kept steady, perhaps because the suppliers found the retailers were trustworthy so that they did not adjust their trust level.

Research question 3 concerns the interaction between the region and the cooperation stage. The results showed that, in the early stage, the Chinese regions and South Korea had a similar level of trust and trustworthiness; in the middle stage, trust and trustworthiness in Hubei were higher than in Liaoning and South Korea; in the late stage, trust in Hubei was higher than trust in Zhejiang. In other words, the influence of region on trust and trustworthiness does not appear in the early cooperation stage but the middle and late stages, reminding people to invest more efforts from the middle stage to maintain the trust level.

6   Conclusion

This study employed a two-tier supply chain experiment to study the trust and trustworthiness in South Korea and Chinese regions including Liaoning, Hubei, and Zhejiang. In the experiment, the supplier solicits demand forecast information from the retailer to plan production, and the loss of overproduction is borne by the supplier. A total of 160 participants took part in the experiment: a half of them played as a retailer, and the other half played as a supplier. The results were as follows. First, trust in Hubei was relatively high among the studied regions and country, whereas no difference was found in the level of trustworthiness. Second, the trust level was steady through the five cooperation stages, whereas the trustworthiness level was lower in the early stage than that in the late stage in Liaoning and Hubei but not in Zhejiang and South Korea. Third, trust and trustworthiness in Hubei were higher than in Liaoning and South Korea in the middle cooperation stage, and trust in Hubei was higher than trust in Zhejiang in the late cooperation stage.

References

1. Talhelm, T., et al.: Large-scale psychological differences within China explained by rice versus wheat agriculture. Science **344**(6184), 603–608 (2014)
2. Özer, Ö., Zheng, Y., Chen, K.-Y.: Trust in forecast information sharing. Manag. Sci. **57**(6), 1111–1137 (2011)
3. Huang, H., Rau, P.L.P.: Cooperative trust and trustworthiness in China and The United States: does guanxi make a difference? Soc. Behav. Pers. Int. J. **47**(5), 1–11 (2019)
4. Özer, Ö., Zheng, Y., Ren, Y.: Trust, trustworthiness, and information sharing in supply chains bridging China and The United States. Manag. Sci. **60**(10), 2435–2460 (2014)
5. Lei, X.: Effect of self-construal and trust on cooperation performance. Tsinghua University, Beijing, China (2016)
6. Alesina, A., La Ferrara, E.: Who trusts others? J. Public Econ. **85**(2), 207–234 (2002)
7. Butler, J.K.: Toward understanding and measuring conditions of trust: evolution of a conditions of trust inventory. J. Manag. **17**(3), 643–663 (1991)
8. Mayer, R.C., Davis, J.H., Schoorman, F.D.: An integrative model of organizational trust. Acad. Manag. Rev. **20**(3), 709–734 (1995)
9. Rousseau, D.M., Sitkin, S.B., Burt, R.S., Camerer, C.: Not so different after all: a cross-discipline view of trust. Acad. Manag. Rev. **23**(3), 393–404 (1998)
10. Flores, F., Solomon, R.C.J.S.T.O.R.: Creating trust. Bus. Ethics Q. **1998**, 205–232 (1998)
11. Caldwell, C., Jeffries, F.L.: Ethics, norms, dispositional trust, and context: components of the missing link between trustworthiness and trust. In: Eighth Annual International Conference on Ethics in Business. De Paul University, Chicago, IL (2001)
12. Caldwell, C., Clapham, S.E.: Organizational trustworthiness: an international perspective. J. Bus. Ethics **47**(4), 349–364 (2003). https://doi.org/10.1023/A:1027370104302

13. Brewer, M.B., Gardner, W.: Who is this "We"? Levels of collective identity and self representations. J. Pers. Soc. Psychol. **71**(1), 83 (1996)

14. Markus, H.R., Kitayama, S.: Culture and the self: implications for cognition, emotion, and motivation. Psychol. Rev. **98**(2), 224 (1991)

15. Kelley, H.H., Thibaut, J.W.: Interpersonal Relations: A Theory of Interdependence. Wiley, New York (1978)

16. Kiyonari, T., Yamagishi, T., Cook, K.S., Cheshire, C.: Does trust beget trustworthiness? Trust and trustworthiness in two games and two cultures: a research note. Soc. Psychol. Q. **69**(3), 270–283 (2006)

17. Colquitt, J.A., Scott, B.A., LePine, J.A.: Trust, trustworthiness, and trust propensity: a meta-analytic test of their unique relationships with risk taking and job performance. J. Appl. Psychol. **92**(4), 909 (2007)

18. Fukuyama, F.: Trust: The Social Virtues and the Creation of Prosperity. Free Press Paperbacks, New York (1995)

19. Bigley, G.A., Pearce, J.L.: Straining for shared meaning in organization science: problems of trust and distrust. Acad. Manag. Rev. **23**(3), 405–421 (1998)

20. Miller, A.S., Mitamura, T.: Are surveys on trust trustworthy? Soc. Psychol. Q. **66**(1), 62–70 (2003)

21. Huff, L., Kelley, L.: Levels of organizational trust in individualist versus collectivist societies: a seven-nation study. Organ. Sci. **14**(1), 81–90 (2003)

22. Oyserman, D.: Is the interdependent self more sensitive to question context than the independent self? Self-Construal Obs. Conversat. Norms (2002)

23. Mamat, M., et al.: Relational self versus collective self: a cross-cultural study in interdependent self-construal between Han and Uyghur in China. J. Cross Cult. Psychol. **45**(6), 959–970 (2014)

24. Li, H.Z., Zhang, Z., Bhatt, G., Yum, Y.-O.: Rethinking culture and self-construal: China as a middle land. J. Soc. Psychol. **146**(5), 591–610 (2006)

25. Chu, J.Y., et al.: Genetic relationship of populations in China. Proc. Natl. Acad. Sci. **95**, 11763–11768 (1998)

26. Wen, B., et al.: Genetic evidence supports demic diffusion of Han culture. Nature **431**(7006), 302–305 (2004)

27. Zhao, T., Lee, T.D.: Gm and Km allotypes in 74 Chinese populations: a hypothesis of the origin of the Chinese nation. Hum. Genet. **83**(2), 101–110 (1989). https://doi.org/10.1007/BF00286699

28. Talhelm, T., Zhang, X., Oishi, S.: Moving chairs in Starbucks: observational studies find rice-wheat cultural differences in daily life in China. Sci. Adv. **4**(4), eaap8469 (2018)

29. Yamagishi, T., Yamagishi, M.: Trust and commitment in The United States and Japan. Motiv. Emot. **18**(2), 129–166 (1994). https://doi.org/10.1007/BF02249397

30. Keser, C.: Trust and Reputation Building in E-commerce (2002)

31. Ganesan, S.: Determinants of long-term orientation in buyer-seller relationships. J. Mark. **58**(2), 1–19 (1994)

32. Anderson, E., Weitz, B.: Determinants of continuity in conventional industrial channel dyads. Mark. Sci. **8**(4), 310–323 (1989)

33. Anderson, E., Weitz, B.: The use of pledges to build and sustain commitment in distribution channels. J. Mark. Res. **29**(1), 18–34 (1992)

34. Cochard, F., Van, P.N., Willinger, M.: Trusting behavior in a repeated investment game. J. Econ. Behav. Organ. **55**(1), 31–44 (2004)

35. Bottom, W.P., Gibson, K., Daniels, S.E., Murnighan, J.K.: When talk is not cheap: substantive penance and expressions of intent in rebuilding cooperation. Organ. Sci. **13**(5), 497–513 (2002)

36. Fischbacher, U.: z-Tree: Zurich toolbox for ready-made economic experiments. Exp. Econ. **10**(2), 171–178 (2007)
37. Hu, X.W.: Quantitative evaluation on Zhejiang family businesses. Zhejiang Sch. J. **2**, 158–162. (in Chinese)

Observing the Influence of Cultural Differences Within India on User Experience of an E-Commerce Application: An Experimental Investigation

P. S. Amalkrishna[(⊠)], Surbhi Pratap, and Jyoti Kumar

Department of Design, Indian Institute of Technology Delhi, New Delhi, India

Abstract. Though several studies have reported differences in online behavior across different cultures where countries have been treated as a single cultural unit, there are limited studies on the influence of subcultural differences on the online behavior of users within a country. Especially within India, there is a dearth of literature investigating the influence of subcultural differences on user preferences on mobile application designs. This paper reports an experimental investigation into consumer preferences for e-commerce mobile application users from three distinct cultural regions within India. Culturally suitable design variations for the three regions were created using user interface markers from literature. This was followed by remote usability testing and structured interviews with participants. It was observed that there were differences between preferences of users from different cultural regions within India and the users from a particular subcultural region preferred the design variation made for that region using culturally suitable user interface elements.

Keywords: Subcultural differences · Mobile user preferences · Online shopping behaviour · E-commerce applications

1 Introduction

With the ubiquity of the internet, geography has ceased to be a limiting factor in the exchange of information [1]. However, the way this information is received depends on several factors like receivers' opinions, beliefs and cultural perspectives [2, 3]. This has led to an attribution of importance to the concept of culture in the design of interactive systems [4, 5]. Even in the context of mobile HCI (Human Computer Interaction), this concept can be viewed from multiple viewpoints, such as accessibility and ergonomics, market advantages and social sustainability [6]. Consequently, cultural gaps have become an important topic in consumer research for mobile devices, and more so for developing countries with their cultural and economic heterogeneity [6]. An experimental study into consumer preferences for e-commerce mobile application users from three distinct cultural regions within India is recorded in this article. Literature has shown that in India there are three distinct cultural areas with different cultural dimensions [7]. It has also

© Springer Nature Switzerland AG 2021
P.-L. P. Rau (Ed.): HCII 2021, LNCS 12771, pp. 475–485, 2021.
https://doi.org/10.1007/978-3-030-77074-7_36

been reported that there is a correlation between the preferred elements of user interface to cultural differences between the users [8, 9].

This study reports findings from usability testing and personal interviews conducted on a total of 24 users from different cultural regions within India, while they made a purchase on an e-commerce mobile application. Each user made the purchase on three different designs for the mobile application (app), leading to a total of 72 usability tests. The design variations in the mobile app were made according to the culturally preferred user interface elements as reported in the literature [10]. Of the 24 users, eight belonged each to South India, North India, and North-East India, which are Indian regions with significant cultural differences [7]. The participants were also interviewed before and after the task was done to get additional insights. The results of this research show that there is a differentiation between the user preferences of three different cultural areas within India. It was also noted that if the app user-interface (UI) is designed in accordance with the cultural aspects of the area of which the user belongs, the user's reaction to using the mobile app is favourable.

2 Background

2.1 Cultural Models

Cultural models generally provide parameters that demonstrate different cultures, but because of their systematic nature, they may be mapped on other structures such as management models [6]. There are several popular theories that have proposed dimensions to a culture like Hall's theory of high or low context cultures [11]; Kluckhohn's five dimensions of attitude to problems, time, nature, form of activity and reaction to compatriots [12]; Trompenaars's seven dimensions of universalistic vs particularistic, individualist vs collectivist, specific vs. diffuse, achievement oriented vs ascriptive and neutral vs affective [13] and Hofstede's 6-D model of power distance, collectivism vs. individualism, femininity vs masculinity, uncertainty avoidance, indulgence vs. restraint and long term vs short term orientation [14]. However, the simple structure of the Hofstede model makes it more accessible in HCI research, especially where a method of calculation or numerical data is needed [6]. The dimensions of this model, for example, have been used to assess performance based on cultural differences [15].

2.2 Distinct Cultural Regions Within India

While the possibility of multiple cultural groups within national borders has been argued in prior studies [16], there is limited research on the design impact of national cultural differences, especially within countries such as India [17]. Based on the anthropological classification of India [18, 19], the following three distinct cultural regions within India have been identified in literature [7]. Figure 1 shows the reported difference in cultural dimensions of these three regions:

1. North India - including the Indian states of Kashmir, Uttarakhand, Delhi, Chandigarh, Punjab, Haryana, Uttar Pradesh, Madhya Pradesh, Chhattisgarh, Gujarat, Rajasthan, Maharashtra, Goa, Daman & Diu, Dadra & Nagar Haveli, Bihar, Jharkhand, Odisha and West Bengal.
2. South India - including the Indian states of Karnataka, Tamil Nadu, Kerala, Andhra Pradesh, Telangana and Puducherry.
3. North-east India - including the Indian states of Sikkim, Assam, Manipur, Arunachal Pradesh, Mizoram, Nagaland, Tripura and Meghalaya.

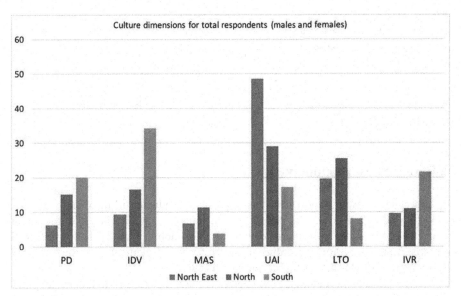

Fig. 1. Difference in Hofstede's culture dimensions for three different Indian regions [7]

2.3 Influence of Culture Differences on UI Elements of HCI Designs

Based on the association reported in literature between the preferred user interface elements and the cultural differences between users [10] and the dimensions of culturally distinct regions within India [7], Table 1 lists the mapping of user-interface design elements for three cultural regions in India:

Table 1. Preferred user interface design elements for three cultural regions in India based on reported literature

Design aspect	South India	North India	North East India
Navigation	Linear, different paths	Linear, restricted	Non-linear, different paths
Error messages	Formal	Formal	Friendly
Structured data	Necessary	Not necessary	Not necessary
Interface level info	High	Low	Low
Support – help, FAQ	Strong	Strong	Not necessary
Image to text ratio	Balanced	Balanced	High
Multimodality	Balanced	Balanced	Low
Color palette	Little saturation	High contrast/bright color	Little saturation
Text content	Friendly	Encouraging words	Friendly
Hierarchy	Flat	Flat	Deep

3   Research Methodology

The experimental study reported in this paper was conducted in three steps. Firstly, an existing popular e-commerce application in India was studied for its information architecture and various user interface elements. The mobile app was then re-designed into three design-variations according to the preferred UI elements for users of three cultural regions of North, North-East and South India (Table 1).

Next, 50 participants belonging to different cultural regions within India were given an online questionnaire to get their demographic data and their availability for the experiment. Out of these, 24 participants of similar age group, gender and educational background were identified, eight from each of the three cultural regions based on how long they had stayed in their native region. The participants were contacted through emails and telephone and were explained the study intent and the methodology in detail. Informed consent to participate in the study and to use the data collected through it was taken through emails.

Further, the users were given a task to make a purchase from the three versions of the mobile application and synchronous remote usability tests [20] were conducted on the 24 users while doing a task on each design variation using a popular remote usability testing tool [21]. Videos of the UT sessions were recorded with the participant's consent and were used for analysis of the data.

Finally, structured telephonic interviews were conducted after the experiment to gather insights about the participants overall experience with the three versions of the websites. The steps of the research methodology are detailed below.

3.1   Redesign of an Existing Popular E-commerce Application

An existing e-commerce mobile application popular in India was studied for its information architecture and its user interface (UI) elements and information architecture (IA) was mapped. The app was then re-designed into 3 variations using the cultural dimensions for North, North-East and South India and their corresponding UI preferences. Figure 2 illustrates the working prototypes that were created for the three versions.

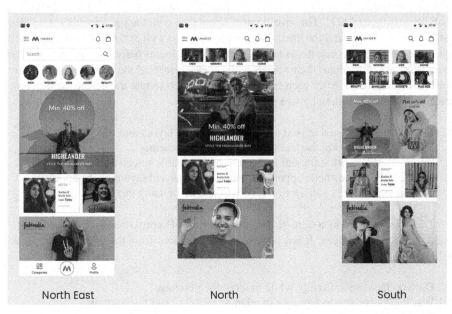

Fig. 2.   Design variations of the e-commerce app for three different cultural regions in India

3.2   Participants

50 participants belonging to different cultural regions within India were given an online questionnaire to get their demographic data and their availability for the experiment. Of these, 24 participants of similar age group, gender and educational background were identified, eight from each of the three cultural regions based on how long they had stayed in their native region. The participants were all males with an undergraduate degree in the discipline of engineering or architecture. The age group of the participants was between 22–28 years with the average age being 25 years and standard deviation of 1.5 years. Each of the participants had stayed in his native place for more than 14 years and the average duration of stay for all the participants was 22 years while the standard deviation was 3.4 years. The participants were contacted through emails and telephone and were explained the study intent and the methodology in detail. Informed consent to participate in the study and to use the data collected through it was taken through emails.

3.3 Experiment Procedure

Firstly, demographic details like gender, age, native place and number of years lived in native place and other states was collected through an online form from each user. Then, the three versions were given randomly to each participant to minimize the 'order effect' using research randomizer [22]. The task given to the participants was: "Your friend from school is getting married. Since it's lockdown and being home is the safest option, you plan to buy a shirt online as a gift for his wedding. Purchase a shirt from the available options in the application". Next, remote moderated testing was conducted using a remote usability testing tool [21]. The investigators connected with the participants via a video call. Then the context and the intent of the experiment as well as the task were explained to the user. Informed consent was taken from the participants before the task began. The users were asked to share their screen while they performed the tasks, and the entire session was recorded. After each task, users were asked to rate the design version on a scale of 1–7 for the following five parameters:

1. Color Scheme - visual appeal to colors used in the layout and images.
2. Ease of Use - perform the tasks effectively and efficiently
3. Information Density - amount of content in the visible screen area
4. Look and Feel - aesthetic appeal of visual elements and layout of the application.
5. Overall experience - satisfaction level of the user in using this application

This was followed by a semi-structured interview of approximately 15 min to understand the user preferences. A few representative questions from the interview are listed below:

1. Did you face any difficulty while making the purchase?
2. Was it easy for you to navigate to other screens? Could you explain?
3. Would you have preferred more items on a single page?
4. How do you usually find a product in an e-commerce application? Do you use the search box to find the item while making a purchase or browse from the home page categories?
5. Which colour do you associate yourself the most? Can you elaborate please.
6. When you think of your native state, what images or thoughts come to your mind? What are your associations with these images?

4 Observations

It was observed overall, that participants from each particular region preferred the design variation made for that region i.e. users from the Northern, North East and Southern regions of India preferred the North Indian, NE Indian and South Indian variations of the application respectively. It was also observed that there was a preference for Content Density and Ease of use by the south Indian participants towards the northern variation, which affected the overall average scores of that region. Figure 3 illustrates the average ratings of the five design parameters for each variation by the participants, while Table 2 enlists a few representative extracts from the verbal reports of the interviews conducted post the usability tests, which provided further insights into the participants' preferences.

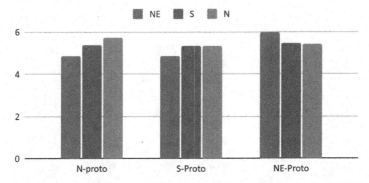

Fig. 3. Average ratings of five design parameters for the three design variations

Table 2. Verbal reports of the interviews conducted post the remote usability testing sessions

Participant code	Cultural region	Verbal report
PS8	South India	"I did not like the hexagons in the third prototype (N version)…"
PS6	South India	"I need all the information to be there. It need not be visible…"
PN1	North India	"Generally i prefer to have bright colors, that feeling of happiness and something cheerful going on…"
PN4	North India	"The second prototype (N version) with hexagonal shapes was very eye-catching. I liked it. I really like these regular shapes (with) sharp edges…"
PNE2	North-east India	"I personally like minimalistic (designs). Minimal content is always preferable for me, be it app, website or anything else…"
PNE7	North-east India	"I go to the categories and then pick a particular option. Also i'm a fan of (using) filters. It's kind of difficult to use a search bar…."

4.1   Observations of Participants from South India

All south Indian users expressed their preference towards more graphical content (high image to text ratio). It was also found that there was a strong use of the search bar for locating a product. All the three versions were found almost equally easy to use by them. However, they preferred the North and North Eastern design versions for their overall experience. While literature had indicated that users from south India would prefer high information density compared to other two regions, it was found that most south Indian participants preferred minimal content. This may arguably be due to the educational background of the participants. These participants also reported preferring low saturation and bright colours. Figure 4 illustrates their average rating of each of the five design parameters for the three design variations.

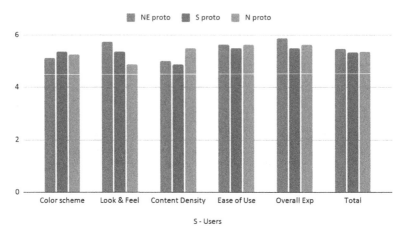

Fig. 4. Average ratings of each parameter by South Indian participants

4.2 Observations of Participants from the North India

Most of the North Indian participants reported that they preferred the content density, colour scheme, ease of use and overall experience of the North Indian design version. However, the look and feel of the North Eastern version was reported to be slightly preferred over the Northern version. The participants reported preferring high contrast bright colours and did not explore alternate paths for navigation reporting multiple paths to be a distraction. Figure 5 illustrates their average rating of each of the five design parameters for the three design variations.

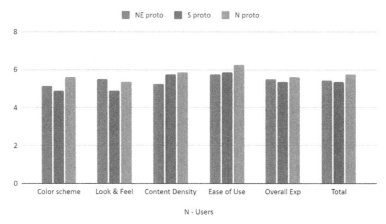

Fig. 5. Average ratings of each parameter by North Indian participants

4.3 Observations of Participants from the North-East India

A majority of North-East Indian participants reported that they preferred the minimal content density, look and feel, colour scheme, ease of use and overall experience of the North-east Indian design version. The participants reported preferring low contrast and less saturated colours and used category-navigation most of the time, rarely using the search-bar. This observation was in contrast to their preference for deep hierarchy as reported in literature. Figure 6 illustrates their average rating of each of the five design parameters for the three design variations.

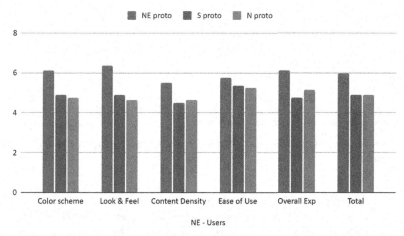

Fig. 6. Average ratings of each parameter by North-east Indian participants

5 Conclusion and Discussion

The contribution of this paper is in identifying and reporting user preferences for mobile application designs at subcultural level within different regions in India. The experimental research results indicate that subcultural disparities in India impact consumer preferences for e-commerce apps, as seen in Fig. 7. The insights from this study can be used in context of user preferences to develop more evolved interactive mobile application designs for sub-cultural Indians.

However, detailed studies are required to establish reasons for the observed differences as well as deviations. Differences for domains other than e-commerce also need to be explored through further studies with larger datasets. There were also limitations to this study, enlisted below; which can be attended to while extending further studies in this area:

1. The study was conducted on a restricted data set (24 participants). Expanding this could yield more conclusive results.

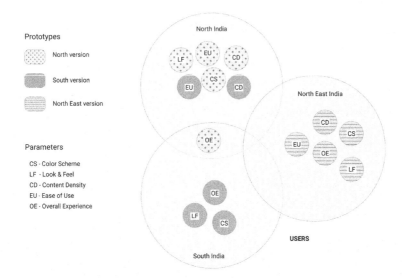

Preference of prototypes by users from three distinct cultural zones   *Pictorial representation. Not actual values

Fig. 7. Experimental observation of subcultural differences in user preferences for an e-commerce mobile application

2. Narrow time intervals between the three prototype tests may have made it difficult for the users to think aloud and could have affected their performance. Also even though randomizer was used to reduce the order effect, by the time the participants used the third design variation, they had become conversed with the task. This could have impacted the overall results.

3. Limited internet connectivity sometimes made it difficult to analyse facial expressions while the remote usability testing was conducted. Also, since the heat maps generated were completely based on mouse clicks in the remote UT tool, precise conclusions could not be made without considering the commonalities in the 5 mentioned parameters.

4. This study is conducted by re-designing an e-commerce mobile application. Different domains of HCI products may be explored further in future studies.

5. External influences of factors like designer's own culture, were not accounted for in this particular study, which can be explored further to ascertain their effect to create culturally suitable designs.

References

1. Burgmann, I., Kitchen, P.J., Williams, R.: Does culture matter on the web? Mark. Intell. Plann. **24**, 62–76 (2006). https://doi.org/10.1108/02634500610641561
2. Fill, C.: Marketing Communications: Contexts, Strategies, and Applications. Prentice Hall, Hoboken (2002)

3. May, P., Ehrlich, Hans-Christian., Steinke, T.: ZIB structure prediction pipeline: composing a complex biological workflow through web services. In: Nagel, W.E., Walter, W.V., Lehner, W. (eds.) Euro-Par 2006. LNCS, vol. 4128, pp. 1148–1158. Springer, Heidelberg (2006). https://doi.org/10.1007/11823285_121

4. Salgado, L., Pereira, R., Gasparini, I.: Cultural issues in HCI: challenges and opportunities. In: Kurosu, M. (ed.) HCI 2015. LNCS, vol. 9169, pp. 60–70. Springer, Cham (2015). https://doi.org/10.1007/978-3-319-20901-2_6

5. Gefen, D., Geri, N., Paravastu, N.: Vive la différence. In: Advances in E-collaboration, pp. 1–12 (2009). https://doi.org/10.4018/978-1-60566-110-0.ch001

6. Aryana, B., Øritsland, T.A.: Culture and mobile HCI: a review. In: Norddesign 2010 Conference, vol. 2, pp. 217–226. 25 August 2010

7. Pratap, S., Kumar, J.: A dimensional analysis across India to study how national cultural diversity affects website designs. In: Chakrabarti, A. (ed.) Research into Design for a Connected World. SIST, vol. 135, pp. 653–664. Springer, Singapore (2019). https://doi.org/10.1007/978-981-13-5977-4_55

8. Reinecke, K., Bernstein, A.: Knowing what a user likes: a design science approach to interfaces that automatically adapt to culture. Miss Q. **37**, 427–453 (2013). https://doi.org/10.25300/misq/2013/37.2.06

9. Marcus, A., Baumgartner, V.-J.: A practical set of culture dimensions for global user-interface development. In: Masoodian, M., Jones, S., Rogers, B. (eds.) APCHI 2004. LNCS, vol. 3101, pp. 252–261. Springer, Heidelberg (2004). https://doi.org/10.1007/978-3-540-27795-8_26

10. Pratap, S., Kumar, J.: CIAM: a new assessment model to measure culture's influence on websites. In: Rau, P.-L. P. (ed.) HCII 2020. LNCS, vol. 12193, pp. 389–408. Springer, Cham (2020). https://doi.org/10.1007/978-3-030-49913-6_33

11. Hall, E.T.: The Silent Language. Doubleday, New York (1969)

12. Kluckhohn, F.R., Strodtbeck, F.L.: Variations in value orientations (1976)

13. Trompenaars, F., Hampden-turner, C.: Riding the Waves of Culture: Understanding Diversity in Global Business. Nicholas Brealey publishing, London (2011)

14. Hofstede, G.: Cultures and Consequences: International Differences in Work-Related Values. Sage Publications, Beverly Hills (1980)

15. Ford, G., Gelderblom, H.: The effects of culture on performance achieved through the use of human computer interaction. In: Proceedings of the 2003 Annual Research Conference of the South African Institute of Computer Scientists and Information Technologists on Enablement through Technology, pp. 218–230, 17 September 2003

16. Thomas, D.C.: Essentials of International Management: A Cross-Cultural Perspective. Sage Publications, Thousand Oaks (2002)

17. Panda, A., Gupta, R.K.: Mapping cultural diversity within India: a meta-analysis of some recent studies. Glob. Bus. Rev. **5**(1), 27–49 (2004)

18. Racial classification of Indian people (by different anthropologist), 19 June 2014. https://www.yourarticlelibrary.com/essay/anthropology/racial-classification-ofindian-people-by-different-anthropologist/41839

19. https://nsdl.niscair.res.in/jspui/bitstream/123456789/339/1/pdf%204.4%20niscair-racial-ethnic-relgious-linguistic-groups-india-text-revised.pdf

20. Andreasen, M.S., Nielsen, H.V., Schrøder, S.O., Stage, J.: What happened to remote usability testing? An empirical study of three methods. In: Proceedings of the SIGCHI Conference on Human Factors in Computing Systems, pp. 1405–1414, 29 April 2007

21. https://maze.co. Accessed 26 Feb 2021

22. https://www.randomizer.org/. Accessed 26 Feb 2021

Competency Model of Chinese Internet Product Managers

Pei-Luen Patrick Rau, Ting-Yu Tony Lin, Hao Chen, and Jian Zheng[(⊠)]

Department of Industrial Engineering, Tsinghua University, Beijing, China
rpl@mail.tsinghua.edu.cn, jzheng23@umd.edu

Abstract. In recent years, the internet industry in China went through dramatic development, to which product managers made a great contribution. This study aimed to define what qualities are required for internet product managers to be competent. This study reviewed the literature on product managers' competency in traditional industries and interviewed three product managers and three recruiters to complement the item pool. Thirty-eight items were selected and then rated on importance by 318 product managers from 173 internet companies in China. Six competency factors were extracted with an exploratory factor analysis: reliability, planning, creativity, leadership, integrity, and user-study. Creativity might be a special requirement for product managers in the internet industry. Internet product managers can use our model to improve their competency. Human resource practitioners can use our model to select, cultivate, and train internet product managers. Previous competency models of product managers focused on traditional industries. This study explored the competency model of internet product managers.

Keywords: Competency · Competency model · Product manager · Internet industry · China

1 Introduction

The number of internet users in China had reached 731 million by the end of 2016, with a penetration rate of over 53.2% nationwide and making up 20% of internet users of the world (CNNIC 2017). The Chinese internet industry has been developing dramatically and becoming ever more worldwide influential. Most Chinese internet organizations are using cross-functional product teams to improve their new product development (NPD) performance. It is estimated that Tencent has more than 500 product teams functioning individually. In a cross-functional environment, obstacles arise from all sides, such as team members' competing identities and loyalties (Webber 2002), conflict and incoherence in decisions (Hong et al. 2004), fragmented and separated operational manners (Rauniar et al. 2008), and barriers to communication (Ashforth and Mael 1989). "The key to project success is to pick the right project manager" (Toney 1997). Similarly, the key to product success is to pick the right or competent product manager.

© Springer Nature Switzerland AG 2021
P.-L. P. Rau (Ed.): HCII 2021, LNCS 12771, pp. 486–495, 2021.
https://doi.org/10.1007/978-3-030-77074-7_37

To select competent product managers, a competency model is needed to lists the competencies required for effective performance in a specific job, organization, or process (Fogg 1999). It can accelerate personal development by clarifying the performance requirements for a job (Brophy and Kiely 2002) and help with the development of improved human resource management (HRM) approaches (El-Sabaa 2001; Nordhaug and Grønhaug 1994).

Previous studies on product manager's competency focused on traditional industries, e.g., the consumer goods industry (Buell 1975) and the construction industry (Dainty et al. 2004; Debrah and Ofori 2005). The competency model of internet product managers has not yet been built. This is odd given that the internet industry is the largest and most established product-based industry. To build such a competency model, we collected competency items from literature and an interview and asked internet product managers to rate the importance of each item. A competency model was built with exploratory factor analysis. The results will help us to learn better about the Chinese internet industry and provide suggestions on the management of relevant industries in both China and other countries.

2 Theory

2.1 Competency

Researchers from different cultures demonstrate the concept of competency differently (Cseh 2003). US literature defined competency by behavior approach: they focused on what distinguishes outstanding workers from typical ones. Boyatzis (1982) defined competency as "an underlying characteristic of a person which results in an effective and/or superior performance in a job", and the underlying characteristic included motives, traits, skills, self-images or social roles, or a body of knowledge. In the UK, the functional approach is more popular. The UK governments introduced a competence-based approach to vocational education and training to establish a national unified system of work-based qualifications. The definition provided by the Manpower Services Commission was "the ability to perform activities in the jobs within an occupation, to the standards expected in employment" (Le Deist and Winterton 2005). One of the obvious differences between the US and UK approaches is that the US approach requires "superior" performance, while the UK approach only requires the ability to perform "to standards".

In the present study, we will employ the functional approach. Competency is defined as knowledge, skill (technical and personal), trait, and behavior related to performing the jobs to the expected standards.

2.2 Competency Model

The competency model is a cluster of related knowledge, attitudes, and skills that affects a major part of one's job (Parry 1996). It serves as an established resource to define employees' successful performance (Brophy and Kiely 2002). Generic models are highly transferable and describe the fundamental competencies required by most jobs (Stasz 1997). For example, Spencer and Spencer (1993) depicted competencies as an iceberg model, which is comprised of five layers (from top to bottom): knowledge, skills, self-concepts, traits, and motives. Specific competency models list the competencies required for a specific job, organization, function, or process (Fogg 1999). For example, the Project Manager Competency Development (PMCD) Framework published by Project Management Institution (PMI 2007) detailed the personal competencies of project managers, including communication, leading, managing, cognitive ability, effectiveness, and professionalism. For more competency models, one can refer to the summary by Bolden et al. (2003).

2.3 Competency of Product Managers in the Traditional Industry

A product manager is responsible for orchestrating various activities and ensuring that the product will meet users' needs. Previous studies about the competency of product managers focused on the traditional industry (e.g. Dainty et al. 2004) or the general industry (Horowitz and Weiden 2010; Rauniar et al. 2008). Following the suggestion of Katz (1974) and El-Sabaa (2001), we classified the competency elements into three typical categories: human skills, organizational skills, and technical skills.

Human Skills
Human skills measure whether a product manager works effectively as a group member and builds a cooperative effort within the team; personality and basic ability are usually categorized into this category (El-Sabaa 2001). Product managers with good human skills are sensitive to others' requirements and thus enhance the potential of success (Belzer 2001; Feger and Thomas 2012).

Basic personality reflects a long-term trend to act in certain ways. The "Big Five" personalities, i.e., extroversion, conscientiousness, agreeableness, openness, and emotional stability, were often used to describe this issue (Costa Jr. and McCrae 1992). There is a large amount of literature supporting that personality can predict job performance (Barrick and Mount 1991). Conscientiousness is one of the most important personalities for managers (e.g. Aitken and Crawford 2008; Bedingfield and Thal 2008; Ebert 2007; Ennis 2008; Fazel and Rashidi 2011; Maronga 2013; Wang et al. 2013). Besides, low neuroticism (Jordan and Cartwright 1998), extroversion (Jordan and Cartwright 1998; Maronga 2013), openness (Bedingfield and Thal 2008; Jordan and Cartwright 1998) also contribute to managers' success.

Communication is also emphasized as a key competency (PMI 2007). Given the prevalence of social components in jobs, employers often evaluated oral communication skills (Robertson et al. 1990), interpersonal skills (Dougherty et al. 1986), and persuasiveness (Dougherty et al. 1986) in recruiting process. Handling relationships within the product team (Crawford 2000; PMI 2007) and professional ethics (Brill et al. 2006; PMI 2007; Sohmen and Dimitriou 2015) are also necessary.

Organizational Skills
Organizational skills refer to the ability to see the cross-function team with a gestalt perspective. It includes understanding the dependence and causality among various functions of a team and extends to visualizing the relationship between a single product and the whole organization (El-Sabaa 2001).

Product managers are leaders of cross-function teams and coordinators of important factors, e.g., company goals and capabilities, customer demand, competition (Horowitz and Weiden 2010). They clarify project targets (Rauniar et al. 2008). The importance of leadership of product managers was well documented in the review by Crawford (2000) and the framework of PMI (2007). The responsibility of product managers as leaders includes team building, member motivation and cultivation, resource allocation, and task distribution and delegation.

Competent managers are good at making plans. They need to understand the goal and focus on it (Ebert 2007; El-Sabaa 2001). Multiple tasks often need to be carried out simultaneously, and some conflicts within the team are inevitable. The manager should understand the priority of tasks (Horowitz and Weiden 2010; PMI 2007) and make wise decisions (Brill et al. 2006; Crawford 2000; Spencer and Spencer 1993). To deliver the product successfully, the manager should agree with the company's strategic direction (Crawford 2000; PMI 2007), and propose both an executable scheme (Brill et al. 2006; Crawford 2000) and a long-term vision of the product (Brill et al. 2006; Crawford 2000; Horowitz and Weiden 2010).

Technical Skills

Technical skill means "an understanding of, and proficiency in, a specific kind of activity, particularly one involving methods, processes, procedures, and techniques. It involves specialized knowledge, the analytical ability within that specialty, and facility in the use of the tools and techniques of the specific discipline" (El-Sabaa 2001).

Researches focused on NPD have discussed technical skills that are useful for product managers, such as analytical thinking (Brill et al. 2006; Cheng et al. 2005), written communication skills (Brill et al. 2006; Horowitz and Weiden 2010), user research and testing methods (Najafi and Toyoshiba 2008), and computer skills (Brill et al. 2006; El-Sabaa 2001; Ennis 2008). Also, to deliver products successfully to the targeted users, a competent product manager needs to understand users' needs (Cheng et al. 2005; Dainty et al. 2004; Kan et al. 2002) and the trends of the industry (Brill et al. 2006; Horowitz and Weiden 2010). A total of 25 items contributing to the competency of product managers were collected.

2.4 Possible Particularity in the Internet Industry

All the items mentioned above are collected from literature in the traditional industry. Probably other competencies are required only, or especially, by the internet industry. To explore such possible items, an interview was carried out.

Behavioral event interview (BEI) is the most commonly used interview method in competency model building (McClelland 1998), and was adopted in this study. A Master's student in management interviewed three product managers and three recruiters, aging from 24 to 36 years old. They all have been working in the Chinese internet industry for 2 to 10 years. They worked in a foreign company, a native listed company, or a native start-up studio.

According to BEI, interviewees were asked to list the three most successful and three worst experiences in work. They described them in detail: cause of the event, people related, solutions, personal feelings, etc. The main questions were: 1) Present your product-related experience. 2) Describe your on-going occupation & responsibility. 3) What is the most successful case you have ever experienced or heard of for a product manager? 4) What about the worst case? 5) Can you think of anyone that you consider a great/bad PM, what makes him so? 6) What is important for a product manager, to your understanding? All interviews were face-to-face and audio recorded. Each interview lasted approximately 60 min. All interviewees were paid 150 CNY afterward.

The linguistic interview data were analyzed to extract the factors concerning product managers' competency model. Following the long table approach (Krueger and Casey 2015), the transcripts of interviews were typed out, followed by a series of cutting and categorizing of the transcripts. In the end, citations reflecting the same item were pasted together and the item was named. Thirteen new elements were added to the list.

3   Method

3.1   Measurement

The questionnaire contains 38 items of competency (Table 1), and questions about participants' gender, position, years of work, etc. The participants were asked to rate "the importance of each item to product managers" on a 7-point scale. One means "*not important at all*", while seven means "*extremely important*". The sample item of human skills was "handling relationship with colleagues, superiors, and subordinates" (Ennis 2008; PMI 2007). The sample item of organizational skills was "understanding the priority of tasks" (Horowitz and Weiden 2010; PMI 2007). The sample item of technical skill was "understanding the future trends of the industry" (Belzer 2001; Edum-Fotwe and McCaffer 2000). The sample item from the interview was "understanding typical user-study methods". To ensure the participants have read the items carefully, one item appeared twice at different positions of the questionnaire.

3.2   Sample

We sent out 420 questionnaires through professional social network applications and collected back 374. Each participant received 30 RMB as remuneration. Four criteria were used to screen the participants: the difference between answers to the two identical items was no more than 2; the time spent on the questionnaire is over 200 s; the same score is rated on less than 90% of the items; correlations with other participants are less than 0.95. Finally, we got 318 pieces of valid feedback. resulting in a subject-item ratio of about 8.4, a generally accepted ratio (Gorsuch 1983; Hatcher and Stepanski 1994).

Among the 318 (112 female) valid respondents, 211 are product managers, 83 are product directors, 15 are product assistant, and 9 holding positions with other titles. As to education, 188 have a degree of bachelor, 129 masters (including 8 MBA), and 1 doctor. Out of the 318 respondents, 214 (76.1%) have been conducting product management for 1 to 3 years. The participants work in 173 companies.

3.3 Data Analysis Method

Items of managers' competency in the traditional industry might not be appropriate for the internet industry. Besides, items from the interview needed to be checked by more subjects. The middle point on the rating scale, i.e., four, is reasonably the cut-off point between unimportant and important. One-sample one-tailed t-tests were used to test whether each of the means is significantly larger than 4.

In the pool of elements selected by the procedure above, some items may be more similar to each other and can be clustered into a set. Explorative factor analysis is used to extract the competency model of product managers in the internet industry, which is more concise than the original elements but can still explain the major variance of them.

4 Results

4.1 Competency Content

The most important item we got is "analysis questions clearly and logically" ($M =$ 6.47), while the least important item is "computer science background" ($M = 3.96$). According to the t-test result, four items are not rated important enough: "Computer science background"; "Design background"; "Operations and business background"; and "International background and foreign languages". They were excluded from further analysis.

4.2 Competency Model

We used the principal components extraction method, analyzed through correlation matrix, extracted factors with eigenvalue over 1, and rotated with Oblimin method with Kaiser Normalization. Items with cross-loading of over .5 on more than one factor and those with loading on all factors below .5 were dropped. The KMO measure of sampling adequacy was 0.884. The Approx. Chi-Square of Bartlett's test of sphericity was 2630.159 ($df = 231, p < .001$). Six factors, i.e., reliability, planning, creativity, leadership, integrity, and user-study, accounted for 63.5% of the total variance. The communalities extractions were above .541. According to the indices above, the construct validity of the model was acceptable. The name of factors was based on the items (Table 1).

5 Discussion

5.1 The Particularity of the Internet Industry

Creativity is the production of "something original and worthwhile" (Sternberg 1996), and involves "the production of a novel, useful products" (Mumford 2003). The importance of creativity is acknowledged by many countries, such as Japan (Zhang and Cantwell 2011), South Korea (Giroud et al. 2012), and multinational corporate (Zhang et al. 2014). Previous research focused more on improving employee's creativity (e.g. Gumusluoglu and Ilsev 2009; Zhang and Bartol 2010), and managers' creativity should

Table 1. Item-factor loading matrix.

Factors	Items
Creativity	Creative ideas in product planning
	Creative ideas in product marketing
	Creative ideas in product's business model
	Understand future trends of the industry
	Sensitivity to competitive products
Reliability	Being steadfast, responsible, and patient
	Stable emotion and stress resistance
	Clear logic when analyzing problems
Leadership	Task distribution and delegation
	Team member cultivation
	Handling relationships with colleagues, superiors, and subordinates
	Team building and motivating team members
	Resource (financial and human) allocation
User-study	Understand typical user study method
	Definition and extraction of user requirement
	Psychology or user study background
Planning	Elaboration on products' executable scheme and features
	Agreement with company's products and strategic direction
	Proposing a long-term vision and development strategy of products
	Understand the priorities of tasks
Integrity	Awareness to protect users' privacy
	Abiding by professional ethics (never plagiary)

be shed more light on. In one study, creativity was rated important only for project managers in novel projects, but not for those in other types of projects (Fazel and Rashidi 2011). A product manager without great creativity may be competent in traditional industries, but not in the internet industry.

The internet industry in China has long been criticized for lacking creativity: practitioners are said to be simply putting established western models into the Chinese market with little modifications. However, the situation is changing. Internet companies are driving technological change and innovating in partnership with universities and businesses worldwide. The internet industry in China values creativity as much as in other countries.

5.2 No Consensus on Educational Background

It was suggested that the Chinese managers' backgrounds affect their competencies (Bu 1994). In the present study, all the four elements not rated important enough were about educational background. Backgrounds like computer science and psychology may be helpful, but they are not indispensable. For example, the minimum qualification for an associate product manager by Google (2016) is a "bachelor's degree in Computer Science, a related field, or *equivalent practical experience.*" Similarly, in most Chinese companies in the internet industry, a degree in CS/design/business will be preferred but not required.

5.3 Managerial Implications

A competency model helps with the development of improved HRM approaches in the industry (El-Sabaa 2001), and is an increasingly versatile and powerful tool in contemporary HRM practice (Collin 1997). According to our model, HR should hunt for those who are conscientious, reliable, emotionally stable, creative, honest. In vocational training, planning, leadership, and user-study experience should be the focuses for product managers.

References

Aitken, A., Crawford, L.: Senior management perceptions of effective project manager behavior: an exploration of a core set of behaviors for superior project managers. Paper Presented at the Proceedings of the PMI Research Conference, Warsaw, Poland (2008)

Ashforth, B.E., Mael, F.: Social identity theory and the organization. Acad. Manag. Rev. **14**(1), 20–39 (1989)

Barrick, M.R., Mount, M.K.: The big five personality dimensions and job performance: a meta-analysis. Pers. Psychol. **44**(1), 1–26 (1991)

Bedingfield, J.D., Thal, A.E.: Project manager personality as a factor for success. Paper Presented at the PICMET 2008 - 2008 Portland International Conference on Management of Engineering and Technology (2008)

Belzer, K.: Project management: still more art than science. Paper Presented at the PM Forum Featured Papers (2001)

Bolden, R., Gosling, J., Marturano, A., Dennison, P.: A review of leadership theory and competency frameworks (2003). https://www.centres.ex.ac.uk/cls/research/abstract.php?id=29

Boyatzis, R.E.: The Competent Manager: A Model for Effective Performance. Wiley, New York (1982)

Brill, J.M., Bishop, M.J., Walker, A.E.: The competencies and characteristics required of an effective project manager: a web-based Delphi study. Educ. Tech. Res. Dev. **54**(2), 115–140 (2006). https://doi.org/10.1007/s11423-006-8251-y

Brophy, M., Kiely, T.: Competencies: a new sector. J. Eur. Ind. Train. **26**(2/3/4), 165–176 (2002)

Bu, N.: Red cadres and specialists as modern managers: an empirical assessment of managerial competencies in China. Int. J. Hum. Resour. Manage. **5**(2), 357–383 (1994). https://doi.org/10.1080/09585199400000022

Buell, V.P.: The changing role of the product manager in consumer goods companies. J. Mark. **39**(3), 3–11 (1975)

Cheng, M.I., Dainty, A.R., Moore, D.R.: What makes a good project manager? Hum. Resour. Manage. J. **15**(1), 25–37 (2005)

CNNIC: Statistical report on internet development in China (2017). https://cnnic.com.cn/IDR/ReportDownloads/201706/P020170608523740585924.pdf

Collin, A.: Learning and development. In: Beardwell, I., Holden, L. (eds.) Human Resource Management: A Contemporary Perspective, pp. 282–344. Pitman, London (1997)

Costa, P.T., Jr., McCrae, R.R.: NEO Personality Inventory-Revised (NEO PI-R). Psychological Assessment Resources, Odessa (1992)

Crawford, L.: Profiling the competent project manager. Paper Presented at the Proceedings of PMI Research Conference (2000)

Cseh, M.: Facilitating learning in multicultural teams. Adv. Dev. Hum. Resour. **5**(1), 26–40 (2003)

Dainty, A.R., Cheng, M.I., Moore, D.R.: A competency-based performance model for construction project managers. Constr. Manage. Econ. **22**(8), 877–886 (2004)

Debrah, Y.A., Ofori, G.: Emerging managerial competencies of professionals in the Tanzanian construction industry. Int. J. Hum. Resour. Manage. **16**(8), 1399–1414 (2005). https://doi.org/10.1080/09585190500220465

Dougherty, T.W., Ebert, R.J., Callender, J.C.: Policy capturing in the employment interview. J. Appl. Psychol. **71**(1), 9–15 (1986). https://doi.org/10.1037//0021-9010.71.1.9

Ebert, C.: The impacts of software product management. J. Syst. Softw. **80**(6), 850–861 (2007). https://doi.org/10.1016/j.jss.2006.09.017

Edum-Fotwe, F.T., McCaffer, R.: Developing project management competency: perspectives from the construction industry. Int. J. Proj. Manage. **18**(2), 111–124 (2000). https://doi.org/10.1016/S0263-7863(98)90075-8

El-Sabaa, S.: The skills and career path of an effective project manager. Int. J. Proj. Manage. **19**(1), 1–7 (2001). https://doi.org/10.1016/S0263-7863(99)00034-4

Ennis, M.R.: Competency models: a review of the literature and the role of the employment and training administration (ETA). US Department of Labor (2008)

Fazel, B., Rashidi, N.: Impact of project managers' personalities on project success in four types of project. Paper Presented at the International Conference on Construction and Project Management (2011)

Feger, A.L.R., Thomas, G.A.: A framework for exploring the relationship between project manager leadership style and project success. Int. J. Manage. **1**(1), 1–19 (2012)

Fogg, C.D.: Implementing Your Strategic Plan: How to Turn "Intent" into Effective Action for Sustainable Change. American Management Association, New York (1999)

Giroud, A., Ha, Y.J., Yamin, M., Ghauri, P.: Innovation policy, competence creation and innovation performance of foreign subsidiaries: the case of South Korea. Asian Bus. Manage. **11**(1), 56–78 (2012). https://doi.org/10.1057/abm.2011.27

Google: Google careers: associate product manager (2016). https://www.google.com/about/careers/jobs#!t=jo&jid=/google/associate-product-manager-university-1600-amphitheatre-pkwy-mountain-view-ca-1029330389

Gorsuch, R.: Factor analysis. In: Hillsdale, N.J.L. (eds) Erlbaum Associates (1983)

Gumusluoglu, L., Ilsev, A.: Transformational leadership, creativity, and organizational innovation. J. Bus. Res. **62**(4), 461–473 (2009)

Hatcher, L., Stepanski, E.J.: A step-by-step approach to using the SAS system for univariate and multivariate statistics. SAS Institute (1994)

Hong, P., Nahm, A.Y., Doll, W.J.: The role of project target clarity in an uncertain project environment. Int. J. Oper. Prod. Manage. **24**(12), 1269–1291 (2004)

Horowitz, B., Weiden, D.: Good product manager/bad product manager. Adv. Res. Econ. Manage. Sci. **15**, 627–633 (2010)

Jordan, J., Cartwright, S.: Selecting expatriate managers: key traits and competencies. Leadersh. Organ. Dev. J. **19**(2), 89–96 (1998). https://doi.org/10.1108/01437739810208665

Kan, S., Jicheng, W., Chaoping, L.: Assessment on competency model of senior managers. J. Chin. Psychol. Acta Psychol. Sin. **3**(34), 193–199 (2002)

Katz, R.L.: Skills of an Effective Administrator. Harvard Business Press, Boston (1974)

Kiessling, T., Harvey, M.: Strategic global human resource management research in the twenty-first century: an endorsement of the mixed-method research methodology. Int. J. Hum. Resour. Manage. **16**(1), 22–45 (2005). https://doi.org/10.1080/0958519042000295939

Krueger, R.A., Casey, M.A.: Focus Groups: A Practical Guide for Applied Research, 5th edn. SAGE, Thousand Oaks (2015)

Le Deist, F.D., Winterton, J.: What is competence? Hum. Resour. Dev. Int. **8**(1), 27–46 (2005)

Maronga, E.: Examination of personality traits as predictors of project manager career success. Int. J. Manage. **2**(2), 1–23 (2013)

McClelland, D.C.: Identifying competencies with behavioral-event interviews. Psychol. Sci. **9**(5), 331–339 (1998). https://doi.org/10.1111/1467-9280.00065

Mumford, M.D.: Where have we been, where are we going? Taking stock in creativity research. Creat. Res. J. **15**(2–3), 107–120 (2003)

Najafi, M., Toyoshiba, L.: Two case studies of user experience design and agile development. Paper Presented at the Agile 2008 Conference, AGILE 2008 (2008)

Nordhaug, O., Grønhaug, K.: Competences as resources in firms. Int. J. Hum. Resour. Manage. **5**(1), 89–106 (1994). https://doi.org/10.1080/09585199400000005

Parry, S.B.: Just what is a competency? (And why should you care?). Training **35**(6), 58 (1996)

PMI: Project manager competency development (PMCD) framework. Newtown Square: Project Management Institute (2007)

Rauniar, R., Doll, W., Rawski, G., Hong, P.: The role of heavyweight product manager in new product development. Int. J. Oper. Prod. Manage. **28**(2), 130–154 (2008)

Robertson, I.T., Gratton, L., Rout, U.: The validity of situational interviews for administrative jobs. J. Organ. Behav. **11**(1), 69–76 (1990). https://doi.org/10.1002/job.4030110109

Sohmen, V.S., Dimitriou, C.K.: Ten core competencies of program managers: an empirical study. Int. J. Health Econ. Dev. **1**(1), 1 (2015)

Spencer, L.M., Spencer, S.M.: Competence at Work: Models for Superior Performance. Wiley, New York (1993)

Stasz, C.: Do employers need the skills they want? Evidence from technical work. J. Educ. Work **10**(3), 205–223 (1997). https://doi.org/10.1080/1363908970100301

Sternberg, R.J.: Cognitive Psychology. Harcourt Brace College Publishers, Fort Worth (1996)

Toney, F.: What the fortune 500 know about PM best practices... and how you can share their knowledge. PM Netw. **11**(2), 30–36 (1997)

Wang, D., Freeman, S., Zhu, C.J.: Personality traits and cross-cultural competence of Chinese expatriate managers: a socio-analytic and institutional perspective. Int. J. Hum. Resour. Manage. **24**(20), 3812–3830 (2013). https://doi.org/10.1080/09585192.2013.778314

Webber, S.S.: Leadership and trust facilitating cross-functional team success. J. Manage. Dev. **21**(3), 201–214 (2002)

Zhang, F., Cantwell, J.A., Jiang, G.: The competence creation of recently-formed subsidiaries in networked multinational corporations: comparing subsidiaries in China and subsidiaries in industrialized countries. Asian Bus. Manage. **13**(1), 5–41 (2014). https://doi.org/10.1057/abm.2013.12

Zhang, X., Bartol, K.M.: Linking empowering leadership and employee creativity: the influence of psychological empowerment, intrinsic motivation, and creative process engagement. Acad. Manag. J. **53**(1), 107–128 (2010)

Zhang, Y., Cantwell, J.: Exploration and exploitation: the different impacts of two types of Japanese business group network on firm innovation and global learning. Asian Bus. Manage. **10**(2), 151–181 (2011). https://doi.org/10.1057/abm.2011.7

The Effectiveness of Scene-Based Icons Inspired by the Oracle Bone Script in Cross-Cultural Communication

Xiaohua Sun[✉], Lin Bao, Weiwei Guo, Yifei Liao, and Xuanye Lu

College of Design and Innovation, Tongji University, Shanghai, China
{xsun,1933629,weiweiguo}@tongji.edu.cn

Abstract. Oracle bone script is an ancient form of writing character used by ancient Chinese. It takes advantage of static pictographic elements to shape scenes, thus conveying dynamic and prosperous messages. The purpose of this study is to demonstrate that scene-based icons inspired by the oracle bone script can be effectively recognized and understood by people from different cultures and thus used to help in cross-cultural communication scenarios. An experiment was conducted with a sample of 16 people from different cultural backgrounds to determine the icons' recognizability. The result indicates that these icons have relatively high recognizability in a cross-cultural context.

Keywords: Oracle bone script · Cross-cultural communication · Scene shaping · Iconic communication

1 Introduction

Icons are small-sized isolated signs [1] that can be seen in everyday life in modern society. Nowadays, we human beings are in an age of visual culture, where people all over the world take advantage of icons or pictorial symbols to understand each other better [2]. Iconic communication makes it possible for understanding across language barriers [3].

Attempts have been made over the years to create universal icons to encourage communication between people worldwide. Credited with being the world's first *true lingua franca* by some media, emoji has played an essential role for people on social media to communicate with each other. Xu compiled icons drawn from public spaces and developed an instant messaging software to help people speaking different languages to communicate with these icons as an intermediary [4].

However, the usability and limitations of these symbol systems remain to be discussed. Emoji serves as a tool to imply or enhance the emotion in text-based communication [5, 6], which means it cannot by itself constitute an intermediary of communication between two parties. Textual messages are translated word by word into a set of icons lined up in a row in Xu's messaging software, which leads to the low efficiency of communication.

© Springer Nature Switzerland AG 2021
P.-L. P. Rau (Ed.): HCII 2021, LNCS 12771, pp. 496–505, 2021.
https://doi.org/10.1007/978-3-030-77074-7_38

In order to enhance the effectiveness of iconic communication, it's a good idea to equip icons with the ability of storytelling. Narrative art like ancient Chinese murals [7, 8], oil paintings of the Middle Age [9] features the power to convey richer messages in a static and limited space. Common narrative methods include the monoscenic narrative, the continuous narrative, and the synoptic narrative [9–11], in which scene-shaping is one of the indispensable steps.

The oracle bone script, as the earliest known form of Chinese writing, is characterized by various scenes and concise pictographs [12]. Therefore, the oracle bone script provides an excellent example for creating an iconic communication system for cross-cultural communication. However, little research has been done in this domain. Our research analyzed the scene-shaping rules of the oracle bone script, designed a set of scene-based icons according to the rules, and finally determined by user testing the effectiveness of the icons used in cross-cultural communication.

2 Scene-Shaping Rules of the Oracle Bone Script

In this section, to find how effectively the oracle bone script conveys messages, we analyzed the scene-shaping rules of the oracle bone script in three aspects, including scene selection, scene organization, and element combination.

2.1 Scene Selection

To tell a story in just one scene, one should first decide which scene of the story to pick. Oracle bone scripts took usage of monoscenic narrative to tell stories. Since every single character of the oracle bone script occupies a relatively fixed small space, it is almost impossible to contain multiple scenes in one character's space.

The first type of scene focuses on a normal moment of the story, which is exceptionally uneventful without any artistic tension. For example, " 𝑌𝑋 " (graze) can be seen as a portrayal of a shepherd and an animal. Another example, the " 𝄞 ", which refers to vegetable plots, portrays how seedlings grow on the ground.

The second type of scene is the climax or the decisive moment of the story. For example, if the story is about a hunter hunting animals with a spear, the story's climax is when the spear pierces the animal, and the animal falls to the ground. Many instances of this type of scene can be found in oracle bone script. Examples include the character " 𝄰 " (blow), which shows one person blows out a breath; the character " 𝄯 " (break off), which indicates precisely the moment a branch was broken.

The third type of scene is what Lessing called the "fruitful moment" [13], which presents the most exciting and tense moment before the climax. Continuing with the hunting story, the most "fruitful moment" refers to when the spear is about to pierce the animal's body. Let's take " 𝄐 " (dawn) as an example. The upper and lower parts are grass, with a sun on the left and a moon on the right, indicating that this is when the sun has risen from the horizon while the moon has not yet disappeared.

2.2   Scene Organization: Selection and Arrangement of the Elements

Based on our analysis of the deciphered oracle bone script, there are two aspects in portraying a scene: selecting the scene elements and visualizing their relations. These two aspects are not separate but complement each other. Song Juan, a Chinese scholar, studied the cognitive associative compounds[1] [14] of the oracle bone script and summarized four construction principles of verbal cognitive associative compounds, which inspired us to develop 3 rules to select and arrange elements.

The first type is "Subject + Object": For scenes that express the relationship of two items. In most cases, the subject is a man. In particular, the subject may become human organs or close-ups of parts of the body. For example, " 🜲 ", refers to "A man is watching a tree", and " 𝕯 ", (an eye) represents the subject (a man) because the action "looking" is the crucial point that the scene highlights. " 🜲 " is a branch meaning a tree.

The second type is "Subject + Object + Environment": For stories happening in a specific environment. In the active voice, it is "Subject + Environment" (the object is omitted), while in the passive voice, it is "Object + Environment" (the subject is omitted). For example, the " 🜲 ", which refers to the mass, portrays three workers under the sun. However, " 🜲 ", means "a cattle sunk in the river". The curves on the left and right side represent the river, and the central element means cattle. When it comes to specific men and items in specific scenes, it is "Subject + Object + Environment". Take " 🜲 ", as an example, it refers to "To raise an insult with two hands in the house". " 🜲 ", means a hand, " 🜲 ", means an insult, " 🜲 ", means an ancient house.

The third type is "Subject + Tool + Object": For sentences composed of a subject, a verb, an object, and a modifier. For example, to portray a man hits another man with a stick—— " 🜲 ", the ancestor use " 🜲 ", (hand) to represent the subject, which is undoubtedly a man, a " 🜲 ", (man) to represent the object, and a " 🜲 ", (stick) to represent the tool.

It is worth noticing that subjects may have multiple variants according to the actions they take. For example, " 🜲 ", (man), as a fundamental element of the oracle bone script, has dozens of variants, such as " 🜲 ", (a dancing man), " 🜲 ", (a standing man), " 🜲 ", (corpse), and " 🜲 ", (a peeing man).

2.3   Element Combination

After selecting the optimal scene and the pictographic elements, combining elements into a complete character also includes scene-shaping rules. We find that the shaping methods of abstracting real-world objects into characters suggested by Xiaohua Li [15] are also applicable in describing scenes of the oracle bone script. But her research is

---

[1] The cognitive associative compounds are compounds of two or more pictographic elements to suggest the meaning of the word to be represented.

limited to the nouns of single pictograms[2]. From the perspective of scene shaping, except for Li's contribution, we also take verbs and simple ideograms[3] into consideration and develop three ways to combine pictographic elements in a scene.

The first one is to outline all the objects which appeared in the scene. For example, in character " 🝔 ",, two people are sitting face to face with a food container between them, which conveys the meaning of treating people to meals.

The second one is to highlight its meaning by exaggerating some parts of the scene. For example, " 𝕌 ,, (tongue) is more evident than real-world situations in character " 𝕤 ",, since the drinking movement is the key point they want to focus on. Furthermore, a specific part is chosen to represent the whole object. The character " 𝕓 ",, describes the scene where a man standing outside the doors asks for help. Here the mouth is used to represent the entire man because the action of asking questions is only related to the mouth. The place where these parts are placed in the character is dependent on the relation of the real-world objects. Taking " 𝕧 ",, (foot) as an example, its position and orientation can be changed as needed. When in 𝕖 (a foot chasing a deer), it is on " 𝕗 " the bottom and pointing upward at the deer, since the man (replaced by the foot) is running after the deer. While in " 𝕙 ",,(a foot into a cave), the foot is on the top and downward-facing to show a man walking into the cave.

The last one is to add marks to highlight the critical parts of the scene. A line is drawn along the knife in character " 𝕜 ",, to emphasize the blade. And the character " 𝕝 ",, has a curve on the man's leg to stress the knee.

3   Test Design

We firstly developed a word list based on cross-cultural scenarios, including highly-used sentences and vocabulary for scenarios such as travel, residential life, etc. We selected some to design pictographic elements and then icons from dozens of sentences and hundreds of words. The selection criteria are to cover the major scenarios of cross-cultural communication and apply all the scene shaping rules mentioned above.

3.1   Pictographic Element Design

We initially designed a set of pictographic elements based on the oracle bone script, which will later be combined into icons according to the scene-shaping rules of the oracle bone script. This set of elements, together with the pictographic elements extracted directly from the oracle, were given to a group of users for testing. They were asked to choose the one with the highest recognition (Table 1).

---

[2] Chinese characters were firstly divided into 6 categories by Shen Xu in *ShuoWenJieZi* according to construction rules. The pictograms and simple ideograms are two of them.
[3] *Ibid.*

Table 1. Examples of 2 types of icon

Element	Type 1	Type 2
Eye		
Man		
Sun		

Their recognizability was evaluated through a user test of 17 samples (aged from 22 to 50 years old, from several countries including China, the UK, and Japan). The test results show that most users (15 out of 17) consider Type 2 more recognizable.

Based on the feedback from the test, we developed a rough guideline to standardize our design:

- Use straight lines, smooth curves, and regular geometric shapes.
- Some frequently-used elements should be flexible enough to produce variants that are suitable for various scenes (e.g., standing, sitting, working man)
- Clearly distinguish people or body parts from inanimate objects to convey the subject-object relationship better in the scene.

3.2 Scene Shaping

After the pictographic element design, we combined different elements into 17 icons (Table 2), which each represent one complete sentence according to the scene-shaping rules of the oracle bone script.

Different scene shaping rules are applied in different icons, and a couple of examples will be introduced in detail as follows.

In Icon No. 1 shown in Table 2, to graphically present "Where is the bathroom/toilet?", we portrayed a complete scene in a normal moment of the story, where a man constructs the scene together with his surroundings: the man stands in the middle of the canvas with a bathroom in the distance. They are connected by a dotted line with a question mark to emphasize the semantic "how to get to someplace".

In Icon No. 9, since the sentence expresses the completion and the result of choice, the moment we chose to manifest is the decisive moment: one chooses the big one and discards the small one. We placed two items, one large and one small, to express the concept of comparison. To emphasize the meaning of choosing, we added an "X" mark on top of the small one and a striking hand towards the large one.

In Icon No. 4, we chose the fruitful moment, where a hand is throwing garbage above two types of trash bins, conveying the message, "Which trash bin should I put the garbage in?". This sentence conforms to the structure of subject-verb-object (SVO), so it's appropriate to apply the "Subject + Tool + Object" here. In particular, we exaggerated the hand to replace the man to highlight the movement of "put".

Table 2. Icons with their meaning and the rules applied

	Icon	Sentence	Scene selection	Scene Organization	Element Combination
1		"Where is the bathroom/toilet?"	1 Normal Moment	2 Subject + Environment	1 Outline all the objects
2		"What is the wifi password here?"	1 Normal Moment	1 Subject + Object	2 Exaggerate one part
3		"How do I use the printer?"	1 Normal Moment	1 Subject + Object	2 Exaggerate one part
4		"Which trash bin should I put the garbage in?"	3 Fruitful Moment	3 Subject + Tool +Object	2 Exaggerate one part
5		"I want a beer"	1 Normal Moment	1 Subject + Object	1 Outline all the objects
6		"I want a red dress"	1 Normal Moment	1 Subject + Object	1 Outline all the objects
7		"I don't want beer"	1 Normal Moment	1 Subject + Object	1 Outline all the objects
8		"I want a shirt, not T-shirt"	1 Normal Moment	1 Subject + Object	1 Outline all the objects
9		"I want a bigger bag"	2 Decisive Moment	1 Subject + Object	3 Add marks to highlight the key point
10		"Which hat should I choose?/Which hat is better?"	1 Normal Moment	1 Subject + Object	1 Outline all the objects
11		"Do you have this dress in other colors?"	2 Decisive Moment	1 Subject + Object	3 Add marks to highlight the key point
12		"Do you have this bag in other sizes?"	2 Decisive Moment	1 Subject + Object	3 Add marks to highlight the key point
13		"Can you lend me a charger ?"	3 Fruitful Moment	1 Subject + Object	2 Exaggerate one part
14		"Can you take a picture for me?"	1 Normal Moment	2 Subject + Object +Environment	1 Outline all the objects
15		"Where can I walk my dog?"	1 Normal Moment	2 Subject + Object +Environment	1 Outline all the objects
16		"When does the library open?"	2 Desicive Moment	2 Object + Environment	1 Outline all the objects
17		"How long does it take to walk to the library?"	1 Normal Moment	2 Subject + Environment	1 Outline all the objects

In Icon No. 16, we depicted the moment when the library opens, which is the climax of the story, in other words, a decisive moment. The rule "Object + Environment" was applied with a passive voice because the door is opened by someone.

4   Evaluation and Results

4.1   Evaluation

To determine the recognizability of the icons, we planned to do user testing with a group of people from diverse cultural backgrounds. The methods used are quantitative data collection and qualitative data collection. Our selection criteria are as follows:

- Speak up to 3 languages. (To avoid them from being used to cross-cultural communication)
- Long-term settlement in one country
- No speech impediment or usage of augmentative communication systems
- Gender balance

Then we experimented with 16 people, whose origins vary from Asia, America to Europe. Most of them are bilingual speakers. The overview of the basic information and the test result is listed in the table below (Table 3):

Table 3. The result of the experiment

Gender	Age	Nation	1	2	3	4	5	6	7	8	9	10	11	12	13	14	15	16	17	Accuracy
M	21	CN	1	1	1	1	1	1	0	1	1	1	1	1	1	1	1	1	0	0.8824
F	20	CN	1	1	1	1	1	1	1	1	0	1	1	1	1	1	1	0	0	0.8235
F	20	CN	1	1	1	1	1	1	1	1	1	1	1	0	1	1	1	0	1	0.8824
F	24	CN	1	1	1	0	1	1	1	1	1	1	0	1	1	1	1	1	1	0.8824
M	20	CN	1	1	1	1	0	1	1	1	1	1	1	0	1	1	0	0		0.7647
M	21	CN	1	1	1	1	1	1	1	1	1	0	1	1	1	1	1	1	1	0.9412
M	24	CN	1	1	1	1	0	1	1	1	0	0	1	1	1	1	1	1	1	0.8235
F	21	CN	1	1	1	1	1	1	1	1	1	0	1	1	1	1	1	1	0	0.8824
F	24	CN	0	1	1	1	1	1	1	1	1	1	0	0	1	1	0	1	1	0.7647
F	23	JP	1	1	1	1	1	1	1	1	1	1	1	1	0	1	0	1	1	0.8824
M	21	CA	1	1	1	1	1	1	1	1	1	1	0	1	0	0	1	1		0.8235
F	21	US	1	1	1	1	1	1	0	1	1	1	0	0	1	1	0	0	1	0.7059
M	22	US	1	1	0	1	1	1	1	1	1	1	1	1	1	1	1	1	0	0.8824
F	22	IT	1	1	1	1	0	1	1	1	1	1	1	0	0	1	1	1	1	0.8235
M	29	AT	1	1	0	1	1	1	1	1	1	1	1	1	1	1	0	1	1	0.8824
F	33	GB	1	0	1	1	0	1	0	1	1	1	1	1	1	1	0	1	1	0.7647

We created a questionnaire to test the 17 icons. Also, we interviewed each participant after the test. We summarized and analyzed the commonalities of their responses, which enabled us to understand the reasons for the cognitive bias of different populations. Each icon was accompanied by a short text as a contextual hint, for example, the identity of the person viewing the icon (e.g., a security guard), the identity of the person using the icon as a communication tool (e.g., a tourist). The participants write down the possible meaning of the 17 icons shown to them one by one (Fig. 1).

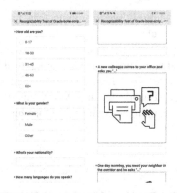

Fig. 1. A screenshot of the questionnaire

4.2 Results

Effectiveness Analysis Between Different Scene Shaping Rules

By calculating the accuracy rate of each icon, we can get a rough idea of how effective the icons are.

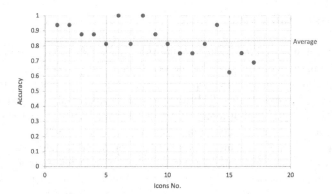

Fig. 2. The average accuracy rate of each icon

We can learn from Fig. 2 that the recognizability of most icons is quite good except No. 15, which is only about 62.5%. Based on the answers they filled in and the data we got from the interviews, we realized that the participants couldn't easily find the relation among all the elements of the icon, so they misunderstood the meaning. Especially in icon No. 15, there are too many icons, including a man, a dog, a dotted circle, a location sign, and a question mark. Some of the participants focused on the location sign, so they thought the man tended to say, "How can I go to…?". Others noticed the circle and guessed, "Please hold your dog around you.". However, the true meaning is "Where can

I walk my dog?". People could rarely see the connection among the question mark, the location sign, and the dotted circle.

The Relation Between Accuracy and Cultural Background
As we can see in Table 4, there is no significant difference in the level of recognition of this icon set among Asians, Europeans, and Americans: the recognition accuracy is between 80% and 86%, which indicates that these icons have good recognizability for people in diverse cultural contexts.

Table 4. The accuracy rate of samples from different regions.

	Region		
	Asia	Europe	America
Percentage of participants	62.50%	18.75%	18.75%
Accuracy	85.29%	82.35%	80.39%

5 Conclusion and Future Work

In this paper, we analyzed the scene-shaping rules of the oracle bone script, designed a set of scene-based icons inspired by the oracle bone script, and took questionnaire investigations and interviews with people from different cultural backgrounds to test the effectiveness of the icons. We learned from the test results that these icons are highly-recognizable for people from different cultural backgrounds, which declares the icons' effectiveness in cross-cultural communication. Future studies will be devoted to the development of an iconic communication system composed of scene-based icons. In this system, we tend to control the scene-based icons with as few elements as possible to avoid information overload and one-sided understanding. Also, we plan to develop a series of principles to unify the iconic pattern for some highly-used phrases.

References

1. Heimbürger, A., Kiyoki, Y.: Pictorial symbols in context—a means for visual communication in cross-cultural environments. In: Proceedings of the IADIS International Conference - Computer Graphics, Visualization, Computer Vision and Image Processing, CGVCVIP 2010, Visual Communication, VC 2010, Web3DW 2010, Part of the MCCSIS 2010, pp. 463–467 (2010)
2. Hariman, R., Lucaites, J.L.: Icons, iconicity, and cultural critique. Sociologica **9**(1) (2015)
3. Beardon, C., Dormann, C., Mealing, S., Yazdani, M.: Talking with pictures: exploring the possibilities of iconic communication. ALT-J **1**(1), 26–39 (1993)
4. Xu, B.: ARTWORK - Book From the Ground. https://www.xubing.com/en/work/details/188?classid=12&type=class. Accessed 21 Jan 2021

5. Bai, Q., Dan, Q., Mu, Z., Yang, M.: A systematic review of emoji: current research and future perspectives. Front. Psychol. **10**(1), 2221 (2019)
6. Danesi, M.: The Semiotics of Emoji: The Rise of Visual Language in the Age of the Internet. Bloomsbury Publishing, London (2016)
7. Zhao, C.: The "narrative" research of portrait art of the Han dynasty. Ph.D., Central Academy of Fine Arts (2007)
8. Zhan, Y.: The image and narration of the story painting of the nine-color deer in Dunhuang Cave 257. Yishu Baijia **26**(03), 196–202 (2010)
9. Ranta, M.: Stories in pictures (and non-pictorial objects): a narratological and cognitive psychological approach. Contemp. Aesthet. J. Arch. **9**(1), 6 (2011)
10. Long, D.: Image narrative: temporalization of space. Jiangxi Soc. Sci. **09**, 39–53 (2007)
11. Mittelstadt, M.G.: Longus: Daphnis and Chloe and Roman narrative painting. Latomus **26**(3), 752–761 (1967)
12. Tang, L.: Introduction to Ancient Chinese Characters (Revised Edition). Qilu Publishing House, Jinan (1981)
13. McClain, J.: Time in the visual arts: lessing and modern criticism. J. Aesthet. Art Crit. **44**(1), 41–58 (1985)
14. Song, J.: Research on the cognitive associative compounds in the inscriptions on bones and tortoise shells. Master degree, Guangzhou University (2010)
15. Li, X.: Study on the pictographic characters in the inscriptions on bones and tortoise shells. Master, Fujian Normal University (2008)

Cultural Discourse in User Behavior: Transfer of Thought in Keyboard Key Practice

Chunyan Wang[1]([envelope]) and Xiaojun Yuan[2]

[1] Luopo Road, Xiaoshan 311200, Zhejiang, China
[2] University at Albany, State University of New York, Albany, NY 12222, USA
xyuan@albany.edu

Abstract. After examining users' keyboard key practice of 95 students in China, we argue that there is a transfer of thought in keyboard key practice of Chinese students. We conceptualize the keys as discourse with cultural factors from the perspective of Cultural Discourse Studies and explore how the users' behaviors deviate the designers' intentions in the key practice. We design a ten-question survey testing the keyboard practice of Chinese students based on the radical behaviorism by Skinner [32], which combines the thoughts and feelings as well as observable behaviors. The survey results indicate that participants generally prefer certain keys and believe each key should not have only one basic function and the keys should be merged according to their operation directions. We then develop a key behavior model by viewing keys as words, and the process of keyboard practice as language comprehension and employment. By dividing the process into designer, discourse, and user; the functions of keys are mapped as designed function and performed function. Results indicate that participants performed poly-functional thought, a universal and powerful thinking method in Chinese language and culture, which may lead to the conflicts with the mono-functional design. We believe the transfer of thought enriches software didactics and inspires a different view of application design.

Keywords: User behavior · Transfer of thought · Transfer of learning · Cultural discourse · Discourse analysis

1 Introduction

Wang and Yuan [1] investigated the characteristics of information communicators in Microsoft Word and concluded that Microsoft prefers the most frequently used English elements to design the user interface and listed them according to western thinking method. Unlike the information speaker/communicator, who often receives full attention and is usually the focus of the discourse research, the receiver/user has not been drawn enough attention [2]. The information or service receiver is considered as proactive and sense-making, who can respond creatively and fairly [2]. From this perspective, information receiver/user is an important part to assess the accuracy of the previous communicator characteristics and whether the current design of word processor and/or keyboard meets the needs of information users or not.

© Springer Nature Switzerland AG 2021
P.-L. P. Rau (Ed.): HCII 2021, LNCS 12771, pp. 506–526, 2021.
https://doi.org/10.1007/978-3-030-77074-7_39

In this study, we introduce the key behavior model, and concept of transfer of thought. A survey was designed to examine users' keyboard key practice, in which 95 students in China answered 10 questions. This paper first discusses existing discourse models that motivate our key behavior model and explores the cultural factor—poly-functional thought of transfer—underlying deviating behaviors of students. The survey and results; discussions and conclusion are then described in a sequential order.

2 Background

2.1 Discourse Models

In the field of computational linguistics, scholars have established a series of discourse models based on speech act theory [3, 4] and other theories [5, 10]. Conversational Roles (COR) Model is well-known as it provides a recursive dialogue between human and computer [6]. Table 1 compares two models.

The COR model is a revised version of CfA model and becomes an integral part of MERIT [13] (with MISSs [7] as its modification) system. These two models, however, ignore errors and cultural factors in the design of information systems.

Table 1. Models of information-seeking dialogues based on discourse theories [6, 7, 11, 13]

Models	Foundations	Features	Pros	Cons
CfA model [8]	Theory of speech acts [3, 9]; Habermas' theory of action [10]	"Describes sequences of dialogue acts" [6]; Acts happen in "progressive dialogue states" [6]	"Basis for the implementation of the Coordinator" [6]; Provides basic frame for other models	A "state-transition network" [6]; "Meta-level dialogues are not addressed" [6]; Commitments can't be dissolved [6]; Being a "merely responding agent", the information provider has no opportunity to clarify their intentions [6]
COR model [11]	Searle's terminology and "taxonomy of illocutionary acts" [9, 12]	All dialogue act labels both dialogue partners [6]; Both partners can fulfill their expectations [6]; The basic dialogue schema contains comprehensive transitions [6]	A "recursive transition network" [6]: It modifies the CfA model, gives both roles of action equal opportunity, and more choices	It is descriptive and ideal [6]. During an information-seeking dialogue, participants should follow "conversational conventions and typical role patterns" to "act cooperatively and negotiate commitments" [6]. However, for beginners or especially the non-English speakers, the conventions or the role patterns may be different

Based on our observation in college teaching, sometimes students do not know how to use keyboard for the lack of knowledge about English conventions. This is against the core assumption of COR model that requires the participants to follow conversational conventions. Even though the MERIT system tries to integrate information-seeking strategies, dialogue structures, scripts, and cases and becomes more practical and prescriptive [13], the computer keyboard or word processors are not smart enough to avoid errors. If users used wrong keys or used keys in a wrong way, the keys could not do anything to stop or correct errors. Besides, as we can see in Sect. 6.3, the functions, such as the "Search" box, work perfectly in one language may not be able to work well in another language.

2.2   Cultural Dimensions and Geography of Thought

Based on value orientation, Hofstede sums up six basic dimensions: power distance, individualism, uncertainty avoidance and masculinity [14], post-increase long-term orientation [15] and laissez-faire and restraint [16]. The cultural dimension theory has a significant impact on the internationalization of interactive design (e.g., the design should be compatible with the language and cultural differences of different countries and regions) and localization (e.g., the software should be adjusted according to local cultural customs when translated into local languages).

Nisbett proves that human behavior is a function of culture in the Geography of Thought. Nisbett [17] points out the main cognitive difference between Asian and the Western thought: Easterners, specifically the East Asians (principally Chinese, Korean, and Japanese) tend to be holistic (Fig. 1, right) while Westerners are more analytic (Fig. 1, left). He lists more differences between the two cultures, as can be seen in Table 2.

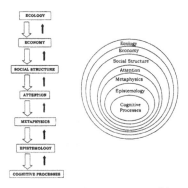

Fig. 1. Schematic model of influences on cognitive processes [17]

Nisbett [17] attempts to explain cultural facts in terms of physical environments. The ecology produces distinct economy and social structure, which forms special attention and folk metaphysics. The folk metaphysics influences tacit epistemology, or beliefs about how to get new knowledge. Social practices influence thinking habits directly.

Table 2. Differences between East Asians and Westerners [17, 18]

Environments	Differences	East Asians	Westerners
Cultural differences	Social structure	Living in a **collective, interdependent** social world with many role prescriptions, they value **harmony** [17]	Living in relatively **independent, individualistic** social circumstances, they like **debating** [17]
	Attitude to the world	**Lack of wonder**, as Chinese civilization outdistanced the others, and they could "predict important occurrences on earth" [17]	Full of **curiosity** [17]
	Social practice, e.g., Philosophy	Evanescence "To understand and appreciate one state of affairs requires the existence of its opposite; what seems to be true now may be the opposite of what it seems to be" [17] "The world is constantly changing and is full of contradictions" [17] "Returning—moving in endless cycles—is the basic pattern of movement of the Tao" [17] The world was "a mass of substances" [17]	Essence Using forms, logic, and separate attributes to understand the fundamental nature of the world [17] "The world is fundamentally static and unchanging" [17] "Aristotle's physics is highly linear" [17] The world was "a collection of discrete objects" [17]
	Science and mathematics	Connection	Contradiction
Perceptual processes	Cognitive process	Holistic	Analytic
	Context	Context-dependent	Context-independent
	Organizing the environment	"Attending to the relationship between the object and the context in which the object is located" [18]	"Focusing on a salient object independently from its context" [18]
	Causes	"Attributing causality to the context or situation" [18]	"Attributing events to causes internal to the object or person" [18]
	Everyday life events	Emphasizing "relationships and similarities" [18]	Using "categorization and rules" [18]

(*continued*)

Table 2. (*continued*)

Environments	Differences	East Asians	Westerners
	Grouping objects	Based on "the relational-contextual information" [18]	"On shared analytic features or shared categories" [18]
	Perceiving similarities	"Holistic judgments of family resemblance" [18]	"Unidimensional rule much more often" [18]
Socialization practices lead to "reproduce a pattern of attention specific to each culture" [18]	Child rearing practices	"Japanese mothers tend to engage their infants in social routines" [18], which "might direct infants' attention to the relationship or to the context in which the object is located" [18]; and to "produce more verbs" [18]	"American mothers label toys and point out their attributes" [18], which "might lead infants to focus on the objects and their appropriate categorizations" [18]; and to "extend to the prevalence of nouns" [18]
	Daily practices	"Japanese perceptual environments are more complex and contain a larger number of objects" [18]	American perceptual environments are simpler and contain fewer objects [18]

Nisbett [17] points out that cognitive and perceptual processes are not fixed or universal but are constructed in part through participation in cultural practices. The cultural environment, both social and physical, shapes perceptual processes.

As we discussed in Sect. 2.1, cultural factors were not included in the existing discourse models, such as COR model. To improve these models with state-of-art culture-based discourses, we choose Cultural Discourse Studies, which layouts the frame for us to select the salient objects – who (communicator/speaker/designer) says what (discourse) to whom (receiver/hearer/user)–in the possess of a communication [19]. After the comparison of the designer intention and the user behavior, we add a new pair of thoughts to the geography of thought.

2.3 Operant Behavior and Deviating Behavior

A person's key/button operation is operant behavior because it is determined primarily by the operation consequences. Operant behavior does not take a certain form as it is selected and shaped by specific consequences [20].

We invent "Deviating behavior" here to refer to the operating behaviors that are not what designers expected in the process of formatting a document. The common deviating behaviors or errors include:

- users prefer "Enter" and "Space bar" while ignore a lot of other keys or buttons;

- users believe "Enter" and "Space bar" are powerful enough to replace "Tab", "Indentation", "Page Break" and others;
- "Enter" and "Space bar" are used to merge the others according to their operation directions. Specifically, they replace "Tab" with "Space bar" as these two both move to the right and replace "Page Break" with "Enter" because they both move down.

2.4 Transfer of Learning and Language Transfer

Transfer of learning involves not only applying knowledge acquired in one situation to another, but also skills acquired in new situations [21].

Fries [22] and Lado [23] stress the negative influence of a learner's native language, which is considered a major cause of failure in second language learning. Lado [23] further claim that learners tend to have positive effects when the two languages are similar, and negative effects if the two are different. Odlin [24] analyzes not only factors among languages (such as grammar, vocabulary, and pronunciation), but also cross-linguistic factors of discourse, individual variation, and sociolinguistic elements can be influenced by the similarities and differences between the native language and the second language.

In this paper, we make an analogy between the language learning and computer practice, and then invent the term "transfer of thought" or "thought transfer" as we view keys as words, and the process of keyboard practice as language comprehension and employment.

2.5 Error Analysis

Errors can give teachers/students strong evidence to proceed in teaching/learning [25]. Corder [25] makes a distinction between random and systematic errors. The random errors are the product of chance circumstances, such as memory lapses, physical states and psychological conditions, which can be corrected with more complete assurance [25]. The systematic errors reveal a learner's underlying knowledge from which teachers are able to reconstruct his/her knowledge [25]. We regard the students' deviating behaviors as systematic errors because they have been repeated by a lot of students for a long time.

To analyze students' behaviors, we first observed their performance in their graduate theses by clicking the "Show/Hide Editing Marks" (Fig. 2). After having seen a lot of spaces and paragraph marks, we were certain that "Space bar" and "Enter" keys were abused. To understand the motivation behind this, we tried to communicate with the students about the designers' thoughts and compared the differences between these two groups. In educational psychology, practice under the guide of theory is better than either practice or theory learning alone [26].

2.6 Quantitative Linguistics

Quantitative linguistics analyzes the rules behind the deviating behaviors. Zipf's law [27] states that not all words are used evenly, instead only a relatively small number

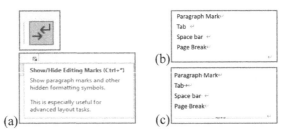

Fig. 2. Hiding and showing editing marks and its cases (English version of Microsoft Word 365, Home Edition, last accessed 2021/2/10).

of them are used frequently, while most words are used only once or not used at all. The highest frequency word is labeled as rank 1, and so on; the proportion between the rank and the frequency is a constant. Zipf's law has been widely used in various fields, including computer science, biology, and physics [28]. It can be applied in the key operation. Zipf [29] explains the reason behind the law is the principle of least effort. After that, dynamic theories, like self-organizing, self-regulation, chaos theory, dynamic system, has begun to find their way in language studies [30].

3 Key Behavior Model

Since the keys are also words, we employ cultural discourse studies (CDS) [2, 19, 31], which regards culture as the main nature of a discourse to design a new model.

In the following, we build the key behavior model. First, we break down the key behavior into three parts: Speaker (a designer who designed the keys), Discourse (keys on keyboard or the related buttons/keys in the Microsoft Word) and Hearer (the receiver of the discourse, or a user). If a speaker and a hearer do not share the same culture, the cross-cultural barriers may appear [31]. The nationality, gender, age or other cultural factors are relevant to the discourse [31]. In this paper, we focus on nationality, as the Microsoft Word was designed by Americans, and its users, are Chinese. The keys are originated in English but are also used in English because at present only English keyboards are available in China. Each key has two levels of functions: the designed function (F1)

Table 3. Key behavior model

	Speaker	Discourse	Hearer
Who (Communicator)	Designer (i.e. nationality, gender, age)	Key (i.e. language)	User (i.e. nationality, gender, age)
What (key function)	Designed function (F1)	Key name or symbol (specific name)	Performed function (F1')
How (Linguistic mapping)	Word meaning (M1)	Word form (specific name)	Word meaning (M1')

and the performed function (F1'). Ideally, F1 is also F1', or at least approximates F1'. Generally, we try to adjust F1' through teaching or training. To explain the relationship between the functions, we map each function with the corresponding word meaning, and then employ linguistic theories to compare the similarities or differences between F1 and F1', M1 and M1' (Table 3).

4 The Survey

In March 2020, a survey was distributed in the class of Computer-aided Translation Technology at Xiamen University of Technology, Fujian province in China, which involved 103 sophomores, majoring in English. Since 8 students did not respond, 95 valid questionnaires were collected at the end. In the survey, we asked participants questions relevant to publicly observable behavior (Group 1 and 2, how do they do…?), and their thoughts and feelings (Group 3, what do they think….?). It was designed by following the radical behaviorism by Skinner [32], which combines the thoughts and feelings as well as observable behaviors, seeks to observe all human behavior [20]. This survey included 10 questions, which were divided into three categories: cognition of functions of common keys, key operation method and the number of key functions. Table 4 shows the details of the survey.

Table 4. Ten questions of the survey

Group 1 Cognition of functions of common keys	Q1	The basic function(s) of "Enter" include(s)
	Q2	The basic function(s) of "Space bar" include(s)
Group 2 Key operation method	Q3	How to have 4 spaces in the beginning of the first line?
	Q4	How to have 4 spaces in the beginning of the other lines except the first line?
	Q5	How to lay 2 options evenly in a row?
	Q6	How to lay 4 options evenly in a row?
Group 3 The number of key functions	Q7	How many functions can a key have?
	Q8	How about a key with only one function?
	Q9	How about a key with 2 functions?
	Q10	How about a key with many functions?

5 Results

Results showed that participants generally preferred certain keys. They believed that each key should have multiple functions, and different keys or buttons in the word processors could be merged according to their operation direction.

5.1 Basic Functions: The Powerful "Enter" and "Space bar"

"Enter" Equals to Downward

To be more specific, "Enter" includes all possible downward functions. The results (Fig. 3) showed that about 69% participants thought the basic function(s) of "Enter" included line feed, paragraph mark and downward movement, 27% supported line feed, and 4% believed only paragraph mark was right.

"Space bar" Equals to Rightward

The "Space bar" encloses rightward functions, like "Tab", "Indent", and "Columns". Regarding the basic functions of "Space bar" (Fig. 4), 74% participants thought it can move cursor to the right, separate spaces between words and move words to the right (like Indent), 22% felt that it represented spaces between words, while 2% thought that moving cursor to the right and 2% thought that moving words to the right were the basic functions of "Space bar."

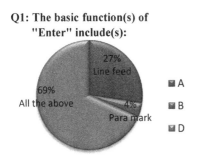

Fig. 3. The basic function(s) of "Enter" (No student chose C. Right-moving)

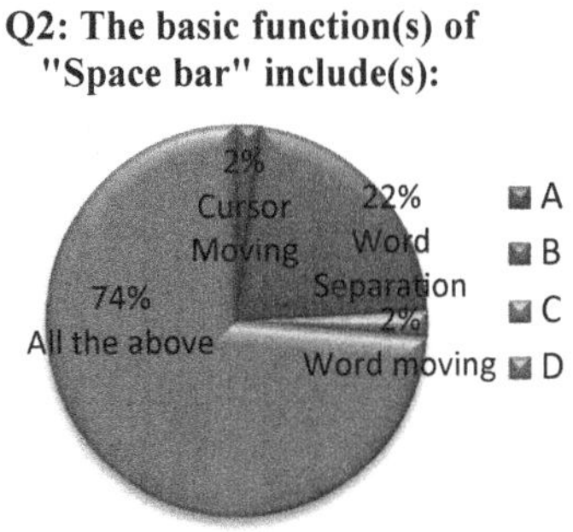

Fig. 4. The basic function(s) of "Space bar"

5.2 Operation Method: Abused "Space bar" in Chinese

To further analyze participants' operation habits, we took the "Space bar" as an example. In Microsoft Word, the "Space bar" is mainly used as a space between words in English and other western languages. Chinese, and some eastern languages, do not use Spaces to separate their characters. The chances of using "Space bar" is little when entering Chinese. However, occupying a large position on the keyboard, the abuse of "Space bar" by Chinese users is common.

"Space bar" Substitutes "Indentation"

Participants generally chose "Space bar" or "All the above" (total 70%, 61%) in these two questions (Fig. 5 and 6). Another 27% selected "Indentation before 1st line" (Fig. 5), and 29% selected "Indentation before other lines except the first line" (Fig. 6). In fact, "Indentation" works only to the whole paragraph (Fig. 7). If we put the cursor at the beginning of the first line, then enter 4 spaces, the whole paragraph moves to the right.

Q3: How to have 4 spaces in the beginning of the 1st line?

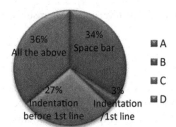

Q4: How to have 4 spaces in the beginning of the other lines except the 1st line?

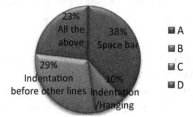

Fig. 5. The first line indentation.

Fig. 6. The hanging indentation.

Fig. 7. Increase Indent function (English version of Microsoft Word 365, Home Edition, last accessed 2021/2/10).

"First line" and "Hanging" are two special layouts in "Indentation" and can be adjusted by the "First line" and "Hanging" functions (Fig. 8). However, only 3% (Fig. 5) and 10% (Fig. 6) students selected these functions.

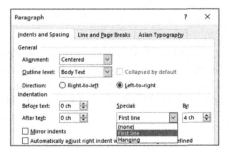

Fig. 8. The place of First line Indentation (English version of Microsoft Word 365, Home Edition, last accessed 2021/2/10).

We should point out that "Space bar" in Microsoft Word and Jinshan WPS has included some "First line" and "Hanging" functions. When there are characters in the paragraph, put the cursor either in the beginning of the first line, or the second line, then move them right, there will be no space symbol when opening "Show Editing Marks" (Fig. 9, lower). However, if there are no characters, for example, using "Space bar" first, then, entering words, there will be still space symbols before the words (Fig. 9, upper).

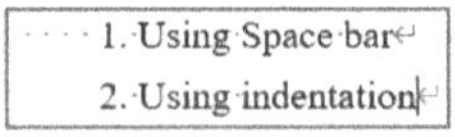

> 1. Using Space bar
> 2. Using indentation

Fig. 9. The effects of "Space bar" in different situations (English version of Microsoft Word 365, Home Edition, last accessed 2021/2/10).

"Space bar" Replaces "Tab" and "Columns"

"Space bar" was also used to sperate options in the survey (Fig. 10 and 11). Interestingly, the more options in a row, the students were more likely to choose "Space bar" (from 45% to 50%) (Fig. 10 and 11). While 22% (Fig. 10) or 16% (Fig. 11) participants adopted "Columns" to arrange the options in one row, only about 7% (Fig. 10) or 8% (Fig. 11) participants knew the "Tab" key on the keyboard, which is specially used for frameless table design or option arrangement. It seems to be a stranger to many Chinese students. Its function is to be replaced by "Space bar" or "Columns".

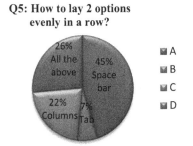

Fig. 10. Two choices laid in a row

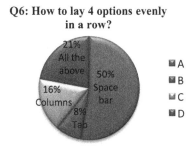

Fig. 11. Four choices laid in a row

"Columns" can split the chosen text into designated columns. If the options are listed in one column first, then they can be split into smaller ones (Fig. 12). However, the "Space bar" is a bad tool here because the position of each option tends to change easily (Fig. 13). "Tab" encloses special spaces for the possible added letters (Fig. 13).

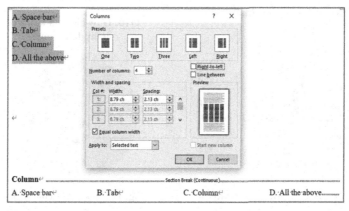

Fig. 12. The effect of "Columns" (English version of Microsoft Word 365, Home Edition, last accessed 2021/2/10).

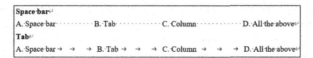

Fig. 13. The effects of "Space bar" and "Tab" (English version of Microsoft Word 365, Home Edition, last accessed 2021/2/10).

5.3 The Ideal Numbers of Functions

To find out the ideal number of functions of a key, group 3 (Fig. 14, 15, 16, 17) includes 4 questions. Though 65% participants (Fig. 14) thought a key could have as many functions as possible, 55% (Fig. 17) admitted that too many functions may increase memory burden or cause confusion easily. Just 2% supported a key with one function (Fig. 14), and about 35% students (Fig. 15) considered a key with a function as a waste of resources. In sum, participants thought the ideal design was a key with two functions (Fig. 16).

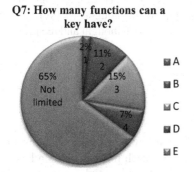

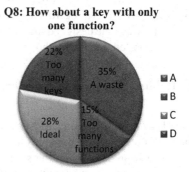

Fig. 14. The numbers of functions of a key.   **Fig. 15.** A key with one function.

Q9: How about a key with 2
functions?

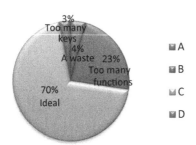

Fig. 16. A key with two functions.

Q10: How about a key with
many functions?

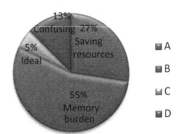

Fig. 17. A key with many functions.

The students' behaviors are common not to young people. Considering that China has popularized basic computer courses in colleges just after Windows 95 or even 98 released [33], most of the college graduates before 1998 have little chance to receive systematic word processing software education. With the popularization of computers and computer courses, the deviating of the key operation is still widespread. A new pattern with a new keyboard or word processing design theory, targeted to eastern Asians is called for. We will discuss the possible theoretic framework in the following section.

6   Discussion

6.1   Case of Key Behavior Model

We fill in the above-mentioned Key Behavior Model with the designed functions and the performed functions (Table 5). As can be seen in the Table 5, $F1 < F1'$, $F2 < F2'$, $M1 \neq M1'$, and $M2 \neq M2'$. F1 has more than one function, but its B and C function are realized by strict conditions. If the cursor was placed in other places, for example, after the second letter of the first line or the second line, "Space bar" immediately became its original shape as a space. Unlike the M1 and M2, where the first meanings are their core meanings, M1' and M2' obviously have hidden hyper-level meanings: right-moving and down-moving. We regard such behavior as transfer of thought, which reflects the conflicts between the designed mono-function-led expansive thought and the preformed poly-function-led organizing thought.

6.2   Cultural Discourse in User Behavior

Zipf's Law and the Theory of Least Effort

Zipf [29] stresses that "the study of words offers a key to an understanding of the entire speech process, while the study of the entire speech process offers a key to an understanding of the personality and of the entire field of biosocial dynamics" (p. 32). Zipf's law describes that in a given-length natural text, not all words are used in the same frequency (Fig. 18) [29]. In our study, the keys on keyboard which can be tapped easily

Table 5. "Enter" and "Space bar" behaviors in word processing layout of Chinese students.

	Designer	Key	User
Who (Communicator)	American corporation	English or blank	Chinese students
What (key function)	Designed function(s)	Key name or symbol	Performed function(s)
	F1: A. Space between English words and/or numbers; B. Indentation of First line (when the cursor is put before the first letter and there is at least one letter); C. Indentation of Hanging (when the cursor is put before the first letter of the second line)	Space bar/blank	F1': A. Space between English and/or Chinese characters and/or numbers; B. Indentation of First line; C. Indentation of Hanging; D. Indentation; E. Tabs; F. Columns; G. Other possible right-moving functions
	F2: A. Paragraph mark;	Enter	F2': A. Paragraph mark; B. Line and Paragraph Spacing; C. Page Break; D. Manual Line Break; E. Other possible down-moving functions
How (Linguistic mapping)	Word meaning	Word form	Word meaning
	M1: A is the core meaning; B and C are parts of users' performed functions	Space bar	M1': Right-moving is the core meaning; A-G are all its specific meanings
	M2: A is the core and only meaning	Enter	M2': Down-moving is the core meaning; A-E are all its specific meanings

are more popular than those buttons scattering below different tabs in the Microsoft Word, and the keys on the right of the keyboard are frequently used than those on the left.

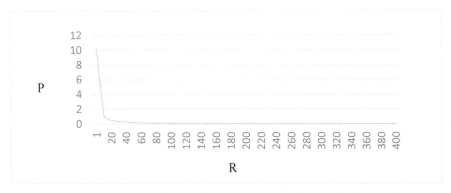

Fig. 18. Zipf's law [29] (Note: we inferred the Top 400 words frequencies data partly from [29], partly according to R × F = C, and C approximates 10% of the total number. In the above Figure, R refers to the rank of frequency, F is short for frequency, P as percentage, equals to (F/Total number) * 100%.)

We cannot infer which key is most frequently used in a document formatting operation, due to the limited sample size of the survey. But the use of commands of Microsoft Word 2003 verified that Zipf's law was executable. We quote the command bar clicks collected by Microsoft in 2005 here [34].

Top 5 Most-Used Commands in Microsoft Word 2003 are 1. Paste, 2. Save, 3. Copy, 4. Undo, and 5. Bold.

Together, these five commands account for around 32% of the total command use in Word 2003. Paste itself accounts for more than 11% of all commands used, and has more than twice as much usage as the #2 entry on the list, Save.

... ...

Beyond the top 10 commands or so, however, the curve flattens out considerably. The percentage difference in usage between the #100 command ("Accept Change") and the #400 command ("Reset Picture") is about the same in difference between #1 and #11 ("Change Font Size")

From these descriptions, we can see that Top 1 "Paste" has the lion's share 11%, one tenth of about 400 commands. The second "Save" goes down sharply about half as the first one. The following three (about 16%) decreased little by little, shaping a curving corner. After 10 commands, the curve flattens out considerably.

Why do people prefer some keys to others? Zipf [29] addresses the principle of least effort. There are two opposite forces – the force of unification and the force of diversification – in the communication. Speakers tend to use as least words as possible to express their ideas, while listeners have to choose the suitable one from a lot of meanings. A word or a few words with a lot of meanings saves the speakers' effort but makes it difficult for the listeners. If listeners are given options, they will like one-word-one-meaning, that is, to decrease the effort. As a compromise, the most used words have a lot of meanings, and on the contrary, many rare words have only one meaning.

The same are the behaviors of keys. The Speaker, the English designer of keyboard or Word, designated each key/button a certain function to complete a task, however, the Hearer, Chinese users, actively rearranged the keys and their functions.

Zipf [29] analyzes the forms and functions of tools, with two main economic principles (1) the Law of Abbreviation and (2) the Law of Diminishing Returns. These two laws deduce another three principles: ① the Principle of Economical Versatility, ② the Principle of Economical Permutation, and ③ the Principle of Economical Specialization. Zipf [29] presumes that there were 10 tools and each of them had one function. The tools were placed within different distances to the user. After a certain time, a tool was likely to absorb the 9 functions of the other tools. This is the Principle of Economical Versatility. The new nearest tool would shrink in shape (the Law of Abbreviation) to fit for the high frequency of use. If the nearest function 1 had to combine with function 3 to realize function 2, which meant the potential replacement (Returns) would not be effective (Diminishing), then the Principle of Economical Permutation determined that function 2 would be kept as it accords to the Principle of Economical Specialization.

In the key behavior model, the keys are both words and tools. The participants tend to use "Enter" and "Space bar" to enclose the functions of other buttons, which are the reflection of the Principle of Economical Versatility. Some of the replaced functions, such as "First line" in "Paragraph", "Page Break" in "Insert", are under the other tabs, which could be easily absorbed by nearer keys; the effects of other functions, for example, "Indent" and "Paragraph and Line Spacing", look similar to keys of "Space bar" and "Enter" respectively, are naturally included by these two keys.

It seems that the Chinese users have expanded the commonly used functions of certain keys. It can be attributed to cultural factors, as Chinese language and culture tended to choose and keep an all-in-one word or tool in their lives.

6.3 Transfer of Thought in Key Practice

Mono-function vs. All-in-One Poly-function

Participants believed that there was more than one basic function of a key, which was a direct manifestation of one thing with multiple functions. This kind of thinking, as we observed, is not an exception, but was ingrained in Chinese daily life, language, and culture, which differs from English one-thing-one-function tradition.

Polysemy is more common in Chinese than in English. Polysemy refers to a word that holds different but related meanings [35]. This can be reflected from the increasing number of English words and the relatively steady Chinese characters. In 1716, Kangxi Dictionary, named after the emperor's era name, was published. It has 47,035 Chinese words [36]. The first English dictionary was compiled by Samuel Johnson and published in 1755, including 42,773 entries [37]. Cihai, the biggest dictionary of Chinese, has approximately 130,000 entries in the 7th Edition [38]. The Oxford Dictionary [39], famous for its comprehensive content, has documented 600,000 words from more than 1,000 years of history across the English-speaking world, which is more than 4 times

Table 6. The words of Chinese and English in the 18th century to now

Century	Chinese characters	English words
18th	47,035	42,773
21st	130,000	600,000

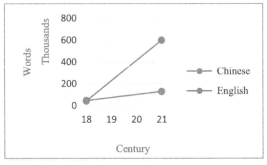

Fig. 19. The total words of English Chinese words from century 18th to now. (Note: To make it clear to see, we change the time from year to century.)

its counterpart (Table 6, Fig. 19). Since China has a much longer history than English-speaking countries, we assume that Chinese words averagely add more meanings to old words while English create more new words to describe new things and express new ideas.

Besides, Chinese characters feature for its homophones, which refers to the words sharing the same pronunciation (tones were excluded) but differing in meanings, origins, or word forms [35]. An extreme case is a story written by 96 characters with the same pronunciation by Zhao Yuanren [40, 41] (Table 7).

Table 7. An example of 96 homophones in a story [40, 41]

The original text by Zhao (Simplified Chinese version)	The Chinese Pinyin (pronunciation) of the text
《施氏食狮史》 1. 石室诗士施氏，嗜狮，誓食十狮。 2. 氏时时适市视狮。 3. 十时，适十狮适市。 4. 是时，适施氏适市。 5. 氏视是十狮，恃矢势，使是十狮逝世。 6. 氏拾是十狮尸，适石室。 7. 石室湿，氏使侍拭石室。 8. 石室拭，氏始试食是十狮。 9. 食时，始识是十狮，实十石狮尸。 10. 试释是事。	« Shī Shì shí shī shǐ » 1. Shíshì shīshì Shī Shì, shì shī, shì shí shí shī. 2. Shì shíshí shì shì shì shī. 3. Shí shí, shì shí shī shì shì. 4. Shì shí, shì Shī Shì shì shì. 5. Shì shì shì shí shī, shì shǐ shì, shǐ shì shí shí shishì. 6. Shì shí shì shí shī shī, shì shíshì. 7. Shíshì shī, Shì shǐ shì shì shíshì. 8. Shíshì shì, Shì shǐ shì shí shì shí shī. 9. Shí shí, shǐ shí shì shí shī, shí shí shí shī shī. 10. Shì shì shì shì.
English version [41]	
Story of Stone Grotto Poet: Eating Lions [40]	
A poet named Shi lived in a stone room. He was fond of lions and swore that he would eat ten lions. He constantly went to the market to look for lions. At ten o'clock, ten lions came to the market, and Shi went to the market. Looking at the ten lions, he used his arrows to make the ten lions dead. Shi picked up the corpses of the ten lions and took them to his stone room. The stone room was damp. Shi ordered a servant to wipe the stone room. As the stone den was being wiped, Shi began to eat the meat of the ten lions. During the meal, he began to realize that the ten lion corpses were in fact ten stone lions. He then tried to write down this story.	

Though this text is an extreme case, there are a lot of words sharing the same pronunciation.

Not only homophones, but there are also homographs in Chinese verbal communication. A word, with the same graph, often same meanings, is pronounced by people from different areas with different accents. When people cannot communicate by speaking, they can understand each other by written words.

In cross-cultural communication, a common and interesting part is about food and cooking. Chopsticks, a typical Chinese tool, is still the main choice for Chinese. Ancient Chinese had made knives and forks with bones thousands of years ago [42] before they changed it into bronze. From the users' point of view, forks were easier to learn than chopsticks. Children in China often use spoons or forks before six, then they begin to use chopsticks clumsily when their fingers become more flexible and powerful. Why do Chinese replace a more convenient tool with a complex one? The least of effort cannot explain it. If we take functions into consideration, it is understandable. Chopsticks is used by one hand and can pick up almost all kinds of food except soup.

New technological tools are still following the poly-function trend. To most Chinese, they begin a day either with WeChat, or Alipay. In WeChat, they can chat with family, friends or business partners, buy most necessities, pay for their goods, learn about the current affairs, watch videos, and order food or taxi, and other everyday life in it. Instead, westerners download different apps to satisfy their needs.

Therefore, to make a document more neatly and orderly, Chinese users rearrange the keys on the keyboard with buttons in the word processors, and automatically categorize them into few types: direction-moving functions. The right-moving function belongs to "Space bar", and the down-moving is to "Enter." Besides, users do not spare extra effort to memorize. The other keys, like "Tap", are naturally kept out of sight. This way, we come to realize why English majors, even teachers teaching English, ignore an English key for so many years and do not bother to look up its meaning in a dictionary when electric dictionary is so easy to access and Chinese users are more curious to know a word's meaning.

From the Chinese users' point of view, the current design of keyboard and Simplified Chinese version of Microsoft Word's layout are not friendly. One thing should be noted is that the "Search" function in Chinese is not workable at all (Fig. 20, Right). Though in Chinese version, the box of "Search" and its function tip are almost the same as its English version, the search results are totally different. When entering "page" in the English "Search" box, we have received a lot of useful information, such as actions with their icons, names, functions, and shortcuts (Fig. 20, Left), but there is no function result at all in simplified Chinese "Search" box (Fig. 20, Right). It seems that when Chinese users do not know which key/button to choose, they cannot get immediate help from Word.

As holistic viewers, Chinese enjoy all the necessary tools in one place. The English designers divide the keys/buttons into different tabs. We list the possible word formatting processes in details here to show what we were expecting to do and where we could find the necessary keys/buttons to finish a text formatting. To separate English words, we tapped "Space bar" on the keyboard. To indent the whole paragraph, we chose "Increase Indent" in "Paragraph" under "Home". To indent the first line, we clicked the "Paragraph

Settings" dialogue, chose "First line" under "Special" in "Indentation", then tapped "OK" and the dialogue box closed, otherwise, we couldn't go on with a dialogue opened. Next time, we reopened the dialogue as we wanted to "Hang" the other lines. The "Columns" belong to "Page Setup", while the "Tabs", is hidden in the "Tabs Dialogue" which is to be opened at the left bottom on "Paragraph Settings Dialogue" box. If a document is to label page numbers, we should "Insert" them. We cannot "Insert" Contents because contents cannot be inserted, but "Referred", so we should go to "References." We chose a module from "Design" and set our favorite as Default. But to modify the details, we had to go back to other places.

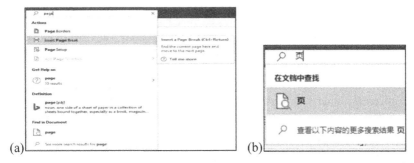

(a)   (b)

Fig. 20. The results list of "Page" (Left: English version; Right: Simplified Chinese version of Microsoft Word 365, Home Edition, last accessed 2021/2/10).

To sum up, Chinese prefer to include multiple functions in one, while westerners are willing to designate the functions to different tools. This poly-functional vs mono-functional cultural differences, like but obviously not holistic vs analytic, is not elaborated by Hofstede or Nisbett. We regard it as a new pair of differing thinking way between East Asians and Westerners. The Chinese students' behaviors, focusing on fewer keys and ignoring others, are just performance of transfer of poly-function thought.

7 Conclusion

In this paper, Chinese participants' keyboard behaviors have been analyzed. They deviated the designer's expectations and showed obvious distinguished cultural factors in them. To analyze the cultural factors, we have reviewed the existing two dialogue models and found none of them involves cultural elements in design. Then we came up with a new key behavior model based on error analysis, geography of thought, transfer of learning, Cultural Discourse Studies, lexicology, and added a new angle between the comparison of Easterners and Westerners. We found that the poly-function may be a common thinking method in Chinese, which can lead to the deviating behaviors.

Poly-function is important in the following aspects: a) If teachers know that systematic errors are part of transfer of thought, they can use it to guide teachers in teaching; b) The keyboard designers should not just keep English words, like "Enter", "Tab" keys. The localization is necessary by considering cultural factors; c) The IT engineers in

China should design products based on behaviors of Chinese users; and d) Designing user interfaces that can be adapted to various mobile applications is critical. We plan to explore the evolution of user interfaces of word processors from desktop to mobile phone to further test the mono- or poly-functional design thought.

References

1. Wang, C., Yuan, X.: Cultural discourse in user interface design: investigating characteristics of communicators in Microsoft word. In: Rau, P.L. (ed.) HCII 2020. LNCS, vol. 12192, pp. 212–227. Springer, Cham (2020). https://doi.org/10.1007/978-3-030-49788-0_16
2. Xu, S.: Chinese Discourse Studies. Palgrave Macmillan, London (2014)
3. Austin, J.: How to Do Things with Words. Clarendon Press, Oxford (1962)
4. Searle, J.R.: A Taxonomy of Illocutionary Acts. University of Minnesota Press, Minneapolis (1975). https://hdl.handle.net/11299/185220
5. Bunt, H.C.: Information dialogues as communicative action in relation to partner modelling and information processing. In: Taylor, M.M., Néel, F., Bouwhuis, D.G. (eds.) The Structure of Multimodal Dialogue, pp. 47–73. Elsevier Science Publishers, Amsterdam (1989)
6. Sitter, S., Stein, A.: Modeling information-seeking dialogues: the conversational roles (COR) model. RIS: Rev. Inf. Sci. (Online J.), 1(1) (1996). https://www.fb10.uni-bremen.de/anglis tik/langpro/webspace/jb/info-pages/misc/sds/cor.pdf
7. Yuan, X.J., Belkin, N.J.: Applying an information-seeking dialogue model in an interactive information retrieval system. J. Doc. 70(5), 829–855 (2014). https://doi.org/10.1108/JD-06-2013-0079
8. Winograd, T., Flores, F.: Understanding Computers and Cognition. Ablex, Norwood (1986)
9. Searle, J.R.: A taxonomy of illocutionary acts. In: Searle, J.R. (ed.) Expression and Meaning. Studies in the Theory of Speech Acts, pp. 1–29. Cambridge University Press, Cambridge (1979)
10. Habermas, J.: The Theory of Communication Action. Fayard, Paris (1981)
11. Sitter, S., Stein, A.: Modeling the illocutionary aspects of information-seeking dialogues. Inf. Process. Manage. 28(2), 165–180 (1992)
12. Searle, J.R., Vanderveken, D.: Foundations of Illocutionary Logic. Cambridge University Press, Cambridge (1985)
13. Belkin, N.J., Cool, C., Stein, A., Thiel, U.: Cases, scripts and information seeking strategies: on the design of interactive information retrieval systems. Expert Syst. Appl. 9(3), 379–395 (1995)
14. Hofstede, G.: Culture's Consequences: International Differences in Work Related Values. Sage Publications, London (1980)
15. Hofstede, G.: Cultures and Organizations: Software of the Mind. McGraw-Hill, London (1991)
16. Hofstede, G., Hofstede, G.J., Minkov, M.: Cultures and Organizations: Software of the Mind, 3rd edn. McGraw-Hill, London (2010)
17. Nisbett, R.E.: The Geography of Thought: How Asians and Westerners Think Differently… and Why. Free Press, Michigan (2003)
18. Nisbett, R.E., Miyamoto, Y.: The influence of culture: holistic vs. analytic perception. Trends Cogn. Sci. 9(10), 467–473 (2005)
19. Xu, S.: Communication of Contemporary China: Studies in a Cultural Discourse Studies Perspective. Peking University Press, Beijing (2010)
20. Cooper, J.O., Heron, T.E., Heward, W.L.: Applied Behavior Analysis, 2nd edn. Wuhan University Press, Wuhan (2012)

21. Slavin, R.E.: Educational Psychology: Theory and Practice, 10th edn. Pearson, Hoboken (2011)
22. Fries, C.: Teaching and Learning English as a Foreign Language. University of Michigan Press, Ann Arbor (1945)
23. Lado, R.: Linguistics Across Cultures: Applied Linguistics for Language Teachers. University of Michigan Press, Ann Arbor (1957)
24. Odlin, T.: Language Transfer: Cross-Linguistic Influence in Language Learning. Cambridge University Press, Cambridge (1989)
25. Corder, S.P.: The significance of learner's errors. Int. Rev. Appl. Linguist. Lang. Teach. **5**, 161–170 (1967). https://doi.org/10.1515/iral.1967.5.1-4.161
26. Judd, C.H.: Psychology: General Introduction. Scribner, New York (1907)
27. Zipf, G.K.: The Psycho-Biology of Language: An Introduction to Dynamic Philology. M.I.T. Press, Cambridge (1935)
28. Jiang, W.Q.: Zipf and the principle of least effort. Tongji Univ. J. Soc. Sci. Sect. **16**(1), 87–95 (2005)
29. Zipf, G.K.: Human Behavior and the Principle of Least Effort: An Introduction to Human Ecology, Kindle Ravenio Books Blackrock, New York (2016)
30. Liu, H.T.: An Introduction to Quantitative Linguistics. The Commercial Press, Beijing (2017)
31. Xu, S.: A Cultural Approach to Discourse. Palgrave Macmillan, London (2005)
32. Skinner, B.F.: The Behavior of Organism. Appleton-Century-Crofts, New York (1938)
33. Lu, X.H.: Computer Applications Guide (Windows 98 Environment). Tsinghua University Press, Beijing (1999)
34. Harris, J.: No Distaste for Paste (Why the UI, Part 7) (2006). https://docs.microsoft.com/en-us/archive/blogs/jensenh/no-distaste-for-paste-why-the-ui-part-7. Accessed 18 Jan 2021
35. Wang, R.P., Wang, Z., Zhu, Y.F.: English Lexicology: A Coursebook. East China Normal University Press, Shanghai (2011)
36. Kangxi Dictionary. https://www.nlc.cn/dsb_zyyfw/mzyk/. Accessed 18 Jan 2021
37. A Dictionary of the English Language. A Digital Edition of the 1775 Classic by Samuel Johnson. https://johnsonsdictionaryonline.com/about-this-project/. Accessed 18 Jan 2021
38. About Cihai. https://www.cihai.com.cn/about/index. Accessed 18 Jan 2021
39. Oxford English Dictionary. https://languages.oup.com/dictionaries/#oed. Accessed 18 Jan 2021
40. Zhao, Y.R. (1930). https://en.wikipedia.org/wiki/Lion-Eating_Poet_in_the_Stone_Den. Accessed 18 Jan 2021
41. Pae, H.K.: Chinese, Japanese, and Korean writing systems: all East-Asian but different scripts. In: Pae, H.K. (ed.) Script Effects as the Hidden Drive of the Mind, Cognition, and Culture, pp. 71–105. Springer International Publishing, Cham (2020). https://doi.org/10.1007/978-3-030-55152-0_5
42. Qinghai Provincial Administration of Cultural Relics: Brief Report on Excavation of Zongri Site in Tongde County, Qinghai. Archaeology 5(1–14, 35) (1998)

Multimodal Features as a Novel Method for Cross-Cultural Studies

Xuhai Xu[1,2], Nan Qie[2], and Pei-Luen Patrick Rau[2(✉)]

[1] Tsinghua University, Beijing, China
[2] University of Washington, Seattle, USA
`rpl@mail.tsinghua.edu.cn`

Abstract. The rise of media and new tools in computer science provide new approaches to study culture. We proposed a novel research method that leverages facial expression and language features from TV series to assist cross-cultural studies. We first compared the statistical results of the features and drew a set of conclusions that can be supported by a number of previous works in cultural studies. Then, we employed the multimodal features to train a classifier to recognize the country that a TV series belongs to. A linear SVM achieved a high average accuracy of 94.4%. We further interpreted the coefficients of the model and obtained new observations from the results. Our method can avoid the drawbacks of the existing cross-cultural research approaches, provide new ideas for media data analysis, and aid culture adaptive product design.

Keywords: Multimodality · Cross-cultural analysis

1 Introduction

Cross-cultural study is a scientific comparative research method that systematically compares among different cultures and aims to answer questions about the incidence and causes of cultural variation, as well as complex problems across a worldwide domain [1]. An improved understanding of people across different cultures can aid transnational corporations development, global team management, individual effectiveness, and product design [2,3].

In the field of human-computer interaction (HCI), the main approaches to cross-cultural investigation involve surveys [4], interviews [5], lab experiments [6], field studies and observational studies [7]. In the era of big data, social media, pop product and various human activity records can provide vivid evidence on culture influence. There is a potential space of leveraging big data analysis methods for cross-cultural research [8]. In this study, we propose a novel research method to conduct cross-cultural study based on big data.

We choose TV series as the data source. TV series is one of the most prevalent entertainment media across the world [9], which contains rich cultural information [10,11]. It can not only reflect some daily routine of a specific subgroup

---

X. Xu—The author was studying at Tsinghua when conducting this research.

© Springer Nature Switzerland AG 2021
P.-L. P. Rau (Ed.): HCII 2021, LNCS 12771, pp. 527–546, 2021.
https://doi.org/10.1007/978-3-030-77074-7_40

(depending on the type of the TV series) that belongs to this culture, but more importantly, TV series can reveal its society's characteristics such as social norm, values and beliefs [12–14]. To the best of our knowledge, there is no previous work conducting cross-cultural studies via exploiting TV series. Our research is an original attempt to do cross-cultural study based on multimodel data extracted from TV series of different cultures.

We believe that our method can avoid the drawbacks of the existing cross-cultural research approaches, provide new ideas for media data analysis, and aid culture adaptive product design. The contributions of this paper are three-fold:

- We proposed a novel cross-cultural research method via comparing multi-modal features extracted from the TV series across different cultures.
- We provided a classifier to automatically classify a TV series into a culture based on its features with high accuracy.
- Our results on the one hand supported a number of previous conclusions and assumptions from traditional cultural studies. On the other hand, the results revealed several new findings that have not been discussed before.

2   Related Work

We reviewed related work on the relationship between media and culture, as well as multimodal analysis method on social media.

2.1   Relationship Between Media and Culture

Media is a strong reflection of the culture [15]. Previous works studied and compared various forms of media among different cultures such as newspaper articles [16], artworks [17], advertisements [18], and street scenes [19]. Pop cultural product such as fiction, music, cartoon, movies and TV series is a collection of media that is more close to contemporary culture [20,21]. Comparison of these media across countries might provide meaningful results that are more relevant to people's life of the day. However, only a few works looked at these media from cross-cultural aspect. Gould et al. compared customers' attitudes towards product placement in movies across three countries [22]. Trepte discussed cultural proximity via comparing TV entertainment evaluation among 325 students from eight countries [21]. These works require enormous human effort [22], or have similar drawbacks as real-life cross-cultural studies such as reference group effect [21,22].

2.2   Multimodal Analysis on Social Media

Since there is few related work on multimodal analysis for TV series, we reviewed another relevant media that has similar data contents such as human faces and natural languages: social media. Social media has become a source of almost infinite information that provides effective new ways of sensing social behaviors and activities via user-generated multimedia contents [23]. Multimodal analysis is one of the powerful methods for social media data mining, which has been employed for various tasks such as sentiment analysis [24], preference-aware recommendation [25], social events detection [26] and public health [23].

The most relevant topic to our research is multimodal sentiment analysis, which was proposed by Morency et al. [24]. They extracted three modalities for sentiment classification (positive, neutral and negative) on a Youtube multimodal dataset. Three modalities included textual features (word polarity), visual features (smile duration and look-away duration), and audio features (pause duration and pitch). They then trained a tri-modal Hidden Markov Model and achieved a F1 score of 0.533 on classification. Cambria et al. blended the output of a facial expression analyzer (Ekman's six universal emotions [27]) and a text-based semantic and sentiment analyzer to generate a continuous stream characterizing a user's sentiment [28]. Poria et al. extended Cambria's work by extracting more detailed facial expression features such as distance between the left eyebrow inner and the outer corner, as well as an improved textual analyzer to extract lexical features [29,30]. They explored both feature-level and decision-level fusion and finally got the best F1 score as 0.776 by an Extreme Learning Model on the same Youtube dataset [24]. You et al. employed a convolutional neural network for image sentiment analysis and trained a paragraph vector model for textual sentiment analysis [31]. They then fused these features, trained a multimodality regression model and achieved an F1 score of 0.800 on a weakly labeled multimodal dataset crawled from websites.

3   Methodology

In order to fill the gaps in the previous research, and overcome the defects mentioned above, we propose a new method for cross-cultural analysis via two consecutive steps: 1) extract visual and textual features respectively from TV series, compare them across different cultures and summarize findings from the comparison; 2) integrate both modalities, train a classifier to recognize a TV series's culture, and find significant features for identifying each culture via model interpretation.

We chose three countries with distinctive cultural characteristics and mature TV industry to exemplify our new method: America, Japan and Korea. For each country, we selected three categories of TV series to cover common daily routine scenarios: family, school and workplace. We collected the video and text (actors' lines) contents of 12 TV series for each category each country according to IMDb leaderboard. Overall, we collected the material of 108 TV series, 36 for each of three countries, which contained over 1,500 h video, approximately 10.8 million figures after downsampling and 8.7 million words after filtering and cleaning.

3.1   Visual Feature Extraction

We employed Affectiva SDK [32], a facial analysis tool trained on large-scale human facial image data, to extract visual features from video contents. The

Affectiva SDK can extract facial information and automatically analyze spontaneous facial expression in video files. It can identify seven emotions and 21 facial
expressions. Table 1 lists these two types of visual features[1]. **Facial expres-**
sions are the fundamental features, which indicate the basic actions appear on
a human face. **Emotions** are the high level features. Each of them is a complex
combination of facial expressions. For example, *Joy* is the combination of *Smile*
(positive expression), *Brow raise* and *Brow forrow* (both negative expressions).
Affectiva outputs a confidence score for each feature, ranging from 0 to 100. The
score can be interpreted as the intensity of the feature.

Table 1. Features extracted by affectiva

Type	Count	Features
Emotion	7	Joy, Fear, Disgust, Sadness, Anger, Surprise, Contempt
Facial expression	21	Smile, Innerbrowraise, Browraise, Browfurrow, Nosewrinkle, Upperlipraise, Lipcornerdepressor, Chinraise, Lippucker, Lippress, Lipsuck, Mouthopen, Smirk, Eyeclosure, Attention, Eyewiden, Cheekraise, Lidtighten, Dimpler, Lipstretch, Jawdrop

We established a pipeline for processing a video file. For each video, we first
converted it to a common resolution (426 * 240, 16:9) to avoid size effect. A
video not at 16:9 ratio was clipped (vertically and horizontally center aligned)
rather than stretched to maintain a normal form of human face. We further
downsampled the video 2 Hz to ease the computation. We then applied Affectiva
SDK on the video. Each figure where a human face was detected would generate a
feature vector at a length of 28 (7 emotions and 21 facial expressions). Therefore,
we could extract a sequence of visual feature vectors via concatenating the results
of all episodes for every TV series.

We further extracted the **occurrence rate** and the average **intensity** of
every feature from the vector sequence. For each feature, we set 5 as the score
threshold to filter the vector sequence where this feature was detected. The occurrence rate is the proportion of the filtered vectors among the whole sequence. And
the average intensity is the mean value of this feature in the filtered sequence.
Eventually, we could obtain a vector at a length of 56 (28 occurrence rate and 28
average intensity) for each TV series. Direct comparison of each visual feature
among countries was an effective approach which would imply the differences
across cultures on facial and emotional expressions.

Note that Affectiva's model was trained on a massive dataset that covered
people from 75 countries and is insensitive to cultural differences [32]. In our
study, we further verified its sensitivity to races by manually selecting 20 pictures
from each country for each emotion. All pictures were agreed by two authors that

---

[1] https://developer.affectiva.com/metrics/.

the target emotion was very obvious. We compared the results of Affectiva among the three cultures and it did not show significant differences for any emotion.

3.2   Textual Feature Extraction

We leveraged two types of textual features: *emotional polarity* and term frequency-inverse document frequency (***TF-IDF***). Due to copyright policies in Japan and Korea, it is difficult to collect Japanese and Korean subtitle files for these two countries' TV series. Therefore, we used Chinese language subtitles for analysis of all three cultural videos. We first merged all episodes' subtitle files of each TV series into one document and got a set of documents that equals the number of all TV series.

For emotional polarity analysis, we utilized BaiDu Artificial Intelligence Open Platform[2] for Chinese textual sentiment analysis. The platform outputs a emotional polarity score ranging from 0 to 1 for each sentence. 0 indicates extreme negative emotion while 1 means extreme positive emotion. Every document, i.e., every TV series, generated a sequence of polarity score. In addition to calculate the average polarity of each document, we counted and calculated the appearance rate of extreme positive emotional sentences (polarity greater than 0.8) and extreme negative emotional sentences (polarity less than 0.2) as two indicators for extreme expressions.

As for TF-IDF, we proposed a word list with 16 semantic categories based on LIWC [33] and relevant literature on linguistics and cultures [34–37]. Table 2 shows several examples of each category. For each TV series, the term frequency (TF) for one category was the sum of the occurrence frequencies in the corresponding document of all words that belongs to this category. Inverse document frequency (IDF) for this category was the logarithmic quotient of the number of all documents and the number of the documents where at least one word in this category appeared. The TF-IDF for one category of one series was calculated by

$$\text{TF-IDF}(T_i, d_k) = \frac{\sum\limits_{t_j \in T_i} n_{t_j, d_k}}{n_{d_k}} \times \log \frac{1 + D}{1 + n_{T_i}}$$

where T_i is the category and d_k is the document, i.e., one TV series. t_j represents every word belonging to the category T_i. n_{d_k} is the overall word count in d_k, and n_{t_i, d_k} is the count of t_i in document d_k. D is the overall number of documents and n_{T_i} is the number of the documents where at least one word in T_i appeared. 1 is used for smoothing. Note that the numbers of t_j differ across different T_i. It makes no sense to compare TF-IDF(T_i, d_k) and TF-IDF(T_j, d_k). We mainly investigated the distinction between series, i.e., TF-IDF(T_k, d_i) versus TF-IDF(T_k, d_j), rather than the categories in one series.

Overall, we extracted three polarity features and 16 TF-IDF features for each document. We concatenated them and obtained a vector with a length of 19 for every TV series.

---

[2] https://ai.baidu.com/.

Table 2. Word list from LIWC and literature (in Chinese)

Type	Count	Words and Phrases
Personal pron (Ppron)	33	I, you, he, she
1st-person pron in Ppron	8	I, me, my. mine
2nd-person pron in Ppron	5	you, your, yours
Kinship terminology	146	son, daughter, father
Approval	31	good, agree, no problem
Disapproval	20	disagree, cannot, deny
Modest	42	humble, my poor self
Respect	226	please, welcome, honor
Plan	23	will, plan, scheme
Rule	38	rule, obey, follow
Sureness	11	must, have to, necessary
Unsureness	17	maybe, might, probably
Musculinity value	43	success, money, fame
Femininity value	39	love, care, concern
Self-expression	7	I think, I want, my opinion
Sex	6	have sex, sleep with

3.3   Multimodal Feature Integration

Besides extracting and comparing visual and textual features between cultures, we also integrated the two modalities and investigated the ability of these features to distinguish which culture a TV series belongs to.

We chose a support vector machine (SVM) classifier with a linear kernel. A simple linear model has advantages over other complex "black-box" models such as deep neural network in interpretability: the coefficient of every feature in a linear SVM indicates the importance of this feature during the classification. Moreover, a high l_1 penalty for a linear SVM could automatically conduct feature selection, therefore, reduce the effect of over-fitting. These properties could provide insightful results about the significance of some features to a culture.

The facial expressions and emotions of the visual features are inner correlated: emotions are complex combinations of facial expressions. Therefore, we chose the fundamental level features, i.e. the facial expression features, for training the classifier to avoid colinearity. The textual features do not have such correlations. Therefore we feed all textual features for classification. Overall, for each TV series, we integrated 42 visual features (21 occurrence rates and 21 intensities) and 19 textual features (3 sentiment features and 16 TF-IDF features), which lead to a vector with a length of 61.

We trained a linear SVM model based on these vectors. Specifically, we used OVR (one-vs-rest) method and trained one binary classifier for one country, taking the rest as the negative instances. For each classifier, we employed LOO (leave-one-out) method for cross-validation to further minimize over-fitting. Note that we chose leave-one-out cross-validation over other techniques such as 10-fold cross-validation due to the bias-variance trade-off [38, 39]. Every instance was classified by a classifier that was trained on the rest of the data. The number of classifiers for one country equaled the number of the instances of this country in the dataset. In order to interpret the models, we averaged the coefficients over all classifiers for each country.

4 Results

We first compared these features one by one across the three countries to study cultural differences. Then, we utilized the 42 fundamental visual features as well as the 19 textual features to train a linear SVM classifier. We further interpreted the coefficients of each features.

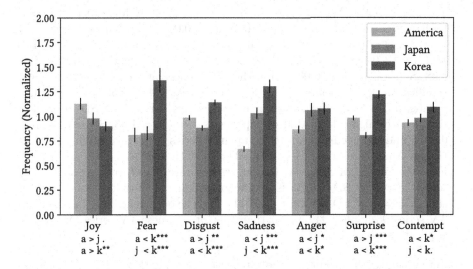

Fig. 1. Occurrence rates of visual emotions of three countries. The texts beneath each emotion show the results of pairwise t-test with Bonferroni adjustment method. The $>$ and $<$ indicates the order between two countries. Note that . for $p < 0.1$, * for $p < 0.05$, ** for $p < 0.01$ and *** for $p < 0.001$. The same for all barplots below.

4.1 Country Comparison

Visual Features: Emotion

As described in the previous section, there are two types of visual features including fundamental facial expression features and high-level emotion features. We directly compared the seven high-level emotion features across three countries to investigate cross-cultural differences. We employed pairwise comparisons using t-test with Bonferroni adjustment method as the post-hoc test to investigate the differences between each pair of countries. Figure 1 visualizes the occurrence rates of the emotions.

Most of the occurrence rates and the intensities of the emotions have significant difference among countries. We first summarized observations from the perspective of the occurrence rate.

- American TV series express positive emotions most frequently. The occurrence rate of *Joy* in American TV series (M = 1.13, SD = 0.35) is significantly higher than that of Korea (M = 0.90, SD = 0.29, p = 0.007) and is marginally significantly higher than that of Japan (M = 0.98, SD = 0.37, p = 0.092).
- Korean TV series tends to express negative emotions, including *Fear, Disgust, Sadness, Anger* and *Contempt*, most frequently. The rates of *Anger* and *Contempt* are significantly higher than those of American series (ps < 0.015). The rate of *Fear* (M = 1.36, SD = 0.75) is significantly higher that those of both American (M = 0.81, SD = 0.43, p < 0.001) and Japanese series (M = 0.83, SD = 0.41, p < 0.001). The rate of *Disgust* in Korean series (M = 1.14, SD = 0.17) is significantly higher than that of American series (M = 0.98, SD = 0.15, p < 0.001), and the occurrence rate in American series is significantly higher than that of Japanese series (M = 0.88, SD = 0.15, p = 0.009). The rate of *Sadness* in Korean series (M = 1.30, SD = 0.40) is significantly higher than that of Japanese series (M = 1.03, SD = 0.34, p = 0.001), and the occurrence rate in Japanese series is significantly higher than that of American series (M = 0.67, SD = 0.17, p < 0.001).
- For the emotion of *Surprise*, the occurrence rate of surprise in Korean TV series (M = 1.22, SD = 0.23) is significantly higher than that of American TV series (M = 0.98, SD = 0.13, p < 0.001), and the occurrence rate in American TV series is significantly higher than that of Japanese TV series (M = 0.80, SD = 0.19, p < 0.001).

Interestingly, the results of the intensities (see Fig. 2) of the emotions are quite different from the observations from occurrence rates.

- Although American TV series has the highest occurrence rate on *Joy*, its intensity of *Joy* (M = 0.99, SD = 0.02) is significantly lower than that of Japanese (M = 1.01, SD = 0.03, p < 0.001) and Korean series (M = 1.00, SD = 0.03, p = 0.028).
- Although Korean TV series has the highest occurrence rates on *Sadness, Fear* and *Disgust*, its *Sadness* intensity does not show any significance versus American and Japanese series; its *Fear* intensity (M = 1.06, SD = 0.14) is almost the same as that of Japanese series (M = 1.07, SD = 0.11), both significantly lower than that in American series (, M = 0.87, SD = 0.11, ps < 0.001); and its *Disgust* intensity (M = 0.94, SD = 0.16) become the lowest among three countries, marginally significantly lower than those of

both Japanese (M = 1.03, SD = 0.19, p = 0.055) and American series (M = 1.03, SD = 0.19, p = 0.072).
- Regardless of the viewpoints of the occurrence rate and the intensities, both the order and the significance level of the *Joy, Anger, Surprise* and *Contempt* between Japanese and Korean series do not change.

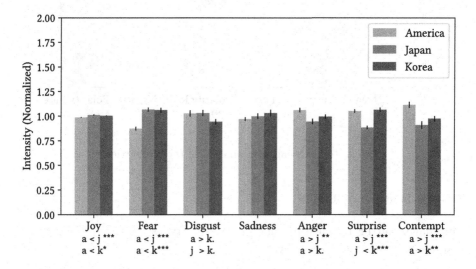

Fig. 2. Intensities of visual emotions of three countries

Textual Features: Polarity

We compared the sentiment polarity score of actors' lines among the three cultures. Figure 3 shows of polarity from the text-based sentiment analysis, which is in consistency with the results of the occurrence rate in the visual emotion analysis.

- The *Mean polarity score* of American series (M = 1.01, SD = 0.03) is significantly higher than that of both Japanese (M = 1.00, SD = 0.03, p = 0.045) and Korean series (M = 1.00, SD = 0.03, p = 0.001).
- American series (M = 1.08, SD = 0.29) has the highest *Positive sentiment occurrence rate*, which is marginally significantly higher than that of both Japanese (M = 0.96, SD = 0.24, p = 0.07) and Korean series (M = 0.96, SD = 0.23, p = 0.06).
- American series (M= 0.88, SD = 0.11) also has the lowest *Negative sentiment occurrence rate*, which is significantly lower than that of both Japanese (M = 1.04, SD = 0.19, p < 0.001) and Korean series (M = 1.09, SD = 0.16, p < 0.001).
- There is no significant difference between Japanese and Korean series.

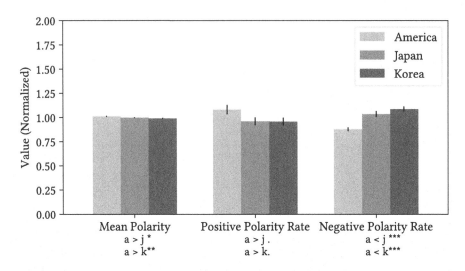

Fig. 3. Textual sentiment analysis of three countries

Textual Features: TF-IDF

The analysis on TF-IDF features of different word types are shown in Fig. 4. We find many interesting observations:

- American series' (M = 1.34, SD = 0.11) lines have significantly higher occurrence frequency on *Personal pronouns* (*Ppron*, same below) than both that of Japanese (M = 0.81, SD = 0.13, p < 0.001) and Korean series (M = 0.85, SD = 0.11, p < 0.001). Meanwhile, its ratio of *1st-person pron in ppron* (M = 1.03, SD = 0.07) is significantly higher than that of Japanese (M = 0.99, SD = 0.09, p = 0.021) and Korean series (M = 0.99, SD = 0.06, p = 0.036). However, American series (M = 0.90, SD = 0.05) have significantly lower occurrence frequencies of *2nd-person pron in ppron* than that of both Japanese (M = 1.00, SD = 0.08, p <0.001), and Japanese series significantly lower than Korean series (M = 1.09, SD = 0.07, p < 0.001). For the occurrence frequency of *Kinship terminology*, both American (M = 0.75, SD = 0.28) and Japanese series (M = 0.94, SD = 0.59) are significantly lower than Korean series (M = 1.31, SD = 0.75, ps < 0.011).
- The occurrence frequency on *Approval* of American series (M = 0.90, SD = 0.05) is significantly higher than that of Japanese (M = 1.00, SD = 0.08, p < 0.001) and Korean series (M = 1.09, SD = 0.07, p < 0.001), while the frequency of *Respect* of American series (M = 0.48, SD = 0.18) is significantly lower than that of Japanese (M = 1.38, SD = 0.66, p < 0.001) and Korean TV series (M = 1.14, SD = 0.35, p < 0.001). The occurrence frequency of as well as *Self expression* of American series (M = 1.72, SD = 0.48) is significantly higher than that of both Japanese (M = 0.71, SD = 0.34, p < 0.001) and Korean series (M = 0.58, SD = 0.27, p < 0.001).

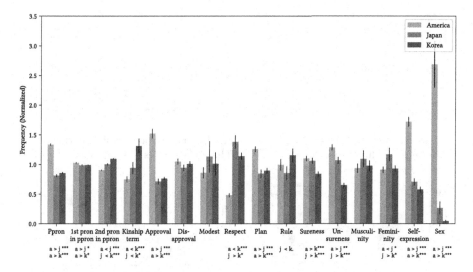

Fig. 4. TF-IDF features of three countries

- American series has significantly higher frequency of *Plan* than both Japanese and Korean series. American series is also highest in the frequency of *Sureness* (M = 1.10, SD = 0.25) and *Unsureness* (M = 1.28, SD = 0.31), with *Sureness* significantly higher than Korean series (M = 0.84, SD = 0.22, p < 0.001), and *Unsureness* significantly higher than both Japanese (M = 1.07, SD = 0.33, p = 0.003) and Korean series (M = 0.65, SD = 0.22, p < 0.001).
- The frequency of *Sex* of American series (M = 2.69, SD = 2.36) is significantly higher than that of Japanese (M = 0.27, SD = 0.62, p < 0.001) and Korean series (M = 0.04, SD = 0.12, p < 0.001).

4.2 Category Comparison

In addition to the comparison between countries, Fig. 5 visualizes the same features analyzed above, but merged the data according to categories instead of countries. Consistent with the ANOVA results that did not show significance on category, most of the features do not show significance from pairwise comparison between three categories. For those features that show significance, the reasons are easy to explain.

For instance, the occurrence rate and intensity of *Joy* in workplace series are significantly lower than those of family series and school series (see Fig. 5b and Fig. 5a). And workplace series' *positive sentiment occurrence rate* is significantly lower than that of the other two categories (see Fig. 5c). This might be explained by the fact that regardless of the countries and cultures, in the workplace, people are tend to be serious and comparatively express less positive emotions. Moreover, workplace series has significantly higher occurrence frequency of *Modest, Respect, Plan, Rule* and *Masculinity value* and family series has significantly

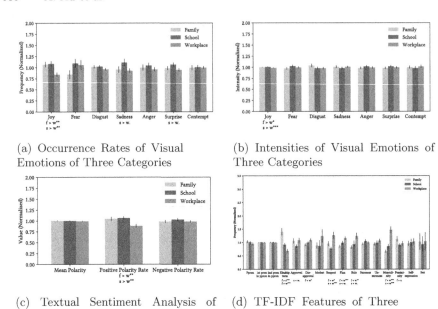

(a) Occurrence Rates of Visual Emotions of Three Categories

(b) Intensities of Visual Emotions of Three Categories

(c) Textual Sentiment Analysis of Three Categories

(d) TF-IDF Features of Three Categories

Fig. 5. Visualization of category comparison

higher frequency of *Kinship terminology*. These TF-IDF features are typically involved in either workplace or family.

The comparison across categories supports the validity of the datasets. These categories expand the coverage of different daily routine scenarios, while do not bias the results of the country comparison.

4.3   Country Classification

Beyond direct comparison on each features individually, we trained a linear SVM model for recognizing the country that a TV series belongs to. We used OVR for classifier training and LOO for cross validation. The model resulted in an average accuracy of 94.4%. Table 3 shows the confusion matrix. Only one series was misclassified as Japanese among 36 American series, while the rest of the misclassified instances were between Japanese and Korean series.

In order to better understand which features have impact on classification for each country, we calculated the classification coefficients of the three classifiers. The bigger the absolute value of a coefficient, the greater impact this feature has on the classification outcome. A feature with a zero coefficient does not contribute to the classifier. We modified the classifiers until the average number of features used for each classifier reaches 20, approximately one-third of all features.

We leveraged l_1 regularization as a feature selection method, and further trained an additional binary SVM model for each country with the features

Table 3. Confusion matrix of the SVM model. Rows are the ground truths and columns are the classification results.

	America	Japan	Korea
America	35	1	0
Japan	0	33	3
Korea	0	2	34

whose coefficients are not zero. We obtained an average accuracy as 96.9% (100%, 93.5% and 97.2% for America, Japan and Korea respectively), and an average F1 score as 0.952 (1.000, 0.899 and 0.957 for America, Japan and Korea respectively). We summarized the features according to their coefficients in Table 4. The features whose coefficient did not equal zero only in one country might indicate the specific characteristics of this country; The features whose coefficient did not equal zero in two countries might imply the distinctions between these two cultures; While the features whose coefficient did not equal zero in all three countries might show the similarities and differences among all three cultures. We discuss our results in more details in the next section.

5 Discussion

We analyzed the multimodal data of American, Japanese and Korean TV series. The results provided vivid evidences of differences among the three cultures based on a fair amount of data, which described behavioral-level culture.

The findings were valuable in providing a practical approach to cross-cultural studies, confirming existing cross-cultural psychological theories, establishing each culture's model for sentimental and oral expressive features, and providing suggestions on application scenarios such as global team management, personal cross-cultural adaptation and cultural related product design.

5.1 Sentimental Features

Positive Sentiment Frequency. Our results indicate that Americans in TV series express positive emotions significantly more frequently, and express negative emotions significantly less frequently than people in the other two cultures. This is in line with the findings of Kitayama et al. [40]. They asked American and Japanese students to report their emotional experience during daily routine. They found that Americans reported considerably higher for positive emotions than for negative emotions, while Japanese did not have such a difference.

Engaging/Disengaging Emotions. In addition to the pleasantness dimension, sentiments can also be described by social orientation dimension. The social orientation dimension divided emotions into engaging emotions and disengaging

Table 4. Multimodal features for each country's classifier

Coefficient type	Visual features	Textual features
Not equals zero only in America classifier	Lid tighten (rate), Lid tighten (intensity)	Modest
Not equals zero only in Japan classifier	Smile (intensity), Brow raise (rate), Smirk (rate), Lip stretch (rate)	–
Not equals zero only in Korean classifier	Lip pucker (rate), Lip stretch (intensity)	–
Not equals zero in American and Japanese Classifiers	–	Ppron, Respect, Plan
Not equals zero in American and Korean Classifiers	Inner brow raise (rate)	2nd-person pron in ppron, Self expression
Not equals zero in Japanese and Korean Classifiers	Smile (rate), Brow furrow (rate), Nose wrinkle (rate), Lip corner depressor (rate), Lip suck (rate), Mouth open (rate), Eye closure (intensity), Eye widen (intensity), Jaw drop (intensity)	Sureness, Unsureness, Musculinity Value, Femininity Value
Not equals zero in all three classifiers	Upper lip raise (rate)	Negative polarity rate, Kinship terminology, Approval, Sex
Equals zero in all three classifiers	Inner brow raise (intensity), Brow raise (intensity), Brow furrow (intensity), Nose wrinkle (intensity), Upper lip raise (intensity), Lip corner depressor (intensity), Chin raise (rate), Chin raise (intensity), Lip pucker (intensity), Lip press (rate), Lip press (intensity), Lip suck (intensity), Mouth open (intensity), Smirk (intensity), Eye closure (rate), Attention (rate), Attention (intensity), Eye widen (rate), Cheek raise (rate), Cheek raise (intensity), Dimpler (rate), Dimpler (intensity), Jaw drop (rate)	Mean polarity, Positive polarity rate, 1st-person pron in ppron, Disapproval, Rule

emotions. Emotions that arise in achieving goals for independent self such as realizing one's own value were disengaging emotions, while emotions arise in meeting interdependent goals such as achieving the connection of the self with others are engaging emotions. Anger, for example, is a negative disengaging emotion that arises in failure to achieve individual goals [40]. Kitayama et al. compared between American and Japanese cultures and their findings showed that Americans reported experience of disengaging emotions significantly more stronger, while Japanese experienced more engaging emotions [41]. This is consistent with the fact that Americans were more independently self-constructed while Japanese were more interdependently self-constructed. Our results confirm this theory in that Americans express *Anger* most strongly among the three cultures.

5.2   Oral Expression Features

Individualism/Collectivism and "I" . Our research finds that in American TV series people use significantly more first-person pronouns, and talk significantly less about kinship. While in Korean TV series people use significantly more second-person pronouns and talk significantly more about kinship. This is in line with the Hofstede's cultural theories. In Hofstede's cultural dimensions, the individualism-collectivism dimension (IDV) describes how much importance a culture attaches to individual independence and individual achievement. People in high individual culture are encouraged to express themselves, while people in high collective culture are encouraged to fit oneself to a group and keep the harmony of the group. In Hofstede's score system, the individualism of countries are ranked with score from 1 to 100 (100 as totally individualism). The American culture is scored 91, ranking as the topmost among 76 countries, indicating American culture as an extremely individual culture. Japanese culture is scored moderately with 46, and Korean cultural is scored 18 as a very collective culture [36]. Our findings on TF-IDF of *1st-person ppron* and *Kinship terminology* confirms the individualism of the three cultures.

Moreover, in American TV series, significantly more *Personal pronouns* are used than the other two countries. Previous research in linguistics has found evidence on the relationship between cultural values and use of personal pronoun: there is less pronoun drop phenomenon in the language of individual culture than collective culture [37]. This is consistent with our findings where more pronouns are dropped in Japanese and Korean cultures.

Power Distance and Respectful Expression. American people use significantly less respectful expressions. Linguistic research has found that among American English, Japanese and Korean languages, American English is the least sensitive to all the power variables and group solidarity [35]. In Hofstede's power distance dimension (PDI, range from 1 to 100), America is scored 40, lower than Japan (54) and Korea (60) [36]. People in high power distance culture emphasize on hierarchy, have a better acceptance of inequality and failing to respect people in higher status will be negatively judged. The results of our research confirm the relatively lower power distance in American culture.

Low/High-Context Culture and Self-expression. People in American TV series express their own thoughts and views significantly more frequently using *Self-expression* terms such as "I think", and uses more *Approval* or *Disapproval* expressions. In Hall's cultural theory, cultures are distributed into low-context and high-context cultures [42]. Low-context cultures require more explicit expression and communication, while in high-context cultures exchanged messages carry implicit meanings with more information than the actually spoken words. Japanese and Korean cultures are considered as representatives of high-context cultures, while American culture is a typical low-context culture. People in high context cultures use more ambiguous and implicit expressions rather than express their views directly. People in low-context culture have higher level of self-disclosure than people in high-context culture [43], which is in line with our findings.

5.3   Inconsonance Between Results and Theories

American people talk more about *Plan, Sureness* and *Unsureness. Plan* is related with long-term orientation dimension, and *Sureness/Unsureness* is related with uncertainty avoidance dimension in Hofstede's theory [36]. In Hofstede's score system, America is scored much lower than Japan and Korea in these two dimensions, indicating that American culture attaches more importance to short-term than long-term benefits, and are more acceptable of unsureness. These indicate that Americans should be the least inclined to emphasize plan and sureness/unsureness. Our results, however, show opposite results that Americans talk more about these topics. On the one hand, the difference may be biased by the high level of self-disclosure in American culture, so that they tend to express their ideas explicitly. On the other hand, the uncertainty avoidance and long-term orientation dimensions are more correlated to behavioral aspects than communicative aspects. The textual analysis is based mainly on dialogue, which can not fully reflect the behavioral facts.

5.4   Features for Culture Classification

For the three cultures, features that contribute to classification vary. For American culture verbal features mainly contribute to the classification, while for Japanese and Korean cultures both facial expression and verbal features have contributions. In low-context culture like America, direct verbal communication plays the main part in expression, while in high-context cultures like Japan and Korea more information is conveyed beyond the verbal part of message [44]. This could be one of the reasons that more facial expression clues contribute to the classification of Japanese and Korean cultures rather than American culture.

Facial expressions carry different messages across cultures [45]. The classifier helps us find typical facial expression features of the three cultures. *Lid tighten* is a facial expression element that only contributes to the classification of American culture. A lower rate of *Lid tighten* and a higher intensity of *Lid tighten* can indicate American culture. Similarly, higher intensity of *Smile*, frequent *Brow raise*, *Smirk* and *Lip stretch* indicate Japanese culture. Frequent *Lip pucker* and *Lip stretch* indicate Korean culture. Previous research claimed that Americans express emotions more through brow and mouth movement, while Japanese people express emotions more through eye movement [46,47]. Our results further find out detailed movement types in each facial part that are typical in each culture, and that lip and brow movements are also important to identify Japanese and Korean cultures.

5.5 Prospect of Application

To our knowledge, this research is the first attempt to use machine learning to culturally analyze TV series. This methodology can be applied to other comparative researches based on media data. The model has good accuracy and can be used to aid design of robot with cultural adaptability, systems that help cross-cultural training, or other artificial intelligence products with cultural analysis.

5.6 Limitation and Future Work

First, we only investigated three countries and three categories. More cultures and scenarios can be involved in the future work to obtain results that covers larger population and scenes. Second, we just leveraged the facial expression and language features. There exists plenty of other visual features such as body poses to be explored in the future. Moreover, other potential modalities such as audio, may provide meaningful results for the cross-cultural studies. Third, we employed a linear model for the sake of interpretability. The linear model can hardly reflect the non-linear or inner relationships among the features for training. In the future work, other machine learning models that have stronger interpretability worth exploration.

6   Conclusion

In this paper, we proposed a new cross-cultural analysis method that leverages multimodal features from TV series. We exemplified our method by analyzing 108 TV series from three countries: America, Japan and Korea, 36 series for each countries. We first extracted visual and textual features from each TV series and compare them directly across countries to investigate the similarities and differences across different cultures. We then trained a linear SVM model to recognize which country a TV series belongs to. We achieved an average accuracy of 94.4%. We further interpreted the model's coefficient to explore the significant features for each country's classification. Our method avoids the drawbacks of the traditional cross-cultural studies approaches, obtains insightful results across different cultures, provides strong evidence for existing literature and meaningful guidelines for future cross-cultural researchers, and has a good prospect in aiding product design.

References

1. Ilesanmi, O.O.: What is cross-cultural research? Int. J. Psychol. Stud. **1**(2), 82 (2009)
2. Black, J.S., Mendenhall, M.: Cross-cultural training effectiveness: a review and a theoretical framework for future research. Acad. Manag. Rev. **15**(1), 113–136 (1990)
3. Holden, N.: Cross-Cultural Management: A Knowledge Management Perspective. Pearson Education, London (2002)
4. Sawaya, Y., Sharif, M., Christin, N., Kubota, A., Nakarai, A., Yamada, A.: Self-confidence trumps knowledge: a cross-cultural study of security behavior. In: Proceedings of the 2017 CHI Conference on Human Factors in Computing Systems, CHI 2017, pp. 2202–2214. ACM, New York (2017). https://doi.org/10.1145/3025453.3025926. http://doi.acm.org/10.1145/3025453.3025926
5. Yarosh, S., Schoenebeck, S., Kothaneth, S., Bales, E.: "Best of both worlds": opportunities for technology in cross-cultural parenting. In: Proceedings of the 2016 CHI Conference on Human Factors in Computing Systems, CHI 2016, pp. 635–647. ACM, New York (2016). https://doi.org/10.1145/2858036.2858210. http://doi.acm.org/10.1145/2858036.2858210
6. Jane, L.E., Ilene, L.E., Landay, J.A., Cauchard, J.R.: Drone & Wo: cultural influences on human-drone interaction techniques. In: Proceedings of the 2017 CHI Conference on Human Factors in Computing Systems, CHI 2017, pp. 6794–6799. ACM, New York (2017). https://doi.org/10.1145/3025453.3025755. http://doi.acm.org/10.1145/3025453.3025755
7. Shen, S., Tennent, H., Claure, H., Jung, M.: My telepresence, my culture?: an intercultural investigation of telepresence robot operators' interpersonal distance behaviors. In: Proceedings of the 2018 CHI Conference on Human Factors in Computing Systems, CHI 2018, pp. 51:1–51:11. ACM, New York (2018). https://doi.org/10.1145/3173574.3173625. http://doi.acm.org/10.1145/3173574.3173625
8. Bail, C.A.: The cultural environment: measuring culture with big data. Theory Soc. **43**(3–4), 465–482 (2014). https://doi.org/10.1007/s11186-014-9216-5
9. Allen, R.C.: To be Continued-: Soap Operas Around the World. Psychology Press, Hove (1995)
10. Fiske, J.: Television Culture. Routledge, London (2010)
11. Kellner, D.: Cultural studies, multiculturalism, and media culture. Gender Race Class Media: Crit. Read. **3**, 7–18 (2011)
12. Brown, M.J., Cody, M.J.: Effects of a prosocial television soap opera in promoting women's status. Human Commun. Res. **18**(1), 114–144 (1991)
13. Browne Graves, S.: Television and prejudice reduction: when does television as a vicarious experience make a difference? J. Soc. Issues **55**(4), 707–727 (1999)
14. Chew, F., Palmer, S., Slonska, Z., Subbiah, K.: Enhancing health knowledge, health beliefs, and health behavior in Poland through a health promoting television program series. J. Health Commun. **7**(3), 179–196 (2002)
15. Lull, J.: Media, Communication, Culture: A Global Approach. Columbia University Press, New York (2000)
16. Morris, M.W., Peng, K.: Culture and cause: American and Chinese attributions for social and physical events. J. Pers. Soc. Psychol. **67**(6), 949 (1994)
17. Nisbett, R.E., Masuda, T.: Culture and point of view. Proc. Natl. Acad. Sci. **100**(19), 11163–11170 (2003)

18. Kim, H., Markus, H.R.: Deviance or uniqueness, harmony or conformity? A cultural analysis. J. Pers. Soc. Psychol. **77**(4), 785 (1999)
19. Miyamoto, Y., Nisbett, R.E., Masuda, T.: Culture and the physical environment: holistic versus analytic perceptual affordances. Psychol. Sci. **17**(2), 113–119 (2006)
20. Rojek, C.: Pop Music, Pop Culture. Polity (2011)
21. Trepte, S.: Cultural proximity in tv entertainment: an eight-country study on the relationship of nationality and the evaluation of us prime-time fiction. Communications **33**(1), 1–25 (2008). https://doi.org/10.1515/COMMUN.2008.001
22. Gould, S.J., Gupta, P.B., Grabner-Kräuter, S.: Product placements in movies: a cross-cultural analysis of Austrian, French and American consumers' attitudes toward this emerging, international promotional medium. J. Advert. **29**(4), 41–58 (2000)
23. Yang, X., Luo, J.: Tracking illicit drug dealing and abuse on instagram using multimodal analysis. ACM Trans. Intell. Syst. Technol. **8**(4), 58:1–58:15 (2017). https://doi.org/10.1145/3011871. http://doi.acm.org/10.1145/3011871
24. Morency, L.P., Mihalcea, R., Doshi, P.: Towards multimodal sentiment analysis: Harvesting opinions from the web. In: Proceedings of the 13th International Conference on Multimodal Interfaces, ICMI 2011, pp. 169–176. ACM, New York (2011). https://doi.org/10.1145/2070481.2070509. http://doi.acm.org/10.1145/2070481.2070509
25. Shah, R.R., Yu, Y., Zimmermann, R.: Advisor: personalized video soundtrack recommendation by late fusion with heuristic rankings. In: Proceedings of the 22nd ACM International Conference on Multimedia, MM 2014, pp. 607–616. ACM, New York (2014). https://doi.org/10.1145/2647868.2654919. http://doi.acm.org/10.1145/2647868.2654919
26. Balasuriya, L., Wijeratne, S., Doran, D., Sheth, A.: Finding street gang members on twitter. In: Proceedings of the 2016 IEEE/ACM International Conference on Advances in Social Networks Analysis and Mining, ASONAM 2016, pp. 685–692. IEEE Press, Piscataway (2016). http://dl.acm.org/citation.cfm?id=3192424.3192554
27. Dalgleish, T., Power, M.: Handbook of Cognition and Emotion. Wiley, Hoboken (2000)
28. Cambria, E., Howard, N., Hsu, J., Hussain, A.: Sentic blending: scalable multimodal fusion for the continuous interpretation of semantics and sentics. In: 2013 IEEE Symposium on Computational Intelligence for Human-like Intelligence (CIHLI), pp. 108–117 (2013). https://doi.org/10.1109/CIHLI.2013.6613272
29. Cambria, E., Olsher, D., Rajagopal, D.: SenticNet 3: a common and commonsense knowledge base for cognition-driven sentiment analysis. In: Proceedings of the Twenty-Eighth AAAI Conference on Artificial Intelligence, AAAI 2014, pp. 1515–1521. AAAI Press (2014). http://dl.acm.org/citation.cfm?id=2892753.2892763
30. Poria, S., Cambria, E., Howard, N., Huang, G.B., Hussain, A.: Fusing audio,visual and textual clues for sentiment analysis from multimodal content. Neurocomputing **174**, 50–59 (2016).https://doi.org/10.1016/j.neucom.2015.01.095. http://www.sciencedirect.com/science/article/pii/S0925231215011297
31. You, Q., Luo, J., Jin, H., Yang, J.: Cross-modality consistent regression for joint visual-textual sentiment analysis of social multimedia. In: Proceedings of the Ninth ACM International Conference on Web Search and Data Mining, WSDM 2016, pp. 13–22ACM, New York (2016). https://doi.org/10.1145/2835776.2835779. http://doi.acm.org/10.1145/2835776.2835779

32. McDuff, D., Mahmoud, A., Mavadati, M., Amr, M., Turcot, J., Kaliouby, R.E.: AFFDEX SDK: a cross-platform real-time multi-face expression recognition toolkit. In: Proceedings of the 2016 CHI Conference Extended Abstracts on Human Factors in Computing Systems, CHI EA 2016, pp. 3723–3726. ACM, New York (2016). https://doi.org/10.1145/2851581.2890247. http://doi.acm.org/10.1145/2851581.2890247

33. Tausczik, Y.R., Pennebaker, J.W.: The psychological meaning of words: LIWC and computerized text analysis methods. J. Lang. Soc. Psychol. **29**(1), 24–54 (2010)

34. Hall, E.T., et al.: The Silent Language, vol. 3. Doubleday, New York (1959)

35. Hijirida, K., Sohn, H.M.: Cross-cultural patterns of honorifics and sociolinguistic sensitivity to honorific variables: evidence from English, Japanese, and Korean. Res. Lang. Soc. Interact. **19**(3), 365–401 (1986)

36. Hofstede, G., Hofstede, G., Minkov, M.: Cultures and organizations: Software of the Mind. Revised and Expanded Third Edition. New York (2010)

37. Kashima, E.S., Kashima, Y.: Culture and language: the case of cultural dimensions and personal pronoun use. J. Cross Cult. Psychol. **29**(3), 461–486 (1998)

38. Burman, P.: A comparative study of ordinary cross-validation, v-fold cross-validation and the repeated learning-testing methods. Biometrika **76**(3), 503–514 (1989)

39. Zhang, Y., Yang, Y.: Cross-validation for selecting a model selection procedure. J. Econ. **187**(1), 95–112 (2015)

40. Kitayama, S., Markus, H.R., Kurokawa, M.: Culture, emotion, and well-being: good feelings in Japan and The United States. Cogn. Emot. **14**(1), 93–124 (2000)

41. Kitayama, S., Karasawa, M., Mesquita, B.: Collective and personal processes in regulating emotions: emotion and self in Japan and the United States. In: The Regulation of Emotion, pp. 251–273 (2004)

42. Hall, E.T.: Beyond Culture, Anchor. Garden City, NY (1976)

43. Chen, G.M.: Differences in self-disclosure patterns among Americans versus Chinese: a comparative study. J. Cross Cult. Psychol. **26**(1), 84–91 (1995)

44. Kim, D., Pan, Y., Park, H.S.: High-versus low-context culture: a comparison of Chinese, Korean, and American cultures. Psychol. Mark. **15**(6), 507–521 (1998)

45. Jack, R.E., Garrod, O.G., Yu, H., Caldara, R., Schyns, P.G.: Facial expressions of emotion are not culturally universal. Proc. Natl. Acad. Sci. **109**(19), 7241–7244 (2012)

46. Jack, R.E., Caldara, R., Schyns, P.G.: Internal representations reveal cultural diversity in expectations of facial expressions of emotion. J. Exp. Psychol. Gen. **141**(1), 19 (2012)

47. Yuki, M., Maddux, W.W., Masuda, T.: Are the windows to the soul the same in the east and west? Cultural differences in using the eyes and mouth as cues to recognize emotions in Japan and the United States. J. Exp. Soc. Psychol. **43**(2), 303–311 (2007)

Comparison of Chinese and Foreign Studies on Skilled Talents Training for Industrial Internet

Ang Zhang[1] and Shuo Guo[2(✉)]

[1] China Academy of Industrial Internet, Beijing 100102, China
[2] Student Affairs Office, Tsinghua University, Beijing 100084, China

Abstract. Skilled talents play an indispensable role in the development of advanced manufacturing industries. The rapid development of the industry is followed by a large gap of skilled talents. This study summarizes the background of the development of the advanced manufacturing industries in the United States, Germany, Japan, and China, as well as the status of talent training. The comparison found that for the development of the Industrial Internet, these countries have successively introduced relevant strategies after 2010, proposing the clear industrial development directions, such as the "Strategy for American Leadership in Advanced Manufacturing", the "Industry 4.0", the "Connected Industries", and the "Industrial Internet". Furthermore, the similarities and differences of talent training in these countries have been compared and analyzed, including training groups, training organization forms, training methods, and training content. The study found that the higher vocational education is the main way to cultivate skilled talents. These countries have formed their vocational education systems that conform to the national conditions and industrial development, and it requires national and local government policies and financial support. The cultivation method of the integration of education with industry is conducive to promoting the adaptation of the talent chain to the development of the industry, but how to stimulate the enthusiasm of the enterprise in the talent training requires more guarantee measures.

Keywords: Manufacturing industry · Industrial internet · Skilled talent training

1 Introduction

Talents are the first resource for the construction of a manufacturing powerhouse. In fast-growing economies, technological innovation is the most important driving force for industrial growth. Technological breakthroughs and high industrialization ability are key to cultivation of emerging industries and upgrading of traditional industries. Talents, as the most active resource for scientific and technological progress, are irreplaceable elements.

As an important cornerstone of the Fourth Industrial Revolution, the Industrial Internet is a key driver of digital transformation and a significant engine for the growth of

© Springer Nature Switzerland AG 2021
P.-L. P. Rau (Ed.): HCII 2021, LNCS 12771, pp. 547–560, 2021.
https://doi.org/10.1007/978-3-030-77074-7_41

new kinetic energy. Since General Electric (GE) in the United States put forward the concept of the Industrial Internet in 2012, some developed countries have laid out the Industrial Internet with a global upsurge, such as the "Advanced Manufacturing American leadership strategy" proposed by the United States [1], the "Industry 4.0" strategy in Germany [2], and the "Connected Industries" strategy in Japan [3]. To accelerate the transformation and upgrading of the manufacturing industry and promote the high-quality development of the industry, China has also put forward a national strategy for implementing the innovative development of the "Industrial Internet" since 2017 [4]. The Industrial Internet is a product of the deep integration of information technology and manufacturing. It effectively connects the supply and the demand sides, and promotes the innovation and development of small and large enterprises and the integrated development of primary, secondary, and tertiary industries. At present, the same or similar concepts as Industrial Internet include Industrial Internet of Things [5] and Smart Manufacturing [6]. Although different countries and institutions have different interpretations of the Industrial Internet, its essence is the application of advanced information technology in the industrial production process.

At present, the manufacturing powers represented by the United States, Japan, Germany, etc., have monopolized the most cutting-edge major equipment and core components, and dominate the global division of labor. The fundamental reason is that they have a world-leading high-quality manufacturing talent team. The United States manufacturing industry is the most active industrial sector for technological innovation, and it is also a place where R&D talents are concentrated, employing 63% of the country's scientists and engineers. At present, countries all over the world have taken manufacturing development as an important strategy to seize the commanding heights of future competition and regard talents as an important support for the implementation of manufacturing development strategies. Take the United States as an example. The National Network for Manufacturing Innovation (NNMI) is the top priority in the United States' manufacturing strategy [7]. The industry and scientific research institutions are connected through the establishment of NNMI, such as additive manufacturing and robotics. NNMI can be used not only as a teaching factory, but also as a postgraduate degree. NNMI helps companies train employees to improve their labor skills and provide technical support for corporate R&D operations. Additionally, NNMI pays equal attention to talent cultivation and technological innovation to enhance the comprehensive competitiveness of the manufacturing industry.

While the Industrial Internet significantly improves the level of productivity, it also puts forward higher demands for training skilled talents. High-quality skilled talents should have superb technology, outstanding creativity, and comprehensive qualities. The traditional education system is difficult to cultivate professional and high-quality skilled talents that could match the depth integration of informatization and industrialization. In the rapid development of the Industrial Internet, the lack of relevant talents has become an important obstacle. According to the proportion of senior skilled workers in Western developed countries, it is estimated that by 2025, the shortage of senior skilled workers in the Chinese manufacturing industry will reach 30–50 million, especially with the lack of high-quality and skilled engineers and skilled workers at the operational level.

Furthermore, the current skilled personnel training cannot support and meet the needs of Industrial Internet development in these developed countries.

This research focuses on the current situation of Industrial Internet skilled talents in China and several other countries. From the perspectives of higher vocational education, corporate internal training, and social training, this paper compares the training methods of high-quality skilled talents for Industrial Internet in various countries, such as the "technology-led manufacturing" in the United States, the "dual-system" in Germany, and "benign interaction between government, colleges and industry" in Japan. Based on the literature and questionnaires, the industrial development background, talent training methods, and targets of different countries are analyzed. Thus, several suggestions are made on the path and direction of cultivating skilled talents.

2 Demand for Manufacturing Talents and Training of Skilled Talents in Developed Countries

2.1 The United States

Background of Talent Demand in Advanced Manufacturing Industry. The United States is the world's largest manufacturing powerhouse, accounting for 18.2% of the world's total production. The United States manufacturing industry is the most active industrial sector for technological innovation, and it is also a place where R&D talents are concentrated, employing 63% of the country's scientists and engineers. The proportion of American manufacturing industry in American GDP is also very large, accounting for about 12%. In 2012, the United States issued the National Strategic Plan for Advanced Manufacturing [8], which began to forcefully promote the return of American manufacturing. At present, 19% of manufacturing workers in the United States are over 54 years old, and only 7% of the total number of workers are under 25 years old, which is a serious age-fault problem. A severe shortage of engineers and skilled workers threatens the American manufacturing. In addition, as manufacturing in the United States has changed dramatically over the past 40 years, specialized skills such as composite materials, precision machining, robotic control and maintenance, radio frequency identification of components, and computer operation have become increasingly important to workers, requiring advanced skills in specific fields such as engineering design and computer programming, but there is a serious shortage of people with these capabilities. In 2018, the total output value of the United States manufacturing industry was US$2.33 trillion, and the manufacturing industry employed approximately 15 million people, with seriously inadequate highly skilled personnel. Of more than 1,600 companies surveyed by Global Talent Management and Rewards, 60% of companies in the Unites States said they had trouble finding and hiring employees with critical skills, especially technical talent in information technology (see Fig. 1) [9].

The United States Adopted a Series of Training Measures to Cultivate Skilled Talents. One measure involved combining vocational education with the universality of general education. There is no separate vocational education system in the school system of American vocational education; it is combined with general education and implemented in the same institutions. Vocational education may be divided into secondary and

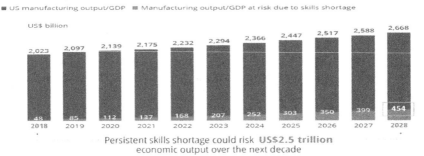

Skills shortage could put US$454 billion of manufacturing GDP at risk in 2028 alone

■ US manufacturing output/GDP   ■ Manufacturing output/GDP at risk due to skills shortage

US$ billion

2018	2019	2020	2021	2022	2023	2024	2025	2026	2027	2028
2,023	2,097	2,139	2,175	2,232	2,294	2,366	2,447	2,517	2,588	2,668
48	85	112	137	168	207	252	303	350	399	454

Persistent skills shortage could risk **US$2.5 trillion** economic output over the next decade

Fig. 1. Impact of manufacturing labor shortages on total manufacturing during 2018–2028 (Data from Deloitte Research Report).

higher education, with secondary vocational education taking place mainly in comprehensive secondary schools, and higher vocational education is mainly implemented in community colleges. Comprehensive middle schools are generally divided into general, academic, and vocational subjects. The courses offered by schools are generally composed of general courses, vocational and technical courses, and practical courses, among which the proportion of theoretical courses and practical courses is 1:2. Comprehensive schools are complemented by regional vocational education centers, where pupils spend half their time in general academic subjects and half their time in vocational subjects, and where the center studies are credited as comprehensive school credits. Community colleges include vocational education, general education, and university transfer education, and their vocational education majors and curriculum settings have obvious locality and diversity. Because the educational purpose of community colleges needs to meet the needs of various industries in the community, their majors and curriculum settings are more practical, which reflects the actual needs and employment trends of the community [10].

This measure also provides a strong guarantee for vocational education from all aspects. In terms of teaching organization, vocational education students need to be able to take up their posts directly, so specialties and courses are determined on the basis of full social investigation, which has strong practicability. In terms of teaching methods, vocational education is flexible and personalized, and the main task is performed by the community college's learning resource centers, which combine library, audio-visual teaching services, and learning laboratories and are staffed by tutors familiar with the various teaching materials and equipment. As regards teacher construction, vocational education especially emphasizes the practical experience of teachers and is equipped with more part-time teachers than full-time teachers. The part-time teachers are generally composed of entrepreneurs in the community, experts in a certain field, or engineering and technical personnel and managers in the production line [11]. Regarding safeguard mechanism, in addition to the above-mentioned legislation, there is direct financial allocation and necessary administrative means management. With regard to funding, state government grants are the main source of funding. At the same time, certain taxes and fees will be levied in the community, as well as enrolling foreign students, obtaining foundation donations, etc., to enrich school funding [12].

The United States aims to improve the technical skills of labor force through vocational and technical education and serve the development of an advanced manufacturing industry by coordinating the government, educators, employers, community colleges, and university courses, which can support the project courses, competency training, and career path planning for skilled talents.

The United States Has Successively Introduced Legislation in Various Periods to Strongly Support and Guarantee the Skilled Personnel Training. In 1862, the United States promulgated the "Morrel Act" to train specialized industrial and agricultural talents, which was the first step in vocational education legislation in the United States. Since then, various legislations such as the "Vocational Education Law," the "Vocational Training Cooperation Law," and the "Education Reform Law" have been successively passed, continuously emphasizing the importance of vocational education.

In July 2018, the President of the United States signed an executive order stating that there are currently more than 6.7 million unfilled jobs in the United States, and these jobs need to be filled through technical skills training [13]. In October 2018, the White House released the "Strategy for American Leadership in Advanced Manufacturing" which pointed out that the United States must strengthen and formulate key human capital strategies to support the next generation of advanced manufacturing technologies, focusing on the development of education. In February 2019, the White House issued the "America Will Dominate the Industries of the Future" [14], which focused on artificial intelligence, advanced manufacturing, quantum information science, and 5G technology as the focus of government support, and planned to expand the apprenticeship system so that skilled personnel can obtain education and skills.

The training of skilled personnel in the United States involves development of attributes like adaptability, pertinence, and flexibility, and imparting knowledge of the economy. The United States attaches great importance to the evaluation of vocational education projects, strengthens the status of evaluation through legislation, and combines evaluation with accountability. Federal government departments, state/local governments, institutions, and various organizations coordinate with each other to develop evaluation indicators and standards, try a variety of evaluation models and methods, and jointly promote the development of vocational education evaluation.

2.2 Germany

Germany Issued "Recommendations for implementing the strategic initiative INDUSTRIE 4.0" in April 2013. It pointed out that Industry 4.0 will bring about new business models, new jobs, and changes in skills, but the existing manufacturing talents and talent structure cannot meet the requirements. According to a 2014 survey of 235 German companies in the fields of information and communications, manufacturing and engineering, automotive and process, electronics and electrical, 30% of those companies believe that highly skilled manufacturing personnel are unable to meet the needs of enterprise development. It is estimated that by 2030, the shortage of German talents with complete vocational training will reach 2.9 million [15]. The implementation of Industry 4.0 has also made the problem of talent structure prominent in Germany. By 2030, the proportion of professionals aged 40 to 44 in Germany will be 24%, while those aged 30 to 34 will be only 19%, which will lead to a serious imbalance in the personnel structure of German professionals [16]. To promote the development of Industry 4.0, "Education 4.0" was launched in 2016, and five key areas of action were put forward to cultivate high-quality digital talents. In 2019, the "National Continuing Education Strategy" was introduced to strengthen personnel training and create a new type of continuing education culture. In November 2019, the German Ministry of Economic Affairs and Energy released the "Nationale Industriestrategie 2030," which is aimed at improving Germany's economic and technological strength and industrial competitiveness. This strategy pointed out that Germany's industrial strength is largely based on excellently trained skilled workers who should be retained and strengthened as the social and international environment changes [17].

"Dual System" Vocational Education System is Known as the Cornerstone of "Made in Germany". It has trained a large number of skilled talents for Germany. It The "Dual System" is a vocational education system in which enterprises and schools cooperate to train skilled talents. The students in vocational schools are also apprentices in enterprises, and students alternately carry out learning and training on both sides. The structure of the German vocational education system is perfect. Three-quarters of the students who have completed junior high school enter various vocational schools, and 70% of them receive "dual system" vocational education [18].

The duality is mainly reflected in the following aspects: (1) Schools and enterprises; (2) Vocational skills training and professional theory teaching; (3) Practical and theoretical teaching materials; (4) Practical and theoretical teachers; (5) Enterprise apprentices and school students; (6) Skills examination and qualification examination; (7) Training certificates and diplomas. The essence of "dual system" education is to take the enterprises as the main body, which play a leading role in the process of enrollment and training. The enterprises not only bear the training costs of students, but also to pay allowances and remuneration for internship instructors. Large and medium-sized enterprises generally have independent training factories; small enterprises also have specific training posts and cross-enterprise factories, which solve the problems of funds, teachers, venues, and equipment, and the students are directly employed after training.

At the same time, the "dual system" vocational education gives full play to the role of autonomous organizations in economic circles, and incorporates chambers of commerce into the whole process of vocational education, including the formulation of training standards, the implementation of training processes, and the assessment of vocational ability. Such a division of labor and cooperation among vocational schools, enterprises, chambers of commerce, and government agencies can not only cultivate skilled and technical talents actually needed by enterprises, but can also make keen adjustments to industrial and occupational changes. It should be pointed out that the concept of enterprise practice and school-enterprise cooperation in "dual system" vocational education also penetrates higher education. Besides research-oriented comprehensive universities, there are also a large number of applied science universities oriented by enterprise practice in Germany.

In the Industry 4.0 area, Germany has paid more attention to cross-border compound talent training represented by information knowledge. The first is to update and optimize the professional knowledge of informatization; the second is to increase the content of informatization in corporate training; and the third is to strengthen investment in digital equipment and environmental construction, etc.

Germany has Mature Models and Mechanisms to Ensure the Promotion of the "Dual System". First, vocational education laws and regulations are comprehensive and strict. The "Basic Law" of the Federal Republic of Germany has provisions on education, which are the most important legal basis for German education. Since the middle of the last century, more than 10 pieces of legislation on vocational education have been promulgated, including the "Federal Vocational Education Act" and the "Regulations on Vocational Education." These laws are rich in content, interlocking, and easy to operate. The school names, training objectives, professional settings, school conditions, sources of funding, and teacher examinations carried out a clear and specific requirement to form a unified and standardized vocational education system. The youth are being trained to do special protection provisions to safeguard the rights and interests of young people. At the same time, it also ensures the relatively balanced development of vocational education in all states [19].

Second, the supporting measures of vocational education are sound. Vocational education in Germany is organically linked with higher vocational education, and young people who embark on vocational education can enjoy the promotion path linked to wages and treatment. The government policy stipulates that the vocational education expenses of enterprises can be included in the production cost and product price, and tax can be reduced or exempted. In Germany, teachers are national civil servants (state employees), with high social status and stable working environment, and vocational education teacher training has a complete system. Germany attaches great importance to career counseling and guidance, providing students with high-quality career services in the complex labor market [20].

2.3 Japan

The Success of "Made in Japan" is Closely Related to the Japanese Government's Emphasis on Talents. Since the 1980s, the pursuit of technological innovation and the persistence of "craftsman spirit" have made "Made in Japan" a synonym for quality excellence. In terms of the amount of investment, the main sources of R&D investment in Japan from 1970 to 2005 were the government and enterprises, with the government accounting for 22% on average, and the enterprises accounting for an average of 77.75%. Second, from the perspective of training institutions, 88.7% of the enterprises acquire manufacturing talents through "self-training and employment of graduates and non-graduates" Japan's manufacturing sector is currently facing a huge turning point, with its own industry's lack of succession leading to low productivity and increasing competition with other emerging countries.

In such a huge environmental change, the Japanese government passed the "Industrial Competitiveness Enhancement Act" in December 2013, and is taking various measures to restore the "strong economy." In order to support the education and R&D of manufacturing base, Japan is promoting the talent training aiming at achieving "Society 5.0," and promoting the "talent revolution" with lifelong learning. Since 2015, the Japanese government has paid more attention to technological innovation and industrial integration, such as the Internet of Things and artificial intelligence, among which the manufacturing industry is the top priority of development. At the Hanover exhibition in 2017, Japanese officials put forward the "Connected Industries," formally declaring that Japan will actively integrate into the new wave of global Industrial Internet development. In the "White Paper on Japanese Manufacturing Industry" in 2018, the importance of "Connected Industries" for Japan to grasp the major opportunities of the Fourth Industrial Revolution was emphasized again [21].

Japan has Formed a Vocational Education System with its own Characteristics. The system includes three major parts: vocational education in schools, public vocational training, and vocational training in enterprises.

Vocational education in schools in Japan includes vocational high schools, comprehensive high schools, vocational subjects, schools, and specialized schools, as well as higher vocational education in short-term universities and higher specialized schools. At present, Japan has broken the traditional subdivision specialty in vocational high schools and recombined vocational disciplines such as natural disciplines, electronic machinery disciplines, and humanities disciplines. Various schools are the general name of all kinds of schools with the advantages of teaching one skill, and specialized schools are part of various schools upgraded, among which students mostly aim at obtaining vocational qualifications and skill appraisal stipulated by the state. Short-term universities are designed to provide general and professional higher education for high school graduates and to train intermediate technical personnel needed by industry. During the period of rapid economic recovery and development after World War II, there was a shortage of senior scientific and technological talents in science and engineering, especially applied talents. Therefore, the Ministry of Education set up a higher specialized school with vocational and technical education as the main part, aiming at teaching specialized knowledge in depth and cultivating the necessary ability for occupation.

Public vocational training in Japan was initially constructed as a system to solve the problem of unemployment, targeting the students who will be employed, people who need to improve their skills and professionals, as well as disabled people and teachers engaged in vocational training. The training is organized by statutory vocational training institutions set up by the state and prefectures, including vocational training centers, vocational training trainees, skills development centers, vocational training universities, etc. Their administrative systems are sound, training standards are uniform, and most of the funds come from central and local government subsidies.

Vocational training in enterprises is the most distinctive vocational education in Japan. The internal training system is very developed, and almost all large enterprises have a complete training system. The training is closely linked to the actual production, lifelong training for employees, and enterprises can use internal training as a "business tool" to win government subsidies. Intra-enterprise vocational training not only improves the production efficiency of the enterprise, but also contributes to vocational education in Japan as a supplement to vocational education in schools.

Japan uses the Law to Regulate the School-running Measures of Vocational Education. Consistent with other developed countries, Japan also attaches great importance to the legislation of vocational education. Since the end of the 19th century, Japan has promulgated the "Vocational Training Law" and the "Higher College Education Law." In addition, Japan has been highly successful in implementing a vocational ability evaluation system, and its skill identification system is the core of the ability evaluation system. More than 600 types of vocational qualification certificates have been developed, and students regard obtaining qualification certificates as the driving force of vocational learning.

In addition to the overall deployment at the strategic level, the Japanese government has also provided large-scale capital investment for the research and development of related technologies, the popularization of small and medium-sized enterprises, and the cultivation of talent technology. In the budget outline for 2019, Japan's budget for science and technology increased by 13.3% to 4.35 trillion yen, of which the budget for AI talents was 13.3 billion yen.

2.4 Summary

It can be seen that the unified characteristics of vocational education in developed countries attach great importance to systems and mechanisms, including **legislative guarantees, connection between secondary and high schools, vocational skills examinations, and lifelong education.** The education methods are flexible, and enterprises have more dominant power in terms of employment, training content, and training subsidies. The guarantee of teachers is comprehensive. Table 1 summarizes several typical effective measures to train senior talents through industry-university cooperation.

Table 1. Several ways of training manufacturing talents through industry-university cooperation in developed countries.

Type	Main features	Representative country or institution
Relying on the project	Using the results of scientific research projects as the medium to achieve complementary advantages	National Science Foundation
Establishment of University Science Park	University construction, government promotion, enterprise support	Stanford Science Park, USA
Establishment of economic entities	Independent legal person, university provides technology, enterprise is responsible for production and operation	Some universities in the United States, Britain, and Canada
Dual system	Aiming at application and based on theoretical knowledge, enterprises (60%) and universities (40%) cooperate to undertake teaching tasks	German University of Applied Technology
Joint establishment of training base	Double tutor system for the graduate students in the base	Japan University

3   China's Industrial Internet Skilled Personnel Training

The training of skilled talents in China includes several main types, such as vocational education, school-enterprise cooperation, and continuing education. The training modes of each type are different in terms of the background of trainees, training objectives, and training methods.

3.1   Background of Skilled Talents Training

In recent years, in order to accelerate the training of senior manufacturing talents and improve the quality of manufacturing talents, talent training development plans have released one after another. The Chinese government has successively issued the "Manufacturing Talent Development Planning Guide," "Skilled Talent Team Construction Implementation Plan (2018–2020)" and other talent plans, pointing out the direction for improving the quality of the manufacturing talent team.

In response to industry needs, China has issued a series of talent training plans such as the "National High-Skilled Talent Revitalization Plan" and the "Advanced Equipment Talent Training Plan." China implemented training and training projects for the shortage of skilled talents in the manufacturing and modern service industries, and the professional and technical talent knowledge updating project. A major talent project to solve the coexistence of the structural surplus and shortage of talent in the manufacturing industry was carried out.

3.2   Higher Vocational Education

In China, higher vocational colleges are the main force in training skilled talents of the Industrial Internet. According to the introduction of the Ministry of Education, in serving the national strategy, more than 1200 majors have been set up in vocational colleges throughout the country, covering all fields of the national economy, and about 10 million high-quality technical and skilled personnel have been trained every year. In the fields of modern manufacturing, strategic emerging industries, and modern service industries, more than 70% of the new front-line employees come from graduates of vocational colleges.

Industrial Internet is interdisciplinary and cross-industry, involving a series of technologies, such as automation, communication technology, etc. "Industrial Internet Talent White Paper" (2020) selected a group of majors that are highly related to industrial Internet from Chinese higher education, involving equipment manufacturing, electronic information and other categories [22]. Some vocational colleges have already carried out the construction of related courses and teaching materials from the perspective of establishing an industrial Internet professional group to promote talent training. The professional knowledge and courses involved in the cultivation of industrial Internet technical talents are not one-dimensional, so continuous learning and training are required.

3.3 Integration of Industry and Education Talents Training

With the successive issuance of the "Implementation Measures for the Construction of Industry-Education Integration Enterprises" and "Implementation Plan for Industry-Education Integration," the system design of industry-education integration enterprises has improved continuously. In April 2019, the Ministry of Education announced a list of 24 recommended enterprises that are integrated with production and education in the early stage of key construction and cultivation, and the construction of enterprises with integration of production and education has taken substantial steps. The Ministry of Education has promoted various industry-university cooperation projects, with a total of 1074 universities participating, and the enterprises provided funds and hardware and software support of 7.744 billion yuan in 2018.

On one hand, universities and enterprises jointly build research institutions. With the acceleration of manufacturing transformation and upgrading, an increasing number of manufacturing giants are paying more attention to innovative talent training methods relying on the advantages of universities. Related research also pointed out that school-enterprise cooperation and the combination of work and study are important and effective ways to cultivate industrial Internet skilled talents. For example, the Haier Group, together with Tsinghua University and Fraunhofer Institute in Germany, established the Industrial Intelligence Research Institute. On the other hand, some enterprises and universities actively promote cooperation in education and training. Manufacturing companies actively cooperate with colleges and universities to carry out education and training programs, launch vocational education, and training courses, in a more flexible way to enable college students to understand the latest demand for talents in the manufacturing industry and provide them with opportunities to participate in production practices.

3.4 Talent Cooperation Projects Between China and Other Countries

China is also actively promoting school-enterprise cooperation with foreign manufacturing giants. A typical example is the demonstration cooperation project of the Sino-French industrial cooperation jointly promoted by the Ministry of Industry and Information Technology and the French Ministry of Economic Affairs in 2017. The local governments, enterprises, and universities of China and France jointly set up experimental classes to formulate talent training programs and teaching plans, jointly implement curriculum teaching, teaching management, and opening and closing, and cultivate high-level professional skills in the Industrial Internet. Proposed by the "High School Discipline Innovation and Intelligence Introducing Plan" jointly implemented by the Ministry of Education and the State Administration of Foreign Experts Affairs since 2006, it promotes the integration of overseas talents with domestic scientific research. The "Intelligent Base" has achieved concrete results [23].

4 Comparative Analysis

By comparing the industrial development and talent training of the aforementioned countries, four characteristics were found. (1) There are laws to follow for the development of industries. (2) Talents in various countries, and the safeguard measures in each country are different. (3) The training measures are highly integrated with social and industrial development. (4) The talent training system and evaluation system were different (Table 2).

Table 2. Comparison of industrial development and talent training in different countries.

Country (Organization)	Industrial development and talents training goal	Representative strategies
China: China Academy of Industrial Internet Alliance of Industrial Internet	Industrial Internet innovation and development strategy Cultivate compound, multi-dimensional and multi-level talents	"Guiding Opinions on Deepening 'Internet + Advanced Manufacturing Industry' to Develop Industrial Internet" "Some Opinions on Deepening the Integration of Industry and Education" "Talent Development Plan for Manufacturing Industry"
America: American Industrial Internet Alliance	Leadership strategy for advanced manufacturing industry Educating, training, and connecting the manufacturing workforce	"National Strategic Plan for Advanced Manufacturing" "Advanced Manufacturing Partnership" "American Leadership Strategy for Advanced Manufacturing"
Germany: German Industry 4.0 Committee	Industry 4.0 plan Multi-faceted industry technical talents who can realize man-machine dialogue throughout the production process	"High-tech Strategy 2020" "Digital Strategy 2025" "Nationale Industriestrategie 2030"
Japan: Japan Industrial Value Chain Promotion Association	Connected Industries plan People First: Actively Promote the Cultivation of Senior Talents Adapted to Digital Technology	"Law on Strengthening Industrial Competitiveness" "Action Plan for Economic Restructuring and Promotion of Creativity" "Program for the Creation of Advanced Research and Technology"

The training of talents in various countries covers all types and levels of personnel, such as corporate managers, leading talents, professional and technical personnel, industrial workers, and university teachers. In the form of training organizations, higher vocational education, integration of production and education, transnational cooperation, and continuing education and training are the main focus. In terms of training methods, a variety of teaching modes are combined, such as intensive face-to-face teaching, on-site workshop teaching, online teaching, and micro-course teaching. As regards training content, flexibly integrates multi-disciplinary and multi-field related professional knowledge involved in the Industrial Internet to form a customized curriculum system. With respect to course organization, industry-specific courses, large-scale public courses, and corporate visit courses are organically integrated, highlighting the transformation of training content to results.

The main education methods in developed countries, such as "dual system," school to work (STW), community college, and enterprise internal training, have something in common with the current education modes in China, such as integration of industry and education, apprenticeship system, and 1 + X certificate, which can be used as a reference. Of course, developed countries are also facing problems such as increasing unemployment rate, lack of exchanges among enterprises, and the decline of vocational education volume brought about by labor-intensive and special technology-intensive industries. In the future, vocational education in various countries still has a certain way to go in terms of informatization, internationalization, and the development of lifelong education.

5 Conclusion

One of the main signs of the deep integration of industry and informatization is the Industrial Internet. The Industrial Internet will greatly change the working mode and skills of workers. The ability to handle multidisciplinary cooperation is becoming increasingly important, and the demand for professional talents is expected to increase. The talent training required for the development Industrial Internet is one of the central tasks that every country should perform well.

Of course, due to the short development time in the Industrial Internet field, as well as the wide range of talents, and the obvious interdisciplinary characteristics, **there are still some urgent problems in the current training of skilled talents, which mainly include the following aspects:** (1) insufficient training capabilities and single training methods; (2) lack of overall training planning; (3) the training evaluation system is not perfect, and needs to be strengthened in training management, training evaluation, and acceptance of training results; (4) insufficient investment in training funds and lack of stable channels. The cultivation of skilled talents lags behind the needs of the development of the industrial Internet industry.

After comparing and analyzing the training situation of skilled talents in several countries, **this study draws some suggestions as follows:** (1) adjust and optimize the discipline and professional structure, establish a multi-level and multi-structure talent training system; (2) strive to achieve a high-skilled talent chain wherein the supply and demand dock with the development of the industrial chain; (3) school-enterprise resource sharing, to promote the complementary advantages of school-enterprise and the deep integration of production and education; and (4) use project cooperation as a link to realize collaborative innovation among universities, enterprises, and research institutes.

References

1. Subcommittee on Advanced Manufacturing Committee on Technology of the National Science & Technology Council, "Strategy for American Leadership in Advanced Manufacturing". Advanced Manufacturing National Program Office, Washington, DC, October 2018
2. Henning, K., Wolfgang, W., Johannes, H.: Recommendations for Implementing the Strategic Initiative INDUSTRIE 4.0. Federal Ministry of Education and Research, Frankfurt (2013)
3. Ministry of Economy, Trade and Industry (METI), CEATEC Japan: Connected Industries. Information Economy Division, Commerce and Information Policy Bureau, May 2018
4. Chen, Z.: Deeply implementing the innovation development strategy of industrial internet. Adm. Reform **6**, 17–20 (2018)
5. Willner, A.: Internet of Things A to Z: Technologies and Applications, pp. 293–318 (2018)
6. Wiesner, S., Wuest, T., Thoben, K.-D.: 'Industrie 4.0' and smart manufacturing - a review of research issues and application examples. Int. J. Autom. Technol. **11**, 4–16 (2017)
7. Singerman, P.: Request for Information on Proposed New Program: National Network for Manufacturing Innovation (NNMI). Federal Register (2012)
8. Executive Office of the President, National Science and Technology Council, "A National Strategic Plan for Advanced Manufacturing", Office of Science and Technology Policy, Washington, DC, February 2012
9. Giffi, C., Wellener, P., Dollar, B.: Deloitte and the manufacturing institute skills gap and future of work study. Deloitte Insights (2018)
10. Cohen, M., Ignash, M.: An overview of the total credit curriculum. New Dir. Commun. Coll. **86**, 13–29 (1994)
11. Snyder, D., Dillow, A.: Digest of Education Statistics 2012. Claitors Publishing, Baton Rouge (2013)
12. Fugate, L., Amey, J.: Career stages of community college faculty: a qualitative analysis of their career paths, roles, and development. Commun. Coll. Rev. **28**(1), 1–22 (2000)
13. The White House: Executive Order Establishing the President's National Council for the American Worker, July 2018
14. Office of Science and Technology Policy: America Will Dominate the Industries of the Future. The White House, February 2019
15. Volkmar, K., Simon K., Reinhard G., Stefan S.: Industry 4.0 Opportunities and Challenges of the Industrial Internet. Germany: Strategy & PWC, p. 37 (2014)
16. Hu, M., Wang, Y., Zhu, M.: The revelations and enlightenment on German vocational education fitting with "Industry 4.0." Mod. Educ. Manag. **10**, 92–97 (2016)
17. Federal Ministry for Economic Affairs and Energy, Germany. National Industrial Strategy 2030, November 2019
18. BIBB: Dual Degree Courses: Plenty of Dynamics (2015)
19. Federal Institute for Vocational Education and Training (BIBB): Industry 4.0 and the Consequences for Labour Market and Economy. Bundesinstitut für Berufsbildung, Germany, p. 8 (2015)
20. Finlay, I., Niven, S., Young, S.: Changing Vocational Education and Training: An International Comparative Perspective. Psychology Press, Hove (1998)
21. The Ministry of Economy, Trade and Industry: White Paper on Japanese Manufacturing, June 2018
22. China Academy of Industrial Internet: White Paper on Industrial Internet Talents, 2020 ed, July 2020. https://www.china-aii.com/index.php?m=content&c=index&a=show&catid=26&id=30
23. Fan, J., Song, Y., Yang, T., Chai, T.: "111 Project"-based cultivation mode of innovative engineering talents with interdisciplinary research ability. Res. High. Eng. Educ. Curric. **6**, 50–56 (2020)

Author Index

Printed in the United States
by Baker & Taylor Publisher Services